The chemistry of the
sulphonium group
Part 2

THE CHEMISTRY OF FUNCTIONAL GROUPS

A series of advanced treatises under the general editorship of
Professor Saul Patai

The chemistry of alkenes (2 volumes)
The chemistry of the carbonyl group (2 volumes)
The chemistry of the ether linkage
The chemistry of the amino group
The chemistry of the nitro and nitroso groups (2 parts)
The chemistry of carboxylic acids and esters
The chemistry of the carbon–nitrogen double bond
The chemistry of amides
The chemistry of the cyano group
The chemistry of the hydroxyl group (2 parts)
The chemistry of the azido group
The chemistry of the acyl halides
The chemistry of the carbon–halogen bond (2 parts)
The chemistry of the quinonoid compounds (2 parts)
The chemistry of the thiol group (2 parts)
The chemistry of the hydrazo, azo and azoxy groups (2 parts)
The chemistry of the amidines and imidates
The chemistry of cyanates and their thio derivatives (2 parts)
The chemistry of diazonium and diazo groups (2 parts)
The chemistry of the carbon–carbon triple bond (2 parts)
Supplement A: The chemistry of double-bonded functional groups (2 parts)
Supplement B: The chemistry of acid derivatives (2 parts)
The chemistry of ketenes, allenes and related compounds (2 parts)
The chemistry of the sulphonium group (2 parts)

The chemistry of the
sulphonium group
Part 2

Edited by

C. J. M. STIRLING

School of Physical and Molecular Sciences
University College of North Wales
Bangor

1981

JOHN WILEY & SONS

CHICHESTER – NEW YORK – BRISBANE – TORONTO

An Interscience® Publication

British Library Cataloguing in Publication Data:

The chemistry of the sulphonium group.—(The
    chemistry of functional groups).
    Part 2
    1. Sulphonium compounds
    I. Stirling, Charles James Matthew
    II. Series
    547'.065   QD341.S8   80-40122
ISBN 0 471 27770 3
ISBN 0 471 27655 3 Set of 2 vols

Typeset by Preface Ltd, Salisbury Wilts.
Printed in the United States of America.

547.065

C HE

FOR E. AND L.—

ANOTHER LINK BETWEEN WALES AND ISRAEL
הקדשה עבור א., ל., והקשר בין ווילס וישראל
CYSWLLT ARALL RHWNG CYMRU AC ISRAEL

Contributing authors

K. K. Andersen   Department of Chemistry, University of New Hampshire, Durham, New Hampshire 03824, USA

M. R. F. Ashworth   Lehrstuhl für Organische Analytik, Universität des Saarlandes, Fachbereich 15.3, 66 Saarbrucken, W. Germany

G. C. Barrett   Oxford Polytechnic, Headington, Oxford OX3 0BP, England

L. F. Blackwell   Department of Chemistry, Biochemistry and Biophysics, Massey University, Palmerston North, New Zealand

E. Block   Department of Chemistry, University of Missouri–St. Louis, St. Louis, Missouri 63121, USA

J. D. Coyle   Chemistry Department, The Open University, Milton Keynes MK7 6AA, England

D. C. Dittmer   Department of Chemistry, Syracuse University, Syracuse, New York 13210, USA

A. Gavezzotti   Istituto di Chimica Fisica e Centro CNR, Università di Milano, Milan, Italy

J. Grimshaw   Department of Chemistry, David Keir Building, Queen's University of Belfast, Belfast BT9 5AG, N. Ireland

A. C. Knipe   School of Physical Sciences, The New University of Ulster, Coleraine, Co. Londonderry, Northern Ireland BT42 1SA

P. A. Lowe   The Ramage Laboratories, Department of Chemistry and Applied Chemistry, University of Salford, Salford M5 4WT, England

G. A. Maw   Department of Biochemistry, Glasshouse Crops Research Institute, Littlehampton, Sussex, England

T. Numata   Department of Chemistry, University of Tsukuba, Sakura-mura, Ibaraki, 305 Japan

S. Oae   Department of Chemistry, University of Tsukuba, Sakura-mura, Ibaraki, 305 Japan

B. H. Patwardhan   Department of Chemistry, Syracuse University, Syracuse, New York 13210, USA

I. C. Paul   Department of Chemistry, School of Chemical Sciences, University of Illinois, Urbana, Illinois 61801, USA

E. F. Perozzi   Ethyl Corporation, 1600 West Eight Mile Road, Ferndale, Michigan 48220, U.S.A.

R. Shaw   Chemistry Department, Lockheed Palo Alto Research Laboratory, Lockheed Missiles & Space Company, Inc., 3251 Hanover Street, Palo Alto, California 94304, USA

H. J. Shine   Department of Chemistry, Texas Tech University, Lubbock, Texas 79409, USA

J. Shorter   Department of Chemistry, The University, Hull HU6 7RX, England

M. Simonetta   Istituto di Chimica Fisica e Centro CNR, Università di Milano, Milan, Italy

T. Yoshimura   Department of Chemistry, University of Tsukuba, Sakura-mura, Ibaraki, 305 Japan

Foreword

Our collaboration on production of this volume on the chemistry of the sulphonium group arose from a sabbatical visit from Wales to Israel.

There is remarkably little review literature on the chemistry of this group, one of the most versatile in organic chemistry, and we have been most fortunate in the group of contributors who have brought this volume to fruition. Chapters on nuclear magnetic resonance spectroscopy and mass spectroscopy did not materialise in time for publication.

We hope that assembly of so much up to date information on this functional group will stimulate further exploration of its wide-ranging and fascinating chemistry.

The preface to the series contains the promise of further volumes on the chemistry of sulphur-containing compounds. We are in the early stages of planning further volumes in the series on the chemistry of other sulphur-containing functional groups, notably those concerned with sulphonyl and sulphinyl groups.

Bangor   May 1980   CHARLES STIRLING
Jerusalem   SAUL PATAI

The Chemistry of Functional Groups
Preface to the series

The series 'The Chemistry of Functional Groups' is planned to cover in each volume all aspects of the chemistry of one of the important functional groups in organic chemistry. The emphasis is laid on the functional group treated and on the effects which it exerts on the chemical and physical properties, primarily in the immediate vicinity of the group in question, and secondarily on the behaviour of the whole molecule. For instance, the volume *The Chemistry of the Ether Linkage* deals with reactions in which the C—O—C group is involved, as well as with the effects of the C—O—C group on the reactions of alkyl or aryl groups connected to the ether oxygen. It is the purpose of the volume to give a complete coverage of all properties and reactions of ethers in as far as these depend on the presence of the ether group but the primary subject matter is not the whole molecule, but the C—O—C functional group.

A further restriction in the treatment of the various functional groups in these volumes is that material included in easily and generally available secondary or tertiary sources, such as Chemical Reviews, Quarterly Reviews, Organic Reactions, various 'Advances' and 'Progress' series as well as textbooks (i.e. in books which are usually found in the chemical libraries of universities and research institutes) should not, as a rule, be repeated in detail, unless it is necessary for the balanced treatment of the subject. Therefore each of the authors is asked *not* to give an encyclopaedic coverage of his subject, but to concentrate on the most important recent developments and mainly on material that has not been adequately covered by reviews or other secondary sources by the time of writing of the chapter, and to address himself to a reader who is assumed to be at a fairly advanced post-graduate level.

With these restrictions, it is realized that no plan can be devised for a volume that would give a *complete* coverage of the subject with *no* overlap between chapters, while at the same time preserving the readability of the text. The Editor set himself the goal of attaining *reasonable* coverage with *moderate* overlap, with a minimum of cross-references between the chapters of each volume. In this manner, sufficient freedom is given to each author to produce readable quasi-monographic chapters.

The general plan of each volume includes the following main sections:

(a) An introductory chapter dealing with the general and theoretical aspects of the group.

(b) One or more chapters dealing with the formation of the functional group in question, either from groups present in the molecule, or by introducing the new group directly or indirectly.

(c) Chapters describing the characterization and characteristics of the functional groups, i.e. a chapter dealing with qualitative and quantitative method of determination including chemical and physical methods, ultraviolet, infrared, nuclear magnetic resonance and mass spectra: a chapter dealing with activating and directive effects exerted by the group and/or a chapter on the basicity, acidity or complex-forming ability of the group (if applicable).

(d) Chapters on the reactions, transformations and rearrangements which the functional group can undergo, either alone or in conjunction with other reagents.

(e) Special topics which do not fit any of the above sections, such as photochemistry, radiation chemistry, biochemical formations and reactions. Depending on the nature of each functional group treated, these special topics may include short monographs on related functional groups on which no separate volume is planned (e.g a chapter on 'Thioketones' is included in the volume *The Chemistry of the Carbonyl Group*, and a chapter on 'Ketenes' is included in the volume *The Chemistry of Alkenes*). In other cases certain compounds, though containing only the functional group of the title, may have special features so as to be best treated in a separate chapter, as e.g. 'Polyethers' in *The Chemistry of the Ether Linkage*, or 'Tetraaminoethylenes' in *The Chemistry of the Amino Group*.

This plan entails that the breadth, depth and thought-provoking nature of each chapter will differ with the views and inclinations of the author and the presentation will necessarily be somewhat uneven. Moreover, a serious problem is caused by authors who deliver their manuscript late or not at all. In order to overcome this problem at least to some extent, it was decided to publish certain volumes in several parts, without giving consideration to the originally planned logical order of the chapters. If after the appearance of the originally planned parts of a volume it is found that either owing to non-delivery of chapters, or to new developments in the subject, sufficient material has accumulated for publication of a supplementary volume, containing material on related functional groups, this will be done as soon as possible.

The overall plan of the volumes in the series 'The Chemistry of Functional Groups' includes the titles listed below:

The Chemistry of Alkenes (two volumes)
The Chemistry of the Carbonyl Group (two volumes)
The Chemistry of the Ether Linkage
The Chemistry of the Amino Group
The Chemistry of the Nitro and Nitroso Groups (two parts)
The Chemistry of Carboxylic Acids and Esters
The Chemistry of the Carbon–Nitrogen Double Bond
The Chemistry of the Cyano Group
The Chemistry of Amides
The Chemistry of the Hydroxyl Group (two parts)
The Chemistry of the Azido Group

The Chemistry of Acyl Halides
The Chemistry of the Carbon–Halogen Bond (two parts)
The Chemistry of Quinonoid Compounds (two parts)
The Chemistry of the Thiol Group (two parts)
The Chemistry of Amidines and Imidates
The Chemistry of the Hydrazo, Azo and Azoxy Groups
The Chemistry of Cyanates and their Thio Derivatives (two parts)
The Chemistry of Diazonium and Diazo Groups (two parts)
The Chemistry of the Carbon–Carbon Triple Bond (two parts)
Supplement A: The Chemistry of Double-bonded Functional Groups (two parts)
Supplement B: The Chemistry of Acid Derivatives (two parts)
The Chemistry of Ketenes, Allenes and Related Compounds (two parts)
The Chemistry of the Sulphonium Group (two parts)

Titles in press:

Supplement E: The Chemistry of Ethers, Crown Ethers, Hydroxyl Groups and their Sulphur Analogs

Future volumes planned include:

The Chemistry of Organometallic Compounds
The Chemistry of Sulphur-containing Compounds
Supplement C: The Chemistry of Triple-bonded Functional Groups
Supplement D: The Chemistry of Halides and Pseudo-halides
Supplement F: The Chemistry of Amines, Nitroso and Nitro Groups and their
   Derivatives

Advice or criticism regarding the plan and execution of this series will be welcomed by the Editor.

The publication of this series would never have started, let alone continued without the support of many persons. First and foremost among these is Dr Arnold Weissberger, whose reassurance and trust encouraged me to tackle this task, and who continues to help and advise me. The efficient and patient cooperation of several staff-members of the Publisher also rendered me invaluable aid (but unfortunately their code of ethics does not allow me to thank them by name). Many of my friends and colleagues in Israel and overseas helped me in the solution of various major and minor matters, and my thanks are due to all of them, especially to Professor Z. Rappoport. Carrying out such a long-range project would be quite impossible without the non-professional but none the less essential participation and partnership of my wife

The Hebrew University
Jerusalem, ISRAEL   SAUL PATAI

Contents

The Chemistry of the Sulphonium Group
Edited by C. J. M. Stirling and S. Patai
© 1981 John Wiley & Sons Ltd.

CHAPTER **13**

Cyclic sulphonium salts

DONALD C. DITTMER and BHALCHANDRA H. PATWARDHAN

Department of Chemistry, Syracuse University, Syracuse, N.Y. 13210, USA

I. INTRODUCTION

The methods of synthesis and the reactions of cyclic sulphonium salts mostly parallel those of their acyclic counterparts. The syntheses involve mainly the reaction of electrophilic reagents with the nucleophilic sulphur atom of a divalent sulphide. The principal reactions of sulphonium salts are displacements (S_N1, S_N2), eliminations, and ylide formation and rearrangements therefrom. However, there are significant differences between cyclic and acyclic sulphonium salts. Three- and four-membered cyclic salts are difficult to prepare because of their enhanced lability caused by ring strain. The reactions of many acyclic and cyclic sulphides containing a good leaving group apparently proceed by cyclic sulphonium intermediates, the reactions of β-halosulphides or mustard gases being illustrative. Cyclic conjugated cationic systems in which the sulphur atom participates have no close analogues in the acyclic systems. These conjugated systems include the dithiolium, thiazolium, oxazolium, and thiapyrylium cations, the thiabenzenes (thiapyrylium ylides), and a variety of mesoionic compounds. Addition reactions are common with such systems.

This chapter covers the synthesis and reactions of cyclic sulphonium salts, starting with the smallest ring size, the three-membered, and proceeding to the four-, five-, six- and higher-membered rings. Cyclic sulphonium cation radicals are discussed in Chapter 14 and will be omitted here. Cyclic sulphoxides which may be considered as cyclic sulphonium zwitterions, because of the polarity of the S—O bond, for the most part will not be discussed.

For a general coverage of the chemistry of cyclic sulphonium salts (included with other topics) the following may be consulted.

Organic Compounds of Sulphur, Selenium and Tellurium, The Chemical Society, London. These biennial Specialist Periodical Reports provide coverage of recent work on all aspects of organosulphur chemistry.

Saturated Heterocyclic Chemistry, Specialist Periodical Reports, The Chemical Society, London.

Aromatic and Heteroaromatic Chemistry, Specialist Periodical Reports, The Chemical Society, London.

J. P. Marino, *Sulphur Containing Cations*, in *Topics in Sulphur Chemistry*, Vol. 1 (Ed. A. Senning), Georg Thieme, Stuttgart, 1976, Chapter 1.

C. J. M. Stirling, *Sulfonium Salts*, in *Organic Chemistry of Sulfur* (Ed. S. Oae), Plenum Press, New York, 1977, p. 473.

E. Block (Editor), *Reactions of Organosulfur Compounds*, Academic Press, New York, 1978.

B. M. Trost and L. S. Melvin, Jr., *Sulfur Ylides*, Academic Press, New York, 1975.

II. THREE-MEMBERED RINGS

A. Thiiranium (Episulphonium) Ions

1. Reviews

Various aspects of the synthesis and chemistry of thiiranium ions have been reviewed[1-9]. In particular, the additions of sulphenyl halides to alkenes may proceed via these intermediates and several reviews cover these reactions[2-6]. The intervention of thiiranium ions in reactions of β-halosulphides has been discussed[7]. The pioneering work of Helmkamp and co-workers on the synthesis and reactions of thiiranium ions has been summarized[8]. Several reviews of thiirans discuss ring-opening reactions which may involve cyclic sulphonium ions[9].

2. Theoretical

Non-empirical SCF-MO calculations on protonated thiirans favour a sulphurane structure **1** over the thiiranium ion structure **2** in the gas phase[10]. In fact, evidence from magnetic resonance measurements support a sulphurane structure **4** for an adduct derived by treatment of the stable thiiranium ion **3** with tetraphenylarsonium chloride (equation 1)[11]. Extended Hückel MO calculations

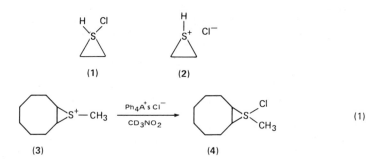

(1)   (2)

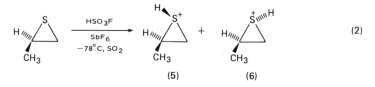

(3)   (4)   (1)

support a trigonal bipyramidal structure for the sulphurane[11]. In solution, the thiiranium structure may be favoured[10], and proton magnetic resonance measurements at low temperatures on 2-methylthiiran[12] (equation 2) and *cis*- and *trans*-2,3-di-*t*-butylthiiran[13] protonated on sulphur by fluorosulphonic acid support the existence of the pyramidal *cis* and *trans* derivatives **5** and **6**. According to MO

calculations, thiiranium ions should have the p configuration in which the substituent on sulphur is perpendicular to the three-membered ring[14,15]. Higher energy configurations at sulphur were sp[3] and sp[2]. It was claimed that the extension of the lowest unoccupied molecular orbital (LUMO) in the region of the ring carbon atoms determined the site at which a nucleophile attacked, the greater the extension of the LUMO the greater the likelihood of attack[15].

3. Stable thiiranium ions

Although stable *S*-alkyl or *S*-acyl thiiranium salts may have been obtained as long ago as 1922[16] or 1958[17], the structures of the compounds were not determined conclusively, and it was not until 1963 that Pettitt and Helmkamp[18] reported stable thiiranium ions **7** derived from cyclooctene (equation 3). These salts were reported to be stable at room temperature.

In general, the preparations of stable thiiranium salts require low temperatures and the absence of nucleophilic reagents or solvents which can cause ring opening.

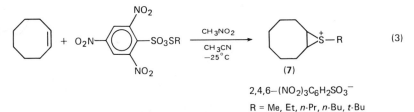

$$2,4,6-(NO_2)_3C_6H_2SO_3^-$$

$$R = Me, Et, n\text{-}Pr, n\text{-}Bu, t\text{-}Bu$$

The preparative methods fall into three categories:

(1) The addition of RS^+ or ArS^+ derived from RSX or ArSX (X = Hal, $ArSO_3$, $R_2\overset{+}{S}$), to an alkene (equation 4).

$$R^1 \quad R^2$$
$$\underset{R^4}{\overset{R^1}{}}C=C\underset{R^3}{\overset{R^2}{}} + RSX \longrightarrow R^1 \underbrace{\qquad}_{R^4} R^2 \tag{4}$$

(2) The reaction of the sulphur atom of an S-alkyl- or arylthio group with an electron-deficient carbon centre appropriately situated for three-membered ring formation via participation of the unpaired electrons on the sulphur atom (equation 5).

$$\tag{5}$$

(3) The alkylation (or protonation) of the sulphur atom in a thiiran (equation 6). The above three methods will be discussed in turn.

$$\tag{6}$$

Reagents RSX, where X represents a good leaving group, are capable of attacking alkenes to give thiiranium ions. The leaving group, X, if it is not to be removed completely from the reaction mixture, should be of low nucleophilicity to avoid ring opening of the thiiranium ion. The system used by Pettitt and Helmkamp[18] illustrates this point. They converted an alkanesulphenyl bromide, in which the bromide ion, while a good leaving group, is also a relatively good nucleophile, to an alkanesulphenium 2,4,6-trinitrobenzenesulphonate in which the arylsulphonate ion is a much poorer nucleophile (equation 7). This stand-in for RS^+ reacted with cyclooctene as described above. Unfortunately, the reaction with cyclohexene resulted only in tar formation.

The reagent, $ArSO_2OSR$, on the basis of conductivity and proton magnetic resonance data, is believed to furnish a complex **8** with acetonitrile, but in the absence of acetonitrile its low conductivity and reactivity suggests the covalent structure (or an intimate ion pair) given in equation 7[19].

Thiiranium ions stable in solution for several hours at -20 to $-50°C$ are obtained by treating alkenes with sulphenyl chlorides or bromides and silver

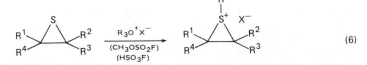

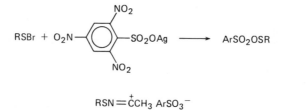

$$RSN = \overset{+}{C}CH_3 \; ArSO_3^-$$

(8)

tetrafluoroborate or hexafluoroantimonate in non-nucleophilic solvents (SO_2, CH_2Cl_2, CH_3NO_2, $ClCH_2CH_2Cl$) (equation 8)[20–22,30]. A hexachloroantimonate salt also was obtained from 2-methylpropene, benzenesulphenyl chloride and antimony pentachloride[23].

$$\underset{R^4}{\overset{R^1}{\diagdown}} C = C \underset{R^3}{\overset{R^2}{\diagup}} + \text{RSHal} + \text{AgX} \quad \xrightarrow[-\text{AgHal}]{-40^\circ C} \quad \underset{R^4}{\overset{R^1}{\diagdown}} \overset{\overset{R}{\underset{|}{S^+ \; X^-}}}{\triangle} \underset{R^3}{\overset{R^2}{\diagup}}$$

(8)

Carbon magnetic resonance spectra of several of the thiiranium hexafluoroantimonates showed considerable deshielding (20–39 p.p.m.) of the ring carbons compared with the respective covalent thiirans.[30] Thiiranium ions also may be made by the addition of alkylthiosulphonium salts ($R_2\overset{+}{S}SR$)[19,24,25] to alkenes (equation 9)[26–28]. The 2,4,6-trinitrobenzenesulphonate[24,26] and hexachloroantimonate[25] salts have been used. Thiiranium hexachloroantimonates are said to be stable for several weeks at $-10^\circ C$ if moisture is excluded. An alternate method of preparation of dimethylmethylthiosulphonium hexachloroantimonate is by treatment of methanesulphenyl chloride and dimethyl disulphide with antimony pentachloride[25].

The proton n.m.r. spectra of these salts show singlets at δ 2.4–2.7 p.p.m. for the S-methyl protons. Protons on the ring carbons appear at δ 3.6–4.3 p.p.m. Only one isomer (trans-S-methyl) is obtained from cis-alkenes.

Another method for the preparation of relatively stable thiiranium ions involves treatment of β-halothioethers (usually prepared by addition of a sulphenyl halide to alkenes) with silver or antimony salts (equations 10 and 11)[18,20–23,29–32].

1-Methyl- and 1-phenylthioniacyclopropane are 'stable' in the gas phase with lifetimes greater than 10^{-5} s[33]. They are generated by electron impact from β-methyl- or β-phenylthioethyl phenyl ether. Their collisional activation spectra have been studied. Cyclic $C_2H_5S^+$ ions (protonated thiirans) have been detected in the collisional activation spectra of cation radicals of thiols[34].

Alkyl iodides, which are used widely to prepare sulphonium ions from sulphides, cannot be used to prepare thiiranium salts by alkylation of thiirans because of the high nucleophilicity of iodide ion, which leads to ring opening (see reference 16 for a possible exception). The development of alkylating agents such as trimethyloxonium salts and methyl fluorosulphonate, which yield poorly nucleophilic anions [e.g. BF_4^- 2,4,6-$(NO_2)_3C_6H_2SO_3^-$, FSO_3^-], has led to the synthesis of several thiiranium ions[13,18]. cis-2,3-Di-t-butylthiiran is methylated by methyl fluorosulphonate (equation 12), but the trans isomer, cyclohexene episulphide and 2,2,3,3-tetramethylthiiran are not[13]. The cis-di-t-butyl salt is very sensitive to moisture and decomposes on standing at room temperature.

R^1	R^2	R^3	R^4	R	X
$-(CH_2)_4-$		H	H	CH_3	BF_4^-
$-(CH_2)_4-$		CH_3	H	Ph	BF_4^-
Ph	H	H	H	CH_3	BF_4^-
t-Bu	H	H	H	Ph	BF_4^-
CH_3	CH_3	H	H	2,4-$(NO_2)_2C_6H_3$ or *p*-ClC_6H_4	SF_6^-
CH_3	H	CH_3	H	2,4-$(NO_2)_2C_6H_3$ or *p*-ClC_6H_4 *p*-$CH_3C_6H_4$	SF_6^-
CH_3	H	H	H	2,4,6-$(CH_3)_3C_6H_2$ *p*-ClC_6H_4 *p*-$CF_3C_6H_4$ *p*-$CF_3C_6H_4$ C_6H_5	SF_6^-
CH_3	H	H	CH_3	2,4,6-$(CH_3)_3C_6H_2$	SF_6^-

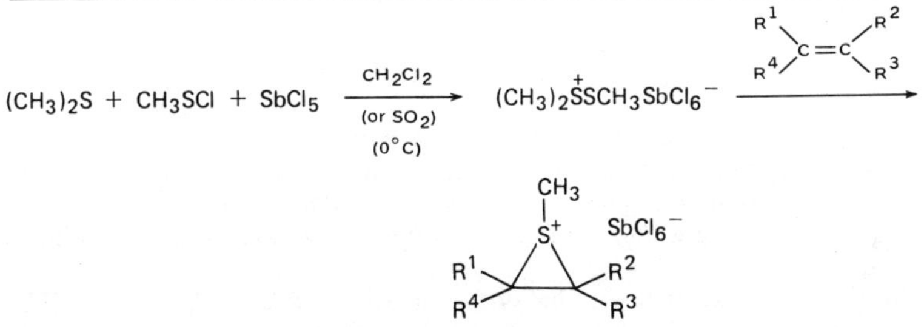

(CH3)2S + CH3SCl + SbCl5
CH2Cl2
(or SO2)
(0°C)
(CH3)2S+SCH3 SbCl6−
R1 R2 C=C R4 R3
CH3
S+
SbCl6−
R1
R2
R4
R3

(9)

R^1	R^2	R^3	R^4
Me	Me	Me	Me
Et	H	Et	H
$-(CH_2)_6-$		H	H
Me	H	H	Me
Me	Me	H	H
Me	H	Me	H
Me	Me	H	Me

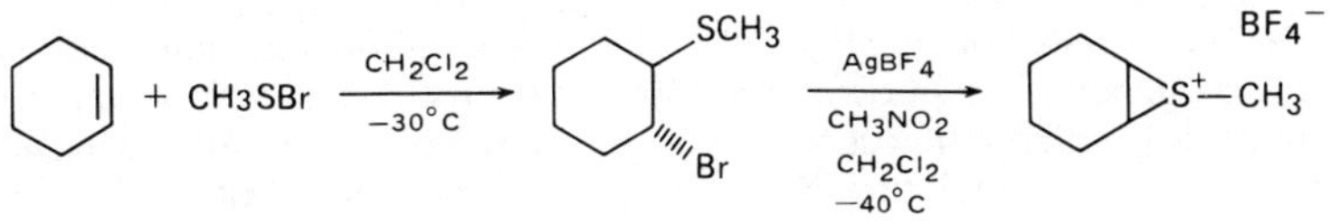

+ CH3SBr
CH2Cl2
−30°C
SCH3
Br
AgBF4
CH3NO2
CH2Cl2
−40°C
BF4−
S+—CH3

(10)

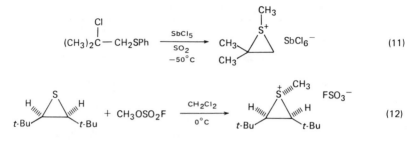

$$(11)$$

$$(12)$$

Cyclooctene episulphide is methylated[18] or ethylated[8] by trimethyl- or triethyloxonium 2,4,6-trinitrobenzenesulphonate.

Protonation of the sulphur atom of 2-methylthiiran, tetramethylthiiran, cis- and trans-1,2-di-t-butylthiiran and the oxygen atom of thiiran 1-oxide by fluorosulphonic acid yields thiiranium ions observable at low temperatures by proton magnetic resonance spectroscopy[12,13]. In the cis-di-t-butylthiiran, protonation occurs 80% at the side opposite the bulky t-butyl groups (equation 13), whereas alkylation apparently occurs 100% at that side[13].

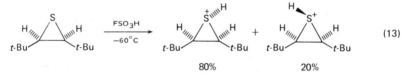

$$(13)$$

4. Thiiranium ions as intermediates in reactions

Thiiranium ions long have been postulated as intermediates in the very reactions which ultimately have yielded stable salts as described in Section 3: additions of RSX to carbon–carbon double bonds, reactions of thioethers substituted in the β-position with a good leaving group, and the reactions of thiirans with electrophilic reagents.

Ionic additions of sulphenyl halides, RSCl, RSBr[35-39], AcSSCl[40], (MeO)$_2$P(O)SCl[41], ClSCN[42], SCl$_2$[40,43,44], Me$_2$NSCl[45]) to alkenes is usually interpreted on the basis of thiiranium ion intermediates[1-5], although under certain conditions sulphuranes may be formed[5c,30,46,47]. The addition of RSX is usually trans[36,48,49], and the thiiranium ion intermediates derived from bicyclic alkenes possess the exo configuration (equation 14)[50-52], except for 7,7-dimethylnorbornene where a methyl group at the 7-position forces endo attack by the sulphenyl chloride[53].

The trans stereospecificity of addition may be lost if an open carbonium ion intervenes in the product-determining step[46,47,54]. Both Markovnikov and anti-Markovnikov regiospecificity for additions of RSX (RS$^+$ and X$^-$) to alkenes is observed (equation 15)[30,35-41,52,55,56]. The anti-Markovnikov product may sometimes be isomerized to the Markovnikov product[52,55,56], particularly if acid is present, and this potential isomerization should be taken into account in the evaluation of isomer ratios.

The regioselectivity of nucleophilic attack on thiiranium ion intermediates (equation 15) depends on steric and electronic factors. A bulky R^1 group results in more anti-Markovnikov product[52,57-59]. Addition of 4-chlorobenzenesulphenyl chloride to cis-1,2-t-butylethylene is ca. 7×10^5 times faster than that to

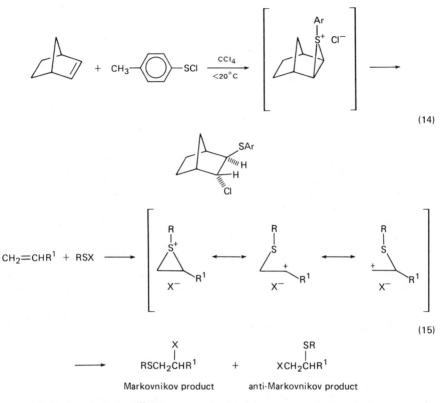

(14)

(15)

Markovnikov product   anti-Markovnikov product

trans-1,2-di-*t*-butylethylene[60]. Less steric hindrance was believed to occur in transition state **9** than in **10** since in **10** the aryl group and a *t*-butyl group are in

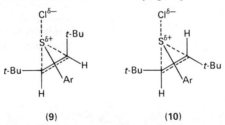

(9)   (10)

close proximity. However, additions to *trans*-1-arylpropenes were *ca.* 5 times faster than to the corresponding *cis* derivatives and a charge-transfer interaction was suggested in order to explain the results[61]. The *cis* isomers of simple alkyl-substituted ethylenes reacted *ca.* 12–59 times faster than the *trans* isomers[61]. These, however, were not as sterically hindered as the 2,3-di-*t*-butylethylenes.

The importance of steric factors in affecting nucleophilic attack on thiiranium ions has perhaps been overemphasized since the addition of acetate ion or other nucleophiles to pre-formed sulphonium ions (equation 16) in liquid sulphur dioxide shows predominantly Markovnikov regioselectivity[30], in contrast to the additions of arenesulphenyl chlorides, ArSCl, which show mainly anti-Markovnikov

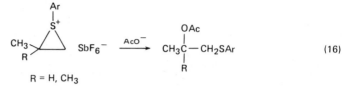

$$R = H, CH_3$$

regioselectivity. Sulphurane intermediates were favoured over thiiranium ions for additions of sulphenyl halides to olefins in non-polar media[30].

A substituent R^1 (equation 15) which is good at stabilizing a positive charge gives more Markovnikov product, either because of greater development of positive charge on the substituted ring carbon in the transition state or because a fully fledged carbonium ion intermediate[46,47,54] is formed from the thiiranium ion (or sulphurane). That 'open' carbonium ions occur after the rate-determining formation of the cyclic thiiranium ion is supported by the observation that while the stereospecificity and regiospecificity of addition of 2,4-dinitrobenzenesulphenyl chloride to *cis*- and *trans*-1-arylpropenes vary depending on the substituents on the aryl ring of the alkene, a linear Hammett plot relating the logarithm of the rate constant for addition with the Hammett substituent constants is obtained ($\rho = -2.7$ for the *cis*- and -3.7 for the *trans*-propenes)[47]. The product stereochemistry with electron-withdrawing substituents is stereospecific but non-regiospecific, as expected for an attack by the nucleophile on a thiiranium ion; that for electron-donating substituents is non-stereospecific but regiospecific, as expected for an 'open' carbonium ion. The Hammett correlation shows that the product-determining steps for these arylpropenes are independent of the rate-determining step (thiiranium ion

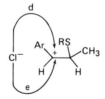

Ar electron-withdrawing
stereospecific—trans attack
non-regiospecific—paths a
and b both likely

Ar electron donating
non-stereospecific—paths
d and e both likely
regiospecific—Ar stabilizes
only one of two possible ions

SCHEME 1

formation), which is a smooth, linear function of the electronic effect of the substituents. The discontinuous nature of the product-forming step as a function of the substituent on the aryl ring is summarized in Scheme I.

The formation of an 'open' carbonium ion intermediate also seems to depend on the counter ion of the thiiranium salt. One commonly used criterion for the intervention of carbonium ions is the observation of rearranged products. Formation of a carbonium ion adjacent to the carbinol carbon in the allylic alcohol **11** would be expected to lead to a rearranged product analogous to that expected from a pinacol–pinacolone rearrangement. Although **11** gave only the products of addition to the double bond under the usual conditions (only a trace amount of rearranged product was formed), in the presence of perchlorate ion up to 51% of rearranged product was obtained (equation 17)[62–64]. Aryl groups substituted with electron-donating substituents (e.g. *p*-CH$_3$OC$_6$H$_4$) gave rearrangement even in the absence of perchlorate ions although more occurred in their presence.

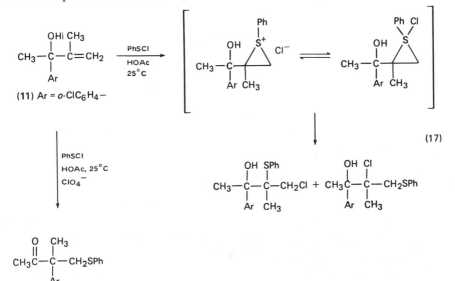

(17)

In the absence of perchlorate ions, it is possible that the more nucleophilic chloride ion reduces the positive charge on the intermediate either by sulphurane formation or electrostatically in an intimate ion pair[5c,8], thus decreasing the tendency for rearrangement. In the presence of perchlorate ions, intimate ion pairs with chloride ion may give way to solvent separated ion pairs and the sulphuranes may dissociate[62–64]. The greater polarity of the medium caused by addition of perchlorate salts would favour dissociation of the sulphurane or intimate ion pair, preparing the way for rearrangements. A concerted process from the trigonal bipyramidal sulphurane to the unrearranged, normal products has been discussed[63]. The favoured sulphurane structure had chlorine and one ring carbon apical[63]. Lithium perchlorate has been used to facilitate the addition of nitriles to the intermediates (equation 18). In the absence of perchlorate ions only the β-chlorosulphide is obtained[65].

Certain electron-withdrawing substituents (R^1, e.g. COCl, COOCH$_3$, CN) cause considerably more of the Markovnikov product to be formed (equation 19)[56,66,67].

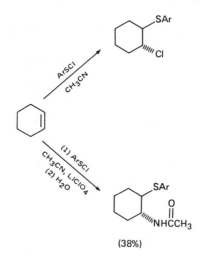

(18)

Ar = 2,4-(NO$_2$)$_2$C$_6$H$_3$−

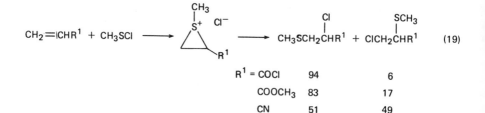

(19)

R^1 = COCl	94	6
COOCH$_3$	83	17
CN	51	49

The result was interpreted on the basis of a special interaction of the substituent with the attacking chloride ion as has been proposed to explain the high reactivity of α-halocarbonyl compounds to nucleophiles[68].

Substituents on sulphur exert an influence on the rates of formation and on the relative stabilities of thiiranium ions formed by addition of sulphenyl derivatives to alkenes. As mentioned previously, the substituent on sulphur tends to be oriented away from bulky substituents on the alkene (see **9** and **10**)[13,60]. Electron-withdrawing substituents on sulphur tend to destabilize its positive charge in the thiiranium ion. The transition state for nucleophilic attack is shifted to one with more positive charge residing on carbon and, thus, more Markovnikov product is formed[40,41,55,58]. Electron-donating substituents which can stabilize a positive charge on sulphur facilitate the addition of sulphenyl halides to alkenes[69,70]. The electron-deficient sulphenyl halide, pentafluorobenzenesulphenyl chloride, does not react with electron-deficient alkenes such as acrylonitrile[71].

Sulphenyl halides can be generated by the action of halogen on disulphides; if the disulphide has a double bond located conveniently for intramolecular reaction with the corresponding sulphenyl halide, cyclizations may occur via thiiranium ion intermediates[72,73]. A synthesis of a thiaprostacyclin derivative was achieved by this route (equation 20)[74].

Similar cyclizations via sulphenyl derivatives have been observed in reactions in the penicillin and cephalosporin series (e.g. equation 21)[75–79].

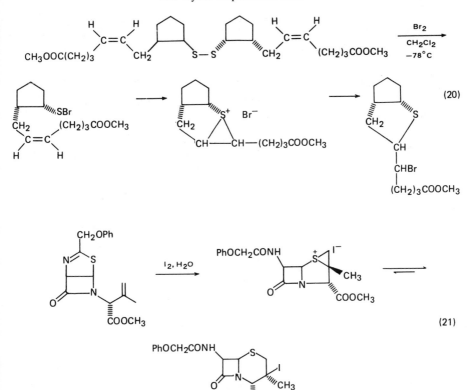

(20)

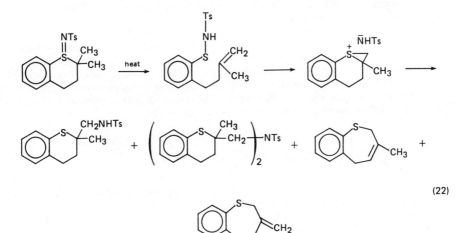

(21)

The intramolecular addition of a sulphenamide to a carbon–carbon double bond may proceed via a thiiranium ion (equation 22)[80].

(22)

Intermediate thiiranium ions may occur upon protonation of a double bond in proximity to a sulphur atom[81-83].

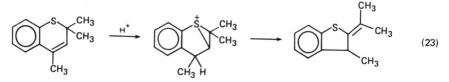

(23)

The second type of reaction in which thiiranium ions are believed to be intermediates involves sulphides with a good leaving group in the β-position which provided some early instances of neighbouring group participation (equation 24)[84,85]. Much interest in these compounds arose because of their physiological properties as mustard gases.

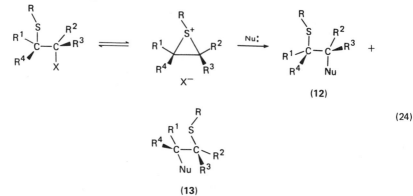

(12)

(24)

(13)

The consequences of β-sulphur atom participation in nucleophilic displacements (equation 24) are (i) usually an increase in rate compared with a non-sulphur-containing model compound, (ii) essentially retained stereochemistry at the carbon atom bearing the leaving group X (as in 12), and (iii) occasional rearrangement (as in 13) or elimination, depending on the structure of the substrate and thiiranium ion intermediate.

Large rate enhancements may attend the formation of a thiiranium intermediate in these systems. In 3-substituted 2-α-5-epithio-5-α-cholestane, which has a sulphur atom bridged between the $C_{(2)}$ and $C_{(5)}$ positions, the rate enhancement of solvolysis of the 3-β isomers (bromide or mesylate) in aqueous dioxan containing sodium acetate may be as great at 10^{11} [86]. Participation by sulphur can occur with the 3-β but not with the 3-α derivative. Retention of configuration at $C_{(3)}$ is observed in solvolysis of the 3-β compounds (equation 25).

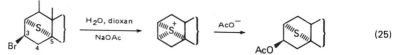

(25)

Other examples of rate enhancements by a neighbouring sulphur atom which may involve thiiranium ions demonstrate that the importance of sulphur participation decreases if the carbonium ion formed by cleavage of the C—X bond is stabilized by other substituents. The solvolyses in 80% aqueous acetone of some

p-nitrobenzoate esters demonstrates this point as the rate differences between **14** and **15** and between **16** and **17** show[87].

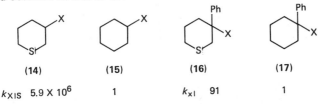

	(14)	**(15)**	**(16)**	**(17)**
k_{XIS}	5.9×10^6	1		
k_{XI}			91	1

Acetolysis of chloromethylthiiran or 2-chloroethylmethyl sulphide shows a common ion rate depression supporting the presence of an intermediate; in the case of chloromethylthiiran, a rearranged product was obtained (equation 26)[88].

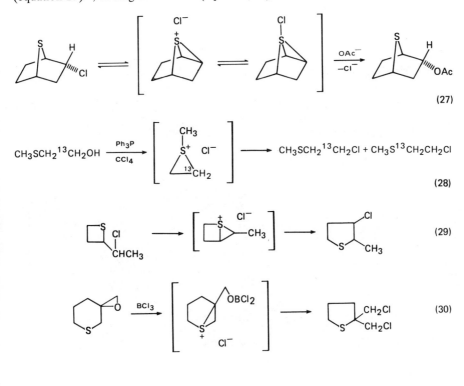

$$(26)$$

However, in the acetolysis of 2-*endo*-chloro-7-thiabicycloheptane, no chloride ion rate depression was observed although the *endo* isomer reacted 5×10^9 times faster than the *exo* isomer. A sulphurane with a bound chloride ion was suggested in order to explain the lack of rate depression by chloride ions (equation 27)[89].

Rearrangements involving thiiranium ion intermediates may manifest themselves as a migration of an alkylthio group (e.g. equation 28)[90–92], ring expansion (equation 29)[93], or ring contraction (equation 30)[94].

$$(27)$$

$$CH_3SCH_2{}^{13}CH_2OH \xrightarrow[CCl_4]{Ph_3P} \left[\text{(thiiranium)} \right] \longrightarrow CH_3SCH_2{}^{13}CH_2Cl + CH_3S{}^{13}CH_2CH_2Cl$$

$$(28)$$

$$(29)$$

$$(30)$$

Eliminations also may occur via a thiiranium ion (equation 31)[95].

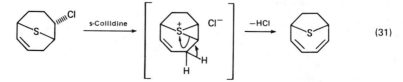

$$(31)$$

Evidence for an unusual sulphurane intermediate **18** was obtained by carbon magnetic resonance spectra of a mixture of 2,6-dichloro-9-thiabicyclo[3.3.1]nonane and fluorosulphonic acid in liquid sulphur dioxide at low temperatures (equation 32)[96]. Only four carbon resonances were observed at $-60°C$; at $-30°C$ formation of the thiiranium ion occurs.

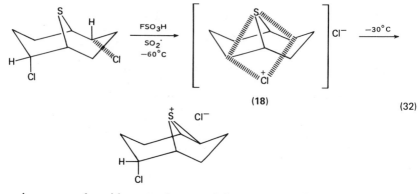

$$(32)$$

A carbene or carbenoid centre β to a sulphur atom can lead to formation of thiiranium ylides. Several reactions have been suggested to involve these intermediates (e.g. equations 33 and 34)[97–100].

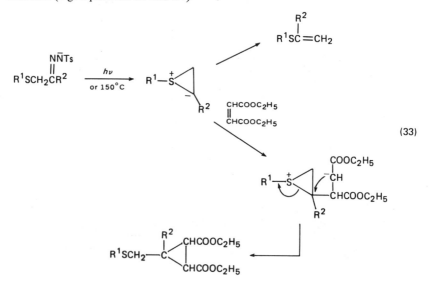

$$(33)$$

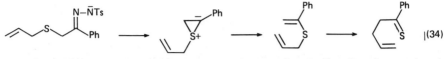

$|(34)$

A thiiranium ion has been suggested as an intermediate in the silver-ion catalysed rearrangement of **19** to **20** (equation 35)[101]. Here, the silver ion plays a role in generating an electron deficiency in proximity to a sulphur atom.

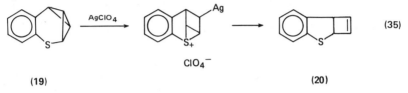

(35)

(19)   (20)

The third type of reaction in which thiiranium ions have been suggested as intermediates is found in reactions of electrophilic reagents with thiirans. Such an intermediate ion may occur in the reaction of methyl iodide or iodine with cis-1,2-dimethylthiiran, which yields cis-2-butene (equation 36)[102–104].

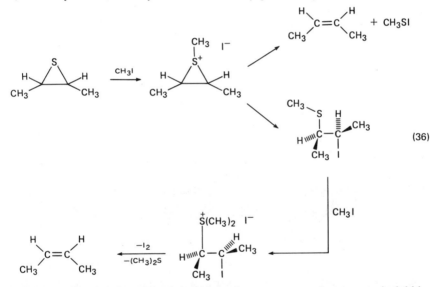

(36)

Addition of trimethyloxonium 2,4,6-trinitrobenzoate to cyclooctene episulphide also yields ring-opened products[26].

The reaction of α-chloroethers[105], α-haloamines[106], and trityl cations[107] with thiirans also may involve thiiranium ions, as may the acid-catalysed ring opening of thiirans[108,109] and the reaction of thiirans with electron-deficient carbonyl compounds (e.g. CF_3COF, equation 37)[110,111]. Thiiran S-oxide undergoes alkylation at oxygen with α-chloroethers to give oxythiiranium ions[112]. Carbenes effect desulphurization of thiirans, possibly via thiiranium ylides (equation 38)[113].

Finally, it may be noted that photolysis of cyclic compounds with a sulphur atom in proximity to an unsaturated site yields products often rationalized on the basis of intermediate thiiranium ions (e.g. equations 39–41)[114–119].

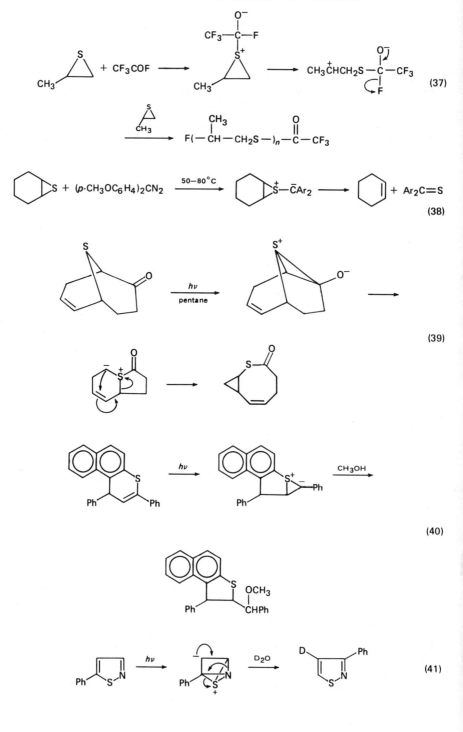

(37)

(38)

(39)

(40)

(41)

5. Reactions of thiiranium ions

In this section reactions at carbon of thiiranium ions will be discussed first, and then reactions at sulphur.

Equation 42 summarizes the ring-opening reactions of thiiranium ions involving reaction at carbon. The regioselectivity and the stereochemistry of reactions of thiiranium ions with nucleophiles have been discussed in Section 4. Occasionally, a fluoride ion originating from the tetrafluoroborate ion intervenes in the capture of a thiiranium ion[20]. Some variation in regioselectivity of attack dependent on the nature of the nucleophile has been observed[22]. Nucleophilic attack apparently can occur readily even at a neopentyl carbon if it is a member of a thiiranium ring[13]. The stereospecificity of the ring opening and lack of rearrangement seems to weigh against an open carbonium ion intermediate (equation 43).

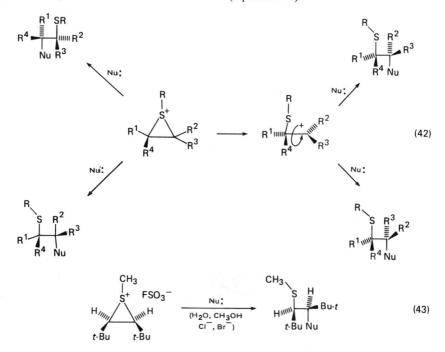

Chloride and bromide ions demethylate S-methylthiiranium salts, derived from adamantylideneadamantane, to give the thiiran[22b]; this reaction is the first of its type.

Rearrangements can occur via the open carbonium ion (e.g., equation 44)[21], and oligomerization can occur by reaction of the open ion with a molecule of alkene (e.g. equations 45 and 46)[20,120].

A neighbouring nucleophilic centre may intramolecularly attack a thiiranium ion to give rearranged products (e.g. equations 47 and 48)[31,43,106,121–124].

Thiiranium ylides are unstable and decompose to a variety of products (e.g. equations 33, 34, 38 and 40)[97–100,113,115] and, as noted in Section 4, alkenes may be produced in eliminations involving thiiranium ions (equation 31)[55,95,125–127].

Sulphur as well as the ring carbons of thiiranium ions may be attacked by

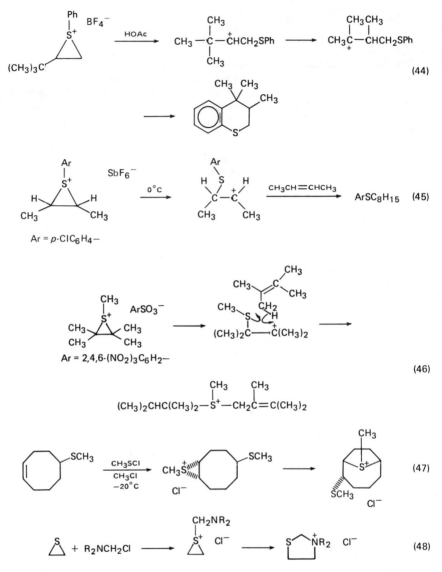

nucleophiles, and sulphurane intermediates may result (e.g. equation 1). The sulphurane **4**, according to the proton magnetic resonance spectrum, is stable at $-5°C$ for at least 30 min[11]. Decomposition to the ring-opened product occurred on warming the solution of **4** to room temperature or on addition of excess of chloride ions, even at $-5°C$ (equation 49).

The stable cyclooctyl thiiranium ion **3** reacts with a variety of nucleophiles to yield cyclooctene (equation 50). Only with acetate ion was attack on carbon observed[128]. This desulphurization has been interpreted as proceeding either through a sulphurane intermediate or an S_N2-like displacement on sulphur[18,128].

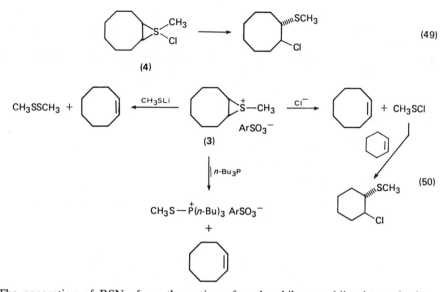

(49)

(4)

(3)

(50)

The generation of RSNu from the action of nucleophiles on thiiranium salts has been demonstrated by isolation or by trapping experiments with cyclohexene and other alkenes[22b,128,129]. The ring size of the parent cycloalkene is important in determining the ease of attack on sulphur. Nine- and eight-membered rings are superior in this respect to seven- and five-membered rings which, in turn, are better than six-membered rings[129]. However, no evidence for sulphurane formation was obtained in the reaction of 1-methyl-cis-2,3-di-t-butylthiiranium fluorosulphonate with chloride ions: no methanesulphenyl chloride was trapped by cyclohexene and there was no reaction between methanesulphenyl chloride and cis-1,2-di-t-butyl-ethylene[13]. The only product was β-chlorosulphide from attack of chloride ion on carbon.

Attack on sulphur of thiiranes by carbenes yields thiiranium ylides which, as mentioned in Section 4, decompose to the alkene and a thiocarbonyl compound (equation 38)[113]. This decomposition can be considered a displacement on sulphur by the pair of electrons on carbon in the thiiranium ylide (equation 51).

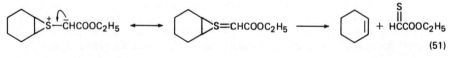

(51)

B. Thiirenium Ions

1. Reviews

Thiirenium ions, both stable and as intermediates in reactions, have been reviewed[1,4,5b,130].

2. Theoretical

An early theoretical paper touched briefly on thiirenium ions as a member of a class of potentially heteroaromatic compounds containing three-membered rings[131a].

Molecular orbital calculations on S-protonated thiiren indicate that a pyramidal configuration is more stable than other configurations[131b]. The barrier to inversion at sulphur was calculated to be 72.9–85 kcal mol^{-1} [132–134]. The barrier to interconversion to an open-chain valence tautomer also was calculated, the thiirenium ion being 66 kcal mol^{-1} more stable (equation 52)[135].

$$\overset{H}{\underset{\triangle}{\overset{+}{S}}} \rightleftharpoons \overset{+}{C}H{=}CHSH \tag{52}$$

3. Stable thiirenium ions

The first stable thiirenium ions were prepared by treatment of acetylenes with methanesulphenyl chloride in the presence of antimony pentachloride at $-120°C$ or by treatment of acetylenes with methyl(bismethylthio)sulphonium hexa-fluoroantimonate at low temperatures (equation 53)[136,137]. The di-t-butyl derivative can be precipitated from pentane as white crystals, m.p. 151–152°C,

$$RC{\equiv}CR + (CH_3S)_2\overset{+}{S}CH_3\ SbCl_6^- \xrightarrow{SO_{2l}} \underset{R \quad R}{\overset{CH_3}{\underset{\triangle}{\overset{|}{S^+}}}} SbCl_6^- \tag{53}$$

$$R = CH_3, C_2H_5, (CH_3)_3C-$$

stable at room temperature[136]. The trimethyl thiirenium salt is unstable above $-40°C$. An X-ray analysis of 1-methyl-2,3-di-t-butyl-2-thiirenium tetrafluoroborate shows the configuration about the sulphur atom to be pyramidal[138]. The stable tetrafluoroborate was prepared by addition of methanesulphenyl chloride to di-t-butylacetylene at $-60°C$ followed by treatment with silver tetrafluoroborate[136]. The initially formed chloride salt was stable for several days in liquid sulphur dioxide but decomposed slowly in methylene chloride. Earlier, stable thiirenium ions were observed by proton and carbon magnetic resonance techniques[139]. In the carbon magnetic resonance spectrum (acetone-d_6, $-60°C$) of the trimethyl-thiirenium cation[139] the ring carbons show resonance at δ 103.1 p.p.m. and the methyl carbon on sulphur shows a resonance at δ 25.1 p.p.m. relative to the tetramethylsilane; the carbon magnetic resonance spectrum (CD$_2$Cl$_2$, room temperature) of the 1-methyl-2,3-di-t-butyl cation[136] shows absorption of the ring carbons at δ 115.1 p.p.m. and the methyl carbon on sulphur absorbs at δ 30.6 p.p.m. and the methyl groups on the ring at δ 9.1 p.p.m. The proton magnetic resonance spectra of the S-methyl protons show absorption at δ 2.51 p.p.m. for the trimethyl derivative[139] and at δ 2.62 p.p.m. for the di-t-butyl derivative[136].

4. Thiirenium ions as intermediates in reactions

Additions of sulphenyl halides (CH$_3$SCl, ArSCl) and sulphur dichloride[140] to acetylenes are often interpreted on the basis of a thiirenium ion as an intermediate although sulphuranes also may be involved. A kinetic study by proton magnetic resonance and conductance methods of the addition of alkyl and aryl sulphenyl chlorides to acetylenes in liquid sulphur dioxide at $-67°C$ indicated a two-step process, a fast step involving formation of a thiirenium ion followed by a slow attack of chloride ion to give *trans* (E) products. The intermediacy of a sulphurane was considered, particularly in less polar solvents (equation 54)[141].

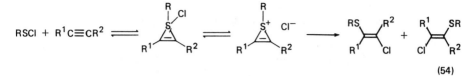

$$\text{(54)}$$

From data on the effect of substituents on the rates of reaction of arylacetylenes with p-chlorophenylsulphenyl chloride, it was concluded that the positive charge in the thiirenium ion was delocalized into the aryl π-orbital system as in **19**[142], although in reactions with 2,4-dinitrobenzenesulphenyl chloride an opposite conclusion was reached, the aryl π-orbital system being conjugated with the double bond of the thiirenium ion as in **20**[143].

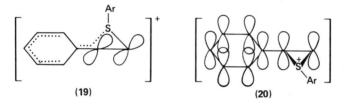

(19)   **(20)**

Olefinic compounds with a good leaving group at one terminus of the carbon–carbon double bond, and a *trans*-thioalkyl or -thioaryl group at the other terminus, can undergo loss of the leaving group with formation either of a vinyl cation or a thiirenium ion (equation 55). Chemical, stereochemical, and kinetic data have been used to support the intervention of thiirenium ions in these reactions[144–150].

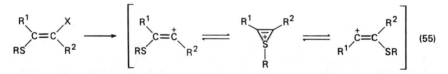

$$\text{(55)}$$

In solvolysis of ^{14}C-labelled vinyl derivatives, the two vinyl carbon atoms were shown to become equivalent, evidence which supported a symmetrical intermediate such as a thiirenium ion (equation 56)[145].

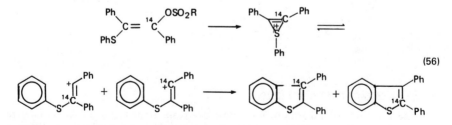

$$\text{(56)}$$

The *trans* (E) stereochemistry of attack by a nucleophile also is consistent with a thiirenium ion intermediate (equation 54).

Large rate enhancements in solvolysis are caused by the presence of a β-*trans*-thioalkyl or -thioaryl group in vinyl sulphonate esters, e.g. the solvolysis of

21 in methanol–nitromethane at 25°C is 37,000 times faster than that of **22**[150]. Such large increases in rate imply neighbouring group participation by sulphur and are consistent with the formation of a thiirenium ion.

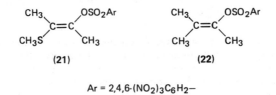

(21)   (22)

Ar = 2,4,6-(NO$_2$)$_3$C$_6$H$_2$−

Another finding in favour of thiirenium intermediates was the obtention of the same product composition (same proportions) from treatment of mixtures of **23** and **24** with hydrogen chloride as from treatment of acetylene **25** with benzenesulphenyl chloride (equation 57). The same intermediate thiirenium ion was postulated for both reactions[151,152].

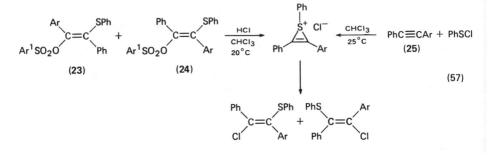

(57)

An example of a 1,2-thiirenium ion postulated in a reaction of a different type from those discussed above is the one proposed as an intermediate in the photochemically induced rearrangement of mesoionic compound **26** (equation 58)[153].

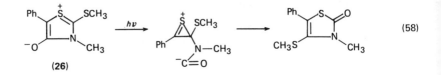

(58)

5. Reactions of thiirenium ions

Some reactions of thiirenium ions have been mentioned above. To summarize, ring-opening reactions occur via attack on either carbon or on sulphur. Sulphurane intermediates (equation 54) and open vinyl cations (equation 56) may also be involved.

The attack of a nucleophile at a ring carbon atom of the thiirenium ion may occur either at the least sterically hindered site[142] (the product being designated as anti-Markovnikov) or at the site most able to stabilize a positive charge[154,155] (the product being designated as Markovnikov). The *trans* (*E*) product is observed. *S*-Aryl derivatives are more reactive to ring opening than *S*-alkyl derivatives, presumably because of resonance of the lone-pair electrons of sulphur in the ring-opened sulphide with the aryl ring[141]. The anti-Markovnikov mode of attack seems to be preferred in weakly polar solvents (in which sulphurane formation may be a complication), in systems in which the substituent on sulphur is capable of stabilizing its positive charge, and in systems in which one carbon site is not hindered sterically. The second factor does not seem particularly important except with *o*-nitro- and 2,4-dinitrobenzenesulphenyl chloride additions to acetylenes in which the destabilizing effect of the electron-withdrawing nitro groups on the positive sulphur atom (and the stabilizing effect on neutral sulphur) leads to more Markovnikov addition[156]. Steric hindrance is not completely effective in blocking attack at a vinyl carbon of a thiirenium ion since 1-methyl-2,3-di-*t*-butylthiirenium chloride in methylene chloride readily yields the vinyl chloride of the *E* configuration (equation 59), although in liquid sulphur dioxide this di-*t*-butyl salt seems much more stable than the trimethylthiirenium derivative[136,139].

$$\text{(59)}$$

The Markovnikov mode of attack is preferred for thiirenium ions in which more positive charge may be developed on carbon or where there is hindrance to nucleophilic attack. Polar solvents seem to favour Markovnikov addition[157] and the extent of this mode of addition of *p*-toluenesulphenyl chloride to phenylacetylene increases more in accord with the measure of polarity of solvent, E_T[158], rather than dielectric constant, *D* (equation 60).

$$\text{ArSCl + PhC}{\equiv}\text{CH} \longrightarrow \text{(60)}$$

Solvent	D	E_T	anti-Markovnikov product	Markovnikov product
EtOAc	6	38.1	100%	—
CHCl$_3$	4.7	39.1	65%	35%
HOAc	6.2	51.9	29%	71%

Nucleophilic attack on sulphur by chloride ion with the formation of sulphenyl chloride and an acetylene has been observed (equation 61)[152]. The sulphenyl

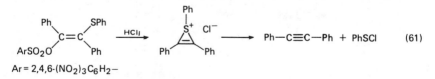

$$\text{Ar = 2,4,6-(NO}_2)_3\text{C}_6\text{H}_2-$$

chloride could be trapped by addition to bis-*p*-tolylacetylene. The ratio of attack at sulphur to attack at carbon was estimated at 1:10[152]. A sulphurane intermediate has not yet been detected in reactions of thiirenium ions, although evidence for them has been obtained in reactions of thiiranium ions[11].

C. 1,2-Dithiiranium Ions

These ions have been suggested as intermediates in several reactions[159,160], e.g. equations 62[161] and 63[162].

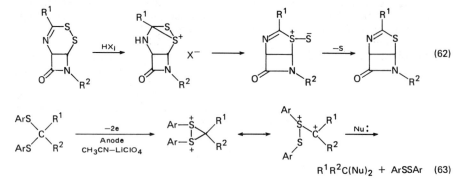

$$R^1R^2C(Nu)_2 + ArSSAr \quad (63)$$

III. FOUR-MEMBERED RINGS

A. Saturated Rings

1. Reviews

Two reviews on thietanes have touched briefly on cyclic, four-membered sulphonium salts[163,164].

2. Stable thietanium salts

Treatment of thietane with methyl iodide originally was believed to yield a bis-methylated thietanium salt[165], but subsequently it was shown that ring opening had occurred and that the compound obtained was dimethyl-(3-iodopropyl)-sulphonium iodide[166].

Relatively stable, simple thietanium salts have been prepared by alkylation of the corresponding thietanes with trimethyloxonium fluoroborate at low temperatures (equation 64). Salt **27a** is reported to be a white solid which decomposes on

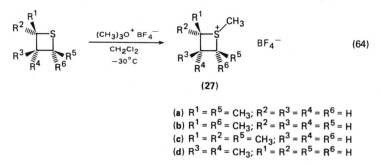

(a) $R^1 = R^5 = CH_3$; $R^2 = R^3 = R^4 = R^6 = H$
(b) $R^1 = R^6 = CH_3$; $R^2 = R^3 = R^4 = R^5 = H$
(c) $R^1 = R^2 = R^5 = CH_3$; $R^3 = R^4 = R^6 = H$
(d) $R^3 = R^4 = CH_3$; $R^1 = R^2 = R^5 = R^6 = H$

warming[167,168], but **27d** is reported to have a melting point of 156–158°C[169]. A brief mention of the 3-ethylthietanium salt was made, but details of its preparation, presumably via triethyloxonium fluoroborate, were not given[169].

The proton magnetic resonance spectra of these salts show absorption for the S-methyl protons at δ 2.75–3.25 p.p.m. and for the ring protons adjacent to sulphur at δ 3.67–4.63 p.p.m.[169]. S-Protonated thietane has been observed at −60°C by n.m.r. techniques. The S—H proton absorbed at δ 7.40 p.p.m. as a multiplet and the ring protons were observed as multiplets at δ 3.20–4.40 p.p.m.[12].

Ionization of sulphurane **28** to a thietanium salt at −40 to −20°C was indicated by the collapse of the two doublets observed in the ^{19}F n.m.r. spectrum (equation 65)[170].

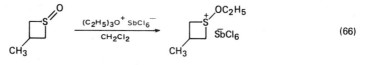

(65)

(28)

Stable S-ethoxy[171], S-methoxy[172], and S-aminosulphonium salts[173] have been obtained (equations 66–68).

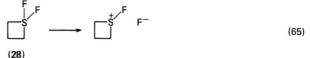

(66)

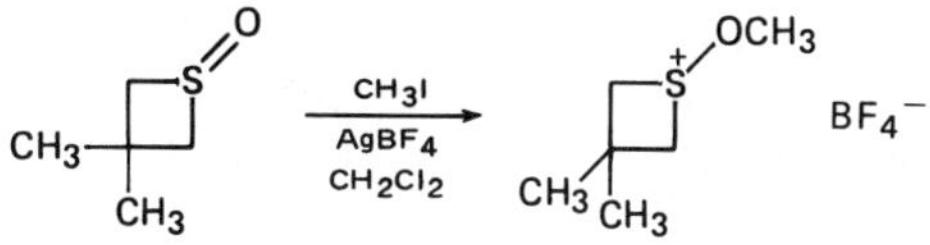

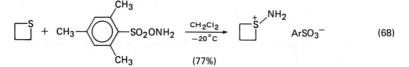

(67)

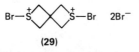

(68)

A stable spirosulphonium dibromide **29** has been reported[174].

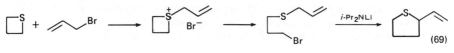

(29)

3. Thietanium ions as reaction intermediates

Treatment of thietanes with electrophilic reagents in the presence of nucleophiles leads to ring-opened products by way of intermediate thietanium ions[163–166,175]. The reaction of allyl bromide with thietane is part of a synthetically useful procedure in the preparation of five- and eight-membered rings (equation 69)[176,177]. The

(69)

reactivity of thietane to the trityl cation is greater than that for the corresponding five- and three-membered heterocycles[107].

Intramolecular examples of the formation of thietanium ions as intermediates have been reported[93,178] (e.g. equation 70)[179].

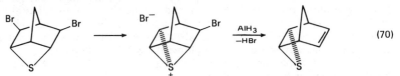

$$(70)$$

The tendency for thietanium ion formation from γ-halosulphides is much less than that for thiiranium or thiolanium ion formation from β-halosulphides and δ-halosulphides, respectively[87,180,181]. Carbonium ion formation γ to a sulphide function or introduction of a superior leaving group at that position may be effective in promoting thietanium ion formation[182,183] (e.g. equation 71)[184].

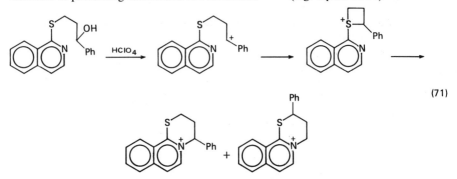

$$(71)$$

The cationic polymerization of thietanes by triethyloxonium fluoroborate involves ring opening of an initially formed S-ethylthietanium ion by nucleophilic attack by the sulphur atom of another molecule of thietane[185]. Intermediate thietanium ions also may undergo ring cleavage by way of an elimination reaction (equation 72)[186]. Chlorine and bromine react with thietanes to yield sulphenyl halides via cyclic S-halosulphonium ions[164].

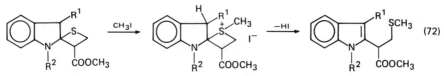

$$(72)$$

Reaction of dimethyl diazomalonate with thietane yields a polymer, probably via the ylide **30**; in the presence of copper(II) sulphate a 26% yield of a tetrahydrothiophene is obtained (equation 73)[187]. Ylides also may be formed by

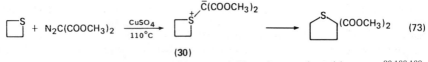

$$(73)$$

$$(30)$$

interaction of a sulphur atom with a neighbouring carbenoid centre[98,188,189]. Formation of a bicyclic ylide from **31** did not appear to occur[190].

Thietanium ion intermediates have been suggested as being involved in several

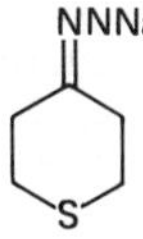

(31)

transformations of cyclic sulphides with a double bond ($>C=O^{191}$, $>C=N^{192}$) in proximity to the sulphur atom (e.g. equation 74).

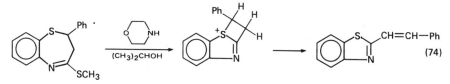

(74)

The photochemistry of several five-membered sulphur-containing cyclic compounds has been interpreted in terms of thietanium ion intermediates[116] (equation 75)[193].

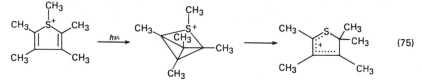

(75)

Intermediates of the type 32 are believed to be involved in the reaction of N-bromosuccinimide or sulphuryl chloride with acyclic β-hydroxysulphoxides[194,195], and the intermediate 33 is believed to be involved in hydrindane formation via an S-phenylthiirane S-oxide[196].

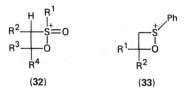

(32)   (33)

4. Reactions of thietanium ions

Nucleophilic reagents may attack thietanium ions either at the α-carbon atoms or at the positive sulphur atom. A number of reactions involving nucleophilic attack at a ring carbon of a thietanium ion intermediate have been described in Section 3. Nucleophilic attack may be intermolecular (e.g. equation 69) or intramolecular, as shown in equations 71[184] and 76[182]. The nucleophilic centre can be an aromatic ring[184] or a carbon–carbon triple bond[196].

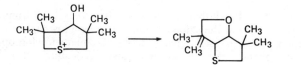

(76)

The relief of strain is presumably the reason why attack by a nucleophile occurs at a carbon atom of the four-membered ring, although this is not true with the alkoxyoxosulphonium ions **32** (equation 77)[194,195]. If R^1 is phenyl, attack then occurs on a ring carbon atom, presumably by an S_N1 process (equation 78)[195].

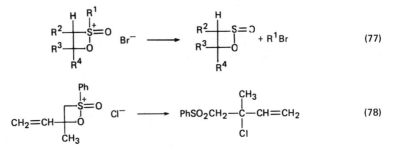

$$(77)$$

$$(78)$$

The propagation of polymerization of thietane by formation of thietanium ions followed by ring opening has been described[12,185]. Ring opening also may occur via elimination reactions (equations 72[186] and 74[192]).

Thietanium ylides, obtained by the action of carbenes centres with a divalent sulphur atom, undergo a variety of reactions[98,188] (equations 79 and 80)[189].

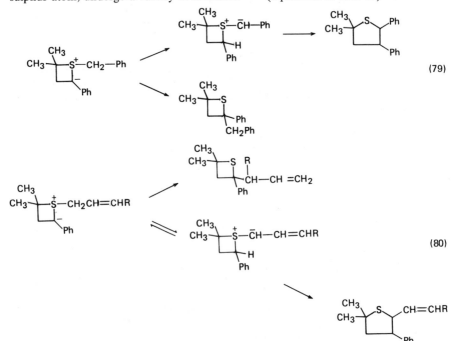

$$(79)$$

$$(80)$$

The attack of *n*-butyllithium on *cis*- or *trans*-1,2,4-trimethylthietanium tetrafluoroborate occurs at the sulphur atom, effecting a low yield but stereospecific desulphurization, the *cis* salt giving trans-1,2-dimethylcyclopropane and the *trans–cis* salt giving the *cis*-cyclopropane (equation 81)[167,168]. Several

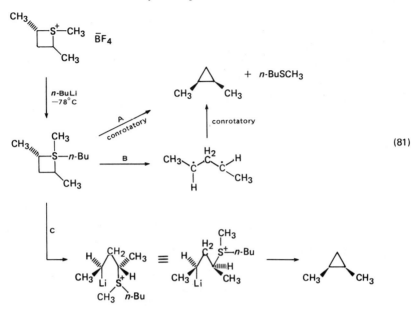

(81)

rationalizations of the results were considered. Pathway C (equation 81) was deemed unlikely because 1,2,2,4-tetramethylthietanium tetrafluoroborate gave 1,2,2-trimethylcyclopropane in a yield (30%) comparable to those from the *cis*- and *trans*-trimethyl salts, despite its seemingly greater steric hindrance to an intramolecular S_N2 displacement reaction. Pathway A was believed unlikely because it involved a strained transition state.

Displacement reactions occur on an *S*-ethoxythietanium ion with inversion at sulphur. Sulphurane intermediates were considered as well as a mechanism involving direct displacement on sulphur[171]. Displacement on sulphur occurs in the hydrolysis of the sulphonium dibromide **29** (equation 82)[174,197]. The salt **29** is said to be unique among thietanium bromides in not undergoing ring cleavage.

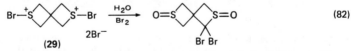

(82)

(29)

The acidity of the protons on the carbon atom adjacent to the positive sulphur atom in several relatively stable thietanium salts has been investigated[169]. In 1,3,3-trimethylthietanium tetrafluoroborate, the 1-methyl protons exchange more than 3×10^4 times more rapidly (D_2O–OD^-) than do the α-methylene protons whose lability to exchange may be affected by the steric effects of the two adjacent methyl groups. Interestingly, the four α-methylene ring protons in 1-ethylthietanium tetrafluoroborate exchange very rapidly but the α-methylene protons of the ethyl group are not readily exchanged. The proton exchange in four-membered sulphonium ions is approximately 100-times faster than exchange in sulphonium ions of larger ring size. The faster rate was attributed to more s-character in the bonding from sulphur to carbon in the four-membered compounds; this increase in s-character was expected to have an acidifying effect on the protons attached to the α-carbon atoms.

B. Unsaturated Rings

1. Theoretical

An MO calculation on the thiacyclobutadiene **34**, which may have a zwitterionic resonance structure, indicates that the most stable conformation has all atoms nearly in the same plane except for the proton bonded to the sulphur atom, the latter having a pyramidal configuration[198].

(34)

2. Stable thiacyclobutenonium salts

The fused thiacyclobutene (thiete) derivative **35** is alkylated by trimethyloxonium fluoroborate to give a stable salt, m.p. 146–147°C (equation 83). The *S*-methyl protons absorb at δ 3.82 p.p.m. in DMSO-d_6[199].

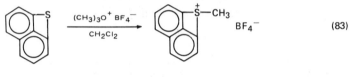

(83)

(35)

Several stable, yellow 1-phenylthiacyclobutenonium ions have been reported to be obtained by treatment of ketosulphides **36** with phosphorus oxychloride and perchloric acid[200] (equation 84). They are relatively high melting and show a single

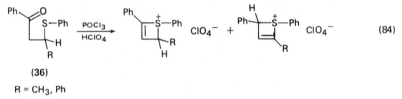

(84)

(36)

R = CH$_3$, Ph

proton absorption in their nuclear magnetic resonance spectra around δ 2.95 or 3.5 p.p.m. The vinyl proton is concealed under the absorptions of the phenyl protons.

3. Thiacyclobutenonium ions as reaction intermediates

These unsaturated sulphonium ions were proposed to account for the apparent interchange of a phenyl group with a hydrogen in the cyclization shown in equation 85[201]. Another kind of ion was suggested as an intermediate in the reaction of thietes with acidic 2,4-dinitrophenylhydrazine (equation 86)[202].

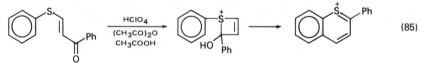

(85)

$$\text{(86)}$$

1-Substituted thiazacyclobutenonium ions have been proposed in the rearrangement of oxime derivatives with a thiomethyl group substituted in the α-position[203] (equation 87)[204].

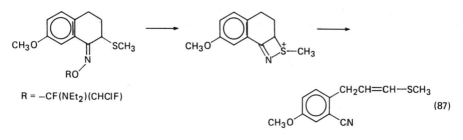

R = −CF(NEt₂)(CHClF)

$$\text{(87)}$$

4. Reactions of thiacyclobutenonium ions

The stable ion obtained by the methylation of **35** (equation 83) undergoes demethylation in refluxing pyridine, cleavage to a sulphoxide on treatment with hydroxide ion, and cleavage to two thioethers on reduction with lithium aluminium hydride (equation 88)[199].

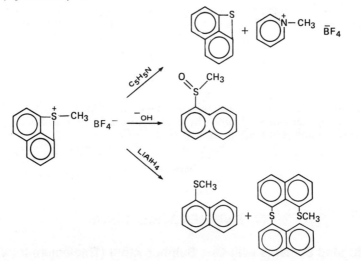

$$\text{(88)}$$

Green S-phenylthiacyclobutadiene derivatives are obtained by treatment of S-phenylthiacyclobutenonium ions with sodium hydride (equation 89)[200]. The salts may isomerize as shown in equation 90[200].

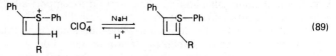

$$\text{(89)}$$

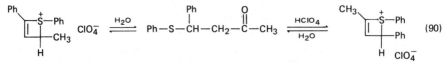

(90)

Nucleophilic attack on thiacyclobutenonium ion intermediates by water causes ring opening (equations 91[202] and 92[203a]). 3-Phenylthiete reacts readily with triethyloxonium tetrafluoroborate to yield the S-ethyl salt which undergoes both ring opening and dealkylation on treatment with nucleophiles[203b].

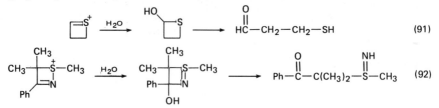

C. Rings with More than One Sulphur Atom

Dithiete radical cations with a variety of substituents have been identified by e.s.r. spectroscopy (equation 93)[205a]. 2-Dialkylamino-1,3-dithietenium salts are obtained by treatment of benzylidenebis-(NN-dialkyldithiocarbamates) with perchloric acid or sulphuric acid[205b]. Electrochemistry of some dithietenium salts has been reported[205c].

A structure containing three sulphur atoms was suggested for the yellow solid obtained by treatment of dithiobenzoic acid with sulphur dichloride. Treatment

$$ R^1 \overset{O}{\underset{\|}{C}} - \overset{OH}{\underset{\|}{C}} H - R^2 \xrightarrow[H_2SO_4]{Na_2S} \left[\begin{array}{c} R^1 \\ R^2 \end{array} \text{S-S} \right]^{+\cdot} $$

(93)

with aniline or morpholine gave the thioamide and sulphur (equation 94)[205d]. A blue, cyclic S_4^{2+} species is obtained by oxidation of sulphur with $S_2O_6F_2$, AsF_5, or SbF_5. It is unstable and tends to form S_8^{2+} [205e].

$$ Ph-\overset{S}{\underset{\|}{C}}-SH + SCl_2 \longrightarrow \overset{Ph}{\underset{S-S}{\diagdown}}\overset{+}{S}\ Cl^- \xrightarrow[-S_2]{PhNH_2} Ph-\overset{S}{\underset{\|}{C}}-NHPh $$

(94)

IV. FIVE-MEMBERED RINGS

A. Saturated Systems with One Sulphur Atom (Thiolanium Ions)

1. Structure

Carbon magnetic resonance spectra have been reported for several S-methylthiolanium salts, and the data are said to support a half-chair conformation with a quasi-equatorial orientation of the S-methyl group[206].

Molecular orbital calculations on ylides derived from thiolanium ions 37–40 confirm that the major stabilizing factor in the ylide is the orbital interaction between the carbanion lone pair orbital and the antibonding sigma orbital of the adjacent $\overset{+}{S}$—$C_{(\alpha)}$ bond[207]. Further details are provided in a publication dealing with hydrogen–deuterium exchange in six-membered cyclic sulphonium ions[208]. The arrangement depicted in **41** is more stable than that depicted in **42**.

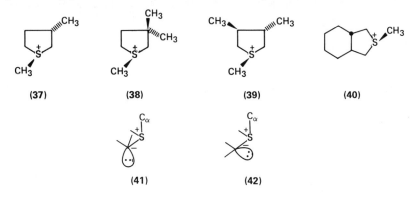

(37)   (38)   (39)   (40)

(41)   (42)

2. Synthesis

Unlike three- and four-membered cyclic sulphonium salts, the synthesis of cyclic five-membered salts poses no particular problems because of their greater stability.

The availability of thiolanes (tetrahydrothiophenes) makes their reactions with electrophilic reagents one of the most convenient methods of preparation. Dimethyl sulphate and alkyl iodides, bromides, and chlorides which are not hindered sterically in displacement reactions (S_N2) are often used. Allyl and propargyl halides and α-haloketones and esters react well. Perchlorate salts may be obtained by exchange of halide by means of silver perchlorate[209] and tetrafluoroborate salts may be obtained similarly[210]. Fluoroborate salts may be converted into the less water-soluble and less hygroscopic hexafluorophosphate salts by treatment with aqueous ammonium hexafluorophosphate[211]. Oxonium and sulphonium salts also alkylate thiolane[211,212]. For long-term storage of these sulphonium salts, a non-nucleophilic anion is preferred over halide. S-Vinylthiolanium salts are made conveniently via β-bromoethyl alcohol (equation 95)[213].

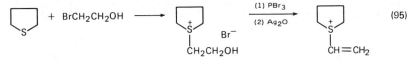

$$\text{(95)}$$

Polymers containing thiolanium ions have been prepared from p-chloromethylstyrene and thiolane[214].

Short-lived carbonium ions can be trapped by thiolane. The sulphate salts obtained as the example of equation 96 shows are converted into their Reineckate salts, $[Cr(SCN)_4(NH_3)_2]^-$, picrates or chlorides for characterization[215].

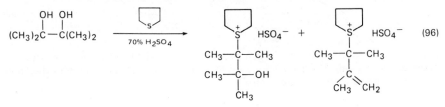

$$\text{(96)}$$

S-(4-Chlorobutyl)thiolanium chloride is one of a number of products obtained by treatment of thiolane with tungsten hexachloride[216].

Displacement by thiolane on a sulphonium ylide in the presence of bistrifluoromethyldisulphide gives a thiolanium ylide (equation 97)[217]. The reaction

(97)

of thiolane with diazoalkanes in the presence of acid yields sulphonium salts (equation 98)[176] or, in the absence of acid, sulphonium ylides[187] (equation 99)[218].

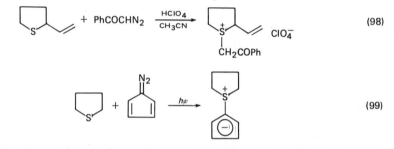

(98)

(99)

Stable ylides of thiolanes are obtained by reaction of thiolane S-oxides with several compounds having acidic hydrogen atoms[219] (equations 100[220] and 101[221]).

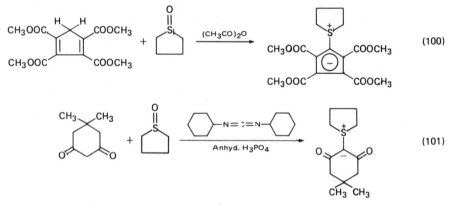

(100)

(101)

S-Arylthiolanium salts can be obtained by treatment of thiolane S-oxide with phenols or anisoles in the presence of Lewis acids[222,223] (equation 102)[224]. Thiolane

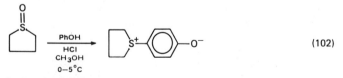

(102)

has been used directly but hydrogen peroxide is added to the reaction mixture to generate the S-oxide[225]. Protonated quinones or quinone imines react with thiolanes to give the corresponding S-aryl salts[226].

Another route to thiolanium ions is by cyclization of a sulphide via attack of the sulphur atom on an electrophilic site (e.g. equations 47, 103[227], 104[228], 105[229], 106[230], 107[222], 108[231], 109[232]). Five-membered ring formation is less favoured than three-membered ring formation when appropriate precursors are compared[233].

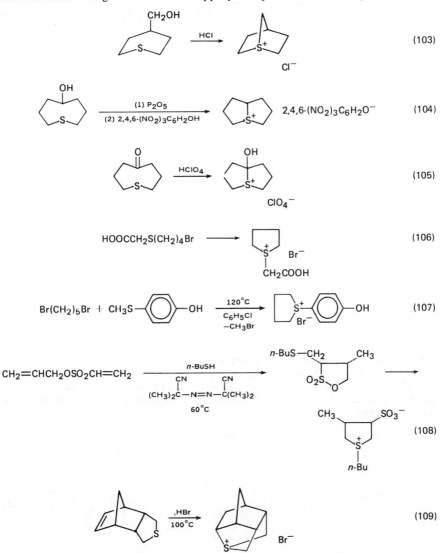

Thiolanium ions in which a sulphur atom is attached to one or more heteroatoms also are known. In some cases, the positive charge on sulphur may be diminished by resonance involving p–d backbonding. Treatment of thiolane with O-mesitylenesulphonylhydroxylamine yields the S-aminosulphonium salt[173]. The sulphoxide reacts similarly (equation 110). Zwitterionic structures may be obtained (equations 110 and 111)[173]. *S*-Amino derivatives may be obtained via *N*-bromoacetamide (equation 112)[234] or via the *S*-chlorothiolanium salts and amines[235] (equation 113)[236].

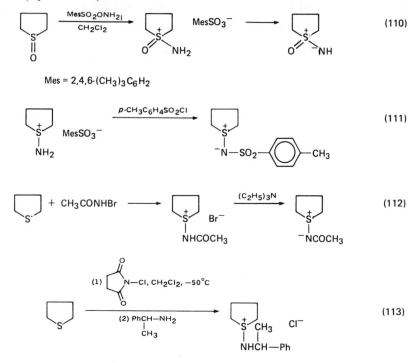

Mes = 2,4,6-(CH$_3$)$_3$C$_6$H$_2$

Salts containing oxygen and nitrogen atoms attached to sulphur have been reported (equations 110 and 114[237]).

S-Alkoxythiolanium ions may be prepared by treatment of the *S*-halo salts with alcohols (equation 115)[235] or by *O*-alkylation of sulphoxides with triethyloxonium tetrafluoroborate[238a].

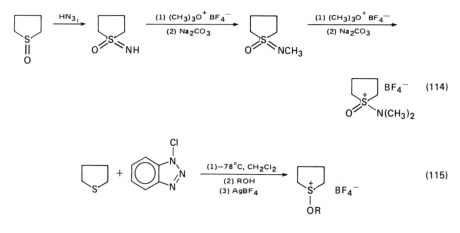

Transannular interaction of a thiolane sulphoxide with a proximate positive centre yields the cation **43** (equation 116)[238b].

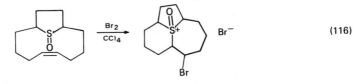

$$\tag{116}$$

(43)

Treatment of thiolane 1,1-dioxide with aryldiazonium fluoroborates or hexafluorophosphates yields aryloxyoxosulphonium salts (equation 117)[239].

$$\tag{117}$$

S-Thioalkylthiolanium ions have been prepared as shown in equation 118[28].

$$\tag{118}$$

The structure of the 1:1 adduct of thiolane with bromine has been investigated by X-ray analysis. The tricoordinate sulphur atom is pyramidal. The proton n.m.r. spectrum is similar to that of the 1-methylthiolanium ion, suggesting that considerable positive charge is localized on sulphur[240].

3. Thiolanium ions as intermediates

The reaction of p-tolyl 4-bromobutyl sulphide with alkoxides proceeds through the thiolanium ion (which can be isolated) followed by a fast reaction with the base to give a mixture of alkene and ether. A common ion retardation is observed in dilute solutions and lithium perchlorate accelerates the reaction. Intimate ion pairs may be involved[241]. Cyclizations to five-membered intermediates are slower than to three-membered but faster than to four-membered intermediates[181,233]. Variations in the transition states for formation of three- and five-membered cyclic ions have been discussed with respect to structural variations and the entropy of activation for three-membered ring formation was determined to be more negative than for five-membered ring formation[233].

A bicyclic thiolanium ion was suggested as an intermediate in a synthesis of d-biotin (equation 119)[242].

The stereospecific cyclopropane formation shown in equation 120 was believed to involve a thiolanium ion intermediate[243]. A Stevens-type rearrangement of an intermediate thiolanium ylide was postulated in the decomposition of the diazo compound **44** (equation 121)[244].

A photochemical reaction of a sulphide in proximity to a carbonyl group has been interpreted as involving a thiolanium ylide (equation 122)[245]. The interaction of tungsten hexachloride with thiolane may involve an S-halosulphonium ion[216].

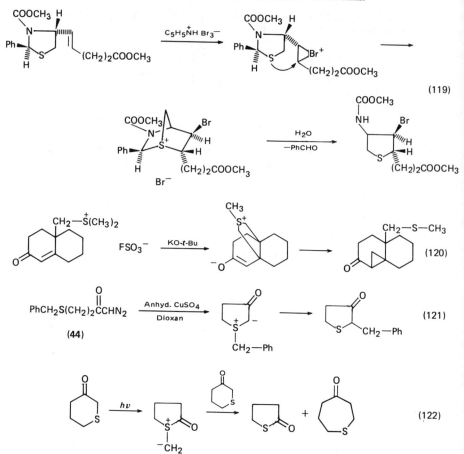

4. Reactions of thiolanium ions

Ring opening of thiolanium ions by nucleophilic attack occurs, but much less readily than with thiiranium and thietanium ions. Azide and methanethiolate ions preferentially attack a ring carbon rather than the methyl group in S-methylthiolanium iodide, although the reverse is true for the six-membered salt[246]. Only when the thiolanium salt is substituted at both the 2- and the 5-positions does attack predominate at the exocyclic S-methyl group[246]. Halide ions, however, under vigorous conditions attack the S-alkyl group rather than a ring carbon atom[247,248]. The ability to break carbon–sulphur bonds in S-arylthiolanium perchlorates was observed to decrease in the order KOH–CH₃OH > Cl⁻ > Br⁻ > I⁻ > 2,4,6(NO₂)₃C₆H₂O⁻. The last two anions did not effect ring opening, and only exchange with perchlorate ion occurred[249]. Amines readily open thiolanium rings at 55°C[223,250]. The bridged sulphonium ion **45** reacts with nucleophiles to yield first a thiiranium–thiolanium ion, followed by rupture of the three-membered ring (equation 123)[232a].

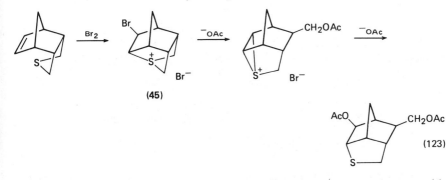

(45)

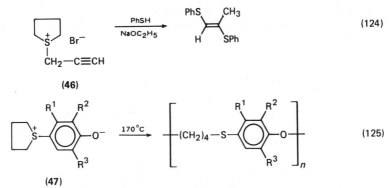

(123)

The thiolanium salt **46** isomerizes to an allenenic salt on treatment with thiophenol and sodium ethoxide followed by elimination of thiolane (equation 124)[251]. The zwitterionic thiolanium salt **47** polymerizes on heating (equation 125)[222,224].

(46)

(124)

(47)

(125)

Ylides derived from thiolanium ions may undergo displacement of thiolane[213,252a] (equation 126)[252b], fragmentation[253] (equation 127)[254], ring expansion (equation 128)[176], or rearrangement (equation 129)[255]. These ring expansions are useful synthetically in the preparation of large rings[177,211,256-258]. Most ylides derived from thiolanium salts show the usual behaviour of sulphur ylides, additions to carbon–carbon double bonds[259-261] giving cyclopropanes and additions to carbon–oxygen double bonds giving oxirans[260]. In certain cases a new, stable ylide

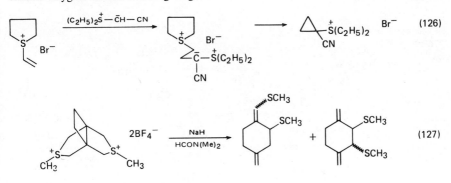

(126)

(127)

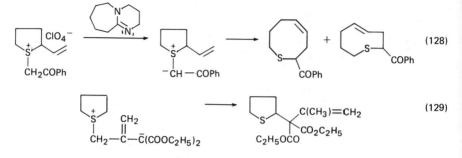

may be formed by addition to a carbon–carbon double bond substituted with highly electron-withdrawing substituents[262].

α-Arylfuran derivatives are obtained in low yield via thiolanium ylides[263]. Phenyl and p-nitrophenyl isocyanates give new ylides when treated with thiolanium ylides **48** (equation 130)[264]. S-Carboxymethylthiolanium bromide gives an amide with p-nitrophenyl isocyanate and a tetrahydropyrimidine with phenyl isocyanate (equation 131)[265]. Silicon and tin derivatives have been obtained from thiolanium ylides (equation 132)[266].

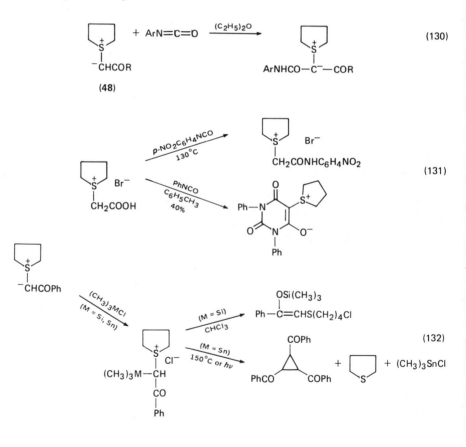

Ylides derived from *S*-amino-substituted thiolanium ions yield cyclopropanes or oxirans with activated alkenes or carbonyl compounds (equation 133)[237].

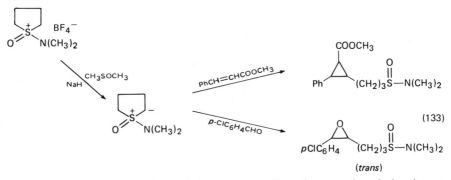

(133)

(*trans*)

A recent process for *ortho*-alkylation of aromatic amines or phenols involves a rearrangement of a thiolanium ion via an ylide (e.g. equation 134)[267].

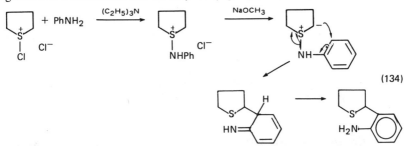

(134)

S-Halosulphonium salts may lose hydrogen halide (HCl, HBr) to give an intermediate ion which is responsible for the various products obtained by halogenation of thiolane[268a]. Displacement on the sulphur atom of *S*-succinimidothiolanium chloride by indole gives indole 3-thioethers (equation 135)[268b].

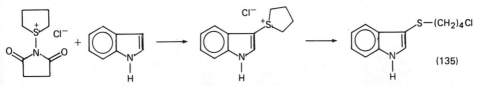

(135)

The rate of pyramidal inversion at the sulphur atom in thiolanium perchlorates **38** and **49** has been determined from the rate of racemization and epimerization[269]. These rates are lower than those of acyclic sulphonium salts, possibly because of added angle strain in the cyclic salts when passing through a planar configuration about sulphur.

Hydrogen–deuterium exchange in thiolanium ions is believed to have a transition state which resembles a carbanion (ylide)[169] and the importance of the anti-bonding σ orbital (σ*) of the S—$C_{(\alpha)}$ bond in stabilizing the carbanion centre has already been mentioned[207,208]. The protons bearing a *cis* relationship to the *S*-methyl group

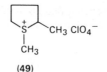

(49)

in the corresponding thiolanium ions exchange most rapidly (NaOD, D_2O). In
S-methylthiolanium iodide, the protons of the S-methyl group exchange fastest[169];
of the protons on carbon atoms α to the sulphur atom, those that are *cis* to the
S-methyl group exchange 12 times faster than those that are *trans*[270,271]. In the
fused, bicyclic thiolanium salt **50**, the four α-protons in the ring exchange at
different rates, the ratio of exchange for $H_{(2)}:H_{(3)}:H_{(4)}:H_{(1)}$ being $200:3:3:1$[272,273].
Again, the fastest exchange rate is found for a proton *cis* to the S-methyl group.

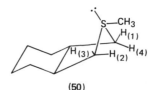

(50)

Oxidation of a bicyclic sulphonium ion gives the corresponding sulphoxonium ion
(equation 136)[274].

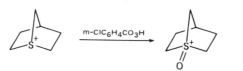

(136)

B. Saturated Systems with One Sulphur Atom and One or More Heteroatoms (O, N)

Five-membered cyclic sulphonium ions which include oxygen as one of the ring
atoms were suggested as intermediates in reactions involving neighbouring group
participation by sulphoxides. One of these intermediates has been isolated
(equation 137)[275]. Attack by water on the sulphur atom (with inversion at sulphur)
gives the iodohydrin. Some reactions may be synchronous, and not involve an
intermediate[276].

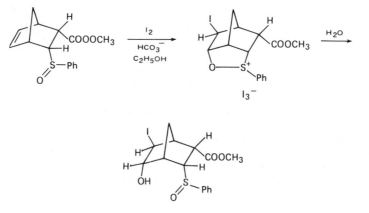

(137)

Oxidation of 4-hydroxythianes yields principally the equatorial sulphoxides and a cyclic intermediate has been suggested (equation 138)[277].

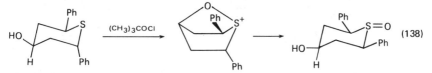

$$(138)$$

Cyclic sulphoxonium intermediates (51) have been proposed for the reactions of β-phenylsulphinylcarboxylic acids, amides[278–282], and alcohols with sulphuryl chloride (e.g. equation 139)[27].

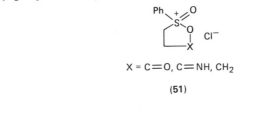

$$X = C{=}O, C{=}NH, CH_2$$

$$(51)$$

$$(139)$$

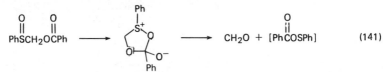

$$Ph-S-CH_2CH_2CONH_2 \xrightarrow{SO_2Cl_2} \quad \xrightarrow{-HCl} \quad PhSO_2CH_2CH_2CN$$

An oxathiolanium ion is believed to be involved in the sensitized photolysis of oxathiolanes in fluorotrichloromethane (equation 140)[279]. Ring cleavage of

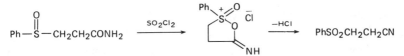

$$(140)$$

oxathiolanes by halogen presumably proceeds via an S-halosulphonium ion[280], and a cyclic zwitterion is proposed as an intermediate in the thermolysis of benzoyloxymethyl phenyl sulphoxide (equation 141)[281].

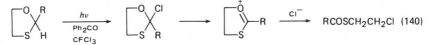

$$(141)$$

The sulphur atom in penicillin derivatives is relatively unreactive to electrophilic reagents and the most powerful alkylating agents [$(CH_3)_3O\ BF_4$, CH_3OSO_2F] are required for the formation of sulphonium salts which readily undergo an elimination reaction with rupture of the five-membered ring[282]. A similar ring opening proceeds via ylides derived from diethyl azomalonate (equation 142)[283].

A bis-salt 52 was obtained by S-alkylation of NN-dialkylthiazolidinium salts with trimethyloxonium tetrafluoroborate[284]. An azasulphoxonium salt has been obtained from a sulphoximine[285], and treatment of aryliminothiadiazolidienediones with triethyloxonium tetrafluoroborate gave unusual sulphonium salts[286a]. Nitrogen analogues of the oxathiolanium ions are known[286b–j]; the nitrogen atom (or atoms) may bear a large share of the positive charge so that the compounds may be considered more as ammonium than sulphonium salts.

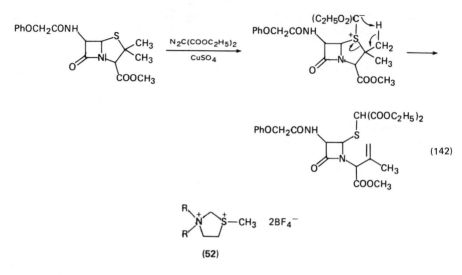

(142)

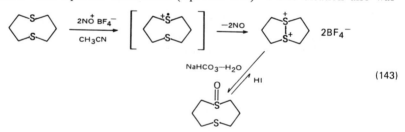

(52)

C. Saturated Systems with Two Sulphur Atoms

S-Methylation of the 1,2-dithiolane ring of lipoic acid with methyl fluorosulphonate at 25°C occurs 37-times faster than the *S*-methylation of diethyl disulphide and 11 times faster than the *S*-methylation of 1,2-dithiane, the increased reactivity being attributed to the greater repulsion of the non-bonded electrons on sulphur in the five-membered ring[287].

A colourless, bicyclic 1,2-dithiolium dication has been obtained by oxidation of 1,5-dithiacyclooctane with nitrosonium tetrafluoroborate. It gave a monosulphoxide on treatment with aqueous bicarbonate (equation 143)[288]. The dication also was

(143)

suggested as an intermediate in the reduction of the sulphoxide by hydrogen iodide[289]. A similar dication in which one sulphur atom has been replaced by methyl-substituted nitrogen also has been prepared[290].

Alkylation of one or both sulphur atoms of 1,3-dithiolanes has been accomplished. Methyl iodide[291,292], dimethyl sulphate[292], and phenacyl bromides[293,294] undergo monoalkylation; both mono- and dialkylation products may be obtained with trimethyl- or triethyloxonium tetrafluoroborates[291,295–298] or methyl fluorosulphonate[292,296,298–299].

Phenacyl ylides derived from 1,3-dithiolanes react with electrophilic alkenes and acetylenes and are acylated by benzoic anhydride and phenyl isocyanate (e.g. equation 144)[293,300].

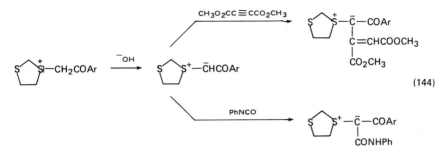

(144)

S-Mono- and SS-disubstituted 1,3-dithiolanium salts are cleaved by nucleophiles [water[295,301], methanol[299], 1,2-diols[298,299], HSCH$_2$CH$_2$OH[299], OH$^-$ [292,297], ammonia solution, or water–copper(I) sulphate[296,297]], the reaction being a useful method of dethioacetal- or dethioketalization (equation 145)[296]; it also has been used to prepare acetals[298,299] (equation 146)[298]. Removal of 1,3-dithiolane protective groups via ditholanium salts also has been accomplished with mesitylsulphonyl-hydroxylamine–water (equation 147)[302] or benzeneseleninic anhydride (equation 148)[303]. Ring expansion of the 1,3-ditholium ring via the S-amino salt has been observed (equation 149)[304].

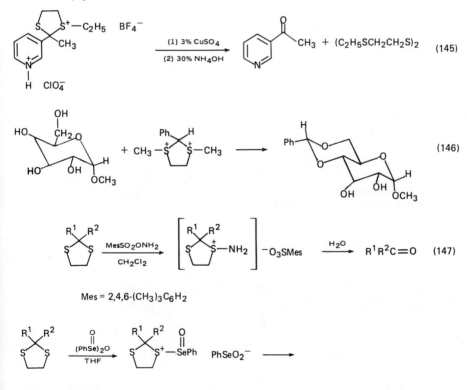

$$R^1R^2CO + (PhSe)_2 + \{SCH_2CH_2SO_2\}_{\overline{n}} \quad (148)$$

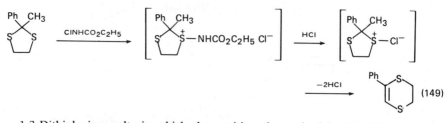

(149)

1,3-Dithiolanium salts in which the positive charge is delocalized between the two sulphur atoms can be prepared in the following ways:

(1) Addition of electrophiles to a 1,3-dithiolane with an exocyclic doubly bonded function at the 2-position (equation 150)[305–310].

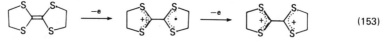

(150)

(2) Removal of an anionic group from the 2-position of a 1,3-dithiolane (equation 151)[311–313].

(151)

(3) Cyclization of appropriately substituted acyclic derivatives[314–321] (e.g. equation 152)[314].

(152)

(4) Removal of electrons from tetrahydrotetrathiafulvalenes and related compounds (equation 153)[322–324].

(153)

These 1,3-dithiolium salts react with a variety of nucleophiles[305–310,313,318,325–327] (e.g. equations 154[306], 155[309], and 156[310]).

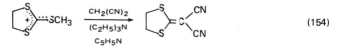

(154)

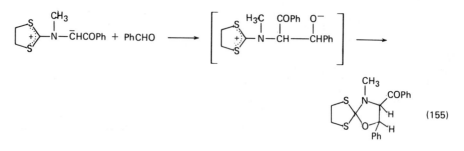

(155)

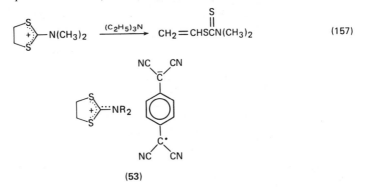

(156)

Bases may remove a proton from the dithiolanium ion[311] (equation 157)[325], and semiconducting (at 20–90°C) complexes **53** with the radical ion of 7,7,8,8-tetracyanoquinodimethane (TCNQ) have been reported[328].

(157)

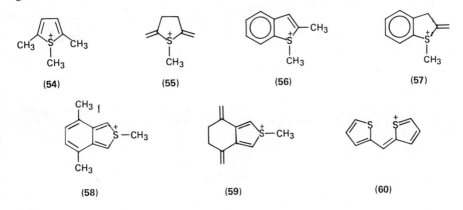

(53)

D. Unsaturated Systems with One Sulphur Atom

1. Theoretical

Barriers for pyramidal inversion of the unsaturated structures **54, 55, 56, 57, 58,** and **59** were calculated to be 19.1, 27.4, 23.2, 26.7, 14.6, and 19.4 kcal mol^{-1}, respectively[132]. Molecular orbital calculations and a discussion of the absorption spectra of polymethine dyes, including **60**, have been given[329]. Related material is given in the theoretical section in Section E which discusses dithiolium salts.

(54)   (55)   (56)   (57)

(58)   (59)   (60)

2. Synthesis

For the most part, unsaturated, five-membered cyclic sulphonium salts (from thiophenes and dihydrothiophenes) can be obtained by alkylation of the sulphur atom by the methods discussed in earlier sections. The low nucleophilicity of the sulphur atom in thiophenes requires the more powerful alkylating agents such as alkyl fluorosulphonates[193,330], trialkyloxonium tetrafluoroborates[331], alkyl halides plus silver perchlorate[331,332] or fluoroborate[330,333] (e.g. equation 158). Yields are low with thiophene itself, but are higher with substituted thiophenes.

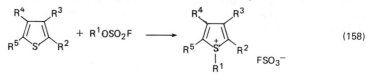

$$+ \ R^1OSO_2F \longrightarrow \tag{158}$$

N.m.r. data for the thiophenium salts indicated that the S-alkyl group is out of the plane of the ring[334]. The carbon n.m.r. spectrum (TMS reference) of 1,2,3,4,5-pentamethylthiophenium fluorosulphonate has been compared with that of the precursor thiophene. A considerable change is observed for the carbon at the 3-position, $\delta_{C(3)}$ (tetramethylthiophene) 132.4 and $\delta_{C(3)}$ (salt) 148.5 p.p.m.[193,330]. The change in chemical shift at $C_{(2)}$ was negligible. The S-methyl carbon resonance is at δ 26.0 p.p.m.[330] and the proton n.m.r. spectrum of the S-methyl group showed a singlet at δ 3.28 p.p.m.[331].

Novel, diamagnetic thiophenium complexes are obtained by treatment of hexafluoro-2-butyne with manganese carbonyl mercaptide complexes (equation 159)[335].

$$CF_3C{\equiv}CCF_3 \ + \ Mn(CO)_4(SR) \longrightarrow \tag{159}$$

S-Aryldibenzothiophenium derivatives have been obtained by the action of aryl Grignard reagents on the corresponding S-alkoxythiophenium salts[336] (equation 160). Other nucleophiles ($C_5H_5^-$, piperidine) react similarly[332].

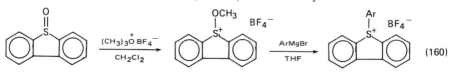

$$\tag{160}$$

Dimethyl diazomalonate reacts with thiophenes to give new ylides[187]; rhodium(III) acetate catalyses the reaction[337]. Ylide **61** is a crystalline solid, stable at room temperature for more than a year[338]. The reaction also is catalysed by light or by copper(II) ions[187,339].

(61)

The synthesis of S-alkyldihydrothiophenium ions poses no particular problems. A typical synthesis is given in equation 161[340]. Ions **62**[341] and **63**[342] (equations 162 and 163, respectively) have fluctuating structures, as indicated by variable-temperature n.m.r. experiments.

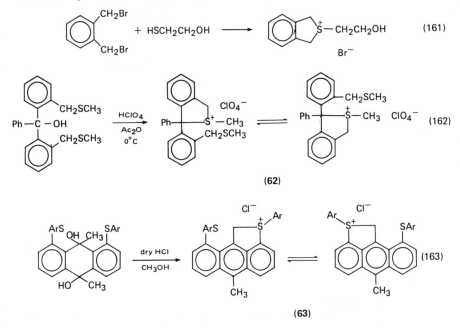

(62)

(163)

(63)

An unusual cyclization of a propargyl sulphide to a cyclic sulphonium ion was observed when the sulphide was treated with methyl iodide (equation 164)[343].

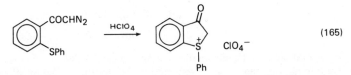

(164)

The formation of S-methylsulphonium salts of benzo[b]dihydrothiophenes by treatment of the S-alkoxysulphonium salts with methylmagnesium bromide or with dimethylcadmium occurs mainly with inversion[344]. An S-phenylbenzodihydrothiophenium salt was obtained via o-thiophenyldiazoacetophenone (equation 165)[345],

(165)

and stable sulphonium ylides were obtained by treatment of sulphides with either dicyanodiazomethane or freshly prepared tetracyanoethylene oxide, the latter giving higher yields (e.g. equation 166)[346].

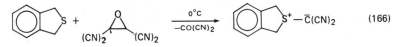

(166)

S-Aminosulphonium salts (via O-mesitylsulphonylhydroxylamine) and ylides (or sulphilimines) have been obtained from benzo[c]-1,3-dihydrothiophene (equation 167)[347]. A transfer of cationic alkylsulphur group to 1,4-diphenylbutadiene to give an S-methyldihydrothiophenium salt has been reported (equation 168)[348].

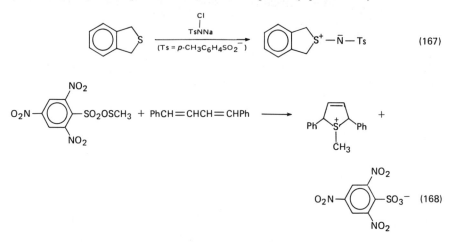

$$(167)$$

$$(168)$$

Stable C-alkylated thiophenium ions have been observed by n.m.r. techniques in the alkylation of thiophenes by alkyl halides (equation 169)[349] or in the protonation of alkylmercaptothiophenes[350a]. The n.m.r. spectra of C-protonated thiophenes has been recorded[350b]. Similar salts have been obtained by oxidation of derivatives of benzo[c]-1,3-dihydrothiophene[351]. The thiophene cation-radical **64**, shown in its allegedly most stable conformation, has been observed by electron-spin resonance spectroscopy in a mixture of thiophene with Lewis acids such as aluminium chloride or boron trifluoride[352].

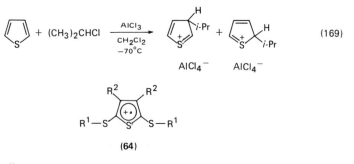

$$(169)$$

(64)

Several polycyclic aromatic cations which have a five-membered sulphur-containing ring have been reported[353–356]. In these, the positive charge may be considerably delocalized (e.g. equation 170)[353].

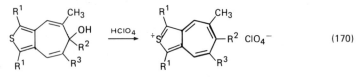

$$(170)$$

Several cations similar to that whose formation is shown in equation 171[357] have been reported[357–359].

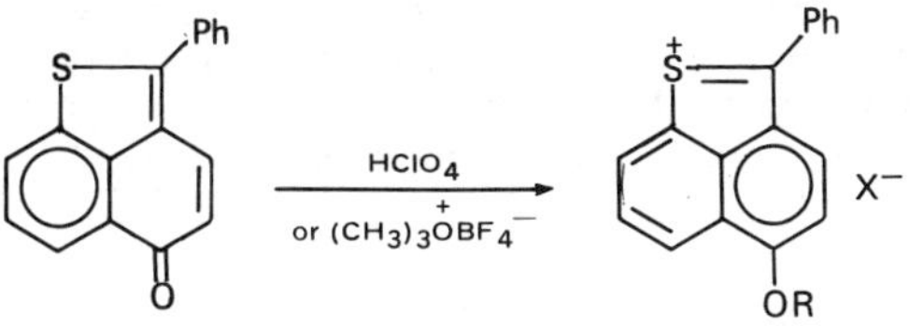

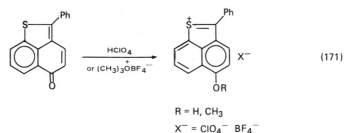

(171)

R = H, CH_3

$X^- = ClO_4^-\ BF_4^-$

3. *Intermediates in reactions*

The photolysis of certain vinyl aryl sulphides may involve the intermediate formation of cyclic sulphonium ylides (thiocarbonyl ylides) as shown in equation 172[360,361].

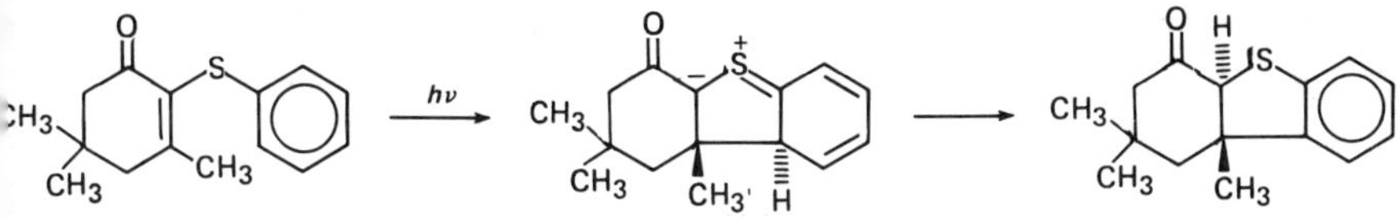

(172)

Ring expansion of 2,5-dihydrothiophene on treatment with dimethyl diazomalonate is believed to occur via a sulphonium ylide (equation 173)[362]. Thiophenium ions, of course, are intermediates in electrophilic substitution reactions of thiophenes such as the detritiation of 2- and 3-tritiothiophenes[363].

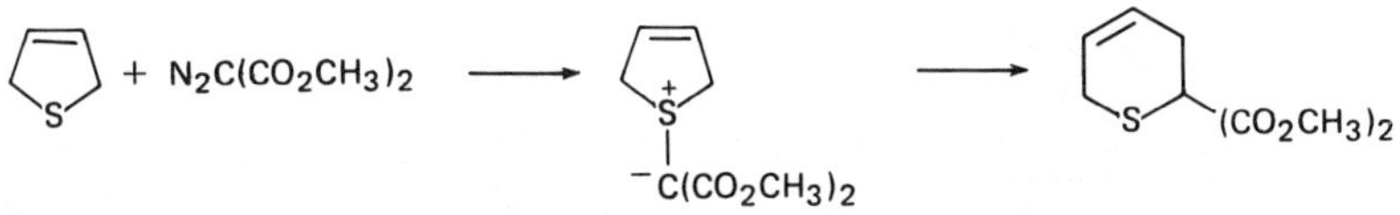

(173)

4. *Reactions*

S-Alkylthiophenium ions are powerful alkylating agents and most decompose readily in hydroxylic solvents to the parent thiophene[333,334,345]. 1-Methylbenzo[*b*]thiophenium salts add bromine; the resulting dibromide can be dehydrohalogenated without affecting the *S*-methyl group (equation 174)[333,334]. Ion **63** reacts with methanol with ring cleavage via elimination and substitution reactions[342b].

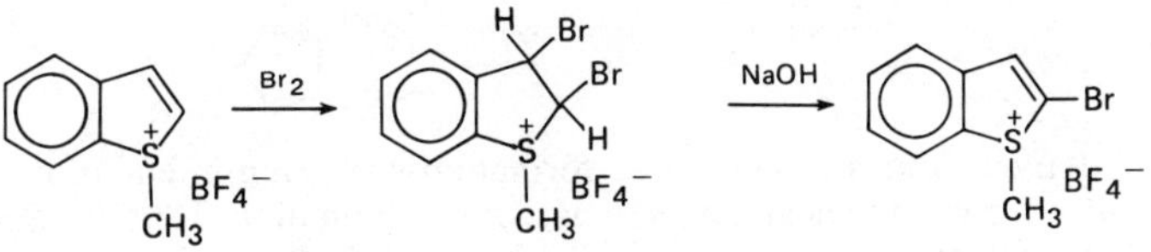

(174)

S-Methylthiophenium salts are inert in Diels–Alder reactions, even with good dienophiles such as hexafluoro-2-butyne. The ylide generated from the *S*-methylthiophenium cation reacts with *p*-nitrobenzaldehyde to give the expected oxirane[330]. Ylide **61** is a useful source of bis(methoxycarbonyl)carbene (equation 175)[338]. The ylide from S-methylbenzo[c]-1,3-dihydrothiophenium iodide undergoes fragmentation with recombination of the fragments from two molecules (equation 176)[364]. Rearrangement of a thiophenium ylide has been observed (equation 177)[365]. Refluxing the ylide derived from the S-phenacyl salt of 2,5-dihydrothiophene results in extrusion of 1,3-butadiene (equation 178)[366].

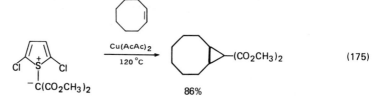

$$\text{(61)} \qquad \qquad \qquad 86\% \qquad \qquad (175)$$

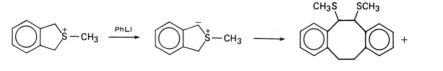

$$(176)$$

$$(177)$$

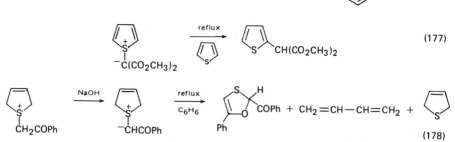

$$(178)$$

Thermolysis of the sulphilimine shown in equation 179 yields the reactive isobenzothiophene, which can be trapped as its Diels–Alder adduct with N-phenylmaleimide[347].

$$\text{230°C} \qquad (179)$$

The ylide derived from the salt whose formation was given in equation 165 reacts with the salt to give a diacylethylene derivative (equation 180)[345]. Some ylides derived photochemically undergo 1,3-dipolar addition reactions (e.g. equation 181)[360].

Additions of nucleophiles to a $-C\!\!=\!\!\overset{+}{S}-$ bond in five-membered rings occurs readily[367]. Facile reductions also are known[354]. Attack on the positive sulphur atom,

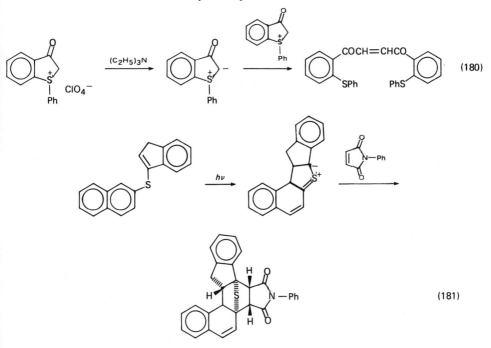

(180)

(181)

in addition to attack on carbon, has been observed particularly in the case of
S-alkoxy cations in which the alkoxy group is displaced by an anionic
group[332,336,344]. Sulphuranes are suggested as intermediates in the reaction of aryl-
or vinyllithium reagents with S-aryldibenzothiophenium tetrafluoroborates
(equation 182)[336]. Path a (equation 182) is favoured by electron-withdrawing
groups in Ar or Ar' and path b by electron-donating groups. Treatment of
S-methylbenzo[c]-1,3-dihydrothiophenium iodide with phenyllithium results in
desulphurization to yield a diene[364]. A similar reaction of a 2,5-dihydrothiophenium

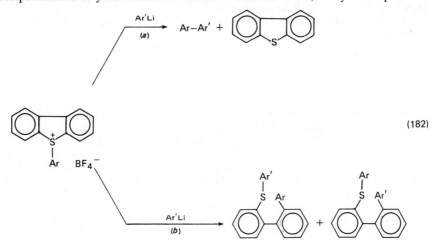

(182)

salt reveals that the stereochemistry of diene formation is a consequence of the disrotatory extrusion of dialkyl sulphide from an intermediate sulphurane (e.g. equation 183)[368]. A β-elimination reaction rather than sulphurane formation is observed to occur with the sterically hindered 2,2,5,5-tetramethyl salt[368].

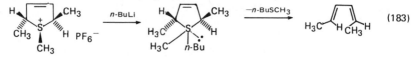

(183)

Loss of a proton occurs easily from C-alkylated thiophenium ions resulting from the attack of electrophiles on thiophenes[349]. Rearrangement of a methyl group has been observed in the photolysis of 1,2,3,4,5-tetramethylthiophenium fluorosulphonate (equation 184)[193].

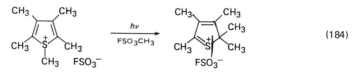

(184)

E. Unsaturated Systems with Two Sulphur Atoms

1. Reviews

A number of reviews of 1,2-dithiolium ions[369–373,374a] and 1,3-dithiolium ions[369–372,374b,375] exist. Precursors of 1,2-dithiolium salts, the 1,2-dithiole-3-thiones, also have been reviewed[374c,376]. 1,3-Dithiolium dyes have been discussed[375].

2. Theoretical treatment and physical properties

Structure and bonding in 1,2-dithiolium and 1,2,4-dithiazolium salts have been discussed in detail[377]. The delocalization energies of the 1,2-dithiolium and 1,3-dithiolium cations are less than that for the thiapyrylium or tropylium ions, the values in β units being 1.57, 1.16, 2.12, and 2.99, respectively[378]. Some variation in these values was obtained depending on the parameters used and whether d-orbitals were involved. The 1,2-dithiolium ion has been treated independently[379]. Calculations predict that nucleophiles and radicals will attack the 3-position of a 1,2-dithiolium ion and the 2-position of a 1,3-dithiolium ion[380,381]. Non-empirical calculations by means of a linear combination of Gaussian orbitals indicated only trivial d-orbital participation[381b]. The positive charge in both 1,2- and 1,3-dithiolium ions was calculated to reside mainly on sulphur and hydrogen atoms[381b].

Proton n.m.r. data support delocalized electronic structures for 1,2- and 1,3-dithiolium ions, greater electron deficiency existing at $C_{(2)}$ of 1,3-dithiolium ions than at $C_{(3)}$ of 1,2-dithiolium ions[369,382]. Aromatic-type anisotropy (diatropy) was observed for 1,2-dithiolium ions[383]. C.m.r. spectra also support aromatic character in 1,3-dithiolium salts[384]. A comparison of proton and carbon chemical shifts for the 1,3-dithiolanium ion **65** and the 1,3-dithiolium ion **66** are given (carbon shifts in parenthesis in p.p.m. downfield from tetramethylsilane)[384]. Binding energies of the sulphur p-electrons in 1,2-dithiolium ions correlate with ^{19}F magnetic resonance chemical shifts observed with 3-(p-fluorophenyl)-1,2-dithiolium ions and are related to the amount of positive charge on sulphur[385].

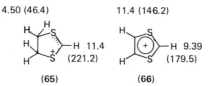

4.50 (46.4)   11.4 (146.2)

(65)   (66)

CNDO/2 calculations on cation **67** indicate that only two localized sulphur—sulphur bonds exist with no appreciable bonding between two sulphur atoms in different rings[386] despite an X-ray analysis of a phenyl-substituted derivative which indicated such an action[387]. Half of the positive charge in **67** resides in each ring although the two sulphur atoms do not share the charge equally[386]. A very short sulphur–oxygen distance (2.18 Å)[388a] in **68** suggests interaction of the electrons on oxygen with a sulphur atom. Molecular orbital calculations (CNDO/2–SCF) support this interaction and indicate that the sulphur d-orbitals play an important role[388b]. An X-ray analysis of 3,5-diamino-1,2-dithiolium iodide gives an S—S bond distance of 2.08 Å[389], larger than that observed in other 1,2-dithiolium salts[390] and which indicates that considerable charge is delocalized on to the nitrogen atoms.

Spectroscopic properties for various 1,2- and 1,3-dithiolium ions have been calculated[329,391]. Calculations of the spectra of 1,2-dithiolium ions which agree well with the observed spectra were carried out using d-orbitals[392]. Calculated and observed ultraviolet absorptions for phenyl derivatives of meso-ionic compounds **69** and **70** were in agreement[393]. In addition to the ultraviolet spectra of 1,2-dithiolium ions, the charge-transfer spectra with iodide ion and polarographic properties have been calculated and compared with measured quantities[391d]. The effects of substituents on the ultraviolet spectra were not determined by the electron distribution in the ground state and the proton n.m.r. spectra correlated with π-electron charges[391d]. Calculations indicate that **71** is more basic (cf. equation 185) than **72**[394].

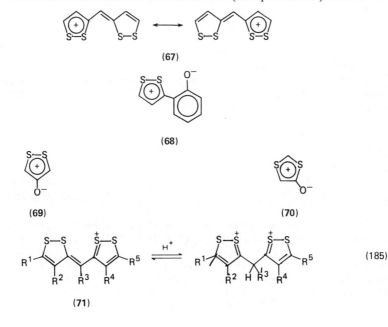

(67)

(68)

(69)   (70)

(185)

(71)

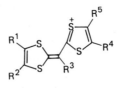

(72)

pK_{R+} values for 4-aryl-1,3-dithiolium salts correlated well with Hammett σ values[395]. A variety of charge-transfer complexes have been observed with 1,2-dithiolium ions[396] and the salt **73** has semiconducting properties[397].

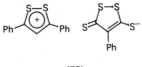

(73)

3. 1,2-Dithiolium salts

a. *Synthesis.* The methods of synthesis of 1,2-dithiolium salts are classified as follows:

(1)

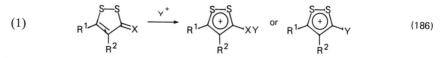

(186)

(2)

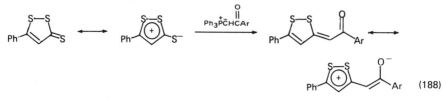

(187)

(3) Ring closure of acyclic intermediates.
(4) Miscellaneous methods.
At the outset, it should be noted that 1,2-dithiolium resonance structures may be written for compounds in which electron delocalization with an exocyclic functional group can occur (e.g. equation 188)[398].

$$\text{(188)}$$

In the first method of synthesis (equation 186), probably the most important one, an exocyclic doubly bonded group, X (S, CR_2, O, NR), either gives up electrons to an electrophilic reagent Y (e.g. R^+, H^+ equations 189–191, 193–195) or is replaced by a group Z (e.g. H, halogen equations 196, 197), which cannot as readily neutralize the positive charge in the dithiolium ion as the displaced group. Some illustrative examples follow (equations 189[399], 190[400], 191[401], 192[402], 193[403], 194[400d], 195[404], 196[396a,405], 197[406], 198[407]).

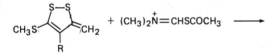

(189)

(190)

(191)

(192)

(193)

(194)

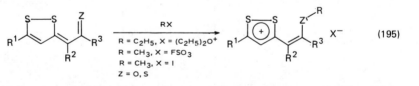

(195)

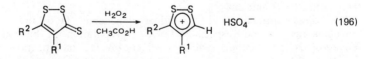

(196)

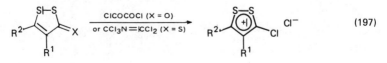

(197)

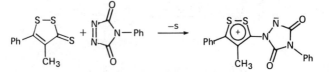

(198)

A less useful method is the removal of hydride ion from 1,2-dithioles (method 2, equation 187)[400m].

The dithiolium ring system may be obtained by cyclization of acyclic intermediates. The cyclization of 1,3-diketones (equation 199) may be accomplished by $H_2S_2-HCl^{408}$, $H_2S-HCl-I_2^{409}$, $H_2S-FeCl_4^{-410}$, $H_2S-I_2-CH_3OH^{411}$, and P_4S_{10} (or $P_4S_{10}-I_2$)[409a,412]. Derivatives of 1,3-diketones also may be converted into 1,2-dithiolium salts (equation 200)[400a,e;413–415]. 1,3-Dithiocarbonyl compounds may be cyclized by I_2^{416}, $FeX_3^{416a,d;417}$, $C_7H_7^+$ $ClO_4^{-\,416a}$ or $H_2O_2-HCl^{416c,d;417}$ (e.g. equation 201). A reaction which also may proceed via a 1,3-dithioketone is shown in equation 202[390c,418].

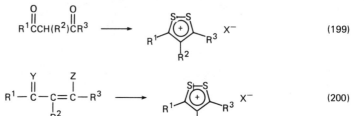

(199)

(200)

Y = O, S, (OC$_2$H$_5$)$_2$, N(CH$_3$)$_2$

Z = OH, Cl, OC$_2$H$_5$, SCH$_3$

(201)

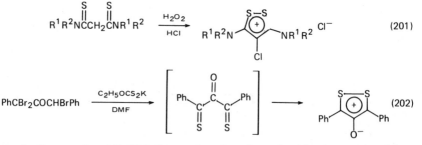

(202)

Several alkenes give 1,2-dithiolium salts when heated with elemental sulphur (equations 203[419] and 204[420]).

The ring contraction of 1,3-dithiacyclohexenes (1,3-dithiins) is a useful miscellaneous method, especially for the preparation of benzodithiolium salts (equation 205)[421]. Other syntheses which involve modification of pre-formed dithiolium salts may be found in the following section on reactions.

$$Cl_2C=C(Cl)CCl_3 \quad \xrightarrow[\text{heat}]{S_8} \quad$$

(203)

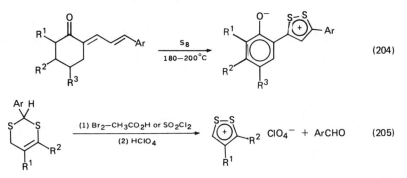

(204)

(205)

b. *Reactions.* The reactions of 1,2-dithiolium ions are categorized as follows: (1) attack by nucleophiles or electrons on a carbon or sulphur atom of the ring; (2) reactions of substituent groups on the ring; and (3) miscellaneous reactions, including rearrangements, dipolar cycloaddition reactions, and the reactions of metal complexes.

The attack of a nucleophile at a ring carbon atom may result in expulsion of another group or in opening of the ring. Nucleophilic attack by the following reagents has been observed: hydride ion (equation 206)[422], oxygen nucleophiles (e.g. equations 207 and 208)[396a,406a,c,d,j,408b,419c], sulphur nucleophiles (e.g. equations 209 and 210)[396a,400a,k,405c,406a,c,423–427], nitrogen nucleophiles (e.g. equations 211–216)[400c,e,k,403a,b,404g,h,405b,f,406c,408b,e,418,423–425,428b,429–450], and carbon nucleophiles (e.g. equations 217–221)[399b,400b,f,g,k–m,403a,404b,408c,420a–c,423,425,432,435,451–473].

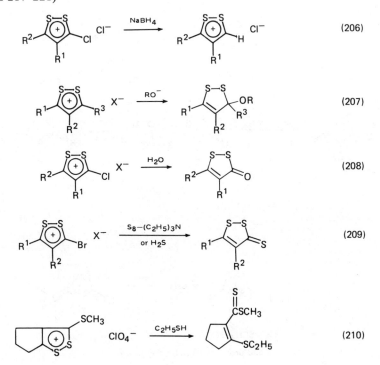

(206)

(207)

(208)

(209)

(210)

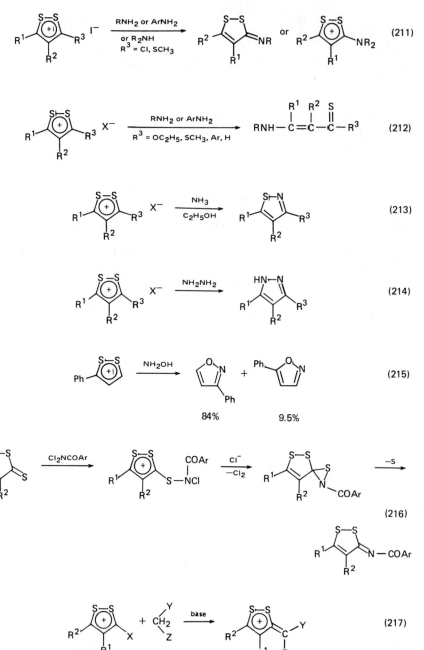

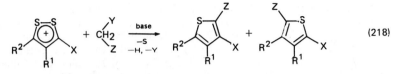

$$Y = H; X = SCH_3, NR_2$$

$$Z = -C(R^3)=C(CN)_2$$
$$\quad -C(R^3)=C(CN)CO_2C_2H_5$$
$$\quad -NO_2$$

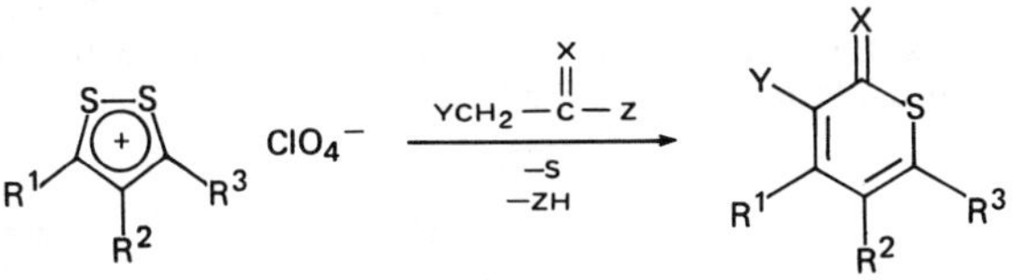

(218)

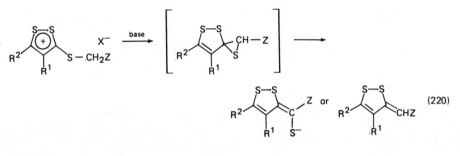

$$X = S, O$$
$$Y = CN, RCO, CO_2C_2H_5$$
$$Z = NH_2, SCH_3, OC_2H_5$$

(219)

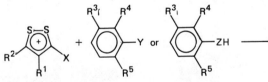

(220)

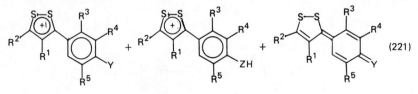

$$X = Cl, SCH_3, H, NR_2$$
$$Y = NR_2, OCH_3; ZH = OH, NH_2$$

(221)

The introduction of a thiono group at the 3-position by reaction of a dithiolium salt with sulphur (equation 209) probably proceeds via a nucleophilic sulphur species (e.g. Nu–S_7S^-) produced by the action of a base or nucleophile on S_8 (but not apparently by a tertiary amine)[474]. In some cases, the action of oxygen nucleophiles (e.g. ROH) on certain 1,2-dithiolium salts (e.g. the 3-chloro derivatives) yields 1,2-dithiole-3-thiones, which probably result from some ring cleavage to give sulphur, the latter reacting with the dithiolium salt as in equation 209 to give the thione[399b,424,427,475]. The reaction of the 4-phenyl-1,2-dithiolium ion with 1,2-diaminoethane or 1,2-diaminopropane (cf. equation 212) has been used in the synthesis of a corrin ring[440].

While a carbon atom in 1,2-dithiolium ions is the principal site of attack by nucleophiles, the sulphur atoms also are susceptible. Intramolecular attack at sulphur is most common (e.g. equations 222–225)[476–479]. Experimental observations and theoretical considerations indicate an intramolecular oxygen–sulphur interaction as shown in equation 226[420a,b,480].

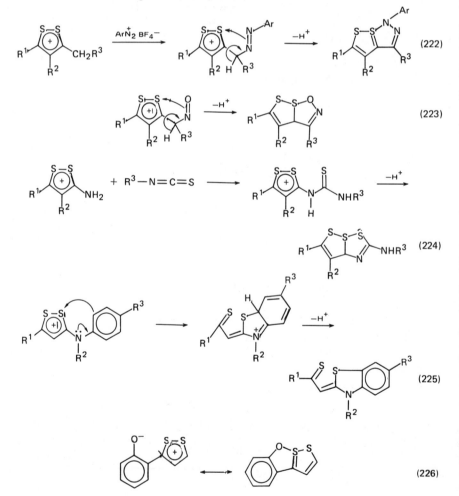

Intermolecular attack on sulphur by hydride ion or hydroxide ion may be involved in the reaction shown in equation 227[416a,416d].

$$Z = O, S, \overset{+}{S}CH_3 \tag{228}$$

$$X = SR, NHR, NR_2$$

Dithiolium salts are one-electron acceptors. One-electron reductions have been accomplished[414,481–483], and the salts also form charge-transfer complexes with a variety of electron donors[396b,c,397,484,485]. Most one-electron reductions of dithiolium salts yield radicals which are in equilibrium with their dimers, but radical **74** is not dimeric from 19 to 95°C[414]. The products of two-electron reductions of 1,2-dithiolium ions are unstable and ring opening occurs[481].

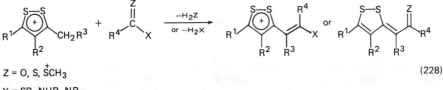

(74)

The protons of a methyl, methylene, or methine group attached directly to the 1,2-dithiolium ring at the 3-(or 5-)position are acidic and can undergo exchange with deuterated protic solvents[486]. The methylene groups also can undergo condensations of the aldol type (e.g. equation 228)[400d,i,k,448,462,487].

Side-chains conjugated with the 1,2-dithiolium ring undergo various reactions, such as the hydrolysis or addition–elimination reaction of enamines (e.g. equation 229)[487,488]. Addition–elimination reactions of thioenol ethers of 1,2-dithiolium salts also occur (equation 230)[404a,c].

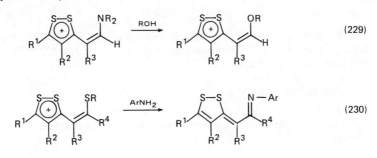

Among the reactions discussed so far in this section were a number of ways electrons could be released to the 1,2-dithiolium ion. A last, more complicated, example is given in equation 231[400b,489].

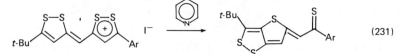

(231)

Derivatives of 4-hydroxy-1,2-dithiolium salts readily form a zwitterion (75)[400a,408a,c]. Demethylation of 3-methylthio-1,2-dithiolium salts to give

(75)

1,2-dithiole-3-thiones may occur in addition to displacement of the methylthio group[400e,403a].

Treatment of 3,4,5-trichloro-1,2-dithiolium chloride with aluminium chloride containing radioactive chlorine resulted in replacement of the non-radioactive halogens[419c]. Protons directly attached to the carbon atoms of the dithiolium ring undergo exchange with D_2O[427,486,490]. 1,2-Dithiolium salts with attached aryl groups undergo electrophilic substitution at the aryl group[405f].

Attempted 1,3-dipolar addition of a mesoionic, dithiolium zwitterion to acetylenes gave thiophenes (from sulphur and the acetylene) and a quinone (equation 232)[390c].

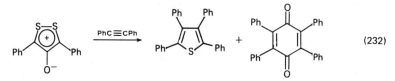

(232)

Thermal fragmentation of 1,2-dithiolium salts has been observed in the ion source of a mass spectrometer; a dithiolyl radical is believed to be an intermediate and loss of a hydrogen atom is common[491,492]. The photochemical behaviour of a 1,2-dithiolium salt apparently yields a dithiolyl radical followed by ring opening (equation 233)[493].

(233)

A number of complex metal halide salts (e.g. $FeCl_4^{2-}$, $CoCl_4^{2-}$, $PtCl_6^{2-}$, $CuCl_4^{2-}$, $SbCl_4^-$, $SnCl_6^{2-}$) of 1,2-dithiolium ions have been reported[396b,494,495]; nickel(II) ions and base cause the ring to break[496].

4. 1,3-Dithiolium salts

a. *Synthesis*. The methods of synthesis parallel those for the preparation of 1,2-dithiolium salts and the principal ones are summarized below (equations 234–242).

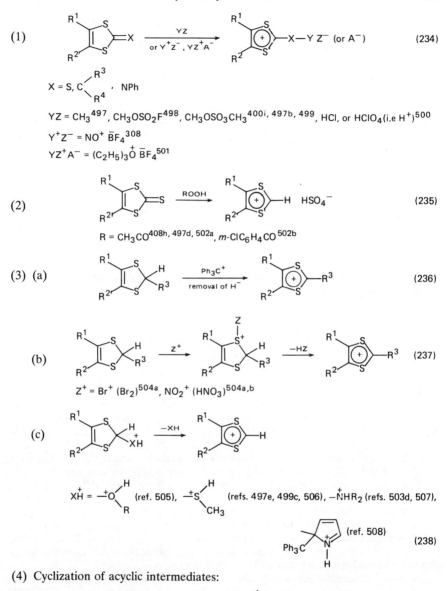

(1)

$$YZ = CH_3{}^{497}, CH_3OSO_2F^{498}, CH_3OSO_3CH_3{}^{400i,\,497b,\,499}, HCl, \text{ or } HClO_4 \text{ (i.e } H^+)^{500}$$

$$Y^+Z^- = NO^+ \, \bar{B}F_4{}^{308}$$

$$YZ^+A^- = (C_2H_5)_3\overset{+}{O} \, \bar{B}F_4{}^{501}$$

(2)

$$R = CH_3CO^{408h,\,497d,\,502a}, \text{ } m\text{-ClC}_6H_4CO^{502b}$$

(3) (a)

(b)

$$z^+ = Br^+ \, (Br_2)^{504a}, NO_2{}^+ \, (HNO_3)^{504a,b}$$

(c)

$$X\overset{+}{H} = -\overset{+}{\underset{R}{O}}{\overset{H}{}} \text{ (ref. 505), } -\overset{+}{\underset{CH_3}{S}}{\overset{H}{}} \text{ (refs. 497e, 499c, 506), } -\overset{+}{N}HR_2 \text{ (refs. 503d, 507),}$$

(ref. 508)

(4) Cyclization of acyclic intermediates:

(a)

$$R^1 = OH^{509}, \text{ alkyl}^{408h,409b,499c,510}, \text{ aryl}^{507a,509a,510}$$
$$R^2 = H^{408h,409b,507a,509a,b,d}, \text{ alkyl}^{499c,509b,e,510}, \text{ aryl}^{507a,509b,c,e}$$
$$R^3 = R_2N^{499c,507a,510}, \text{ RS}^{509a,c}, \text{ alkyl}^{408h,409b,509b,c}, \text{ aryl}^{509,510}$$

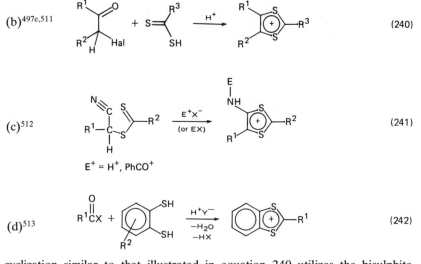

(b)[497e,511]   (240)

(c)[512]   (241)

E$^+$ = H$^+$, PhCO$^+$

(d)[513]   (242)

A cyclization similar to that illustrated in equation 240 utilizes the bisulphite addition product of 2-chloroacetaldehyde (equation 243)[514]. Diselenium salts have been prepared by the method shown in equation 240[515].

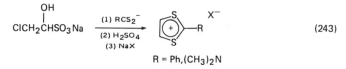

$$\underset{\text{CICH}_2\text{CHSO}_3\text{Na}}{\overset{\text{OH}}{|}} \xrightarrow[\substack{(2)\ H_2SO_4 \\ (3)\ NaX}]{(1)\ RCS_2^-} \qquad (243)$$

R = Ph,(CH$_3$)$_2$N

Several other methods have been used for the preparation of particular 1,3-dithiolium ions. Cycloaddition of carbon disulphide to benzyne intermediates followed by protonation yields benzo-1,3-dithiolium salts[516]. Elimination of hydrogen bromide from a 1,3-dithiolanium salt either by heat or with triethylamine has been observed[314,316]. Treatment of tetrathiafulvalenes with oxidizing agents gives bis-1,3-dithiolium salts via intermediate cation radicals[517]; the latter are relatively good semiconductors and their electron-spin resonance spectra have been studied. The oxidation also can be accomplished electrochemically[518].

b. *Reactions.* The principal reactions of 1,3-dithiolium ions are (1) attack by nucleophiles or electrons on the ring, nucleophiles adding to or displacing a group at C$_{(2)}$, and electrons affecting dimerization through C$_{(2)}$, (2) reactions of substituents on the ring (e.g. aldol condensations), and (3) miscellaneous reactions such as the formation of carbenes and ylides, photochemical transformations, dipolar cycloaddition reactions, and the formation of complexes.

The types of compounds which add to the 2-position of the 1,3-dithiolium ring are as follows: hydride ion sources (BH$_4^-$, AlH$_4^-$, 5:1 acetone–*i*-PrOH)[497e,499a,500c, 503a,d,506,507] (e.g. equation 244[499a]), oxygen nucleophiles[395,423,499a,503d,504a,b,505c, 516b,519,520a–c] (e.g. equations 245[505c], 246[520b], 247[520a]), sulphur nucleophiles[497e,f, 503d,514,516b,520a,c,521a,b] (e.g. equations 247[520a], 248[497f]), selenium nucleophiles[515a–c], nitrogen nucleophiles[423,497b,499a,503d,522a–c] (e.g. equations 249[503d], 250[522c], 251[522b]), and carbon nucleophiles[400i,420b,451,464b,497b–d,498,499a–c,500b,502a,b,503b–e,504b,506,507b,508,513c,d, 516a,517a,518,519,520a,523a–d,524a,b,525a–c,526] (e.g. equations 252[503e], 253[526], 254[524b], 255[525a], 256[451], 257[503c], 258[497d,506,525c]).

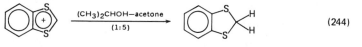

(244)

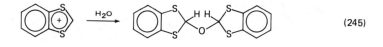

(245)

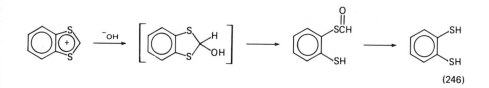

(246)

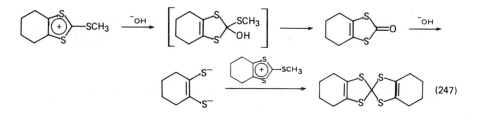

(247)

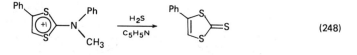

(248)

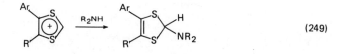

(249)

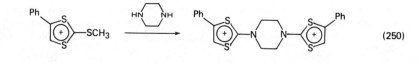

(250)

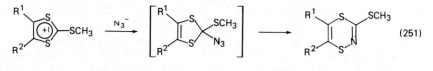

(251)

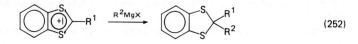

(252)

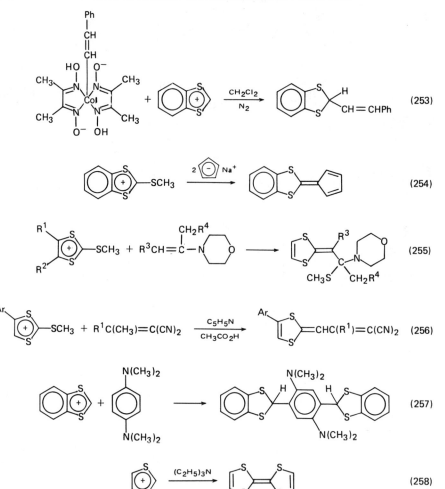

As the above examples show, a 2-S-methyl group is easily displaced by nucleophiles from a 1,3-dithiolium salt. Displacement reactions also may occur directly on the methyl carbon of an S-methyl group to give the thione (equation 259)[497c].

Electrons also may be added to 1,3-dithiolium salts either electrochemically or via a reducing metal (e.g. zinc)[500c,501,505a,511b,518,527a,b,528,529]. The usual result is formation of a dimer from the radicals initially formed (e.g. equation 260[528]). Cleavage of carbon–sulphur bonds was observed when sodium was used (equation 261[505a]).

Protons on a carbon atom at the 2-position of a 1,3-dithiolium salt are acidic,

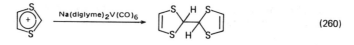

(260)

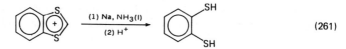

(261)

and several reactions of the aldol type have been demonstrated[408h,499b,513d] (e.g. equation 262)[499b].

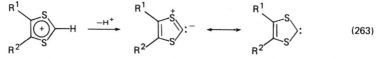

(262)

1,3-Dithiolium ylides are formed readily by treatment with weak bases[497d,499a,c, 502b,503d,507b,517a,519,520b,532a–c,525b,c] (equation 263). Hydrogen–deuterium exchange is observed[523d].

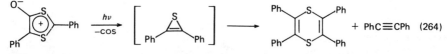

(263)

Dimerization reactions involving these intermediates may be viewed either as nucleophilic substitutions on a 1,3-dithiolium salt or as carbene dimerizations. The tetrathiafulvalenes from 4-substituted 1,3-dithiolium salts are mixtures of isomers[523d].

Photolysis of a mesoionic 1,3-dithiolium salt is believed to proceed via a diphenylthiirene (equation 264)[530].

$$O^-$$

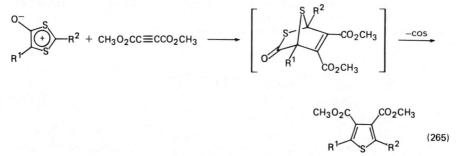

The cycloaddition reactions of similar mesoionic compounds have been investigated[509c,d,531a–d] (e.g. equation 265).

(265)

Complexes of tetrathiafulvalenes (TTF) with electron acceptors may involve the 1,3-dithiolium ions, at least formally[497d,502b,507b,532a–c]. Such complexes are of interest because in a number of cases high electrical conductivity is observed. The reader is referred to a recent summary for more information on this subject[533].

F. Unsaturated Systems with One Sulphur Atom and One or More Heteroatoms (N, O)

1. Reviews

Discussions of thiazolium, isothiazolium, and oxathiazolium salts may be found in reviews on thiazoles[534], isothiazoles[535], and mesoionic compounds[536].

2. Theoretical

Molecular orbital calculations have been made for thiazolium, selenazolium, and oxazolium salts[537–539]. As expected, $C_{(2)}$ is the most electron-deficient carbon atom[537]. Dipole moments, charge distributions, bond lengths, polarographic half-wave potentials, and spectra of various derivatives have been calculated[537–539].

3. Thiazolium salts

Thiazoles invariably alkylate at the nitrogen atom and thiazolium salts are therefore probably more ammonium than sulphonium salts. For this reason, the discussion of these compounds will be limited. The thiazolium moiety is important as a component of thiamin (vitamin B_1) and of certain cyanine dyes.

The two principal methods of syntheses are alkylation of thiazoles[540] and cyclizations analogous to those used in the synthesis of thiazoles proper, e.g. equations 266[541] and 267[542].

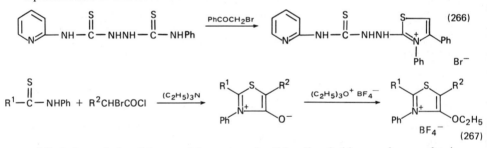

Alkylation of the thione sulphur atom in thiazoline-2-thiones also results in thiazolium ion formation (equation 268)[543], as does the addition of other electrophiles to an exocyclic double bond (equation 269)[544].

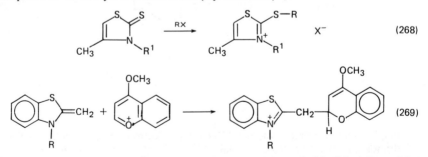

Treatment of 5-acylamino-Δ^4-thiazoline-2-thiones with methyl iodide followed by alkali yields mesoionic compounds[545]. Other syntheses of mesoionic thiazolium derivatives have been described[546]. For example, 1,3-polar cycloaddition of mesoionic oxazolium salts to carbon disulphide yields mesoionic thiazolium salts[286e,547].

An *N*-aminothiazolium salt has been obtained by treatment of a thiazole with *O*-mesitylenesulphonylhydroxylamine (MesSO$_2$ONH$_2$)[548].

Oxidation of thiazoline-2-thiones also yields thiazolium salts[549-551] (equation 270)[550].

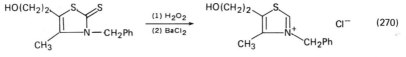

$$(270)$$

The three most important reactions of thiazolium salts are their reactions with nucleophiles, the reactions of the aldol type involving active methylene side-groups, and the reactions of thiazolium ylides, the last being involved in the biological action of thiamine. Nucleophilic additions of hydroxide ion occur readily, frequently with ring opening (equation 271 and 272)[548,552] or with displacement of a good leaving group (via addition–elimination)[553] (equation 273)[553a]. Sodium borohydride reduces the 2,3-double bond of the salts[554].

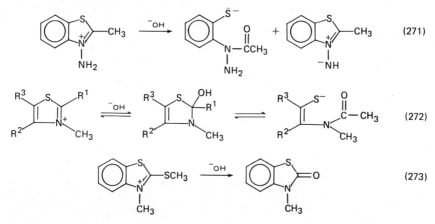

$$(271)$$

$$(272)$$

$$(273)$$

Condensation of carbonyl compounds with the active methylene groups of 2-alkylthiazolium salts is a useful method for extending the side-chain[555] as, for example, in the synthesis of the photochromic spiropyran shown in equation 274[556].

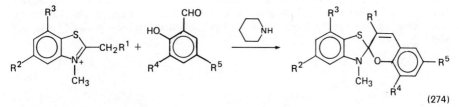

$$(274)$$

Various thiazolium ylides have been used to model biological reactions involving thiamine catalysis. They have been used to effect acyloin-type condensations[557], and asymmetric induction of the benzoin condensation by asymmetric thiazolium salts gave products of relatively high optical purity (up to 51%)[558]. Thiazolium salts catalyse the addition of aliphatic aldehydes to activated double bonds[559]. A variety of products are obtained by addition of thiazolium ylides to carbodiimides[560a] (equation 275) and isothiocyanates[560b].

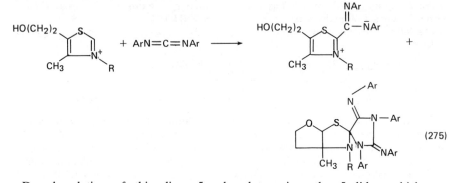

(275)

Decarboxylation of thiazolium 5-carboxylates gives the 5-ylides, which are generated 10^3 times more readily than ylides from imidazolium ions but 250 times less readily than 5-ylides from oxazolium ions[561]. The formation of the 2-ylide from 3,4-dimethylthiazolium ions is about 3×10^3 times faster than for the corresponding imidazolium ion and 100 times slower than for the oxazolium ion[562]. Ylides are involved in the dimerization of benzothiazolium salts (equation 276)[563], and extensive polarographic studies on the dimer have been carried out[564].

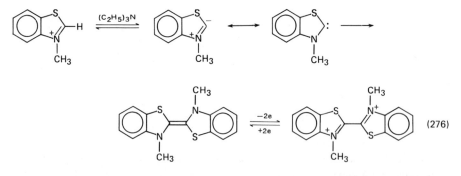

(276)

Another useful reaction which proceeds via the ylide is the ring expansion of thiazolium salts to 1,4-thiazines (equation 277), the so-called Takamizawa reaction[565]. Treatment of the 3-methylbenzothiazolium ion with phenacyl bromide and triethylamine gave 2-benzoyl-4-methyl-1,4-benzothiazine, probably via an ylide[566].

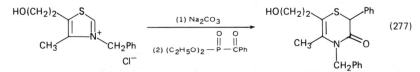

(277)

Among miscellaneous reactions may be mentioned the 1,3-dipolar cycloaddition reactions of mesoionic thiazolium salts[531b,c,542,546b,c,567] and their photochemical transformations[153].

4. Isothiazolium salts

The N-alkylation of isothiazoles by alkyl halides[428b,568], tosylates[428b,568a,c,569], sulphates[568c,570], methyl fluorosulphonate[428b,571], or triethyloxonium tetrafluoroborate[572] is generally a satisfactory method of preparation for isothiazolium salts, although the reaction of alkyl iodides with simple isothiazoles was slow and led to decomposition upon heating[568a]. Yields may be low in some cases[560]. Alkylation of benzo-1,2- and benzo-2,1-isothiazoles with methyl iodide is faster than the alkylation of the corresponding benzoxazoles[568b].

N-Substituted isothiazol-3-thiones may be alkylated at the thione sulphur atom to yield isothiazolium salts[434d]. Treatment of 6-substituted 1,6a-dithia-6-azapentalenes with methyl iodide gives 5-(2-methylthiovinyl)isothiazolium salts (e.g. equation 278)[573]. An X-ray analysis has confirmed the structure of one of the products[574].

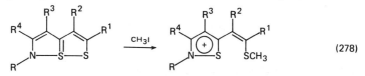

(278)

Oxidative cyclizations have been used to prepare isothiazolium salts[428b,433,441,575] (e.g. equation 279)[433].

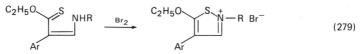

(279)

Various transformations of isothiazolium salts result from treatment with nucleophiles. Rupture of the ring is common[576] (e.g. equation 280)[428b].

Loss of an N-alkyl group can occur via acyclic intermediates (equations 281 and 282)[428b,572a], and displacements of good leaving groups can occur (equation 283)[572b]. Nucleophiles may attack the N-alkyl group directly to remove the group (equation 283).

The 3- or 5-thione function may be introduced into isothiazolium salts by a mixture of sulphur and pyridine[434d,441,446b], but 3-unsubstituted N-aryl (but not N-alkyl) salts give 1,2-dithiole-3-imines[446b].

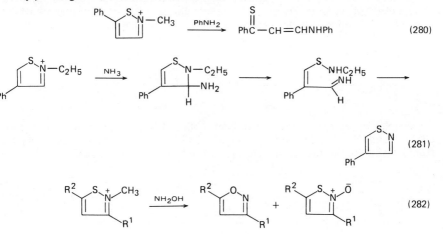

(280)

(281)

(282)

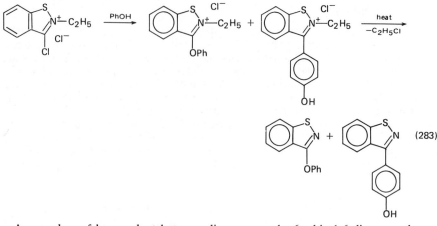

A new class of hypervalent heterocyclic compounds, 6a-thia-1,6-diazapentalenes, is obtained by treatment of 2-methyl-5-(2-methylthiovinyl)isothiazolium salts with methylamine (equation 284)[573].

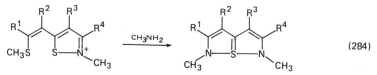

$$(284)$$

The hydrogen–deuterium exchange (via the ylides) of isothiazolium and related salts has been studied[577].

The 3-methyl group of the 1,3-dimethyl-2,1-benzoisothiazolium ion condenses with aldehydes, ketones, orthoesters, and diazonium salts to give new types of styryl cyanine and azo dyes[578].

5. Oxathiolium salts and related compounds

The cations to be discussed are the S-substituted salts of $5H$-1,2-oxathiole **76**, $3H$-1,2-oxathiolium salts **77**, and 1,3-oxathiolium salts **78**. In the latter two ions, there may be considerable delocalization of the positive charge.

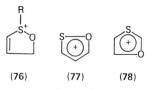

Examples of the first type **76** have been obtained from sulphuranes[579] (equation 285)[579a] and allene sulphoxides (equation 286)[580]. As expected, these salts are readily hydrolysed to the sulphoxides, presumably with inversion at sulphur[579a].

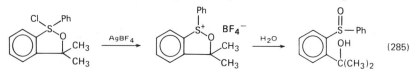

$$(285)$$

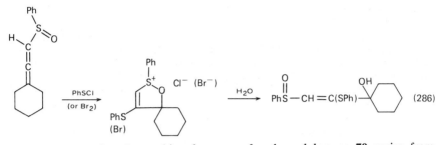

(286)

The infrared carbonyl stretching frequency for the sulphurane **79** varies from 1609 and 1832 cm^{-1} depending on the ligand L and the varying importance of several resonance structures in the representation of the hybrid[581].

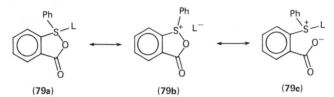

(79a)   (79b)   (79c)

A cation analogous to **79b** was proposed as an intermediate in the reduction and racemization of 2-methylsulphinyl benzoic acid by halide ions[582], and a similar resonance structure was proposed for an intermediate in the free-radical decomposition of *t*-butyl 2-phenylthioperoxybenzoate[583]. A nitrogen-containing variant of **76** has been reported (equation 287)[584].

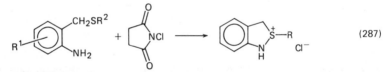

(287)

The first example of a cation of type **77** was observed by n.m.r. methods in the protonation of 1,6-dioxa-6a-thiapentalenes (equation 288)[585]. Previously, such cations had been suggested as intermediates in the nitrosation of 1,6-dioxa-6a-thiapentalenes[477].

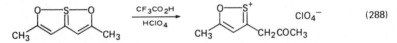

(288)

The 1,3-oxathiolium salts **78** are more common. They may be prepared by various cyclization methods (e.g. equations 289[588] and 290[589–592])[511a,586–594].

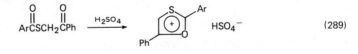

(289)

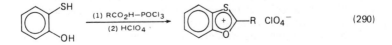

(290)

Alternatively, they can be generated by removal of an anionic group (H⁻, RO⁻) from an oxathiole[589,595].

The principal reaction of 1,3-oxathiolium salts is nucleophilic attack at the 2-position. Addition of hydride ion (via NaBH₄ or LiAlH₄) followed by hydrolysis results in a method for conversion of carboxylic acids[589–592] or nitriles[593] into aldehydes (equation 291)[590,592]. Ketones can be prepared in a similar manner by first treating the salt with a Grignard reagent instead of a hydride ion source before hydrolysis, although some reduction and dimerization is observed in the process.

$$ \underset{}{\text{(291)}} $$

A variety of other nucleophiles (H_2O[587,596], HO^-[596], C_2H_5OH[596], $PhSH$[596], morpholine[596], $Ph\bar{C}HCOPh$[588]) have been added to the 2-position of 1,3-oxathiolium ions. Zinc metal causes reductive dimerization and manganese dioxide effects rupture of the ring[596]. A de-ethoxylation of a 2-ethoxybenzo-1,3-dithiole was observed (equation 292)[596] and the reaction with hydroxide ion leads to cleavage of the ring and another nucleophilic addition[596] (equation 293). Water can also effect ring cleavage[587].

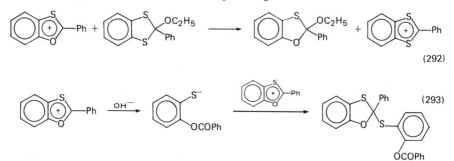

$$ \text{(292)} $$

$$ \text{(293)} $$

Thiophenes can be obtained by treatment of 1,3-oxathiolium salts with active methylene compounds[586,597–600] (e.g. equation 294) and aminothiazoles are obtained by treatment with cyanamide[598].

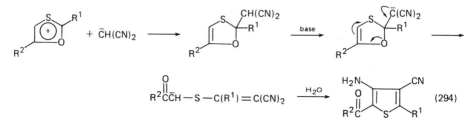

$$ \text{(294)} $$

6. Thiadiazolium salts and related compounds

A number of these cyclic salts with one sulphur atom and two nitrogen atoms exemplify mesoionic derivatives, which have been reviewed[601], and molecular orbital calculations have been carried out on them[393].

Thiadiazolium salts are similar to thiazolium or isothiazolium salts and are more like ammonium than sulphonium salts. Thiadiazoles[602] and thiatetrazoles[603] have been reviewed.

1,3,4-Thiadiazolium derivatives can be prepared by cyclization reactions from acyclic thiocarbonyl compounds[604–611] (e.g. equations 295[604], 296[607–609], 297[611]).

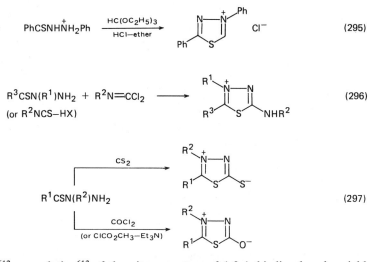

Alkylation[612] or acylation[613] of the nitrogen atom of 1,3,4-thiadiazoles also yield salts. Mesoionic 1,3,4-thiadiazolium salts may be obtained by disproportionation of the 2,2′-disulphides[614].

The dipole moments of a number of mesoionic compounds including derivatives of 1,3,4-thiadiazoles have been obtained[615], the ultraviolet absorption bands of 1,3,4-thiadiazolium 2-thiolate zwitterions have been assigned, and correlations of the n → π* transition at 330–440 nm with solvent polarity have been made[616].

Nucleophiles add readily to the 2-position of 1,3,4-thiadiazolium salts (e.g. equation 298)[617] and the kinetics of the ring cleavage by water of 2-methylthio-4,5-diphenyl-1,3,4-thiadiazolium iodide to N-benzoyl-N-phenyldithio-carbamate have been studied[618]. Alkyl hydrazines cleave the salts; recyclization then occurs to give tetrazines[619]. Aryl hydrazines give triazolium salts[619].

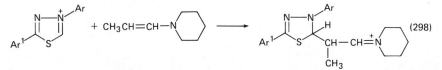

An ylide is a likely intermediate in the conversion of 4-substituted 1,3,4-thiadiazolium iodides into thiadiazine derivatives by treatment with dialkylbenzoylphosphonates (equation 299)[620].

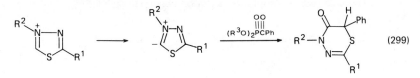

Photolysis of mesoionic 1,3,4-thiadiazolium derivatives results in ring opening (equation 300)[621], and these mesoionic compounds also condense with dimethyl acetylenedicarboxylate to yield the tetramethyl ester of thiophenetetracarboxylic acid (equation 301)[622].

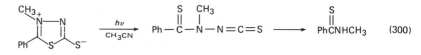

$$(300)$$

$$(301)$$

1,2,4-Thiadiazolium compounds, isomers of the 1,3,4-salts discussed above, may be obtained by cyclization reactions (e.g. equation 302)[623] or by alkylation of 1,2,4-thiadiazoles[612,624]. Methylation may occur at both $N_{(2)}$ or $N_{(4)}$ (the ratio for attack is 4:1 in equation 303)[624].

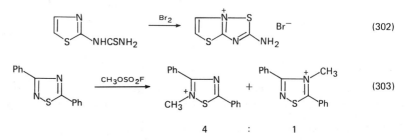

$$(302)$$

$$(303)$$

These salts react readily with nucleophiles to give ring-opened products (equation 304)[624]. The attack may occur at $C_{(5)}$ [RO^-, RNH_2, R_2NH, $(NC)_2C^-$] or at the sulphur atoms (RS^-, BH_4^-, CN^-).

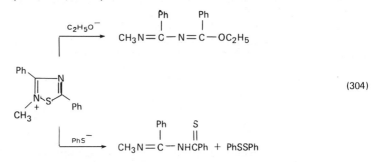

$$(304)$$

1,2,3-Thiadiazolium ions can be prepared by alkylation, which occurs at the 3-position (equation 305)[625]. Unlike the salts of 1,2,4-thiadiazoles, no ring cleavage occurs with nucleophiles, simple dealkylation occurring instead[625].

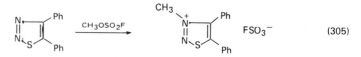

$$(305)$$

Treatment of a mesoionic 3-aryl-1,2,3-oxadiazolium ion with thionyl chloride–dimethylformamide gave a 1,2,3-thiadiazolium salt, which was prepared independently from a mesoionic 3-aryl-1,2,3-thiadiazolium ion (equation 306)[626]. Oxidation gave a new mesoionic derivative[626].

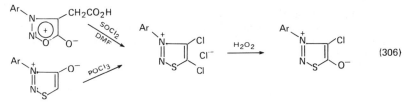

(306)

A 1,2,3-thiadiazolium ion intermediate was suggested in order to account for the formation of thiatetrazapentalenes and their reversible exchange with aryldiazonium tetrafluoroborates shown in equation 307[627].

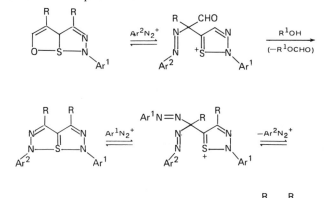

(307)

A number of methods for the preparation of mesoionic derivatives of thiadiazolium salts have been discussed by Ollis and Ramsden[601].

2,1,3-Benzothiadiazolium salts and their selenium analogues have been obtained by treatment of N-methyl-o-phenylenediamine with thionyl chloride or with H_2SeO_3 (equation 308)[628].

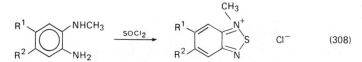

(308)

Isomerization of 1,2,3,4-oxatriazolium-5-thiolates by treatment with ammonia solution was reported to give the corresponding thiatriazolium zwitterions (equation 309)[629]. Treatment of the ethoxythiatriazolium zwitterion in equation 309 with primary aromatic amines, sodium sulphide, or the dicyanomethide ion results in displacement of the ethoxy group with formation of new mesoionic compounds[630].

The interesting sulphur-trinitrogen ylide **80** loses nitrogen on treatment with

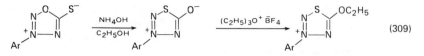

(309)

dilute acid or on heating (equation 310)[631]. Mesoionic oxathiazolium compounds have been prepared; they react with certain 1,3-dipolarophiles and alkenes to give isothiazoles. Photolysis (R = Ph) gives carbon dioxide and benzonitrile sulphide (equation 311)[632].

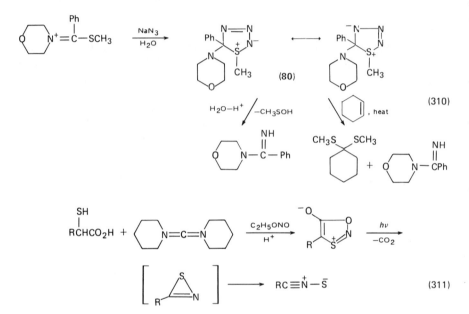

(310)

(311)

G. Unsaturated Systems with Two Sulphur Atoms and One or More Nitrogen Atoms

The alkylation of phenylthiuret is believed to occur at the sulphur atom to give an intermediate sulphonium ion which rearranges to a 1,2,4-thiadiazoline (equation 312)[633].

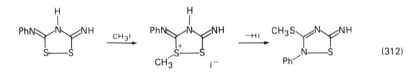

(312)

1,2,4-Dithiazolium salts can be obtained by alkylation of the thione sulphur atom of 1,2,4-dithiazole-3-thiones[634] (equation 313)[634a] or the imino nitrogen atom of 3-imino-1,2,4-dithiazoles[635].

$$\text{R}_2\text{N} \overset{\text{S}\text{---}\text{S}}{\underset{\text{N}}{\diagdown\diagup}}\text{S} \quad \xrightarrow[\text{(2) HClO}_4]{\text{(1) (CH}_3)_2\text{SO}_4} \quad \text{R}_2\text{N}\overset{\text{S}\text{---}\text{S}^+}{\underset{\text{N}}{\diagdown\diagup}}\text{SCH}_3 \quad \text{ClO}_4^- \tag{313}$$

Oxidation of dithiobiurets[416c,634c,636] (equation 314)[634c], arylthioamides, and their *N*-aminomethylene or *N*-hydroxymethylene derivatives also yields 1,2,4-dithiazolium salts (equation 315)[637].

$$\text{R}_2\text{NCSNHCSNR}_2 \quad \xrightarrow[\text{HX}]{\text{H}_2\text{O}_2} \quad \text{R}_2\text{N}\overset{\text{S}\text{---}\text{S}^+}{\underset{\text{N}}{\diagdown\diagup}}\text{NR}_2 \quad \text{X}^- \tag{314}$$

$$\text{ArCSNH}_2 \quad \xrightarrow[\text{HX}]{\text{H}_2\text{O}_2} \quad \text{Ar}\overset{\text{S}\text{---}\text{S}^+}{\underset{\text{N}}{\diagdown\diagup}}\text{Ar} \quad \text{X}^- \tag{315}$$

A novel formation of 1,2,4-thiadiazolium ions by alkylation of certain substituted dithiazoles has been observed (equation 316)[635a].

Hydride ion abstraction from a 1,2,4-dithiazole by triphenylmethyl tetrafluoroborate gives the cation (equation 316a)[638].

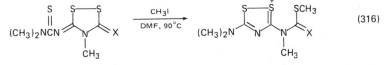

$$\begin{array}{c}\text{S}\\\|\\(\text{CH}_3)_2\text{NCN}\end{array}\overset{\text{S}\text{---}\text{S}}{\underset{\underset{\text{CH}_3}{|}}{\diagdown\diagup}}\text{X} \quad \xrightarrow[\text{DMF, 90°C}]{\text{CH}_3\text{I}} \quad (\text{CH}_3)_2\text{N}\overset{\text{S}\text{---}\overset{+}{\text{S}}}{\underset{\underset{\text{CH}_3}{|}}{\diagdown\diagup}}\overset{\text{SCH}_3}{\underset{\text{X}}{}} \tag{316}$$

X = NCSN(CH₃)₂, NCH₃, S

$$\text{Ph}\overset{\text{S}\text{--}\text{S}}{\underset{\text{N}}{\diagdown\diagup}} \quad \xrightarrow{\text{Ph}_3\overset{+}{\text{C}}\ \text{BF}_4^-} \quad \text{Ph}\overset{\text{S}\text{--}\text{S}}{\underset{\text{N}}{\diagdown(+)\diagup}} \quad \text{BF}_4^- \tag{316a}$$

These 1,2,4-dithiazolium salts react with nucleophilic reagents either with cleavage or alteration of the ring[635a,636a,637,639] (equations 317[637], 318[639b], 319[636a]) or with displacement of a group[634b,c] (e.g. equation 320)[634c].

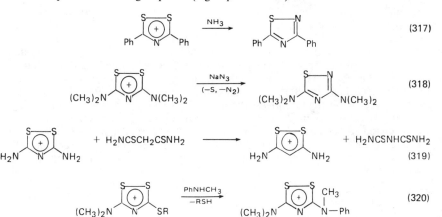

$$\text{Ph}\overset{\text{S}\text{---}\text{S}}{\underset{\text{N}}{\diagdown(+)\diagup}}\text{Ph} \quad \xrightarrow{\text{NH}_3} \quad \text{Ph}\overset{\text{S}\text{---}\text{N}}{\underset{\text{N}}{\diagdown\diagup}}\text{Ph} \tag{317}$$

$$(\text{CH}_3)_2\text{N}\overset{\text{S}\text{---}\text{S}}{\underset{\text{N}}{\diagdown(+)\diagup}}\text{N(CH}_3)_2 \quad \xrightarrow[(-\text{S},\ -\text{N}_2)]{\text{NaN}_3} \quad (\text{CH}_3)_2\text{N}\overset{\text{S}\text{---}\text{N}}{\underset{\text{N}}{\diagdown\diagup}}\text{N(CH}_3)_2 \tag{318}$$

$$\text{H}_2\text{N}\overset{\text{S}\text{---}\text{S}}{\underset{\text{N}}{\diagdown(+)\diagup}}\text{NH}_2 + \text{H}_2\text{NCSCH}_2\text{CSNH}_2 \longrightarrow \text{H}_2\text{N}\overset{\text{S}\text{---}\text{S}}{\underset{\text{N}}{\diagdown(+)\diagup}}\text{NH}_2 + \text{H}_2\text{NCSNHCSNH}_2 \tag{319}$$

$$(\text{CH}_3)_2\text{N}\overset{\text{S}\text{---}\text{S}}{\underset{\text{N}}{\diagdown(+)\diagup}}\text{SR} \quad \xrightarrow[-\text{RSH}]{\text{PhNHCH}_3} \quad (\text{CH}_3)_2\text{N}\overset{\text{S}\text{---}\text{S}}{\underset{\text{N}}{\diagdown(+)\diagup}}\overset{\text{CH}_3}{\underset{\text{N—Ph}}{}} \tag{320}$$

1,2,3-Dithiazolium ions have been obtained by treatment of arylamines[640], nitrosobenzenes[640c,641], or *N*-sulphinylanilines[640c,642] with disulphur dichloride or by treatment of *o*-aminothiophenols with thionyl chloride[643] (e.g. equation 321)[643a].

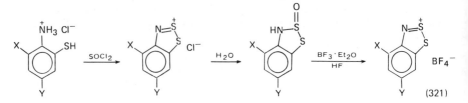

(321)

The ions are hydrolysed by water to *S*-oxide derivatives, which can be converted into the more stable perchlorate or tetrafluoroborate 1,2,3-dithiazolium salts[643a].

A 1,3,2,4-dithiadiazolium salt was prepared by treating *p*-bromobenzoylimino-1,3,2,4-dithiadiazole with methyl fluorosulphonate (equation 322)[644].

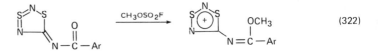

(322)

V. SIX-MEMBERED RINGS

A. Thiapyrylium Salts and Related Compounds

1. Reviews

Among six-membered cyclic sulphonium salts, the thiapyrylium salts and their ylides, known as 'thiabenzenes', are most important; they have been reviewed more or less extensively[645–647]. A comparison of the electronic structure of thiapyrylium ions with other sulphur-containing 6-π-electron systems has been made[648].

2. Theoretical

Because the thiapyrylium ring is a 6-π-electron system, considerable theoretical interest in it has been generated. A number of molecular orbital calculations have been reported[378,380,381b,391d,e,649]. Experimental and theoretical studies indicate that the thiapyrylium ion **81** has the charge more localized on the heteroatom than does the pyrylium ion **82**, the latter being more reactive to nucleophilic attack at carbon than the former[649a]. The energies of the occupied molecular orbitals are generally lower in **82** than in **81**[649a]. Delocalization energies, charge densities, and spectroscopic properties of thiapyrylium salts have been obtained in the various theoretical treatments. In thiapyrylium ions the α- and the γ-positions to sulphur are predicted to be most electron deficient[378b,391d,649a,b] and n.m.r. spectra support the prediction[649b,650]. The n.m.r. signals for the β- and γ-protons for the thiapyrylium cation are superimposed, unlike those for the oxygen and nitrogen analogues[649b].

(81)   (82)

The extent of d-orbital participation in thiapyrylium ions is controversial, some workers claiming that it is trivial[381b,649c] and others supporting the involvement of d-orbitals[378b,380,649b]. Calculations on mesoionic thiapyrylium compounds indicate that d-orbitals can be neglected[393].

Molecular orbital calculations of the spectra of polymethine dyes, including some containing thiapyrylium moieties, have been reported[329,391c].

Barriers to pyramidal inversion at the sulphur atom in 'thiabenzene' have been calculated as 42.5[651] or 56[198] kcal mol^{-1}.

3. Synthesis

The principal methods of synthesis of thiapyrylium salts are summarized as follows.

a. *Treatment of pyrylium salts with sulphide ion.* This reaction involves ring opening and recyclization (equation 323)[649g,h,652–657]. The reaction apparently does not always succeed[658].

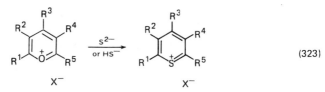

(323)

b. *Cyclization reactions*[659–671]. (E.g. equations 324–331). Some of these (*viz.* the cyclizations of 1,5-diketones)[659,660] are analogous to steps involved in the first method above. Disproportionation reactions occur frequently (e.g. equations 324[659e] and 326[669f]). Cyclizations of the Friedel–Crafts type are frequently used to prepare thiapyrylium salts[200,201,669,670,671] (e.g. equations 326[669f] and 327[671]).

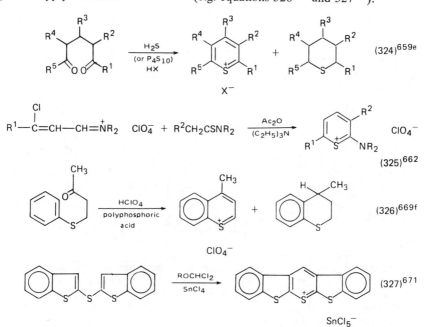

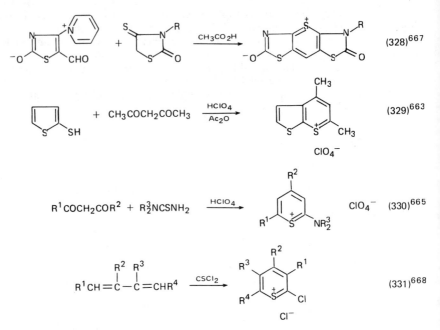

$$(328)^{667}$$

$$(329)^{663}$$

$$(330)^{665}$$

$$(331)^{668}$$

c. *Via thiapyrans or their dihydro or tetrahydro derivatives.* Hydride ion removal from thiapyran derivatives or dehydration of an hydroxythiapyran is involved in many of the cyclization reactions (method b) (e.g. equations 332^{659j} and 333^{670b}).

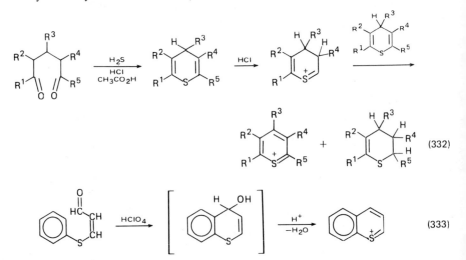

$$(332)$$

$$(333)$$

The direct removal of hydride ion from a thiapyran can be accomplished by the use of the triphenylmethyl cation[650a,658,665,672-682] (e.g. equation 334)[674], 2,3-dichloro-5,6-dicyano-*p*-benzoquinone (DDQ)–perchloric acid[673], or carbonium ions derived from the same thiapyrans by protonation (i.e. the disproportionation reaction exemplified by equation 332, in which ultimately some tetrahydrothiapyran

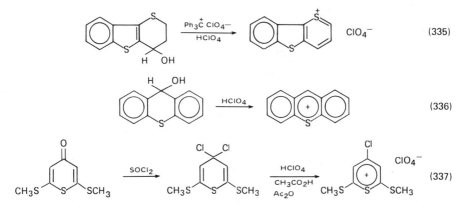

is obtained as well as the thiapyrylium ion)[83,659d,j,683]. In some cases, the triphenylmethyl cation, DDQ–HClO$_4$, or o-chloranil–HClO$_4$ also effects both hydride ion removal and elimination of water[649e,650a,684] (equation 335)[649e]. Simple dehydrations and similar elimination reactions involving appropriate thiapyran derivatives can give thiapyrylium salts[652a,657b,685-692] (equations 336[685] and 337[688]).

Several thiapyran derivatives have been observed to aromatize to thiapyrylium salts on treatment with methyl iodide[693], triethyloxonium tetrafluoroborate[693], dimethyl sulphate[693], sulphuryl chloride–HClO$_4$[681,685,694-696] (equation 338)[685,694], or phosphorus pentachloride[690].

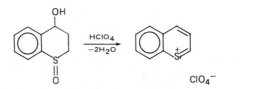

Six-membered cyclic sulphoxides have been converted into thiapyrylium ions (usually concomitantly with dehydration of an alcohol function) on treatment with strong acids[681,695,697-699] (equation 339)[695]. Debenzylation of a 4H-thiapyran by

treatment with perchloric acid–iron(III) chloride gives a thiapyrylium salt in addition to a product arising from rearrangement of the benzyl group[659h,700]. A copper(II) perchlorate oxidation procedure yields bisthiapyrylium ions (equation 340)[701].

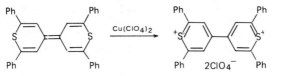

d. *Aromatization of exocyclic thiapyran derivatives.* Thiapyrans with an exocyclic doubly bonded sulphur[702], oxygen[703], nitrogen[661], or carbon[704] atom can often be converted into thiapyrylium salts by treatment with electrophilic reagents (equation 341). Examples are given in equations 342[702f], 343[703b], and 344[704b].

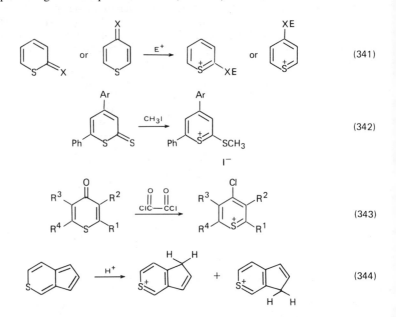

(341)

(342)

(343)

(344)

e. *Miscellaneous methods.* The thiapyrylium ion itself was first prepared by ring expansion of thiophene (equation 345)[705].

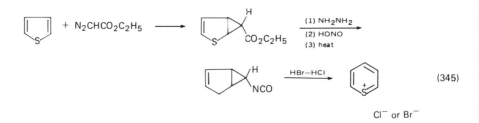

(345)

Cl⁻ or Br⁻

Thiapyrylium ions previously had been suggested as occurring in the mass spectra of alkylthiophenes[706].

4. Reactions

a. *Nucleophilic attack on carbon.* Nucleophiles attack either the $C_{(2)}$, $C_{(4)}$, or $C_{(6)}$ atoms of the thiapyrylium ring. Frequently, a group (SR, Cl, NR_2) at these positions is displaced or eliminated (e.g. equations 346[707], 347[708], 348[702b]) and ring opening may occur (e.g. equation 349[709]). Intermediates resulting from nucleophilic attack may react further to give structures distantly removed from those of the original salts (e.g. equations 350[710] and 351[711]).

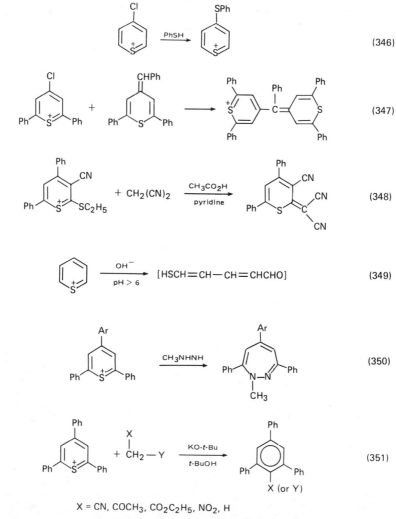

$$X = CN, COCH_3, CO_2C_2H_5, NO_2, H$$
$$Y = CN, COCH_3, CO_2C_2H_5, NO_2, CONH_2, H$$

Reduction with lithium aluminium hydride or sodium borohydride gives 2H- and 4H-thiapyrans[659h,669f,683i,693,709,712–714]. Trichlorosilane[676] and phosphonium iodide[683e] have also been used, although not as widely.

Oxygen[687,707,709,715–717], sulphur[668,687,707], selenium[702a], nitrogen[657a,702d,e,707,710,718–721], phosphorus[722,723], and carbon (organometallic reagents[654,659b,d,h,672,683i,694,696,707,709,723–726], cyanide ion[727], active methylene compounds[469,658,662,688,702b,708,711,728–734], activated aromatic rings[695,735]) nucleophiles have been investigated. Examples of these reactions are given in equations 352–356. The K_a value for hydrolysis $(R^+ + H_2O = ROH = H^+)$ of the thiapyrylium cation is 1.87×10^{-9} as compared with a value of 1.8×10^{-5} for the tropylium ion[736]. The phenazathionium ion generated *in situ* from phenothiazine is attacked by nucleophiles at the 3-position (equation 357)[737].

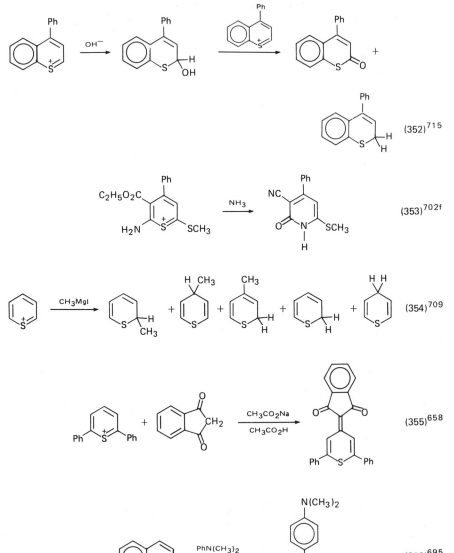

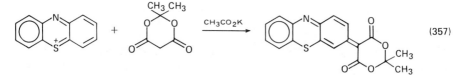

(357)

b. *Nucleophilic attack on sulphur to give 'thiabenzenes'.* Attack on the sulphur atom of thiapyrylium salts by carbanionic reagents (e.g. aryllithium compounds) may occur to give substances called thiabenzenes[693,738]. Recent evidence shows that the reaction products are better formulated as ylides[676,739,740], which formulation was considered by the early workers as part of a resonance representation (equation 358) in which the major emphasis was on covalent structures involving expansion of the valence shell of sulphur[693,738]. Many of what were previously called stable thiabenzenes were identified as oligomers[739]. Thiabenzenes which are relatively more stable contain groups which can stabilize a negative charge on carbon[727c,739a,741–743].

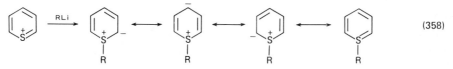

(358)

In addition to the use of organolithium reagents to prepare thiabenzenes[654,676,693,694,739,744,745], they also are obtained by treatment of thiapyrylium salts with Grignard reagents[725,739,746,747b] or by deprotonation of 2*H*- or 4*H*-thiinium salts[676,694,727c,739,740b,741,743,747c,d] (e.g. equation 359)[741].

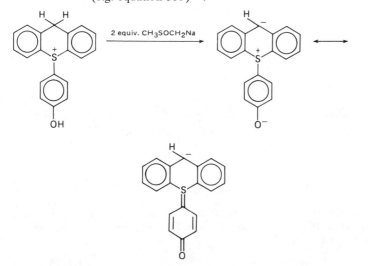

(359)

Thiabenzene derivatives undergo the Stevens rearrangement[693,738a,b,739,741,743,744,746,747a,c,d,748] (e.g. equation 360)[693] and act as dipolar reagents in additions to dimethyl acetylenedicarboxylate, diphenylcyclopropenone, and diphenylcyclopropenethione[742,743] (e.g. equation 361)[743].

Treatment of thiinium ion **83** with methyllithium gave the 4*H*-thioxanthene and methylene which was trapped with cyclohexene[747d] (equation 362).

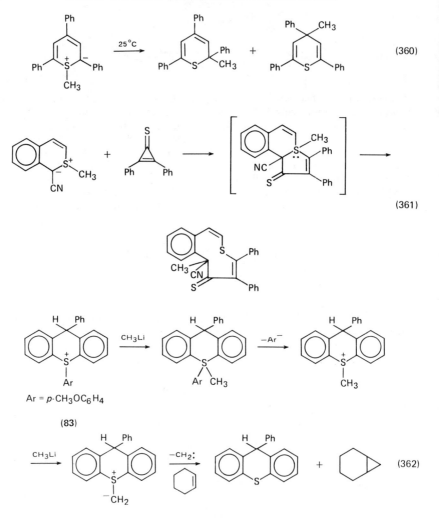

(360)

(361)

(83)

(362)

Thiabenzenes are susceptible to oxygen[654,727c,738a] (e.g. equations 363[727c] and 364[738a]).

The configuration of the sulphur atom in thiabenzenes is pyramidal, and an optically active thiabenzene was obtained by deprotonation of the sulphonium salt precursor with brucine[739c].

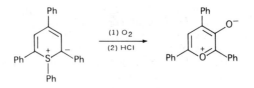

(363)

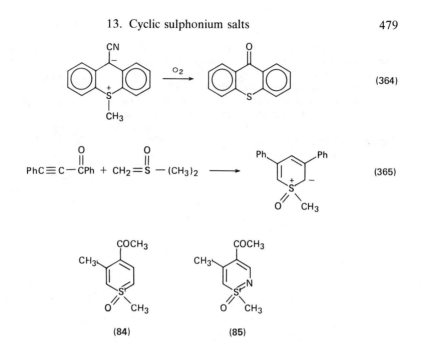

(364)

(365)

(84)   (85)

Thiabenzene oxides (see equation 365)[740c] also have ylide character[676,740a,c,749,750], and an X-ray analysis of **84** and **85** indicates that they are in a half-boat conformation with an appreciable contribution of pπ–dπ bonding to the S—C and S—N bonds[751].

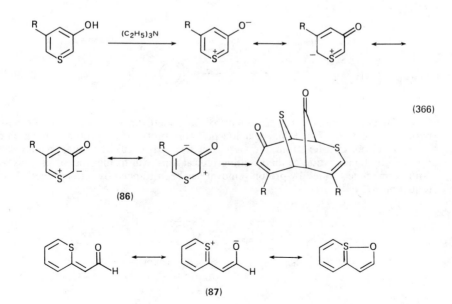

(366)

(86)

(87)

Although not strictly thiabenzenes, the thiapyrylium betaines, e.g. **86**, dimerize via sulphonium ylides[682,752] (equation 366). Finally, a thiabenzene structure was proposed as resulting from an interaction of a negative oxygen atom with a positive sulphur atom in a thiapyrylium structure **87**[753].

c. *Reactions of side groups.* α-Methyl or α-methylene groups in thiapyrylium salts possess acidic hydrogen atoms and a number of aldol-type condensations have been recorded[754–756]. Such reactions have been used to prepare cyanine dyes[657b,658,734]. Equations 367–370 exemplify these reactions.

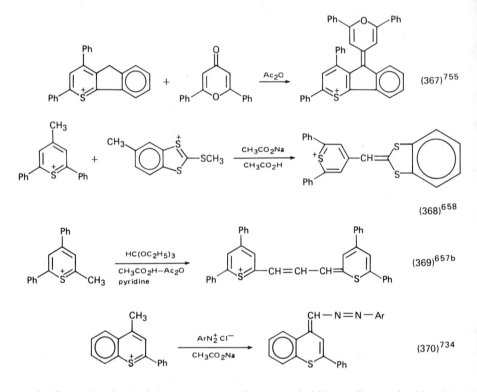

A dimer is obtained by treatment of a γ-methylthiapyrylium salt **88** with pyridine (equation 371)[701]; the reaction probably proceeds by proton removal and a one-electron transfer from the deprotonated species to the thiapyrylium salt followed by dimerization of the resulting radical ions with further loss of protons[757]. An intermediate formed by proton removal has been trapped by tetracyanoethylene (equation 372)[758].

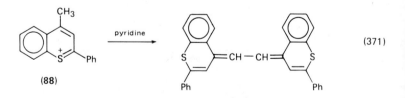

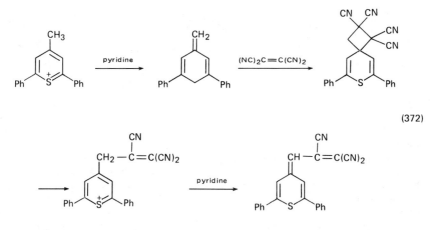

(372)

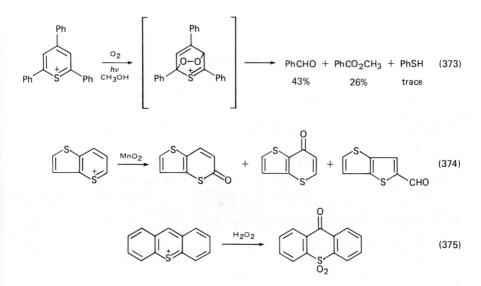

1-Methylthiabenzene 1-oxides can be alkylated or acylated at the methyl group[740a].

d. *Oxidation.* Oxygen[649f] (equation 373), manganese dioxide[687,709,759,760] (equation 374)[760], and hydrogen peroxide[683e] (equation 375) oxidize some thiapyrylium salts. Photo-oxygenation via triplet oxygen leads to rupture of the thiapyrylium ring, but singlet oxygen does not react[649f]. The hydrogen peroxide oxidation probably proceeds by initial addition of the reagent to the thiapyrylium ring.

PhCHO + PhCO$_2$CH$_3$ + PhSH (373)
43%   26%   trace

(374)

(375)

e. *Addition of electrons (one-electron reductions).* Treatment of several thiapyrylium salts with one-electron reducing agents (zinc[674,683e,761], cobaltocene[683e], dithionite[683e], phosphonium iodide[683e], and alkali metals[683e]) yields dimers via

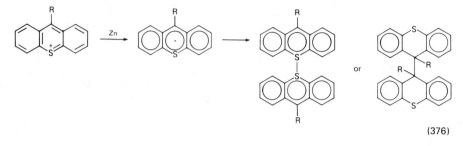

(376)

intermediate radicals. When steric hindrance inhibits dimerization at carbon, dimerization at sulphur occurs, at least initially (equation 376)[683e]. The radical intermediates react with oxygen to give thiapyrones. Electron-spin resonance has been used to investigate the symmetrical 4,4'-bithiopyrylium radical cation[761]. Polarographic studies on thiapyrylium ions have revealed two reversible one-electron transfers[702d,762,763]. *t*-Butylmagnesium chloride has been reported to yield a dimer as well as the expected addition compound and a product of a two-electron reduction[694].

f. *Miscellaneous reactions.* Charge-transfer complexes of thiapyrylium ions with a variety of electron donors have been investigated[655,764–767] and the photo- and semiconductive properties of complexes with anions of cyanocarbon acids (resistivities $10^{10}–10^{12}$ ohm cm at room temperature) have been reported[764b].

An unusual exchange of an ethyl group in thiopyrylium iodide **89** with the methyl group of solvent acetonitrile has been described (equation 377)[702h]. The tetrafluoroborate salt did not undergo alkyl group exchange.

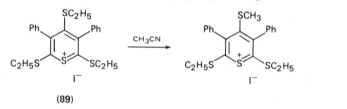

(377)

(89)

B. Other Six-membered Sulphonium Salts with One Sulphur Atom

1. Structure and conformation

Conformational energies of all possible conformers of the thianium salt **90** have been calculated, and the high pharmacological activity of the *trans* compound is believed to be related to its conformational stability[768]. A study of 34 *S*-alkylthianium salts established that $\Delta G^0_{axial \rightleftarrows equatorial}$ for *S*-methyl was small (0.0–0.3 kcal mol^{-1}) because of flattening of the six-membered ring at the sulphur end[769]. Equatorial substituents at $C_{(2)}$ or $C_{(6)}$ which inhibit the flattening increase ΔG^0 for

(90)

S-methyl. An axial methyl group at $C_{(4)}$ stabilizes the conformation of an axial *S*-methyl group. Groups larger than methyl favour an equatorial *S*-alkyl group.

Carbon magnetic resonance substituent effects are a blend of α, β, γ, and δ effects; values of chemical shifts for more or less conformationally fixed systems may be used (in addition to proton n.m.r.) to assess the position of conformational equilibrium in more mobile systems[770,771]. For instance, the *trans* to *cis* ratio for 4-isopropyl-1-methylthianium tetrafluoroborate in water at 100°C is 1.32[206]; the *trans* to *cis* ratio for 4-*t*-butyl-1-methylthianium perchlorate in chloroform at 100°C is 1.45[772a]. C.m.r. spectra of equatorial and axial *S*-methylthianium yields have been reported[772b].

Rates of alkylation of *cis*- and *trans*-3,5-di-*t*-butylthianes and 4-phenylthiane have been determined. Both kinetic and thermodynamic stereoselectivities favour equatorial *S*-methyl groups[773]. Twist conformations of thianes were believed to be unimportant. *S*-Protonated thianes have the proton in the axial position[774].

Activation parameters have been determined for pyramidal inversion at sulphur for the bicyclic sulphonium salt **91** and its ylide[775]. The equatorial conformer is favoured in the salt and the axial conformer in the ylide[775].

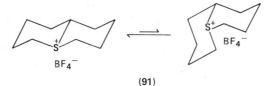

(91)

The energy barrier for pyramidal inversion in 9,9-dimethyl-10-phenylthioxan-thylium perchlorate is 25.4 kcal mol^{-1} [699].

An X-ray analysis of sulphonium salt **92** indicates a greater contribution from the zwitterionic structure than from the thiabenzene structure[776].

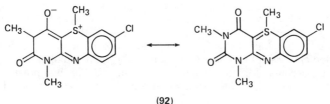

(92)

2. Synthesis

The preparation of six-membered cyclic sulphonium salts can be accomplished by alkylation methods (previously described) via alkyl halides, alkyl halide–silver ion[332], alkyl halide–mercury(II) iodide[777], alcohols or ethers plus HX, alkenes–HX, trialkyloxonium salts, alkyl fluorosulphonates, or alkyl trifluoromethanesulphonates (triflates)[256]. Alkylation of the sulphur atom of a cephalosporin with methyl fluorosulphonate results in epimerization at $C_{(6)}$[778].

Cyclization of halosulphides to thianium salts is facile[779] (e.g. equation 378)[780]. In alkylation reactions assisted by silver perchlorate, the use of its acetonitrile complex is said to be safer than use of the salt alone in dichloromethane[332]. An improved synthesis of 1-thioniabicyclo[4.4.0]decyl bromide has been reported (equation 379)[775,781]. Thianes have been *t*-butylated by treatment with *t*-butanol and three equivalents of perchloric acid[770].

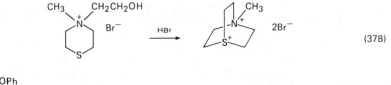

$$(378)$$

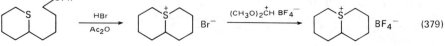

$$(379)$$

Bis-thiazinium dications can be prepared by cyclizations of dithiooxamides (equation 380)[782].

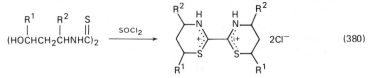

$$(380)$$

A tricyclic salt was obtained by a transannular ring closure (equation 381)[783]. Several zwitterionic S-arylthianium salts have been prepared, either directly from the thiane and a phenol[225] or from 1,5-dibromopentane and 4-(methylthio)-phenol[222,224].

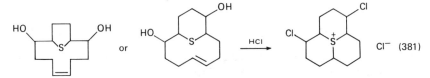

$$(381)$$

S-Chlorothianium salts are intermediates in the reactions of thianes with t-butyl hypochlorite[784–786], and S-aminothianium salts are obtained by treatment of thianes with O-mesitylenesulphonylhydroxylamine[173a,b,347] or by N-alkylation of sulphilimines[787]. Sulphiliminyl sulphonium salts have been prepared from thianes and N-halodiphenylsulphilimines (equation 382)[788]. A sulphurane structure is proposed to account for the behaviour of the salt in non-polar solvents[788]. S-Azathianium salts are intermediates useful in the alkylation of aromatic amines[267c,d].

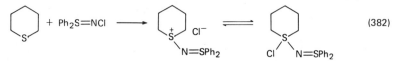

$$(382)$$

Ylides can be obtained directly from thianes and diazoalkanes[187,339,362,789] (e.g. equation 383)[362]. They are frequently unstable and rearrangements occur[362,790–792]. A cephalosporin has been converted into a penicillin via this diazoalkane rearrangement route[791]. Attack of carbenoid species on the sulphur atom of

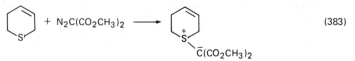

$$(383)$$

4-t-butylthiane gives mainly the axial ylide, whereas nitrenes are less specific[789]. There is a substantial barrier to inversion of the sulphonium ylides[789].

Thianes and other sulphides react with tetracyanoethylene oxide to give zwitterionic sulphonium salts (or ylides) as intermediates in the formation of dicyanomethylides[346,793].

Interaction of electrophilic reagents with a thiocarbonyl group can give six-membered cyclic sulphonium salts[794] (e.g. equation 384)[639a]. Dimethylamino-1-oxathioniacyclohexane tetrafluoroborates have been obtained by N-alkylation of an $-S(O)=NCH_3$ group[237].

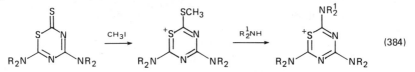

$$(384)$$

Removal of hydride ion from the α-position of a thiane yields a salt (equation 385)[673]. Protonation, alkylation, or acylation of the oxygen atom of a ketone function of certain cyclic ketone sulphoxides yields bicyclic sulphoxonium ions (e.g. equation 386)[795].

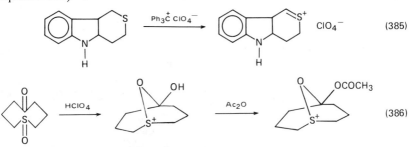

$$(385)$$

$$(386)$$

3. Reactions

As is the case with the sulphonium salts discussed previously, the principal reactions are those with nucleophilic reagents, those involving ylide formation (acidity of the α-protons) and those involving inversion at sulphur. Reactions which are closely similar to reactions discussed earlier will not be reviewed in detail.

Displacement reactions with strong bases as nucleophiles frequently result in considerable elimination[241,779,796] which is particularly apparent in thianium salts (equation 387)[779]. Nucleophilic attack on the S-methylthianium ion occurs mainly at the methyl group (equation 388), in contrast to results with five-membered cyclic salts in which the greater ring strain favours ring opening[246].

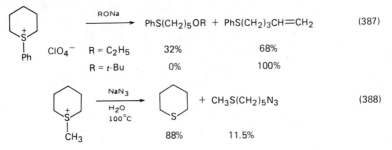

$$(387)$$

$$(388)$$

Determination of the stereochemistry and regioselectivity of hydrogen–deuterium exchange in cyclic sulphonium salts has been the object of a number of investigations. Conformational analysis in heterocyclic systems has been reviewed[797]. Variation in the kinetic acidity (exchange rates) of axial and equatorial protons in S-methylthianium salts and the results of molecular orbital calculations have led to the suggestion that overlap of the carbanionic carbon orbital with the antibonding (σ^*) orbitals of the groups attached to sulphur plays a major role in stabilization of the carbanion, which in turn is important in determining the exchange rates[208]. The protons of the methyl group were observed to exchange 700 times faster than the protons of the α-methylene groups[169,798], and the successive exchange rates of the four α-methylene protons indicate little selectivity of one disastereotopic pair of α-hydrogens over the other pair, which contrasts with the greater selectivity in the five-membered rings[270]. The overall exchange rate of the α-methylene protons was slower than in the five- or seven-membered cyclic sulphonium salts[271]. Exchange in the bicyclic sulphonium salts **93** indicates that

(93)

there is no particularly strong geometrical constraint on formation of an α-carbanion[799]. The pK$_a$ values for a number of conjugate acids of phenacyl sulphonium ylides, including a thianium derivative, have been determined[800a]. Methylation of 6-membered cyclic ylides is reported to be highly stereoselective[800b].

Rearrangements of ylides derived from six-membered cyclic sulphonium salts are common[209,256,258,362,366,790,792,801] (equations 389[362], 390[790], 391[258]). The ylide derived from 1-dimethylamino-1-oxathioniacyclohexane tetrafluoroborate reacts with p-chlorobenzaldehyde to give an epoxide (equation 392)[237].

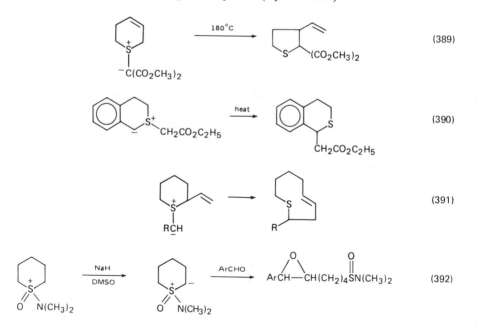

Cyclic sulphilimines react with methoxide ion to yield α-methoxysulphides (equation 393)[802] and the sulphilimine derivative **94** decomposes thermally with elimination (equation 394)[803]. Ring expansions of certain sulphilimines have been observed[803d]. The ylide **95** is alkylated at oxygen and not at carbon (equation 395) and no evidence could be obtained for formation of a stable thianaphthol[804].

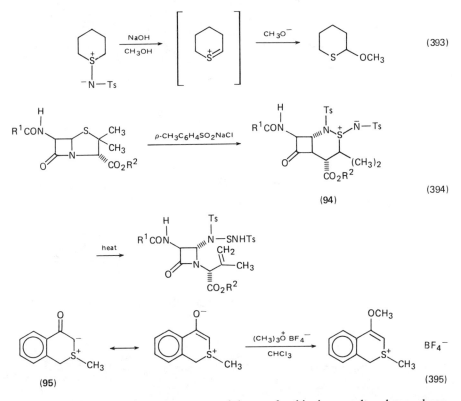

The barriers to inversion at sulphur of thianium salts have been investigated[269,775]. The rates of stereomutation of some five- and six-membered cyclic sulphonium salts are considerably slower than the rates for acyclic analogues, presumably because of angle strain in the planar intermediate state[269]. Displacements on the sulphur atom of S-alkoxythianium[805] and S-aminothianium[787] salts result in inversion at sulphur.

S-Alkoxythianium tetrafluoroborate is rapidly and quantitatively reduced to the sulphide by sodium sulphite or sodium hydrogen sulphite[806]. Photolysis of a β-ketosulphonium salt in methanol yields a variety of products (equation 396)[807]. Photolysis of pyrimido-1,4-benzothiazinesulphonium ylides gave ring expansion

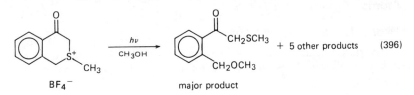

(equation 397)[808]. Attempted conversion of an *S*-amino salt into a sulphilimine by treatment with a basic ion-exchange resin resulted in the formation of a dimer (equation 398)[173b].

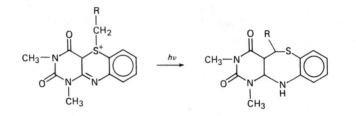

(397)

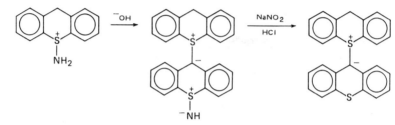

(398)

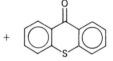

The zwitterionic salt **96** decomposes thermally with the loss of carbon oxysulphide (equation 399)[809].

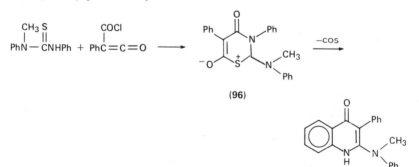

(399)

C. Six-membered Cyclic Sulphonium Salts with Two or More Sulphur Atoms

1. Structure and conformation

The stereochemistry of 1,3-bis-dithianium salts has been investigated by n.m.r. techniques[810]. The 1,3-diethyl salt was a mixture of *cis* and *trans* isomers (62:28), and rapid ring inversion occurred[810]. The $C_{(2)}$ hydrogen atoms of the cyclic salts (S—CH$_3$, S—C$_2$H$_5$) are less acidic than analogous hydrogen atoms in acyclic systems. The relative acidities of the diastereotopic $C_{(2)}$ hydrogen atoms depend on the medium and the configuration and conformation of the molecule[810]. The conformations and configurations of both mono- and di-S-ethyl- or S-methyl-sulphonium salts of 1,3-dithiane have been studied by n.m.r. measurements[811]. Molecular orbital calculations on **97** and **98** have been reported[378c].

(97)   (98)

The mono- and diprotonation of 1,3-dithiane and the exchange of the protons in fluorosulphonic acid have been investigated by proton and carbon n.m.r. techniques; protonation gives the equatorial conjugate acid[812]. Fast >SH proton exchange and fast ring reversal is observed[812]. Mono- and diprotonated forms are in the ratio 85:15. Proton exchange (equilibration of axial and equatorial protons) does not interconvert the 4- and the 6-positions, the base which delivered a proton to the sulphur atom remaining in proximity to that atom[812].

2. Synthesis

Either or both sulphur atoms in 1,3- or 1,4-dithianes can be alkylated by the usual variety of reagents: alkyl halides[291,800,813–816], oxonium salts[291,295,810,817–820], sulphate[821], or fluorosulphonate[292,299] esters. Apparently, the more powerful alkylating agents, e.g. (C$_2$H$_5$)$_3$O BF$_4$ or CH$_3$OSO$_2$F, are required for disubstitution (equation 400)[295]. 1,4-Dithiins are monoalkylated only (equation 401)[819].

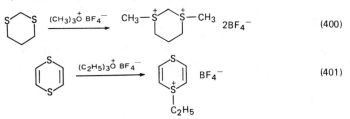

1,2-Dithiane can be monoalkylated by trimethyloxonium 2,4,6-trinitrobenzene-sulphonate (equation 402)[822]. 1,2-Dithianes also are monoalkylated by methyl fluoro-sulphonate at rates only slightly greater than for diethyl disulphide[287].

Treatment of 1,3-dihalopropanes with salts of NN-dimethyldithiocarbamic acid gives dithienium ions (equation 403)[320,321], and dimerization of β-chloroethyl sulphides gives 1,4-dithianium salts[823].

A sulphur atom in 1,3-dithianes can be aminated with O-mesitylenesulphonyl hydroxylamine[173b,302]; a dihydro-1,4-dithiin can be aminated with ethyl N-chlorocarbamate[304] to give S-aminosulphonium salts. 1,4-Dithiane reacts with

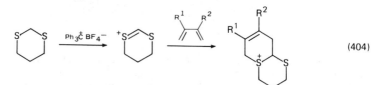

$$Br(CH_2)_3Br + [(CH_3)_2NCS_2^-]_2(CH_3)_2Sn^{2+} \longrightarrow \qquad\qquad (403)$$

tetracyanoethylene oxide to give the S-dicyanomethylene ylide[346]. An ylide is also obtained by treatment of dihydro-1,4-dithiin with biscarbomethoxycarbene[339].

Trityl fluoroborate removes a hydride ion from 1,3-dithiane to give the 1,3-dithienium ion, which reacts with dienes to give bicyclic 1,3-dithianium salts (equation 404)[824]. The same dithienium ion can be obtained from 1,3-dithiane

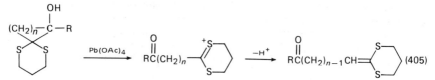

$$(404)$$

and N-chlorosuccinimide followed by ionization of the 2-chlorine atom in liquid sulphur dioxide[312]. A dithienium ion has been proposed as an intermediate in reactions of a bromotetrathiaadamantane derivative[825]. Protonation of the exocyclic double bond in 3-methylene-1,3-dithianes also gives 1,3-dithienium intermediates[826]. Oxidation of 3,3-disubstituted 1,3-dithianes, in which one substituent bears an α-hydroxy group, with lead tetraacetate gives 1,3-dithienium intermediates (equation 405)[827].

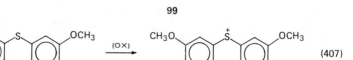

$$(405)$$

The bis-salt of dithiabicyclodecane **99** is obtained by oxidation of 1,6-dithiacyclodecane with the nitrosonium cation (equation 406)[288]. Oxidation of certain 1,4-dithiins electrochemically[828,829] or with $PhICl_2$[830], $HClO_4$[828], or O_2–$AlCl_3$[828] has given bis-salts and radical cations (equation 407).

$$\qquad\qquad \xrightarrow{NO\,BF_4^-} \qquad\qquad 2BF_4^- \qquad\qquad (406)$$

99

$$\xrightarrow{(OX)} \qquad\qquad (407)$$

3. Reactions

Six-membered cyclic sulphonium salts with two sulphur atoms are, as expected, susceptible to nucleophilic attack. They are reduced by sources of hydride ion[326,826b] $[NaAlH_2(OCH_2CH_2OCH_3)_2$[326] (equation 408), $NaBH_4$[326], $LiAlH_4$[326], $(C_2H_5)_3SiH$[826b]$], and they undergo either dealkylation[816,831,832] (equation 409)[832], ring cleavage[292,295,299,320,810,817,820,831,833] (equations 410[810] and 411[820]), or addition reactions[312,320,326] (dithienium ions, e.g. equation 412[312]) with other nucleophiles.

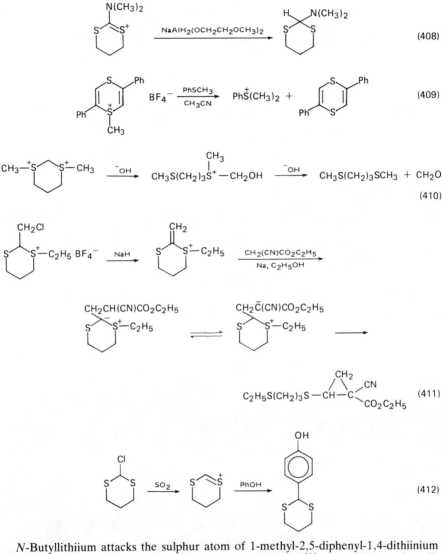

$$\text{(408)}$$

$$\text{(409)}$$

$$\text{(410)}$$

$$\text{(411)}$$

$$\text{(412)}$$

N-Butyllithiium attacks the sulphur atom of 1-methyl-2,5-diphenyl-1,4-dithiinium tetrafluoroborate to effect ring cleavage (equation 413)[816]. Attack by a phenolic oxygen atom on the positive sulphur atom of the S-phthalimido-1,3-dithianium ion

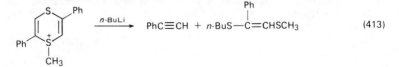

$$\text{(413)}$$

is believed to occur in the first step of a novel synthesis of o-phenol aldehydes (equation 414)[834].

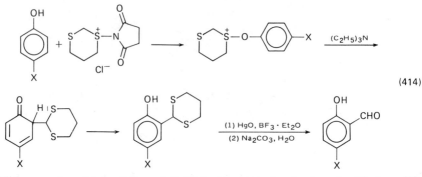

(414)

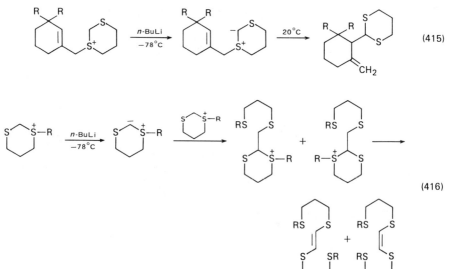

Ylides can be obtained from 1,3-dithianium salts by treatment with bases[295, 810,814,815,817,824,833,835], although with strongly nucleophilic bases ring opening may occur. Rearrangement of the ylides is frequently observed and has uses in synthesis (equation 415)[814]. Ylides derived from mono-S-alkyl-1,3-dithianium ions react with the salt precursor to give dimers (equation 416)[836]. The pK_a of the 1-(4-bromophenacyl)-1,4-dithianium ion is 6.63, compared with 7.00 for the monosulphur (thianium) ion[800].

(415)

(416)

The formal carbon–sulphur double bond in 1,3-dithienium salts shows dienophilic (equation 404)[824] and electrophilic activity towards alkenes (equation 417)[837]. The S-methyl-1,2-dithianium ion reacts with 1,3-dienes and with ethylene, propene, 1-pentene, cis- and trans-2-butene, and cyclohexene to give ring-expanded products (equation 418)[822].

The cyclic sulphonium salts also undergo elimination reactions[816,820,824,833,836] (e.g. equations 413[816], 416[836], 419[820]).

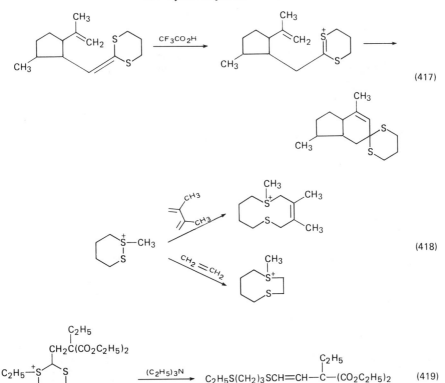

(417)

(418)

(419)

The intermediate N-ethoxycarbonyliminosulphonium chloride, obtained from 2-phenyldihydro-1,4-dithiin and ethyl N-chlorocarbamate, in the presence of hydrogen chloride yields 2-chloro-3-phenyl-5,6-dihydro-1,4-dithiin[304].

VI. SEVEN- AND HIGHER-MEMBERED RINGS

A. Reviews

Seven-membered sulphur heterocycles, including their salts, have been reviewed[838,839], and the Stevens rearrangement of macrocyclic sulphonium salts has been surveyed in connection with the various methods of preparing 2,2-cyclophanes[840].

B. Synthesis

Alkylation of larger ring sulphides usually proceeds readily, and difficult cases yield to the use of the more powerful alkylating agents. The use of alkyl halides is common[229,257,841–843], iodides being preferred for less reactive sulphides. The silver ion assisted alkylation by halides frequently is advantageous[844,845]. Alkyl and other trifluoromethanesulphonates have been used as alkylating agents for cyclic sulphides, the alkylation being an intermediate step in ring expansions[256].

Dibenzothiepinium salts can be obtained from 2,2'-bisbromomethylbiphenyl and thiols[340]. The dimethoxycarbocation[846] (e.g. equation 420)[846j], trimethyloxonium tetrafluoroborate[847], or methyl fluorosulphonate[846d,848] have been used to prepare the large-ring cyclic salts used in cyclophane syntheses. Benzyne has been used as a phenylating agent; it gives an ylide directly (equation 421)[849].

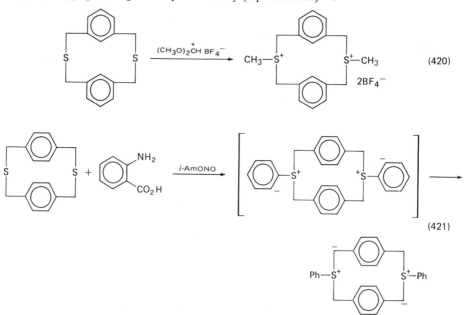

(420)

(421)

A thiepanium ylide is obtained as a colourless solid by treatment of thiepan with dimethyldiazomalonate[187], and 1-dimethylamino-1-oxathioniacycloheptane and its ylide have been prepared[237].

Thiacycloheptane reacts with protonated quinones to give sulphonium salts of hydroquinones (equation 422)[226].

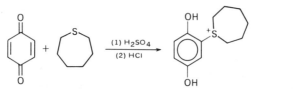

(422)

The thiepin derivative **100** disproportionates presumably via thiepinium intermediates (equation 423)[850].

Cyclization of ethyl 6-chlorohexylsulphide to S-ethylthiepanium chloride occurs less readily than cyclizations to five- and six-membered rings[851]. A bicyclic sulphonium salt has been proposed as an intermediate in the rearrangement of an eight-membered cyclic sulphide (equation 424)[852]. A seven-membered sulphonium salt intermediate was proposed in order to account for the rearrangement of alcohol **101** (equation 425)[853], and similar intermediates involving the corresponding ketone have been suggested[854].

Eight- and ten-membered cyclic sulphonium salts with two sulphur atoms are

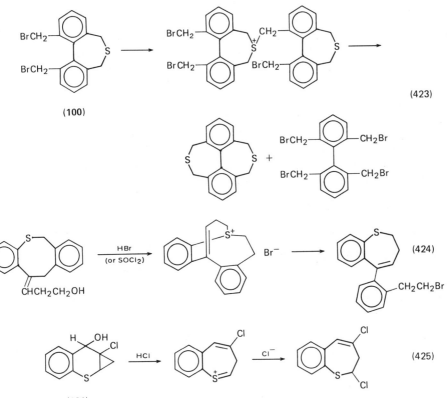

(423)

(100)

(424)

ĊHCH₂CH₂OH

(101)

(425)

obtained by treatment of the *S*-methyl salts of 1,2-dithiane with alkenes and dienes (equation 418)[822].

The *S*-methyl salt of dibenzothiepin was not resolvable[841], and a study of the nuclear Overhauser effect on the conformations of another dibenzothiepin salt has been reported[845]. The proton and carbon n.m.r. spectra of salt **102** show that for *n* = 6 two conformations can be detected, the aromatic carbon atoms becoming non-equivalent for this ring size[855].

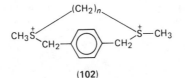

(102)

Macrocyclic sulphoxides can be converted into *S*-hydroxy or *S*-alkoxy salts on protonation or alkylation[289,856]. The former are intermediates in the reduction of sulphoxides by hydrogen iodide (equation 426)[289]. An *O*-*t*-butyl salt is obtained on treatment of thiepane with *t*-butyl hypochlorite[857].

Thiepane is *N*-substituted with diphenyl *N*-bromosulphilimine to give a salt[858]. An *S*-chlorosulphonium salt of a benzothiepane has been obtained by treatment of the sulphoxide with acetyl chloride–antimony pentachloride[859].

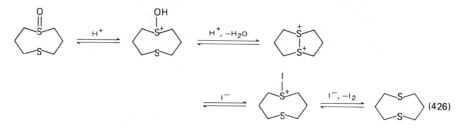

C. Reactions

The larger ring sulphonium salts are stable, although prolonged heating of S-methylthiepanium iodide causes ring opening[860].

Sulphonium ylides derived from large-ring sulphides are used in ring expansions to give macrocyclic derivatives[176,256,257] (e.g. equation 427)[256]. The ylides of a number of macrocyclic sulphonium salts have found utility in cyclophane and annulene synthesis via Stevens rearrangements[846,847,848a] (e.g. equation 428[848a]).

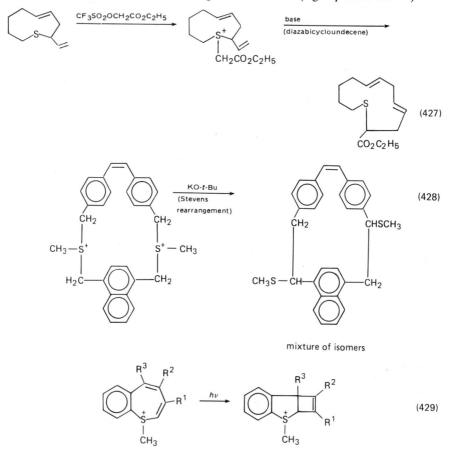

The photochemical rearrangement of 1-methylbenzothiepinium salts yields heterobicycloheptadienes (equation 429)[861].

VII. REFERENCES

1. W. H. Mueller, *Angew. Chem. Int. Ed. Engl.*, **8**, 482 (1969).
2. N. Kharasch, in *The Chemistry of Organic Sulfur Compounds* (Ed. N. Kharasch), Vol. 1, Pergamon Press, New York, 1961, Chapter 32.
3. P. B. de la Mare and R. Bolton, *Electrophilic Additions to Unsaturated Systems*, Elsevier, New York, 1966, p. 166.
4. R. C. Fahey, in *Topics in Stereochemistry* (Eds. E. L. Eliel and N. L. Allinger), Vol. III, Wiley, New York, 1968, p. 299.
5. (a) G. H. Schmid and D. G. Garrat, in *The Chemistry of Double-Bonded Functional Groups* (Ed. S. Patai), Wiley, New York, 1977, p. 828; (b) G. H. Schmid, in *Topics in Sulfur Chemistry* (Eds. A. Senning and P. S. Magee), Vol. 3, Georg Thieme, Stuttgart, 1977, p. 101; (c) W. A. Smit, N. S. Zefirov, I. V. Bodrikov and M. Z. Krimer, *Acc. Chem. Res.*, *Chem. Res.*, **12**, 289 (1979); (d) E. Kühle, *Synthesis*, 567 (1971).
6. L. Rasteikiene, D. Greiciute, M. G. Lin'kova and I. L. Knunyants, *Russ. Chem. Rev.*, **46**, 548 (1977).
7. K. D. Gundermann, *Angew. Chem. Int. Ed. Engl.*, **2**, 674 (1963).
8. G. K. Helmkamp and D. C. Owsley, *Mech. React. Sulfur Compd.*, **4**, 37 (1969).
9. A. V. Fokin and A. F. Kolomiets, *Russ. Chem. Rev.*, **45**, 25 (1976); M. Sander, *Chem. Rev.*, **66**, 297 (1966); L. Goodman and E. J. Reist, in *The Chemistry of Organic Sulfur Compounds* (Ed. N. Kharasch and C. Y. Meyers), Vol. 2, Pergamon Press, New York, 1966, p. 93.
10. V. M. Csizmadia, G. H. Schmid, P. G. Mezey and I. G. Csizmadia, *J. Chem. Soc. Perkin Trans. II*, 1019 (1977).
11. D. C. Owsley, G. K. Helmkamp and M. F. Rettig, *J. Amer. Chem. Soc.*, **91**, 5239 (1969).
12. G. A. Olah and P. J. Szilagyi, *J. Org. Chem.*, **36**, 1121 (1971).
13. P. Raynolds, S. Zonnebelt, S. Bakker and R. M. Kellogg, *J. Amer. Chem. Soc.*, **96**, 3146 (1974).
14. R. D. Bach and H. F. Henneike, *J. Amer. Chem. Soc.*, **92**, 5589 (1970).
15. Y. Kikuzono, T. Yamabe, S. Nagata, H. Kato and K. Fukui, *Tetrahedron*, **30**, 2197 (1974).
16. P. P. Budnikoff and E. A. Schilow, *Chem. Ber.*, **55**, 3848 (1922).
17. L. Goodman, A. Benitez and B. R. Baker, *J. Amer. Chem. Soc.*, **80**, 1680 (1958).
18. D. J. Pettitt and G. K. Helmkamp, *J. Org. Chem.*, **28**, 2932 (1963); **29**, 2702 (1964).
19. G. K. Helmkamp and D. C. Owsley, *Q. Rep. Sulfur Chem.*, **2**, 303 (1967).
20. E. A. Vorob'eva, M. Z. Krimer and V. A. Smit, *Bull. Acad. Sci. USSR, Div. Chem. Sci.*, **25**, 2553 (1976).
21. E. A. Vorob'eva, M. Z. Krimer and V. A. Smit, *Bull. Acad. Sci. USSR, Div. Chem. Sci.*, **25**, 1267 (1976).
22. (a) V. A. Smit, M. Z. Krimer and E. A. Vorob'eva, *Tetrahedron Lett.*, 2451 (1975); (b) J. Bolster and R. M. Kellogg, *J. Chem. Soc. Chem. Commun.*, 630 (1978).
23. M. Oki, W. Nakanishi, M. Fukunaga, G. D. Smith, W. L. Duax and Y. Osawa, *Chem. Lett.*, 1277 (1975).
24. G. K. Helmkamp, H. N. Cassey, B. A. Olsen and D. J. Pettitt, *J. Org. Chem.*, **30**, 933 (1965).
25. G. Capozzi, V. Lucchini, G. Modena and F. Rivetti, *J. Chem. Soc. Perkin Trans. II*, 900 (1975); A. S. Gybin, W. A. Smit, V. S. Bogdanov, M. Z. Krimer and J. B. Kalyan, *Tetrahedron Lett.*, 383 (1980).
26. G. K. Helmkamp, B. A. Olsen and D. J. Pettitt, *J. Org. Chem.*, **30**, 676 (1965).
27. G. Capozzi, O. De Lucchi, V. Lucchini and G. Modena, *Tetrahedron Lett.*, 2603 (1975).
28. G. Capozzi, O. De Lucchi, V. Lucchini and G. Modena, *Synthesis*, 677 (1976).

29. E. A. Vorob'eva, M. Z. Krimer and V. A. Smit, *Bull. Acad. Sci. USSR, Div. Chem. Sci.*, **24**, 113 (1975).
30. W. A. Smit, A. S. Gybin, V. S. Bogadanov, M. Z. Krimer and E. A. Vorob'eva, *Tetrahedron Lett.*, 1085 (1978).
31. W. H. Mueller, *J. Amer. Chem. Soc.*, **91**, 1223 (1969).
32. O. V. Kil'disheva, M. G. Lin'kova, L. P. Rasteikene, V. A. Zabelaite, N. K. Potsyute and I. L. Knunyants, *Proc. Acad. Sci. USSR, Chem. Sect.*, **203**, 331 (1972).
33. K. Levsen, H. Heimbach, C. C. Van De Sande and J. Monstrey, *Tetrahedron*, **33**, 1785 (1977).
34. B. van de Graaf, P. P. Dymerski and F. W. McLafferty, *J. Chem. Soc. Chem. Commun.*, 978 (1975).
35. N. Kharasch and C. M. Buess, *J. Amer. Chem. Soc.*, **71**, 2724 (1949).
36. N. Kharasch and A. J. Havlik, *J. Amer. Chem. Soc.*, **75**, 3734 (1953).
37. W. L. Orr and N. Kharasch, *J. Amer. Chem. Soc.*, **75**, 6030 (1953).
38. N. Kharasch, C. M. Buess and W. King, *J. Amer. Chem. Soc.*, **75**, 6035 (1953).
39. D. R. Hogg and N. Kharasch, *J. Amer. Chem. Soc.*, **78**, 2728 (1956).
40. W. H. Mueller and P. E. Butler, *J. Org. Chem.*, **32**, 2925 (1967).
41. W. H. Mueller, R. M. Rubin and P. E. Butler, *J. Org. Chem.*, **31**, 3537 (1966).
42. R. G. Guy and I. Pearson, *Bull. Chem. Soc. Jap.*, **49**, 2310 (1976).
43. (a) F. Lautenschlaeger, *J. Org. Chem.*, **31**, 1679 (1966); (b) *Q. Rep. Sulfur Chem.*, **2**, 331 (1967).
44. F. Lautenschlaeger and N. V. Schwartz, *J. Org. Chem.*, **34**, 3991 (1969).
45. W. H. Mueller and P. E. Butler, *J. Org. Chem.*, **33**, 2111 (1968).
46. G. H. Schmid and V. J. Nowlan, *J. Org. Chem.*, **37**, 3086 (1972).
47. G. H. Schmid and V. J. Nowlan, *Can. J. Chem.*, **54**, 695 (1976).
48. D. J. Cram, *J. Amer. Chem. Soc.*, **71**, 3883 (1949).
49. G. H. Schmid and V. M. Csizmadia, *Can. J. Chem.*, **44**, 1338 (1966).
50. H. Kwart and R. K. Miller, *J. Amer. Chem. Soc.*, **78**, 5678 (1956).
51. S. J. Cristol, R. P. Arganbright, G. D. Brindell and R. M. Heitz, *J. Amer. Chem. Soc.*, **79**, 6035 (1957).
52. W. H. Mueller and P. E. Butler, *J. Amer. Chem. Soc.*, **88**, 2866 (1966).
53. H. C. Brown, J. H. Kawakami and K. T. Liu, *J. Amer. Chem. Soc.*, **95**, 2209 (1973).
54. K. Toyoshima, T. Okuyama and T. Fueno, *J. Org. Chem.*, **43**, 2789 (1978).
55. W. H. Mueller and P. E. Butler, *J. Amer. Chem. Soc.*, **90**, 2075 (1968).
56. W. A. Thaler, W. H. Mueller and P. E. Butler, *J. Amer. Chem. Soc.*, **90**, 2069 (1968).
57. G. M. Beverly and D. R. Hogg, *J. Chem. Soc. Chem. Commun.*, 138 (1966).
58. G. M. Beverly, D. R. Hogg and J. H. Smith, *Chem. Ind. (London)*, 1403 (1968).
59. W. A. Thaler, *J. Org. Chem.*, **34**, 871 (1969).
60. C. L. Dean, D. G. Garratt, T. T. Tidwell and G. H. Schmid, *J. Amer. Chem. Soc.*, **96**, 4958 (1974).
61. K. Izawa, T. Okuyama and T. Fueno, *Bull. Chem. Soc. Jap.*, **47**, 1480 (1974).
62. I. V. Bodrikov, L. G. Gurvich, N. S. Zefirov, V. R. Kartashov and A. L. Kurts, *J. Org. Chem. USSR*, **10**, 1553 (1974).
63. V. R. Kartashov, I. V. Bodrikov, E. V. Skorobogatova and N. S. Zefirov, *J. Org. Chem. USSR*, **12**, 289 (1976).
64. I. V. Bodrikov, T. S. Ganzhenko, N. S. Zefirov and V. R. Kartashov, *Dokl. Acad. Nauk SSSR*, **226**, 831 (1976); *Chem. Abstr.*, **84**, 150088 (1976).
65. N. S. Zefirov, N. K. Sadovaya, A. M. Magarramov, I. V. Bodrikov and V. R. Kartashov, *J. Org. Chem. USSR*, **12**, 904 (1976).
66. H. Chartier and R. Vessiere, *C.R. Acad. Sci., Ser. C*, **270**, 646 (1970).
67. M. G. Lin'kova, D. I. Greichute, L. P. Rasteikene and I. L. Knunyants, *Bull. Acad. Sci. USSR, Div. Chem. Sci.*, **20**, 2390 (1971).
68. P. D. Bartlett and E. N. Trachtenberg, *J. Amer. Chem. Soc.*, **80**, 5808 (1958).
69. D. R. Hogg, *Q. Rep. Sulfur Chem.*, **2**, 339 (1967).
70. C. Brown and D. R. Hogg, *J. Chem. Soc. Chem. Commun.*, 357 (1965).
71. T. S. Leong and M. E. Peach, *J. Fluorine Chem.*, **6**, 145 (1975).
72. H. Morita and S. Oae, *Heterocycles*, **5**, 29 (1976).
73. S. Ikegami, J. Ohishi and Y. Shimizu, *Tetrahedron Lett.*, 3923 (1975).

74. M. Shibasaki and S. Ikegami, *Tetrahedron Lett.*, 4037 (1977).
75. R. G. Micetich and R. B. Morin, *Tetrahedron Lett.*, 979 (1976).
76. S. Kukolja, S. R. Lammert, M. R. Gleissner and A. I. Ellis, *J. Amer. Chem. Soc.*, **97**, 3192 (1975).
77. S. Kukolja and S. R. Lammert, *J. Amer. Chem. Soc.*, **94**, 7169 (1972).
78. D. H. R. Barton, F. Comer, D. G. T. Greig, P. G. Sammes, C. M. Cooper, G. Hewitt and W. G. E. Underwood, *J. Chem. Soc. C*, 3540 (1971).
79. D. H. R. Barton, F. Comer, D. G. T. Greig, G. Lucente, P. G. Sammes and W. G. E. Underwood, *J. Chem. Soc. Chem. Commun.*, 1059 (1970).
80. M. Kise, M. Murase, M. Kitano, T. Tomita and H. Murai, *Tetrahedron Lett.*, 691 (1976).
81. D. D. MacNicol and J. J. McKendrick, *J. Chem. Soc. Perkin Trans. I*, 2493 (1974); *Tetrahedron Lett.*, 2593 (1973).
82. F. H. M. Deckers, W. N. Speckamp and H. O. Huisman, *J. Chem. Soc. Chem. Commun.*, 1521 (1970).
83. E. R. de Waard, W. J. Vloon and H. O. Huisman, *J. Chem. Soc. Chem. Commun.*, 841 (1970).
84. R. C. Fuson, C. C. Price and D. M. Burness, *J. Org. Chem.*, **11**, 475 (1946).
85. P. D. Bartlett and C. G. Swain, *J. Amer. Chem. Soc.*, **71**, 1406 (1949).
86. T. T. Tsuji, T. Komeno, H. Itani and H. Tanida, *J. Org. Chem.*, **36**, 1648 (1971).
87. S. Ikegami, T. Asai, K. Tsuneoka, S. Matsumura and S. Akaboshi, *Tetrahedron*, **30**, 2087 (1974).
88. O. N. Nuretdinova and B. A. Arbuzov, *Mater. Nauch. Konf. Inst. Org. Fiz. Khim. Akad. Nauk SSSR*, 1 (1970); *Chem. Abstr.*, **76**, 71679 (1972).
89. I. Tabushi, Y. Tamaru, Z. Yoshida and T. Sugimoto, *J. Amer. Chem. Soc.*, **97**, 2886 (1975).
90. D. C. Billington and B. T. Golding, *J. Chem. Soc. Chem. Commun.*, 208 (1978).
91. A. M. Jeffery and D. M. Jerina, *J. Amer. Chem. Soc.*, **97**, 4427 (1975).
92. J. F. King, K. Abikar, D. M. Deaken and R. G. Pews, *Can. J. Chem.*, **46**, 1 (1968); J. F. King and K. Abikar, *Can. J. Chem.*, **46**, 9 (1968).
93. J. V. Cerny and J. Polacek, *Coll. Czech Chem. Commun.*, **31**, 1831 (1966).
94. S. Ikegami, J. I. Ohishi and S. Akaboshi, *Chem. Pharm. Bull.*, **23**, 2701 (1975).
95. P. H. McCabe and C. M. Livingston, *Tetrahedron Lett.*, 3029 (1973).
96. J. A. J. M. Vincent, P. Schipper, A. de Groot and H. M. Buck, *Tetrahedron Lett.*, 1989 (1975).
97. J. H. Robson and H. Shechter, *J. Amer. Chem. Soc.*, **89**, 7112 (1967).
98. K. Kondo and I. Ojima, *J. Chem. Soc. Chem. Commun.*, 62 (1972).
99. I. Ojima and K. Kondo, *Bull. Chem. Soc. Jap.*, **46**, 1539 (1973).
100. S. S. Hixon and S. H. Hixon, *J. Org. Chem.*, **37**, 1279 (1972); A. Kortmann and A. Bhattacharjya, *J. Amer. Chem. Soc.*, **98**, 7081 (1976).
101. I. Murata, T. Tatsuoka and Y. Sugihara, *Tetrahedron Lett.*, 199 (1974).
102. G. K. Helmkamp and D. J. Pettitt, *J. Org. Chem.*, **25**, 1754 (1960).
103. G. K. Helmkamp and D. J. Pettitt, *J. Org. Chem.*, **27**, 2942 (1962).
104. G. K. Helmkamp and D. J. Pettitt, *J. Org. Chem.*, **29**, 3258 (1964).
105. E. Vilsmaier and W. Schalk, *Justus Liebigs Ann. Chem.*, **750**, 104 (1971).
106. H. Böhme and G. Dahler, *Chem. Ber.*, **103**, 3058 (1970).
107. S. Sakai, Y. Asai and Y. Ishii, *Kogyo Kagaku Zasshi*, **70**, 2036 (1967); *Chem. Abstr.*, **68**, 86644 (1968).
108. H. Kwart and J. A. Herbig, *J. Amer. Chem. Soc.*, **85**, 1508 (1963).
109. N. V. Schwartz, *J. Org. Chem.*, **33**, 2895 (1968).
110. A. V. Fokin, A. F. Kolomets, T. I. Fedyushina and V. I. Shevchenko, *Proc. Acad. Sci. USSR, Chem. Sect.*, **237**, 672 (1977).
111. M. G. Lin'kova, L. D. Parshina, Z. K. Stumbrevichute, O. V. Kildisheva and I. L. Knunyants, *Proc. Acad. Sci. USSR, Chem. Sect.*, **196**, 124 (1971).
112. E. Vilsmaier and B. Hloch, *Synthesis*, 590 (1971).
113. Y. Hata, M. Watanabe, S. Inoue and S. Oae, *J. Amer. Chem. Soc.*, **97**, 2553 (1975).
114. A. Padwa and A. Battisti and E. Shefter, *J. Amer. Chem. Soc.*, **91**, 4000 (1969); **93**, 1304 (1971); **94**, 521 (1972).
115. A. G. Schultz and R. H. Schlessinger, *Tetrahedron Lett.*, 4787, 4791 (1973).

116. M. Maeda, A. Kawahara, M. Kai and M. Kojima, *Heterocycles*, **3**, 389 (1975).
117. J. G. Pacifici and C. Diebert, *J. Amer. Chem. Soc.*, **91**, 4595 (1969).
118. M. F. Semmelhack, S. Kunkes and C. S. Lee, *J. Chem. Soc. Chem. Commun.*, 698 (1971).
119. A. Couture, A. Delevallee, A. Lablache-Combier and C. Parkanyi, *Tetrahedron*, **31**, 785 (1975).
120. E. Carbin, G. K. Helmkamp, W. M. Barnes and M. Sundaralingam, *Int. J. Sulfur Chem. A*, **2**, 129 (1972).
121. J. G. Traynham, G. R. Franzen, G. A. Knesel and D. J. Northington, Jr., *J. Org. Chem.*, **32**, 3285 (1967).
122. E. J. Corey and E. Block, *J. Org. Chem.*, **31**, 1663 (1966).
123. E. D. Weil, K. J. Smith and R. J. Gruber, *J. Org. Chem.*, **31**, 1669 (1966).
124. F. Lautenschlaeger, *Can. J. Chem.*, **44**, 2813 (1966).
125. K. Undheim and K. R. Reistad, *Acta Chem. Scand.*, **24**, 2949 (1970).
126. J. Alexander, G. Lowe, N. K. McCullum and G. K. Ruffles, *J. Chem. Soc. Perkin Trans. I*, 2092 (1974).
127. P. Brownbridge and S. Warren, *J. Chem. Soc. Perkin Trans. I*, 1131 (1977).
128. D. C. Owsley, G. K. Helmkamp and S. N. Spurlock, *J. Amer. Chem. Soc.*, **91**, 3606 (1969).
129. G. H. Schmid and P. H. Fitzgerald, *J. Amer. Chem. Soc.*, **93**, 2547 (1971).
130. (a) G. Modena and G. S. Scorrano, *Mech. React. Sulfur Compd.*, **3**, 115 (1968); (b) G. Capozzi, V. Lucchini and G. Modena, *Rev. Chem. Int.*, **2**, 347 (1979).
131. (a) M. E. Volpin, Y. D. Koreshkov, V. G. Dulova and D. N. Kursanov, *Tetrahedron*, **18**, 107 (1962); (b) I. G. Csizmadia, F. Bernardi, V. Lucchini and G. Modena, *J. Chem. Soc. Perkin Trans. II*, 542 (1977).
132. I. G. Csizmadia, A. J. Duke, V. Lucchini and G. Modena, *J. Chem. Soc. Perkin Trans. II*, 1808 (1974).
133. D. T. Clark, *Int. J. Sulfur Chem. C*, **7**, 711 (1972).
134. J. B. Andose, A. Rank, R. Tang and K. Mislow, *Int. J. Sulfur Chem. A*, **1**, 66 (1971).
135. A. S. Denes, I. G. Csizmadia and G. Modena, *J. Chem. Soc. Chem. Commun.*, 8 (1972).
136. G. Capozzi, V. Lucchini, G. Modena and P. Scrimin, *Tetrahedron Lett.*, 911 (1977).
137. G. Capozzi, V. Lucchini and G. Modena, *Abstracts of Papers, Eighth International Symposium on Organic Sulphur Chemistry*, Portoroz, Yugoslavia, June 18–23, 1978, p. 54.
138. R. Destro, T. Pilati and M. Simonetta, *J. Chem. Soc. Chem. Commun.*, 576 (1977).
139. G. Capozzi, O. D. Lucchi, V. Lucchini and G. Modena, *J. Chem. Soc. Chem. Commun.*, 248 (1975).
140. T. J. Barton and R. G. Zika, *J. Org. Chem.*, **35**, 1729 (1970).
141. G. Capozzi, V. Lucchini, G. Modena and P. Scrimin, *Nouv. J. Chim.*, **2**, 95 (1978).
142. G. H. Schmid, A. Modro, D. G. Garratt and K. Yates, *Can. J. Chem.*, **54**, 3045 (1976).
143. T. Okuyama, K. Izawa and T. Fueno, *J. Org. Chem.*, **39**, 351 (1974).
144. (a) G. Capozzi, G. Melloni and G. Modena, *J. Chem. Soc. C*, 2621 (1970); (b) 2625 (1970); (c) 3018 (1971).
145. G. Capozzi, G. Melloni, G. Modena and U. Tonellato, *J. Chem. Soc. Chem. Commun.*, 1520 (1969).
146. G. Modena and U. Tonellato, *J. Chem. Soc. B*, 381 (1971).
147. G. Capozzi, G. Modena and U. Tonellato, *J. Chem. Soc. B*, 1700 (1971).
148. G. Modena and U. Tonellato, *J. Chem. Soc. B*, 374 (1971).
149. L. Di Nunno, G. Melloni, G. Modena and G. Scorrano, *Tetrahedron Lett.*, 4405 (1965).
150. A. Burighel, G. Modena and U. Tonellato, *J. Chem. Soc. Chem. Commun.*, 1325 (1971).
151. G. Scorrano and U. Tonellato, *Boll. Sci. Fac. Chim. Ind. Bologna*, **27**, 377 (1969); *Chem. Abstr.*, **72**, 110504 (1970).
152. G. Modena, G. Scorrano and U. Tonellato, *J. Chem. Soc. Perkin Trans. II*, 493 (1973).
153. O. Buchardt, J. Domanus, N. Harrit, A. Holm, G. Isaksson and J. Sandström, *J. Chem. Soc. Chem. Commun.*, 376 (1974).

154. A. Dondini, G. Modena and G. Scorrano, *Boll. Sci. Fac. Chim. Ind. Bologna*, **22**, 26 (1974).
155. G. H. Schmid and M. Heinola, *J. Amer. Chem. Soc.*, **90**, 3466 (1968).
156. N. Kharasch and C. N. Yiannios, *J. Org. Chem.*, **29**, 1190 (1964).
157. V. Calo', G. Modena and G. Scorrano, *J. Chem. Soc. C*, 1339 (1968).
158. C. Reichardt, *Angew. Chem. Int. Ed. Engl.*, **4**, 29 (1965).
159. H. Tanida, R. Muneyuki and T. Tsushima, *Tetrahedron Lett.*, 3063 (1975).
160. H. Matsuyama, H. Minato and M. Kobayashi, *Bull. Chem. Soc. Jap.*, **48**, 3287 (1975).
161. R. G. Micetich, C. G. Chin and R. B. Morin, *Tetrahedron Lett.*, 975 (1976).
162. J. G. Gourcy, G. Jeminet and S. Simonet, *J. Chem. Soc. Chem. Commun.*, 634 (1974).
163. Y. Etienne, R. Soulas and H. Lumbroso, in *The Chemistry of Heterocyclic Compounds, Heterocyclic Compounds with Three- and Four-Membered Rings*, Part 2 (Ed. A. Weissberger), Interscience, New York, 1964, Chapter 5.
164. M. Sander, *Chem. Rev.*, **66**, 341 (1966).
165. E. Grishkevich-Trokhimovski, *J. Russ. Phys. Chem. Soc.*, **48**, 880 (1916); *Chem. Abstr.*, **11**, 784 (1917).
166. G. M. Bennett and A. L. Hock, *J. Chem. Soc.*, 2496 (1927).
167. B. M. Trost, W. L. Schinski and I. B. Mantz, *J. Amer. Chem. Soc.*, **91**, 4320 (1969).
168. B. M. Trost, W. L. Schinski, F. Chen and I. B. Mantz, *J. Amer. Chem. Soc.*, **93**, 676 (1971).
169. G. Barbarella, A. Garbesi and A. Fava, *Helv. Chim. Acta*, **54**, 2297 (1971).
170. D. B. Denney, D. Z. Denney and Y. F. Hsu, *J. Amer. Chem. Soc.*, **95**, 4064 (1973).
171. R. Tang and K. Mislow, *J. Amer. Chem. Soc.*, **91**, 5644 (1969).
172. W. O. Siegl and C. R. Johnson, *Tetrahedron*, **27**, 341 (1971).
173. (a) Y. Tamura, K. Sumoto, J. Minamikawa and M. Ikeda, *Tetrahedron Lett.*, 4137 (1972); (b) Y. Tamura, H. Matsushina, J. Minamikawa, M. Ikeda and K. Sumoto, *Tetrahedron*, **31**, 3035 (1975); (c) Y. Tamura, J. Minamikawa and M. Ikeda, *Synthesis*, 1 (1977).
174. H. J. Backer and K. J. Keuning, *Recl. Trav. Chim. Pays-Bas*, **53**, 798 (1934); 52, 499 (1933).
175. E. Vilsmaier and W. Schalk, *Synthesis*, 429 (1971).
176. E. Vedejs and J. P. Hagen, *J. Amer. Chem. Soc.*, **97**, 6878 (1975).
177. E. Vedejs, J. P. Hagen, B. L. Roach and K. L. Spear, *J. Org. Chem.*, **43**, 1185 (1978).
178. P. M. Weintraub and A. D. Sill, *J. Chem. Soc. Chem. Commun.*, 784 (1975).
179. S. D. Ziman and B. M. Trost, *J. Org. Chem.*, **38**, 649 (1973).
180. D. J. Anderson, *Dissertation Abstr.*, **21**, 2468 (1961).
181. H. Böhme and K. Sell, *Chem. Ber.*, **81**, 123 (1948).
182. A. de Groot, J. A. Boerma and H. Wynberg, *Tetrahedron Lett.*, 2365 (1968).
183. A. de Groot, J. A. Boerma and H. Wynberg, *Recl. Trav. Chim. Pays-Bas*, **88**, 994 (1969).
184. H. Singh, V. K. Vij and K. Lal, *Indian J. Chem.*, **12**, 1242 (1974).
185. W. Drijvers and E. J. Goethals, *Makromol. Chem.*, **148**, 311 (1971); *Chem. Abstr.*, **76**, 25681 (1972); E. J. Goethals, W. Drijvers, D. Van Ooteghem and A. M. Buyle, *J. Macromol. Sci. Chem.*, **7**, 1375 (1973); E. J. Goethals and W. Drijvers, *Makromol. Chem.*, **165**, 329 (1973); *Chem. Abstr.*, **78**, 160178 (1973).
186. C. Marazona, J.-L. Fourrey and B. C. Das, *J. Chem. Soc. Chem. Commun.*, 742 (1977).
187. W. Ando, T. Yagihara, S. Tozune, I. Imai, J. Suzuki, T. Toyama, S. Nakaido and T. Migata, *J. Org. Chem.*, **37**, 1721 (1972).
188. K. Kondo and I. Ojima, *Chem. Lett.*, 119 (1972).
189. K. Kondo and I. Ojima, *Bull. Chem. Soc. Jap.*, **48**, 1490 (1975).
190. P. Y. Johnson, E. Koza and R. E. Kohrman, *J. Org. Chem.*, **38**, 2967 (1973).
191. P. Y. Johnson and M. Berman, *J. Chem. Soc. Chem. Commun.*, 779 (1974); *J. Org. Chem.*, **40**, 3046 (1975).
192. M. Wilhelm and P. Schmidt, *Helv. Chim. Acta*, **53**, 1697 (1970).
193. H. Hogeveen, R. M. Kellogg and K. A. Kuindersma, *Tetrahedron Lett.*, 3929 (1973).
194. F. Jung, N. K. Sharma and T. Durst, *J. Amer. Chem. Soc.*, **95**, 3420 (1973).

195. P. J. R. Nederlof, M. J. Moolenaar, E. R. DeWaard and H. O. Huisman, *Tetrahedron*, **34**, 447 (1978).
196. P. T. Lansbury, T. R. Demmin, G. E. DuBois and V. R. Haddon, *J. Amer. Chem. Soc.*, **97**, 394 (1975).
197. J. P. Marino, in *Topics in Sulfur Chemistry* (Ed. A. Senning), Vol. 1, G. Thieme, Stuttgart, 1976, p. 53.
198. F. Bernardi, N. D. Epiotis, S. Shaik and K. Mislow, *Tetrahedron*, **33**, 3061 (1977).
199. J. Meinwald, S. Knapp, S. K. Obendorf and R. E. Hughes, *J. Amer. Chem. Soc.*, **98**, 6643 (1976).
200. R. S. Devdhar, V. N. Gogte and B. D. Tilak, *Tetrahedron Lett.*, 3911 (1974).
201. N. Engelhard and A. Kolb, *Justus Liebigs Ann. Chem.*, **673**, 136 (1964).
202. D. C. Dittmer, P.-L. Chang, F. A. Davis, M. Iwanami, I. K. Stamos and K. Takakashi, *J. Org. Chem.*, **37**, 1111 (1972).
203 (a) C. A. Grob and J. Ide, *Helv. Chim. Acta,* **57**, 2571 (1974); (b) D. C. Dittmer, B. H. Patwardhan, E. J. Parker and T. C. Sedergran, Abstracts of Papers, Ninth International Symposium on Organic Sulfur Chemistry, Riga, USSR, June 9–14, 1980, p. 93.
204. R. L. Autrey and P. W. Scullard, *J. Amer. Chem. Soc.*, **90**, 4924 (1968).
205. (a) G. A. Russell, R. Tanikaga and E. R. Talaty, *J. Amer. Chem. Soc.*, **94**, 6125 (1972); (b) Y. Ueno and M. Okawara, *Chem. Lett.*, 863 (1973); (c) P. R. Moses and J. Q. Chambers, *J. Electroanal. Chem. Interfacial Electrochem.*, **49**, 105 (1974); *Chem. Abstr.*, **80**, 70056 (1974); (d) E. Campaigne, M. Pragnell and F. Haaf, *J. Heterocyclic Chem.*, **5**, 141 (1968); (e) R. J. Gillespie, J. Passmore, P. K. Ummat and O. C. Vaidya, *Inorg. Chem.*, **10**, 1327 (1971).
206. G. Barbarella, P. Dembech, A. Garbesi and A. Fava, *Org. Magn. Reson.*, **8**, 108 (1976).
207. G. Barbarella, P. Dembech, A. Garbesi, G. D. Andreetti, G. Bocelli, F. Bernardi, A. Bottoni and A. Fava, *Abstracts of Papers, Eighth International Symposium on Organic Sulphur Chemistry*, Portoroz, Yugoslavia, June 18–23, 1978, p. 37.
208. G. Barbarella, P. Dembech, A. Garbesi, F. Bernardi, A. Bottoni and A. Fava, *J. Amer. Chem. Soc.*, **100**, 200 (1978).
209. Y. Yano, T. Matoyoshi, K. Misu and W. Tagaki, *Phosphorus Sulfur*, **1**, 25 (1976).
210. M. Umehara, K. Kanai, H. Kitano and K. Fukui, *Nippon Kagaku Zasshi*, **83**, 1060 (1962); *Chem. Abstr.*, **59**, 11398 (1963).
211. V. Cere', S. Pollicino, E. Sandri and A. Fava, *J. Amer. Chem. Soc.*, **100**, 1516 (1978).
212. D. Van Ooteghem, R. Deveux and E. J. Goethals, *Int. J. Sulfur Chem.*, **8**, 31 (1973).
213. R. Manske and J. Gosselck, *Tetrahedron Lett.*, 2097 (1971).
214. M. J. Hatch, F. J. Meyer and W. D. Lloyd, *J. Appl. Polym. Sci.*, **13**, 721 (1969).
215. H. Bosshard, M. E. Baumann and G. Schetty, *Helv. Chim. Acta*, **53**, 1271 (1970); H. Bosshard, *Helv. Chim. Acta*, **55**, 37 (1972).
216. P. M. Boorman, T. Chivers and K. N. Mahadev, *Can. J. Chem.*, **55**, 869 (1977).
217. H. Matsuyama, H. Minato and M. Kobayashi, *Bull. Chem. Soc. Jap.*, **46**, 3828 (1973).
218. W. Ando, Y. Saiki and T. Migata, *Tetrahedron*, **29**, 3511 (1973); W. Ando, J. Suzuki, Y. Saiki and T. Migita, *J. Chem. Soc. Chem. Commun.*, 365 (1973).
219. R. Gompper and H. Euchner, *Chem. Ber.*, **99**, 527 (1966).
220. G. Seitz, *Chem. Ber.*, **101**, 585 (1968).
221. A. F. Cook and J. G. Moffatt, *J. Amer. Chem. Soc.*, **90**, 740 (1968).
222. D. L. Schmidt, H. B. Smith, M. Yoshimine and M. J. Hatch, *J. Polym. Sci. Part A1*, **10**, 2951 (1972).
223. E. N. Karaulova, T. S. Bobriuskaya, G. D. Gal'pern, V. D. Nikitina, L. A. Shekhoyan and A. V. Koshevnik, *Chem. Heterocyclic Compd.*, **11**, 659 (1975).
224. M. J. Hatch, M. Yoshimine, D. L. Schmidt and H. B. Smith, *J. Amer. Chem. Soc.*, **93**, 4617 (1971).
225. E. N. Karaulova, G. D. Gal'pern, V. D. Nikitina, T. A. Bardina and L. M. Petrova, *Chem. Heterocyclic Compd.*, **9**, 1337 (1973).
226. H. Bosshard, *Helv. Chim. Acta*, **55**, 32 (1972).
227. V. Prelog and E. Cerkovnikov, *Justus Liebigs Ann. Chem.*, **537**, 214 (1939).
228. C. G. Overberger and A. Lusi, *J. Amer. Chem. Soc.*, **81**, 506 (1959).
229. N. J. Leonard, T. W. Milligan and T. L. Brown, *J. Amer. Chem. Soc.*, **82**, 4075 (1960).

230. E. Larsson, *Acta. Univ. Lund, Sect. II*, No. 22 (1965); *Chem. Abstr.*, **64**, 12635 (1966).
231. E. DeWitte and E. J. Goethals, *J. Macromol. Sci., Chem.*, **A5**, 73 (1971).
232. (a) P. Wilder, Jr., and R. F. Gratz, *J. Org. Chem.*, **35**, 3295 (1970); (b) P. Wilder, Jr., and L. A. Feliu-Otero, *J. Org. Chem.*, **30**, 2560 (1965).
233. R. Bird and C. J. M. Stirling, *J. Chem. Soc. Perkin Trans. II*, 1221 (1973).
234. H. Kise, G. F. Whitfield and D. Swern, *J. Org. Chem.*, **37**, 1121 (1972).
235. C. R. Johnson, C. C. Bacon and W. D. Kingsbury, *Tetrahedron Lett.*, 501 (1972).
236. A. D. Dawson and D. Swern, *J. Org. Chem.*, **42**, 592 (1977).
237. C. R. Johnson and L. J. Pepoy, *J. Org. Chem.*, **37**, 671 (1972).
238. (a) F. Lautenschlaeger, *J. Org. Chem.*, **31**, 1679 (1966); (b) N. N. Novitskaya, R. V. Kunakova and G. A. Tolstikov, *Chem. Heterocyclic Compd.*, **9**, 797 (1973).
239. M. C. Whiting, G. R. Chalkley, D. J. Snodin and G. Stevens, *J. Chem. Soc. C*, 682 (1970).
240. G. Allegra, G. E. Wilson, E. Benedetti, C. Pedone and R. Albert, *J. Amer. Chem. Soc.*, **92**, 4002 (1970).
241. C. J. W. Stirling and A. C. Knipe, *J. Chem. Soc. B*, 1218 (1968).
242. P. N. Confalone, G. Pizzolato, E. G. Baggiolini, D. Lollar and M. R. Uskokovic, *J. Amer. Chem. Soc.*, **97**, 5936 (1975).
243. R. S. Matthews and T. E. Meteyer, *J. Chem. Soc. Chem. Commun.*, 1576 (1971).
244. K. Kondo and I. Ojima, *J. Chem. Soc. Chem. Commun.*, 860 (1972).
245. G. Berchtold and K. K. Maheshwari, *J. Chem. Soc. Chem. Commun.*, 13 (1969).
246. E. I. Eliel, R. O. Hutchins, R. Mebane and R. L. Willer, *J. Org. Chem.*, **41**, 1052 (1976).
247. G. M. Bennett and A. L. Hock, *J. Chem. Soc.*, **130**, 477 (1927).
248. C. Leroy, M. Martin and L. Bassery, *Bull. Soc. Chim. Fr.*, 590 (1974).
249. E. N. Karaulova, G. D. Gal'pern, T. S. Bobruiskaya, L. R. Barykina, A. Y. Koshevnik and L. K. Il'ina, *Tezisy Dokl. Nauchn. Sess. Khim. Tekhnol. Org. Soedin. Ser. Sernistykh Neftei, 14th*, 174 (1975); *Chem. Abstr.*, **88**, 190544 (1978); E. N. Karaulova, G. D. Gal'pern and L. R. Barykina, *Abstracts of Papers, Eighth International Symposium Organic Sulphur Chemistry*, Portoroz, Yugoslavia, June 18–23, 1978, p. 129.
250. E. N. Karaulova, G. D. Gal'pern, T. S. Bobruiskaya and V. D. Nikitina, *Proc. Acad. Sci. USSR, Chem. Sect.*, **216**, 289 (1974).
251. J. W. Batty, P. D. Howes and C. J. M. Stirling, *J. Chem. Soc. Perkin Trans. I*, 59 (1973).
252. (a) I. Ernest, *Tetrahedron*, **33**, 547 (1977); (b) D. Jeckel and J. Gosselck, *Tetrahedron Lett.*, 2101 (1972).
253. F. Weygand and H. Daniel, *Chem. Ber.*, **94**, 3145 (1961).
254. I. Lantos and D. Ginsberg, *Tetrahedron*, **28**, 2507 (1972).
255. G. Griffiths, P. D. Howes and C. J. M. Stirling, *J. Chem. Soc. Perkin. Trans. I*, 912 (1977).
256. E. Vedejs, M. J. Mullins, J. M. Renga and S. P. Singer, *Tetrahedron Lett.*, 519 (1978).
257. R. Schmid and H. Schmid, *Helv. Chim. Acta*, **60**, 1361 (1977).
258. E. Sandri, V. Cere, C. Paolucci, S. Pollicino, L. Lunazzi and A. Fava, *Abstracts of Papers, Eighth International Symposium Organic Sulphur Chemistry*, Portoroz, Yugoslavia, June 18–23, 1978, p. 205.
259. G. Schmidt and J. Gosselck, *Tetrahedron Lett.*, 3445 (1969).
260. M. Hetschko and J. Gosselck, *Chem. Ber.*, **106**, 996 (1973).
261. M. Yoshimoto, N. Ishida and Y. Kishida, *Chem. Pharm. Bull.*, **20**, 2593 (1972).
262. M. Watanabe, T. Kinoshita and S. Furukawa, *Chem. Pharm. Bull.*, **23**, 82 (1975).
263. M. Watanabe, M. Baba, T. Kinoshita and S. Furukawa, *Chem. Pharm. Bull.*, **24**, 2421 (1976).
264. H. Wittmann and F. A. Petio, *Z. Naturforsch.*, **29B**, 765 (1974).
265. H. Wittmann, D. Sobhi and K. Dehghani, *Z. Naturforsch.*, **29B**, 414 (1974).
266. K. Itoh, S. Kato and Y. Ishii, *J. Organometal. Chem.*, **34**, 293 (1972).
267. (a) P. G. Gassman, T. J. van Bergen and G. Gruetzmacher, *J. Amer. Chem. Soc.*, **95**, 6508 (1973); (b) P. G. Gassman and D. R. Amick, *Tetrahedron Lett.*, 889 (1974); (c) P. G. Gassman and G. D. Gruetzmacher, *J. Amer. Chem. Soc.*, **96**, 5487 (1974); (d) P. G. Gassman, G. Gruetzmacher and T. J. van Bergen, *J. Amer. Chem. Soc.*, **96**, 5512 (1974).

268. (a) G. E. Wilson, Jr., and R. Albert, *Tetrahedron Lett.*, 6271 (1968); *J. Org. Chem.*, **38**, 2156, 2160 (1973); (b) K. Tomita, A. Terada and R. Tachikawa, *Heterocycles*, **4**, 729 (1976).
269. A. Garbesi, N. Corsi and A. Fava, *Helv. Chim. Acta*, **53**, 1499 (1970).
270. O. Hofer and E. L. Eliel, *J. Amer. Chem. Soc.*, **95**, 8045 (1973).
271. G. Barbarella, A. Garbesi and A. Fava, *Helv. Chim. Acta*, **54**, 341 (1971).
272. G. Barbarella, A. Garbesi, A. Boicelli and A. Fava, *J. Amer. Chem. Soc.*, **95**, 8051 (1973).
273. G. Barbarella, A. Garbesi and A. Fava, *J. Amer. Chem. Soc.*, **97**, 5883 (1975).
274. M. Kobayashi, K. Okuma and H. Takeuchi, *Abstracts of Papers, Eighth International Symposium Organic Sulphur Chemistry*, Portoroz, Yugoslavia, June 18–23, 1978, p. 132.
275. F. Montanari, R. Danieli, H. Hogeveen and G. Maccagnani, *Tetrahedron Lett.*, 2685 (1964).
276. M. Cinquini, S. Colonna and F. Montanari, *J. Chem. Soc. C*, 572 (1970); 1213 (1967).
277. J. Klein and H. Stollar, *Tetrahedron*, **30**, 2541 (1974); H. Stollar and J. Klein, *J. Chem. Soc. Perkin Trans. I*, 1763 (1974).
278. T. Durst, K.-C. Tin and M. J. V. Marcil, *Can. J. Chem.*, **51**, 1704 (1973).
279. J. W. Hartgerink, L. C. J. van der Laan, J. B. F. N. Engberts and T. J. de Boer, *Tetrahedron*, **27**, 4323 (1971).
280. G. E. Wilson, Jr., and M.-G. Huang, *J. Org. Chem.*, **41**, 966 (1976); G. E. Wilson, Jr., *J. Amer. Chem. Soc.*, **87**, 3785 (1965).
281. T. J. Maricich, R. A. Jourdenais and C. K. Harrington, *J. Amer. Chem. Soc.*, **95**, 2378 (1973).
282. P. M. Denerley and E. J. Thomas, *Tetrahedron Lett.*, 71 (1977).
283. M. Numata, Y. Imashiro, I. Minamida and M. Yamaoka, *Tetrahedron Lett.*, 5097 (1972).
284. H. Böhme, G. Dahler and W. Krack, *Justus Liebigs Ann. Chem.*, 1686 (1973).
285. C. R. Johnson, G. F. Katekar, R. F. Huxol and E. R. Janiga, *J. Amer. Chem. Soc.*, **93**, 3771 (1971).
286. (a) R. Neidlein and P. Leinberger, *Chem. Ztg.*, **99**, 465 (1975); *Chem. Abstr.*, **84**, 43947 (1976); (b) S. P. McManus, J. T. Carroll and C. U. Pittman, Jr., *J. Org. Chem.*, **35**, 3768 (1970); (c) R. D. Westland, M. H. Lin and J. M. Vandenbelt, *J. Heterocyclic Chem.*, **8**, 405 (1971); (d) A. P. Sineokov and V. S. Kutyreva, *Chem. Heterocyclic Compd.*, **7**, 1534 (1971); (e) E. Funke, R. Huisgen and F. C. Schaefer, *Chem. Ber.*, **104**, 1550 (1971); (f) T. Nakai, K. Hiratani and M. Okawara, *Bull. Chem. Soc. Jap.*, **47**, 398 (1974); (g) A. Eidem, K. Undheim and K. R. Reistad, *Acta Chem. Scand.*, **25**, 1 (1971); (h) V. N. Gogte, R. N. Sathe and B. D. Tilak, *Indian J. Chem.*, **11**, 1115 (1973); (i) G. A. Ulsaker and K. Undheim, *Acta Chem. Scand.*, **31B**, 917 (1977); (j) G. A. Ulsaker, F. G. Evans and K. Undheim, *Acta Chem. Scand.*, **31B**, 919 (1977).
287. R. F. Hudson and F. Filippini, *J. Chem. Soc. Chem. Commun.*, 726 (1972).
288. W. K. Musker and P. B. Roush, *J. Amer. Chem. Soc.*, **98**, 6745 (1976).
289. J. T. Doi and W. K. Musker, *J. Amer. Chem. Soc.*, **100**, 3533 (1978).
290. W. K. Musker, A. S. Hirschon and J. T. Doi, *J. Amer. Chem. Soc.*, **100**, 7754 (1978).
291. H. Böhme and W. Krack, *Justus Liebigs Ann. Chem.*, **758**, 143 (1972).
292. T. L. Ho and C. M. Wòng, *Synthesis*, 561 (1972).
293. M. Hetschko and J. Gosselck, *Tetrahedron Lett.*, 1691 (1972).
294. S. Kato, H. Ishihara, M. Mizuta and Y. Hirabayashi, *Bull. Chem. Soc. Jap.*, **44**, 2469 (1971).
295. I. Stahl, M. Hetschko and J. Gosselck, *Tetrahedron Lett.*, 4077 (1971).
296. T. Oishi, H. Takechi, K. Kamemoto and Y. Ban, *Tetrahedron Lett.*, 11 (1974).
297. T. Oishi, H. Kamemoto and Y. Ban, *Tetrahedron Lett.*, 1085 (1972).
298. R. M. Munavu and H. H. Szmant, *Tetrahedron Lett.*, 4543 (1975).
299. E. J. Corey and T. Hase, *Tetrahedron Lett.*, 3267 (1975).
300. M. Hetschko and J. Gosselck, *Chem. Ber.*, **107**, 123 (1974).
301. M. Fetizon and M. Jurion, *J. Chem. Soc. Chem. Commun.*, 382 (1972).
302. Y. Tamura, K. Sumoto, S. Fujii, H. Satoh and M. Ikeda, *Synthesis*, 312 (1973).

303. D. H. R. Barton, N. J. Cussans and S. V. Ley, *J. Chem. Soc. Chem. Commun.*, 751 (1977).
304. H. Yoshino, Y. Kawazoe and T. Taguchi, *Synthesis*, 713 (1974).
305. R. Gompper and E. Kutter, *Angew. Chem. Int. Ed. Engl.*, **2**, 687 (1963).
306. R. Gompper and E. Kutter, *Chem. Ber.*, **98**, 1365 (1965).
307. R. Gompper, E. Kutter and R. R. Schmidt, *Chem. Ber.*, **98**, 1374 (1965).
308. M. P. Doyle and D. M. Hedstrand, *J. Chem. Soc. Chem. Commun.*, 643 (1977).
309. Y. Ueno and M. Okawara, *Bull. Chem. Soc. Jap.*, **45**, 1797 (1972).
310. Y. Ueno and M. Okawara, *Bull. Chem. Soc. Jap.*, **47**, 1033 (1974).
311. D. L. Coffen and P. E. Garrett, *Tetrahedron Lett.*, 2043 (1969).
312. K. Arai and M. Oki, *Tetrahedron Lett.*, 2183 (1975).
313. K. Hiratani, T. Nakai and M. Okawara, *Bull. Chem. Soc. Jap.*, **47**, 904 (1974).
314. T. Nakai, K. Hiratani and M. Okawara, *Bull. Chem. Soc. Jap.*, **49**, 827 (1976).
315. K. Hiratani, T. Nakai and M. Okawara, *Chem. Lett.*, 1041 (1974).
316. K. Hiratani, H. Shiono and M. Okawara, *Chem. Lett.*, 867 (1973).
317. H. R. Kricheldorf, *Angew. Chem. Int. Ed. Engl.*, **10**, 726 (1971).
318. T. Nakai, Y. Ueno and M. Okawara, *Bull. Chem. Soc. Jap.*, **43**, 156 (1970).
319. T. Tanaka and T. Abe, *Inorg. Nucl. Chem. Lett.*, **4**, 569 (1968).
320. T. Nakai, Y. Ueno and M. Okawara, *Bull. Chem. Soc. Jap.*, **43**, 3175 (1970).
321. T. Tanaka, K. Tanaka and T. Yoshimitsu, *Bull. Chem. Soc. Jap.*, **44**, 112 (1971).
322. J. Q. Chambers, N. D. Canfield, D. R. Williams and D. L. Coffen, *Mol. Phys.*, **19**, 581 (1970).
323. (a) N. D. Canfield, J. Q. Chambers and D. L. Coffen, *J. Electroanal. Chem. Interfacial Electrochem.*, **24**, A7 (1970); *Chem. Abstr.*, **72**, 96051 (1970); (b) P. R. Moses, R. M. Harnden and J. Q. Chambers, *J. Electroanal. Chem. Interfacial Electrochem.*, **84**, 187 (1977); *Chem. Abstr.*, **88**, 29574 (1978); (c) R. M. Harnden, P. R. Moses and J. Q. Chambers, *J. Chem. Soc. Chem. Commun.*, 11 (1977).
324. D. L. Coffen, J. Q. Chambers, D. R. Williams, P. E. Garrett and N. D. Canfield, *J. Amer. Chem. Soc.*, **93**, 2258 (1971).
325. T. Nakai and M. Okawara, *Bull. Chem. Soc. Jap.*, **43**, 1864 (1970).
326. K. Hiratani, T. Nakai and M. Okawara, *Bull. Chem. Soc. Jap.*, **46**, 3872 (1973).
327. T. Nakai and M. Okawara, *Bull. Chem. Soc. Jap.*, **43**, 3882 (1970).
328. S. Araki and T. Tanaka, *Bull. Chem. Soc. Jap.*, **51**, 1311 (1978).
329. J. Fabian and H. Hartmann, *Tetrahedron*, **29**, 2597 (1973).
330. R. F. Heldeweg and H. Hogeveen, *Tetrahedron Lett.*, 75 (1974).
331. G. C. Brumlik, A. I. Kosak and R. Pitcher, *J. Amer. Chem. Soc.*, **86**, 5360 (1964).
332. R. M. Acheson and J. K. Stubbs, *J. Chem. Soc. Perkin Trans. 1*, 899 (1972).
333. R. M. Acheson and D. R. Harrison, *J. Chem. Soc. Chem. Commun.*, 724 (1969).
334. R. M. Acheson and D. R. Harrison, *J. Chem. Soc. C*, 1764 (1970).
335. M. J. Barrow, J. L. Davidson, W. Harrison, D. W. A. Sharp, G. A. Sim and F. B. Wilson, *J. Chem. Soc. Chem. Commun.*, 583 (1973).
336. B. M. Trost and H. C. Arndt, *J. Amer. Chem. Soc.*, **95**, 5288 (1973); R. W. Larochelle and B. M. Trost, *J. Amer. Chem. Soc.*, **93**, 6077 (1971).
337. R. J. Gillespie, J. Murray-Rust, P. Murray-Rust and A. E. A. Porter, *J. Chem. Soc. Chem. Commun.*, 83 (1978).
338. J. Cuffe, R. J. Gillespie and A. E. A. Porter, *J. Chem. Soc. Chem. Commun.*, 641 (1978).
339. W. Ando, H. Higuchi and T. Migita, *J. Org. Chem.*, **42**, 3365 (1977).
340. E. A. Steck and E. H. Wilson, *J. Heterocyclic Chem.*, **12**, 1065 (1975).
341. R. Breslow, S. Garratt, L. Kaplan and D. LaFollette, *J. Amer. Chem. Soc.*, **90**, 4051 (1968); R. Breslow, L. Kaplan and D. LaFollette, *J. Amer. Chem. Soc.*, **90**, 4056 (1968).
342. (a) R. J. Basalay and J. C. Martin, *J. Amer. Chem. Soc.*, **95**, 2565 (1973); (b) J. C. Martin and R. J. Basalay, *J. Amer. Chem. Soc.*, **95**, 2572 (1973).
343. A. Terada and Y. Kishida, *Chem. Pharm. Bull.*, **17**, 974 (1969).
344. K. K. Andersen, R. I. Caret and I. Karup-Nielsen, *J. Amer. Chem. Soc.*, **96**, 8026 (1974).

506   Donald C. Dittmer and Bhalchandra H. Patwardhan

345. W. T. Flowers, G. Holt and M. A. Hope, *J. Chem. Soc. Perkin Trans. I*, 1116 (1974).
346. K. Friedrich and J. Rieser, *Justus Liebigs Ann. Chem.*, 641 (1976).
347. Y. Tamura, H. Matsushima, M. Ikeda and K. Sumoto, *Synthesis*, 277 (1974).
348. G. K. Helmkamp, D. C. Owsley and B. R. Harris, *J. Org. Chem.*, **34**, 2763 (1969).
349. L. I. Belen'kii, A. P. Yakubov and I. A. Bessonova, *J. Org. Chem. USSR*, **13**, 329 (1977).
350. (a) A. P. Yakubov, N. V. Grigor'eva and L. I. Belen'kii, *J. Org. Chem. USSR*, **14**, 593 (1978); (b) H. Hogeveen, *Recl. Trav. Chim. Pays-Bas*, **85**, 1072 (1966).
351. T. G. Melent'eva, I. P. Soloveichik, D. A. Oparin and L. A. Pavlova, *J. Org. Chem. USSR*, **8**, 1341 (1972); I. P. Soloveichik, T. G. Melent'eva, D. A. Oparin and L. A. Pavlova, *J. Org. Chem. USSR*, **10**, 615 (1974).
352. C. M. Camaggi, L. Lunazzi and G. Placucci, *J. Chem. Soc. Perkin Trans II*, 1491 (1973).
353. M. Winn and F. G. Bordwell, *J. Org. Chem.*, **32**, 1610 (1967); A. V. El'tsov and A. A. Ginesina, *J. Org. Chem. USSR*, **3**, 184 (1967).
354. A. V. El'tsov, A. A. Ginesina and L. N. Kivokurtseva, *Tetrahedron Lett.*, 735 (1968).
355. A. A. Ginesina, L. N. Kivokurtseva and A. V. El'tsov, *J. Org. Chem. USSR*, **5**, 559 (1969).
356. G. Seitz and H. Mönnighoff, *Angew. Chem.*, **82**, 938 (1970).
357. R. Neidlein and A. D. Kraemer, *Tetrahedron Lett.*, 4713 (1976).
358. E. Campaigne and D. R. Knapp, *J. Heterocyclic Chem.*, **7**, 107 (1970).
359. R. Neidlein and M. H. Salzl, *J. Chem. Res., S*, 118 (1977).
360. A. G. Schultz and M. B. DeTar, *J. Amer. Chem. Soc.*, **96**, 296 (1974).
361. A. G. Schultz, W. Y. Fu, R. D. Lucci, B. G. Kurr, K. M. Lo and M. Boxer, *J. Amer. Chem. Soc.*, **100**, 2140 (1978).
362. W. Ando, S. Kondo, K. Nakayama, K. Ichibori, H. Kohoda, H. Yamato, I. Imai, S. Nakaido and T. Migita, *J. Amer. Chem. Soc.*, **94**, 3870 (1972).
363. R. S. Alexander and A. R. Butler, *J. Chem. Soc. Perkin Trans. II*, 1998 (1977).
364. J. Bornstein, J. E. Shields and J. H. Supple, *J. Org. Chem.*, **32**, 1499 (1967).
365. R. J. Gillespie, A. E. A. Porter and W. E. Willmott, *J. Chem. Soc. Chem. Commun.*, 85 (1978).
366. S. Mageswaran, W. D. Ollis and I. O. Sutherland, *J. Chem. Soc. Chem. Commun.*, 656 (1973).
367. R. Neidlein and M. H. Salzl, *Justus Liebigs Ann. Chem.*, 1938 (1977).
368. B. M. Trost and S. D. Ziman, *J. Amer. Chem. Soc.*, **93**, 3825 (1971).
369. H. Prinzbach and E. Futterer, *Adv. Heterocyclic Chem.*, **7**, 39 (1966).
370. E. Klingsberg, in *Organosulfur Chemistry* (Ed. M. J. Janssen), Interscience, New York, 1967, p. 171.
371. R. D. Hamilton and E. Campaigne, in *The Chemistry of Heterocyclic Compounds. Special Topics in Heterocyclic Chemistry* (Eds. A. Weissberger and E. C. Taylor), Wiley-Interscience, New York, 1977, p. 271.
372. E. Campaigne and R. D. Hamilton, *Q. Rep. Sulfur Chem.*, **5**, 275 (1970).
373. E. Klingsberg, *Lec. Heterocyclic Chem.*, **1**, S 19 (1972).
374. (a) Eds. D. S. Breslow and H. Skolnik, *The Chemistry of Heterocyclic Compounds. Multi-Sulfur and Sulfur and Oxygen 5- and 6-Membered Heterocycles, Part 1*, Interscience, New York, 1966, p. 405; (b) p. 551; (c) p. 347.
375. R. Wizinger-Aust, *Q. Rep. Sulfur Chem.*, **5**, 191 (1970).
376. P. S. Landis, *Chem. Rev.*, **65**, 237 (1965).
377. A. Hordvik, *Q. Rep. Sulfur Chem.*, **5**, 21 (1970).
378. (a) R. Zahradnik and J. Koutecky, *Coll. Czech. Chem. Commun.*, **28**, 1117 (1963); (b) J. Koutecky, *Coll. Czech. Chem. Commun.*, **24**, 1608 (1959); (c) R. Zahradnik, *Adv. Heterocyclic Chem.*, **5**, 1 (1965).
379. G. Bergson, *Ark. Kemi*, **19**, 181 (1962); *Chem. Abstr.*, **57**, 15085 (1962).
380. R. Zahradnik and J. Koutecky, *Tetrahedron Lett.*, 632 (1961).
381. (a) J. Koutecky, J. Paldus and R. Zahradnik, *Coll. Czech. Chem. Commun.*, **25**, 617 (1960); (b) M. H. Palmer and R. H. Findlay, *Tetrahedron Lett.*, 4165 (1972).
382. H. Prinzbach, E. Futterer, H. Berger and A. Lüttringhaus, *Angew. Chem. Int. Ed. Engl.*, **4**, 88 (1965).

383. B. Bock, M. Kuhr and H. Musso, *Chem. Ber.*, **109**, 1184 (1976).
384. (a) K. Sakamoto, N. Nakamura, M. Oki, J. Nakayama and M. Hoshino, *Chem. Lett.*, 1133 (1977); (b) G. A. Olah and J. L. Grant, *J. Org. Chem.*, **42**, 2237 (1977).
385. B. J. Lindberg, R. Pinel and Y. Mollier, *Tetrahedron*, **30**, 2537 (1974).
386. J. Sletten, *Acta Chem. Scand.*, **30A**, 397 (1976).
387. A. Hordvik, *Acta Chem. Scand.*, **19**, 1253 (1965).
388. (a) I. C. Paul, quoted in (b) J. P. DeBarbeyrac, D. Gonbeau and G. Pfister-Guillouzo, *J. Mol. Struct.*, **16**, 103 (1973).
389. A. Hordvik, *Acta Chem. Scand.*, **19**, 1037 (1965); **17**, 1809 (1963).
390. (a) H. C. Freeman, G. H. W. Milburn, C. E. Nockolds, R. Mason, G. B. Robertson and G. A. Rusholme, *Acta Crystallogr. Sect. B*, **30**, 886 (1974); (b) R. Mason, G. B. Robertson and G. A. Rusholme, *Acta Crystallogr. Sect. B*, **30**, 894, 906 (1974); (c) J. R. Cannon, K. T. Potts, C. L. Raston, A. F. Sierakowski and A. H. White, *Aust. J. Chem.*, **31**, 297 (1978).
391. (a) R. Zahradnik and C. Parkanyi, *Coll. Czech. Chem. Commun.*, **30**, 3016 (1965); (b) A. Mehlhorn, J. Fabian and R. Mayer, *Z. Chem.*, **5**, 23 (1965); (c) J. Fabian, H. Hartmann and K. Fabian, *Tetrahedron*, **29**, 2609 (1973); (d) K. Fabian, H. Hartmann, J. Fabian and R. Mayer, *Tetrahedron*, **27**, 4705 (1971); (e) J. Fabian, K. Fabian and H. Hartmann, *Theor. Chim. Acta*, **12**, 319 (1968).
392. C. Guimon, D. Gonbeau and G. Pfister-Guillouzo, *Tetrahedron*, **29**, 3399 (1973).
393. J. Fabian, *J. Prakt. Chem.*, **315**, 690 (1973).
394. J. Fabian and H. Hartmann, *Z. Chem.*, **12**, 349 (1972).
395. A. Takamizawa and K. Hirai, *Chem. Pharm. Bull.*, **18**, 865 (1970).
396. (a) M. G. Voronkov and T. V. Lapina, *Chem. Heterocyclic Compd.*, **6**, 416 (1970); (b) E. A. Luksha, T. V. Lapina, M. G. Voronkov and Y. A. Bankovskii, *Chem. Heterocyclic Compd.*, 549; (c) G. Kiel, U. Reuter and G. Gattow, *Chem. Ber.*, **107**, 2569 (1974).
397. N. Loayza and C. T. Pedersen, *J. Chem. Soc. Chem. Commun.*, 496 (1975).
398. R. S. Tewan and K. G. Gupta, *Synth. Commun.*, **8**, 315 (1978).
399. (a) J. Faust and J. Fabian, *Z. Naturforsch.*, **B24**, 577 (1969); (b) C. Bouillon and J. Vialle, *Bull. Soc. Chim. Fr.*, 4560 (1968).
400. (a) D. Barillier, P. Rioult and J. Vialle, *Bull. Soc. Chim Fr.*, 659 (1977); (b) C. Lemarie-Retour, M. Stavaux and N. Lozac'h, *Bull. Soc. Chim. Fr.*, 1659 (1973); (c) F. Boberg and W. V. Gentzkow, *J. Prakt. Chem.*, **315**, 965, 970 (1973); (d) C. Metayer, G. Duguay and H. Quiniou, *Bull. Soc. Chim. Fr.*, 4576 (1972); (e) G. Duguay and H. Quiniou, 637; (f) G. Caillaud and Y. Mollier, 147 (1972); (g) G. Caillaud and Y. Mollier, 2326 (1971); (h) G. Caillaud and Y. Mollier, 2018 (1970); (i) E. Klingsberg, *Synthesis*, 29 (1972); (j) M. G. Voronkov and T. V. Lapina, *Chem. Heterocyclic Compd.*, **6**, 319 (1970); (k) R. Mayer and H. Hartmann, *Chem. Ber.*, **97**, 1886 (1964); (l) H. Behringer, M. Ruff and R. Wiedenmann, *Chem. Ber.*, **97**, 1732 (1964); (m) E. Klingsberg and A. M. Schreiber, *J. Amer. Chem. Soc.*, **84**, 2941 (1962); (n) J. Teste and N. Lozac'h, *Bull. Soc. Chim. Fr.*, 437 (1955); (o) A. Lüttringhaus and U. Schmidt, *Chem. Ztg.*, **77**, 135 (1953); (p) B. Böttcher and F. Bauer, *Justus Liebigs Ann. Chem.*, **568**, 227 (1950); (q) B. Böttcher and A. Lüttringhaus, *Justus Liebigs Ann. Chem.*, **557**, 89 (1947); (r) F. Challenger, E. A. Mason, E. C. Holdsworth and R. Emmott, *J. Chem. Soc.*, 292 (1953); (s) A. Marei and M. M. A. El Sukkary, *UAR J. Chem.*, **14**, 101 (1971); *Chem. Abstr.*, **77**, 126473 (1972).
401. V. N. Drozd, G. S. Bogomolova and Y. M. Udachin, *J. Org. Chem. USSR*, **14**. 833 (1978).
402. C. T. Pedersen and V. D. Parker, *Tetrahedron Lett.*, 771 (1972).
403. (a) Y. Mollier and N. Lozac'h, *Bull. Soc. Chim. Fr.*, 614 (1961); (b) C. Paulmier, Y. Mollier and N. Lozac'h, *Bull. Soc. Chim. Fr.*, 2463 (1965).
404. (a) H. Behringer and J. Falkenberg, *Chem. Ber.*, **102**, 1580 (1969); (b) D. Festal, J. Tison, N. Kim Son, R. Pinel and Y. Mollier, *Bull. Soc. Chim. Fr.*, 3339 (1973); (c) E. Klingsberg, *J. Org. Chem.*, **33**, 2915 (1968).
405. (a) J. L. Charlton, S. M. Loosmore and D. M. McKinnon, *Can. J. Chem.*, **52**, 3021 (1974); (b) C. Metayer and G. Duguay, *C. R. Acad. Sci. Ser. C*, **273**, 1457 (1971); (c) A. Grandin, C. Bouillon and J. Vialle, *Bull. Soc. Chim. Fr.*, 4555 (1968); (d) H. Quiniou and N. Lozac'h, *Bull. Soc. Chim. Fr.*, 1167 (1973); (e) W. Walter and J. Curts,

Justus Liebigs Ann. Chem., **649**, 88 (1961); (f) E. Klingsberg, *J. Amer. Chem. Soc.*, **83**, 2934 (1961).

406. (a) J. Faust and R. Mayer, *Angew. Chem. Int. Ed. Engl.*, **2**, 326 (1963); (b) J. Faust and R. Mayer, *Justus Liebigs Ann. Chem.*, **688**, 150 (1965); (c) J. Faust, H. Spies and R. Mayer, *Naturwissenschaften*, **54**, 537 (1967); (d) F. Boberg and W. v. Gentzkow, *Justus Liebigs Ann. Chem.*, **766**, 1 (1972); (e) R. S. Spindt, D. R. Stevens and W. E. Baldwin, *J. Amer. Chem. Soc.*, **73**, 3693 (1951); (f) P. S. Landis and L. A. Hamilton, *J. Org. Chem.*, **25**, 1742 (1960); (g) R. Wiedermann, W. v. Gentzkow and F. Boberg, *Justus Liebigs Ann. Chem.*, **742**, 103 (1970).

407. V. N. Drozd and G. S. Bogomolova, *J. Org. Chem. USSR*, **13**, 1871 (1977).

408. (a) D. Barillier, P. Rioult and J. Vialle, *Bull Soc. Chim. Fr.*, 444 (1976); (b) S. Coen, J. C. Poite and J. P. Roggero, *Bull Soc. Chim. Fr.*, 611 (1975); (c) D. Barillier, P. Rioult and J. Vialle, *Bull. Soc. Chim. Fr.*, 3031 (1973); (d) D. Barillier, C. Gy, P. Rioult and J. Vialle, *Bull. Soc. Chim. Fr.*, 277 (1973); (e) J. C. Poite, S. Coen and J. Roggero, *Bull. Soc. Chim. Fr.*, 4373 (1971); (f) M. Schmidt and H. Schulz, *Chem. Ber.*, **101**, 277 (1968); (g) M. Schmidt and H. Schulz, *Z. Naturforsch.*, **B23**, 1540 (1968); (h) D. Leaver, W. A. H. Robertson and D. M. McKinnon, *J. Chem. Soc.*, 5104 (1962); (i) K. Inouye, S. Sato and M. Ohta, *Bull. Chem. Soc. Jap.*, **43**, 1911 (1970); (j) A. Chinone, K. Inouye and M. Ohta, *Bull. Chem. Soc. Jap.*, **45**, 213 (1972).

409. (a) J. P. Guemas and H. Quiniou, *Bull. Soc. Chim. Fr.*, 592 (1973); (b) D. Leaver and W. A. H. Robertson, *Proc. Chem. Soc.*, 252 (1960).

410. K. Knauer, P. Hemmerich and J. D. W. Van Voorst, *Angew. Chem. Int. Ed. Engl.*, **6**, 262 (1967).

411. A. R. Henrickson and R. L. Martin, *J. Org. Chem.*, **38**, 2548 (1973).

412. (a) M. Stavaux and N. Lozac'h, *Bull. Soc. Chim. Fr.*, 2082 (1967); (b) J. P. Biton, G. Duguay and H. Quiniou, *C. R. Acad. Sci., Ser. C*, **267**, 586 (1968).

413. H. Hartmann, K. Fabian, B. Bartho and J. Faust, *J. Prakt. Chem.*, **312**, 1197 (1970).

414. R. C. Haddon, F. Wudl, M. L. Kaplan, J. H. Marshall and F. B. Bramwell, *J. Chem. Soc. Chem. Commun.*, 429 (1978).

415. R. Maignan and J. Vialle, *Bull. Soc. Chim. Fr.*, 2388 (1973).

416. (a) M. B. Kolesova, L. I. Maksimova and A. V. El'tsov, *J. Org. Chem. USSR*, **6**, 606 (1970); (b) G. Barnikow, *Chem. Ber.*, **100**, 1389 (1967); (c) U. Schmidt, *Chem. Ber.*, **92**, 1171 (1959); (d) K. A. Jensen, H. R. Baccaro and O. Buchardt, *Acta Chem. Scand.*, **17**, 163 (1963); (e) A. D. Grabenko, L. N. Kulaeva and P. S. Pel'kis, *Chem. Heterocyclic Compd.*, **10**, 806 (1974).

417. L. Menabue and G. C. Pellacani, *J. Chem. Soc. Dalton Trans.*, 455 (1976).

418. A. Schönberg and E. Frese, *Chem. Ber.*, **103**, 3885 (1970).

419. (a) F. Boberg, *Angew. Chem.*, **72**, 629 (1960); (b) F. Boberg, *Justus Liebigs Ann. Chem.*, **679**, 109 (1964); (c) F. Boberg, R. Wiedermann and J. Kresse, *J. Labeled Compd.*, **10**, 297 (1974).

420. (a) R. Pinel, Y. Mollier and N. Lozac'h, *Bull. Soc. Chim. Fr.*, 856 (1967); (b) Y. Poirier and N. Lozac'h, *Bull. Soc. Chim. Fr.*, 865 (1967); (c) R. Pinel, Y. Mollier and N. Lozac'h, *C. R. Acad. Sci., Ser. C*, **260**, 5065 (1965).

421. A. Lüttringhaus, M. Mohr and N. Engelhard, *Justus Liebigs Ann. Chem.*, **661**, 84 (1963).

422. J. Faust and R. Mayer, *Z. Naturforsch.*, **B22**, 790 (1967).

423. D. Leaver, D. M. McKinnon and W. A. H. Robertson, *J. Chem. Soc.*, 32 (1965).

424. E. Klingsberg, *J. Org. Chem.*, **28**, 529 (1963).

425. G. Purrello, *Gazz. Chim. Ital.*, **96**, 1000 (1966).

426. G. J. Wentrup, M. Koepke and F. Boberg, *Synthesis*, 525 (1975).

427. H. Newman and R. B. Angier, *J. Chem. Soc. Chem. Commun.*, 353 (1967).

428. (a) J. Faust, H. Spies and R. Mayer, *Z. Naturforsch.*, **B22**, 789 (1967); (b) P. Sykes and H. Ullah, *J. Chem. Soc. Perkin Trans. I*, 2305 (1972).

429. F. Boberg, G. J. Wentrup and U. Puttins, *Abstracts of Papers, Eighth International Symposium on Organic Sulfur Chemistry*, Portoroz, Yugoslavia, June 18–23, 1978, p. 48.

430. G. J. Wentrup and F. Boberg, *Justus Liebigs Ann. Chem.*, 387 (1978).

431. F. Boberg, G. J. Wentrup and M. Koepke, *Synthesis*, 502 (1975).
432. B. Bartho, J. Faust, R. Pohl and R. Mayer, *J. Prakt. Chem.*, **318**, 221 (1976).
433. J. Faust, *Z. Chem.*, **15**, 478 (1975).
434. (a) G. Le Coustumer and Y. Mollier, *Bull Soc. Chim. Fr.*, 3349 (1973); (b) 2958 (1971); (c) 3076 (1970); (d) G. Coustumer and Y. Mollier, *C. R. Acad. Sci., Ser. C*, **270**, 433 (1970).
435. A. Reliquet and F. Reliquet-Clesse, *C. R. Acad. Sci., Ser. C*, **275**, 689 (1972).
436. F. Clesse, A. Reliquet and H. Quiniou, *C. R. Acad. Sci., Ser. C*, **272**, 1049 (1971).
437. G. Duguay and H. Quiniou, *Bull. Soc. Chim Fr.*, 1918 (1970).
438. C. Metayer and G. Duguay, *C. R. Acad. Sci., Ser. C*, **273**, 1457 (1971).
439. J. Bignebat, H. Quiniou and N. Lozac'h, *Bull. Soc. Chim. Fr.*, 127 (1969); 1699 (1966).
440. S. C. Tang, G. N. Weinstein and R. H. Holm, *J. Amer. Chem. Soc.*, **95**, 613 (1973).
441. D. M. McKinnon and E. A. Robak, *Can. J. Chem.*, **46**, 1855 (1968).
442. Y. N'Guessan and J. Bignebat, *C. R. Acad. Sci., Ser. C*, **280**, 1323 (1975).
443. U. Schmidt, A. Lüttringhaus and F. Hübinger, *Justus Liebigs Ann. Chem.*, **631**, 138 (1960).
444. S. Tamagaki, K. Sakaki and S. Oae, *Bull. Chem. Soc. Jap.*, **48**, 2983, 2985 (1975).
445. J. C. Poite, A. Perichaut and J. Roggero, *C. R. Acad. Sci., Ser. C*, **270**, 1677 (1970).
446. (a) R. A. Olofson, J. M. Landesberg, R. O. Berry, D. Leaver, W. A. H. Robertson and D. M. McKinnon, *Tetrahedron*, **22**, 2119 (1966); (b) G. E. Bachers, D. M. McKinnon and J. M. Buchshriber, *Can. J. Chem.*, **50**, 2568 (1972).
447. A. Grandin and J. Vialle, *Bull. Soc. Chim. Fr.*, 4002 (1971).
448. C. Metayer, G. Duguay and H. Quiniou, *Bull. Soc. Chim. Fr.*, 163 (1974).
449. J. L. Adelfang, *J. Org. Chem.*, **31**, 2388 (1966).
450. E. Klingsberg, *Chem. Ind. (London)*, 1813 (1968).
451. J.-M. Catel and Y. Mollier, *Abstracts of Papers, Eighth International Symposium on Organic Sulfur Chemistry*, Portoroz, Yugoslavia, June 18–23, 1978, p. 57.
452. D. Festal and Y. Mollier, *Tetrahedron Lett.*, 1259 (1970).
453. (a) R. J. S. Beer, R. P. Carr, D. Cartwright, D. Harris and R. A. Slater, *J. Chem. Soc. C*, 2490 (1968); (b) R. J. S. Beer, D. Cartwright and D. Harris, *Tetrahedron Lett.*, 953 (1967).
454. Y. Poirier and N. Lozac'h, *Bull. Soc. Chim. Fr.*, 2090 (1967).
455. E. Klingsberg, *J. Org. Chem.*, **31**, 3489 (1966).
456. E. Klingsberg, *J. Amer. Chem. Soc.*, **85**, 3244 (1963).
457. Y. Mollier, F. Terrier, R. Pinel, N. Lozac'h and C. Menez, *Bull. Soc. Chim. Fr.*, 2074 (1967).
458. E. Klingsberg, *J. Heterocyclic Chem.*, **3**, 243 (1966).
459. H. Behringer and R. Wiedenmann, *Tetrahedron Lett.*, 3705 (1965).
460. Y. Mollier, F. Terrier and N. Lozac'h, *Bull. Soc. Chim. Fr.*, 1778 (1964).
461. Y. Mollier and N. Lozac'h, *Bull. Soc. Chim. Fr.*, 157 (1963).
462. D. Leaver and D. M. McKinnon, *Chem. Ind. (London)*, 461 (1964).
463. E. Futterer, A. Lüttringhaus and H. Prinzbach, *Tetrahedron Lett.*, 1209 (1963).
464. (a) U. Schmidt, R. Scheuring and A. Lüttringhaus, *Justus Liebigs Ann. Chem.*, **630**, 116 (1960); (b) E. Campaigne and R. D. Hamilton, *J. Org. Chem.*, **29**, 1711 (1964).
465. (a) N. K. Son, R. Pinel and Y. Mollier, *Bull. Soc. Chim. Fr.*, 1356 (1974); (b) R. Pinel, N. K. Son and Y. Mollier, *C. R. Acad. Sci., Ser. C*, **271**, 955 (1970).
466. F. Clesse, J. P. Pradere and H. Quiniou, *Bull. Soc. Chim. Fr.*, 586 (1973).
467. (a) G. Caillaud and Y. Mollier, *Bull. Soc. Chim. Fr.*, 151 (1972); (b) 331 (1971).
468. S. Davidson and D. Leaver, *J. Chem. Soc. Chem. Commun.*, 540 (1972).
469. E. I. G. Brown, D. Leaver and D. M. McKinnon, *J. Chem. Soc. C*, 1202 (1970).
470. B. Bartho, J. Faust and R. Mayer, *Z. Chem.*, **15**, 440 (1975).
471. J. Faust, B. Bartho and R. Mayer, *Z. Chem.*, **15**, 395 (1975).
472. (a) G. A. Reynolds, *J. Org. Chem.*, **33**, 3352 (1968); (b) N. Lozac'h and C. T. Pedersen, *Acta Chem. Scand.*, **24**, 3189 (1970).
473. R. Pinel and Y. Mollier, *Bull. Soc. Chim. Fr.*, 1032 (1973).
474. P. D. Bartlett, E. F. Cox and R. E. Davis, *J. Amer. Chem. Soc.*, **83**, 103 (1961).
475. J. Faust, H. Spies and R. Mayer, *Z. Chem.*, **7**, 275 (1967).

510   Donald C. Dittmer and Bhalchandra H. Patwardhan

476. R. M. Christie and D. H. Reid, *J. Chem. Soc. Perkin Trans. I*, 228 (1976).
477. R. M. Christie, A. S. Ingram, D. H. Reid and R. G. Webster, *J. Chem. Soc. Perkin Trans. I*, 722 (1974).
478. H. Behringer and D. Bender, *Chem. Ber.*, **100**, 4027 (1967).
479. B. Bartho, J. Faust and R. Mayer, *Tetrahedron Lett.*, 2683 (1975).
480. R. Pinel, Y. Mollier, J. P. de Barbeyrac and G. Pfister-Guillouzo, *C. R. Acad. Sci., Ser. C*, **275**, 909 (1972).
481. K. Bechgaard, V. D. Parker and C. T. Pedersen, *J. Amer. Chem. Soc.*, **95**, 4373 (1973).
482. C. T. Pedersen, K. Bechgaard and V. D. Parker, *J. Chem. Soc. Chem. Commun.*, 430 (1972).
483. C. T. Pedersen and V. D. Parker, *Tetrahedron Lett.*, 767 (1972).
484. J. M. Catel and Y. Mollier, *C. R. Acad. Sci., Ser. C*, **280**, 673 (1975).
485. M. G. Voronkov and T. V. Lapina, *Chem. Heterocyclic Compd.*, **6**, 416 (1970).
486. G. Duguay and H. Quiniou, *C. R. Acad. Sci., Ser. C*, **283**, 495 (1976).
487. R. Mayer and H. Hartmann, *Z. Chem.*, **6**, 312 (1966).
488. D. H. Reid, J. G. Dingwall and S. McKenzie, *J. Chem. Soc. C*, 2543 (1968).
489. C. Retour, M. Stavaux and N. Lozac'h, *Bull. Soc. Chim. Fr.*, 3360 (1971).
490. H. Prinzbach, E. Futterer and A. Lüttringhaus, *Angew. Chem. Int. Ed. Engl.*, **5**, 513 (1966).
491. C. T. Pedersen and J. Moller, *Tetrahedron*, **30**, 553 (1974).
492. C. T. Pedersen, N. L. Huaman and J. Moller, *Acta Chem. Scand.*, **28B**, 1185 (1974).
493. C. T. Pedersen and C. Lohse, *Tetrahedron Lett.*, 5213 (1972).
494. G. A. Heath, R. L. Martin and I. M. Stewart, *J. Chem.*, **22**, 83 (1969).
495. M. Nakatani, Y. Takahashi and A. Ouchi, *J. Inorg. Nucl. Chem.*, **31**, 3330 (1969).
496. E. Uhlemann and K. H. Uteg, *Z. Chem.*, **10**, 468 (1970).
497. (a) E. Fanghänel, *Z. Chem.*, **4**, 70 (1964); (b) R. Mayer and B. Gebhardt, *Chem. Ber.*, **97**, 1298 (1964); (c) E. Klingsberg, *J. Amer. Chem. Soc.*, **86**, 5290 (1964); (d) D. Dolphin, W. Pegg and P. Wirz, *Can. J. Chem.*, **52**, 4078 (1974); (e) P. Calas, J. M. Fabre, M. Khalife-El-Saleh, A. Mas, E. Torreiles and L. Giral, *C. R. Acad. Sci., Ser. C*, **281**, 1037 (1975); (f) H. Spies, K. Gewald and R. Mayer, *J. Prakt. Chem.*, **313**, 804 (1971).
498. N. F. Haley, *J. Org. Chem.*, **43**, 678 (1978).
499. (a) G. Scherowsky and J. Weiland, *Justus Liebigs Ann. Chem.*, 403 (1974); (b) L. Soder and R. Wizinger, *Helv. Chim. Acta*, **42**, 1733 (1959); (c) A. Mas, J. M. Fabre, E. Torreilles, L. Giral and G. Brun, *Tetrahedron Lett.*, 2579 (1977).
500. (a) R. Mayer and H. Kröber, *J. Prakt. Chem.*, **316**, 907 (1974); (b) A. Lüttringhaus, H. Berger and H. Prinzbach, *Tetrahedron Lett.*, 2121 (1965); (c) W. Kirmse and L. Horner, *Justus Liebigs Ann. Chem.*, **614**, 4 (1958); (d) E. Campaigne and F. Haaf, *J. Org. Chem.*, **30**, 732 (1965).
501. S. Wawzonek and S. M. Heilmann, *J. Org. Chem.*, **39**, 511 (1974).
502. (a) E. Klingsberg, *J. Amer. Chem. Soc.*, **84**, 3410 (1962); (b) H. K. Spencer, M. P. Cava, F. G. Yamagishi and A. F. Garito, *J. Org. Chem.*, **41**, 730 (1976).
503. (a) I. Degani and R. Fochi, *Synthesis*, 759 (1976); (b) K. Hirai, *Tetrahedron*, **27**, 4003 (1971); (c) J. Nakayama, K. Ueda, M. Hoshino and T. Takemasa, *Synthesis*, 770 (1977); (d) A. Takamizawa and K. Hirai, *Chem. Pharm. Bull.*, **17**, 1931 (1969); (e) I. Degani and R. Fochi, *J. Chem. Soc. Perkin Trans. I*, 1886 (1976).
504. (a) W. R. H. Hurtley and S. Smiles, *J. Chem. Soc.*, 1821 (1926); (b) W. R. H. Hurtley and S. Smiles, *J. Chem. Soc.*, 534 (1927).
505. (a) I. Degani and R. Fochi, *Synthesis*, 471 (1976); (b) J. Nakayama, E. Seki and M. Hoshino, *J. Chem. Soc. Perkin Trans. I*, 468 (1978); (c) W. R. H. Hurtley and S. Smiles, *J. Chem. Soc.*, 2263 (1926); (d) J. Nakayama, K. Fujiwara and M. Hoshino, *Chem. Lett.*, 1099 (1975).
506. F. Wudl, M. L. Kaplan, E. J. Hufnagel and E. W. Southwick, *J. Org. Chem.*, **39**, 3608 (1974).
507. (a) A. Takamizawa and K. Hirai, *Chem. Pharm. Bull.*, **17**, 1924 (1969); (b) P. Calas, J. M. Fabre, E. Torreilles and L. Giral, *C. R. Acad. Sci., Ser. C*, **280**, 901 (1975).
508. J. Nakayama, Y. Watabe and M. Hoshino, *Bull. Chem. Soc. Jap.*, **51**, 1427 (1978).

509. (a) E. Campaigne, R. D. Hamilton and N. W. Jacobsen, *J. Org. Chem.*, **29**, 1708 (1964); (b) H. Gotthardt, M. C. Weisshuhn and B. Christl, *Chem. Ber.*, **109**, 740 (1976); (c) H. Gotthardt, C. M. Weisshuhn, O. M. Huss and D. J. Brauer, *Tetrahedron Lett.*, 671 (1978); (d) K. T. Potts, D. R. Choudhury, A. J. Elliott, and U. P. Singh, *J. Org. Chem.*, **41**, 1724 (1976); (e) H. Gotthardt and B. Christl, *Tetrahedron Lett.*, 4743 (1968).

510. E. Campaigne and N. W. Jacobsen, *J. Org. Chem.*, **29**, 1703 (1964).

511. (a) K. Fabian and H. Hartmann, *J. Prakt. Chem.*, **313**, 722 (1971); (b) Y. Ueno, M. Bahry and M. Okawara, *Tetrahedron Lett.*, 4607 (1977).

512. M. Ohta and M. Sugiyama, *Bull. Chem. Soc. Jap.*, **36**, 1437 (1963).

513. (a) I. Degani and R. Fochi, *Synthesis*, 263 (1977); (b) L. Soder and R. Wizinger, *Helv. Chim. Acta*, **42**, 1779 (1959); (c) R. Wizinger and L. Soder, *Chimia*, **12**, 79 (1959); (d) R. Wizinger and D. Dürr, *Helv. Chim. Acta*, **46**, 2167 (1963).

514. Y. Ueno, A. Nakayama and M. Okawara, *Synthesis*, 277 (1975).

515. (a) K. Bechgaard, D. O. Cowan and A. N. Bloch, *J. Chem. Soc. Chem. Commun.*, 937 (1974); (b) K. Bechgaard, D. O. Cowan, A. N. Bloch and L. Henriksen, *J. Org. Chem.*, **40**, 746 (1975); (c) J. R. Andersen and K. Bechgaard, *J. Org. Chem.*, **40**, 2016 (1975).

516. (a) J. Nakayama, E. Seki and M. Hoshino, *J. Chem. Soc. Perkin Trans. I*, 468 (1978); (b) J. Nakayama, *J. Chem. Soc. Perkin Trans. I*, 525 (1975).

517. (a) F. Wudl, G. M. Smith and E. J. Hufnagel, *J. Chem. Soc. Chem. Commun.*, 1453 (1970); (b) F. Wudl, D. Wobschall and E. J. Hufnagel, *J. Amer. Chem. Soc.*, **94**, 670 (1972).

518. S. Hunig, G. Kiesslich, H. Quast and D. Scheutzow, *Justus Liebigs Ann. Chem.*, 310 (1973).

519. H. D. Hartzler, *J. Amer. Chem. Soc.*, **95**, 4379 (1973); **92**, 1412 (1970).

520. (a) E. Fanghänel and R. Mayer, *Z. Chem.*, **4**, 384 (1964); (b) D. Buza, A. Gryffkeller and S. Szymanski, *Rocz. Chem.*, **44**, 2319 (1970); (c) D. Buza and S. Szymanski, *Rocz. Chem.*, **45**, 501 (1971).

521. (a) K. Hirai, H. Sugimoto and T. Ishiba, *J. Org. Chem.*, **42**, 1543 (1977); (b) E. Fanghänel, *Z. Chem.*, **7**, 58 (1967).

522. (a) E. Campaigne, T. Bosin and R. D. Hamilton, *J. Org. Chem.*, **30**, 1677 (1965); (b) E. Fanghänel, *Z. Chem.*, **5**, 386 (1965); (c) E. Campaigne and R. D. Hamilton, *J. Org. Chem.*, **29**, 2877 (1964).

523. (a) H. Anzai, *Denshi Gijutsu Sogo Kenkyujo Iho*, **39**, 667 (1975); *Chem. Abstr.*, **84**, 164 654 (1976); (b) J. M. Fabre, E. Torreilles, J. P. Gibert, M. Chanaa and L. Giral, *Tetrahedron Lett.*, 4033 (1977); (c) H. Prinzbach, H. Berger and A. Lüttringhaus, *Angew. Chem. Int. Ed. Engl.*, **4**, 435 (1965).

524. (a) G. Seitz and H. G. Lehmann, *Arch. Pharm. (Weinheim)*, **307**, 853 (1974); *Chem. Abstr.*, **82**, 72737 (1975); (b) R. Gompper and E. Kutter, *Chem. Ber.*, **98**, 2825 (1965).

525. (a) E. Fanghänel, *J. Prakt. Chem.*, **317**, 137 (1975); (b) K. M. Pazdro and W. Polaczkowa, *Rocz. Chem.*, **45**, 811 (1971); *Chem. Abstr.*, **75**, 110213 (1971); (c) L. R. Melby, H. D. Hartzler and W. A. Sheppard, *J. Org. Chem.*, **39**, 2456 (1974).

526. K. Miura and M. Tada, *Chem. Lett.*, 1139 (1978).

527. (a) E. Fanghänel, L. Van Hinh and G. Schukat, *Z. Chem.*, **16**, 317 (1976); (b) A. Kruger and F. Wudl, *J. Org. Chem.*, **42**, 2778 (1977).

528. A. R. Siedle and R. B. Johannesen, *J. Org. Chem.*, **40**, 2002 (1975).

529. P. R. Moses and J. Q. Chambers, *J. Amer. Chem. Soc.*, **96**, 945 (1974).

530. H. Kato, M. Kawamura, T. Shiba and M. Ohta, *J. Chem. Soc. Chem. Commun.*, 959 (1970).

531. (a) H. Gotthardt, M. C. Weisshuhn and B. Christl, *Chem. Ber.*, **109**, 753 (1976); (b) H. Matsukubo and H. Kato, *J. Chem. Soc. Perkin Trans. I*, 632 (1975); (c) H. Matsukubo and H. Kato, *J. Chem. Soc. Chem. Commun.*, 840 (1975); (d) K. T. Potts and U. P. Singh, *J. Chem. Soc. Chem. Commun.*, 569 (1969).

532. (a) J. Ferraris, D. O. Cowan, V. Walatka and J. H. Perlstein, *J. Amer. Chem. Soc.*, **95**, 948 (1973); (b) F. Wudl, C. H. Ho and A. Nagel, *J. Chem. Soc. Chem. Commun.*, 923 (1973); (c) F. Wudl, *J. Amer. Chem. Soc.*, **97**, 1962 (1975).

512   Donald C. Dittmer and Bhalchandra H. Patwardhan

533. J. S. Miller and A. J. Epstein (Eds.), *Synthesis and Properties of Low-Dimensional Materials*, *Ann. N.Y. Acad. Sci.*, **313** (1978).
534. (a) R. H. Wiley, D. C. England and L. C. Behr, *Org. React.*, **6**, 367 (1951); (b) J. M. Sprague and A. H. Land, in *Heterocyclic Compounds*, (Ed. R. C. Elderfield), Vol. 5, Wiley, New York, 1957, p. 638.
535. (a) K. R. H. Wooldridge, *Adv. Heterocyclic Chem.*, **14**, 2 (1972); (b) M. Davis, *Adv. Heterocyclic Chem.*, 43.
536. (a) W. D. Ollis and C. A. Ramsden, *Adv. Heterocyclic Chem.*, **19**, 3 (1976); (b) K. T. Potts, in *Lectures in Heterocyclic Chemistry* (Eds. R. N. Castle and I. Lalezari), Vol. 4, *J. Heterocyclic Chem.*, S 35 (1978).
537. Y. Ferré, E. J. Vincent, H. Larivé and J. Metzger, *Bull. Soc. Chim. Fr.*, 3862 (1972).
538. Y. Ferré, E. J. Vincent, H. Larivé and J. Metzger, *Bull. Soc. Chim. Fr.*, 1003 (1973).
539. Y. Ferré and E. J. Vincent, *C. R. Acad. Sci., Ser. C*, **272**, 1916 (1971).
540. e.g. P. J. Nigrey and A. F. Garito, *J. Chem. Eng. Data*, **22**, 451 (1977).
541. I. M. Bazavova, R. G. Dubenko and P. S. Pel'kis, *J. Org. Chem. USSR*, **14**, 181 (1978).
542. K. T. Potts, S. J. Chen, J. Kane and J. L. Marshall, *J. Org. Chem.*, **42**, 1633 (1977).
543. Y. Gelernt and P. Sykes, *J. Chem. Soc. Perkin Trans. I*, 2610 (1974).
544. A. I. Tolmachev, E. F. Karaban and L. M. Shulezhko, *Chem. Heterocyclic Compd.*, **13**, 137 (1977).
545. T. Shiba and H. Kato, *Bull. Chem. Soc. Jap.*, **46**, 964 (1973).
546. (a) M. Baudy and A. Robert, *J. Chem. Soc. Chem. Commun.*, 23 (1976); (b) K. T. Potts, J. Baum, E. Houghton, D. N. Roy and U. P. Singh, *J. Org. Chem.*, **39**, 3619 (1974); (c) K. T. Potts, E. Houghton and U. P. Singh, *J. Org. Chem.*, **39**, 3627 (1974).
547. R. Huisgen and T. Schmidt, *Justus Liebigs Ann. Chem.*, 29 (1978).
548. Y. Tamura, H. Hayashi and M. Ikeda, *Synthesis*, 126 (1974).
549. I. Ito, T. Ueda, Y. Kuroyanagi and K. Suzuki, *Chem. Pharm. Bull.*, **25**, 171 (1977).
550. A. Takamizawa and Y. Sato, *Chem. Pharm. Bull.*, **14**, 742 (1966).
551. H. Neef, K. D. Kohnert and A. Schellenberger, *J. Prakt. Chem.*, **315**, 701 (1973).
552. P. Haake and M. Duclos, *Tetrahedron Lett.*, 461 (1970).
553. (a) P. Sohar, G. H. Denny and R. D. Babson, *J. Heterocyclic Chem.*, **7**, 1369 (1970); (b) K. Dimroth and K. Severin, *Justus Liebigs Ann. Chem.*, 380 (1973).
554. G. M. Clarke and P. Sykes, *J. Chem. Soc. Chem. Commun.*, 370 (1965).
555. A. V. Kazymov and E. P. Shchelkina, *Chem. Heterocyclic Compd.*, **7**, 1287 (1971).
556. (a) J. Kister, A. Blanc, E. Davin and J. Metzger, *Bull. Soc. Chim. Fr.*, 2297 (1975); (b) A. Samat, J. Kister, F. Garnier, J. Metzger and R. Guglielmetti, *Bull. Soc. Chim. Fr.*, 2627 (1975).
557. (a) F. Jordan and Y. H. Mariam, *J. Amer. Chem. Soc.*, **100**, 2534 (1978); (b) H. Stetter and G. Dämbkes, *Synthesis*, 403 (1977); (c) H. Stetter, R. Y. Rämsch and H. Kuhlmann, *Synthesis*, 733 (1976); (d) J. Castells, H. Llitjos and M. Moreno-Manas, *Tetrahedron Lett.*, 205 (1977); (e) R. C. Cookson and R. M. Lane, *J. Chem. Soc. Chem. Commun.*, 804 (1976).
558. J. C. Sheehan and T. Hara, *J. Org. Chem.*, **39**, 1196 (1974).
559. H. Stetter and H. Kuhlmann, *Angew. Chem. Int. Ed. Engl.*, **13**, 539 (1974).
560. (a) A. Takamizawa, S. Matsumoto and S. Sakai, *Chem. Pharm. Bull.*, **22**, 299 (1974); (b) A. Takamizawa, S. Matsumoto and S. Sakai, *Chem. Pharm. Bull.*, **22**, 293 (1974).
561. P. Haake, L. P. Bausher and J. P. McNeal, *J. Amer. Chem. Soc.*, **93**, 7045 (1971).
562. P. Haake, L. P. Bausher and W. B. Miller, *J. Amer. Chem. Soc.*, **91**, 1113 (1969).
563. (a) H. Quast and S. Hünig, *Angew. Chem. Int. Ed. Engl.*, **3**, 800 (1964); (b) H. W. Wanzlick and H. J. Kleiner, *Angew. Chem. Int. Ed. Engl.*, **3**, 65 (1964).
564. S. Hünig, D. Scheutzow and H. Schlaf, *Justus Liebigs Ann. Chem.*, **765**, 126 (1972).
565. (a) A. Takamizawa, Y. Hamashima and H. Sato, *J. Org. Chem.*, **33**, 4038 (1968); (b) A. Takamizawa and H. Harada, *Chem. Pharm. Bull.*, **21**, 770 (1973).
566. J. A. Van Allan, J. D. Mee, C. A. Maggiulli and R. S. Henion, *J. Heterocyclic Chem.*, **12**, 1005 (1975).
567. (a) K. T. Potts and D. McKeough, *J. Amer. Chem. Soc.*, **96**, 4268 (1974); (b) K. T. Potts, J. Baum and E. Houghton, *J. Org. Chem.*, **39**, 3631 (1974); (c) S. Nakazawa, T. Kiyosawa, K. Hirakawa and H. Kato, *J. Chem. Soc. Chem. Commun.*, 621 (1974).

568. (a) P. Chaplen, R. Slack and K. R. H. Wooldridge, *J. Chem. Soc.*, 4577 (1965); (b) M. Davis, L. W. Deady and E. Homfield, *Aust. J. Chem.*, **27**, 1221 (1974); (c) M. Davis, E. Homfeld and K. S. Lal Srivastava, *J. Chem. Soc. Perkin Trans. I*, 1863 (1973).
569. H. P. Benschop, A. M. Van Oosten, D. H. J. M. Platenburg and C. van Hooidonk, *J. Med. Chem.*, **13**, 1208 (1970).
570. D. M. McKinnon and M. E. Hassan, *Can. J. Chem.*, **51**, 3081 (1973).
571. G. V. Boyd and T. Norris, *J. Chem. Soc. Perkin Trans. I*, 1028 (1974).
572. (a) J. M. Landesberg and R. A. Olofson, *Tetrahedron*, **22**, 2135 (1966); (b) H. Böshagen and W. Geiger, *Chem. Ber.*, **107**, 1667 (1974).
573. A. S. Ingram, D. H. Reid and J. D. Symon, *J. Chem. Soc. Perkin Trans. I*, 242 (1974).
574. F. Leung and S. C. Nyburg, *Can. J. Chem.*, **50**, 324 (1972).
575. (a) F. Boberg and W. V. Gentzkow, *Justus Liebigs Ann. Chem.*, 247 (1973); (b) J. Liebscher and H. Hartmann, *Z. Chem.*, **14**, 189 (1974).
576. (a) H. Ullah and P. Sykes, *Chem. Ind. (London)*, 1162 (1973); (b) R. F. Meyer, B. L. Cummings, P. Bass and H. O. J. Collier, *J. Med. Chem.*, **8**, 515 (1965).
577. R. A. Olofson and J. M. Landesberg, *J. Amer. Chem. Soc.*, **88**, 4263 (1966).
578. N. F. Haley, *J. Org. Chem.*, **43**, 1233 (1978).
579. (a) J. C. Martin and T. M. Balthazor, *J. Amer. Chem. Soc.*, **99**, 152 (1977); (b) G. W. Astrologes and J. C. Martin, *J. Amer. Chem. Soc.*, **99**, 4400 (1977); (c) L. J. Adzima and J. C. Martin, *J. Amer. Chem. Soc.*, **99**, 1657 (1977); (d) T. M. Balthazor and J. C. Martin, *J. Amer. Chem. Soc.*, **97**, 5634 (1975).
580. L. Horner and V. Binder, *Justus Liebigs Ann. Chem.*, **757**, 33 (1972).
581. P. Livant and J. C. Martin, *J. Amer. Chem. Soc.*, **99**, 5761 (1977).
582. D. Landini, F. Rolla and G. Torre, *Int. J. Sulfur Chem. A*, **2**, 43 (1972); *Chem. Abstr.*, **77**, 139372 (1972).
583. D. L. Tuleen, W. G. Bentrude and J. C. Martin, *J. Amer. Chem. Soc.*, **85**, 1938 (1963).
584. P. K. Claus, in *Organic Sulphur Chemistry*, (Ed. C. J. M. Stirling), Butterworths, London, 1975, p. 449.
585. D. H. Reid and R. G. Webster, *J. Chem. Soc. Perkin Trans. I*, 2097 (1975).
586. Y. Tominaga, Y. Matsuda and G. Kobayashi, *Heterocycles*, **4**, 9 (1976).
587. K. T. Potts, J. Kane, E. Carnahan and U. P. Singh, *J. Chem. Soc. Chem. Commun.*, 417 (1975).
588. K. Hirai and T. Ishiba, *Heterocycles*, **3**, 217 (1975).
589. I. Degani, R. Fochi and P. Tundo, *Ann. Chim. (Rome)*, **62**, 570 (1972).
590. L. Costa, I. Degani, R. Fochi and P. Tundo, *J. Heterocyclic Chem.*, **11**, 943 (1974).
591. I. Degani, R. Fochi and P. Tundo, *Gazz. Chim. Ital.*, **105**, 907 (1975).
592. I. Degani and R. Fochi, *J. Chem. Soc. Perkin Trans. I*, 323 (1976).
593. I. Degani and R. Fochi, *Synthesis*, 757 (1976).
594. (a) H. Hartmann and F. Mohn, *J. Prakt. Chem.*, **313**, 737 (1971); (b) H. Hartmann, *J. Prakt. Chem.*, 730.
595. H. Wevers and W. Drenth, *Recl. Trav. Chim. Pays-Bas*, **93**, 99 (1974).
596. I. Degani, R. Fochi and P. Tundo, *J. Heterocyclic Chem.*, **11**, 507 (1974).
597. H. Hartmann, *Z. Chem.*, **11**, 421 (1971).
598. H. Hartmann, H. Schäfer and K. Gewald, *J. Prakt. Chem.*, **315**, 497 (1973).
599. K. Hirai and T. Ishiba, *Chem. Pharm. Bull.*, **19**, 2194 (1971).
600. K. Hirai and T. Ishiba, *Chem. Pharm. Bull.*, **20**, 2384 (1972).
601. W. D. Ollis and C. A. Ramsden, *Adv. Heterocyclic Chem.*, **19**, 50 (1976).
602. (a) L. M. Weinstock and P. I. Pollak, *Adv. Heterocyclic Chem.*, **9**, 107 (1968); (b) J. Sandstrom, *Adv. Heterocyclic Chem.*, **9**, 165 (1968); (c) F. Kurzer, *Adv. Heterocyclic Chem.*, **5**, 119 (1965).
603. K. A. Jensen and C. Pedersen, *Adv. Heterocyclic Chem.*, **3**, 263 (1964).
604. (a) G. Scherowsky, *Chem. Ber.*, **107**, 1092 (1974); (b) G. Scherowsky, *Tetrahedron Lett.*, 4985 (1971).
605. G. V. Boyd and A. J. H. Summers, *J. Chem. Soc. C*, 2311 (1971).
606. A. R. McCarthy, W. D. Ollis and C. A. Ramsden, *J. Chem. Soc. Perkin Trans. I*, 627 (1974).
607. R. Grashey, M. Baumann and R. Hamprecht, *Tetrahedron Lett.*, 2939 (1972).

608. (a) W. D. Ollis and C. A. Ramsden, *J. Chem. Soc. Chem. Commun.*, 1222 (1971); (b) W. D. Ollis and C. A. Ramsden, *J. Chem. Soc. Perkin Trans. I*, 633 (1974).
609. A. Y. Lazaris, S. M. Shmuilovich and A. N. Egorochkin, *Chem. Heterocyclic Compd.*, 9, 1216 (1973).
610. A. Y. Lazaris and A. N. Egorochkin, *J. Org. Chem. USSR*, 6, 2351 (1970).
611. (a) R. Grashey, M. Baumann and W. D. Lubos, *Tetrahedron Lett.*, 5881 (1968); (b) R. Grashey, M. Baumann and W. D. Lubos, *Tetrahedron Lett.*, 5877 (1968).
612. T. J. Curphey and K. S. Prasad, *J. Org. Chem.*, 37, 2259 (1972).
613. A. Alemagna and T. Bacchetti, *Gazz. Chim. Ital.*, 102, 1068, 1077, 1084 (1972).
614. (a) A. M. Kiwan and H. M. N. H. Irving, *J. Chem. Soc. Chem. Commun.*, 928 (1970); (b) M. Busch, *Chem. Ber.*, 28, 2635 (1895); (c) M. Busch, *J. Prakt. Chem.*, 60, 25 (1899).
615. R. N. Hanley, W. D. Ollis, C. A. Ramsden, G. Rowlands and L. E. Sutton, *J. Chem. Soc. Perkin Trans. I*, 600 (1978).
616. A. M. Kiwan and H. M. Marafie, *J. Heterocyclic Chem.*, 13, 1273 (1976).
617. G. Scherowsky and H. Matloubi, *Justus Liebigs Ann. Chem.*, 98 (1978).
618. P. B. Talukdar and A. Chakraborty, *J. Indian Chem. Soc.*, 51, 600 (1974); 52, 893 (1975); *Chem. Abstr.*, 82, 72336 (1975); 84, 73302 (1976).
619. O. P. Shvaika and V. I. Fomenko, *J. Org. Chem. USSR*, 10, 376 (1974).
620. A. Takamizawa and H. Sato, *Chem. Pharm. Bull.*, 18, 1201 (1970).
621. (a) R. M. Moriarty, R. Mukherjee, O. L. Chapman and D. R. Eckroth, *Tetrahedron Lett.*, 397 (1971); (b) R. M. Moriarty, J. M. Kliegeman and R. B. Desai, *J. Chem. Soc. Chem. Commun.*, 1255 (1967).
622. R. M. Moriarty and A. Chin, *J. Chem. Soc. Chem. Commun.*, 1300 (1972).
623. G. Barnikow and J. Bödeker, *J. Prakt. Chem.*, 313, 1148 (1971).
624. S. Crook and P. Sykes, *J. Chem. Soc. Perkin Trans. I*, 1791 (1977).
625. S. Crook, P. G. Jones, O. Kennard and P. Sykes, *Chem. Ind. (London)*, 840 (1977).
626. S. I. Burmistrov and V. A. Kozinskii, *J. Org. Chem. USSR*, 10, 906 (1974).
627. R. M. Christie, D. H. Reid, R. Walker and R. G. Webster, *J. Chem. Soc. Perkin Trans. I*, 195 (1978).
628. G. I. Eremeeva, B. K. Strelets and L. S. Efros, *Chem. Heterocyclic Compd.*, 11, 238 (1975).
629. R. N. Hanley, W. D. Ollis and C. A. Ramsden, *J. Chem. Soc. Chem. Commun.*, 306 (1976).
630. R. N. Hanley, W. D. Ollis and C. A. Ramsden, *J. Chem. Soc. Chem. Commun.*, 307 (1976).
631. S. I. Mathew and F. Stansfield, *J. Chem. Soc. Perkin Trans. I*, 540 (1974).
632. H. Gotthardt, *Tetrahedron Lett.*, 1277, 1281 (1971); *Chem. Ber.*, 105, 188, 196 (1972).
633. G. Bhaskaraiah, *Indian J. Chem.*, 12, 134 (1974).
634. (a) J. E. Oliver, R. T. Brown and N. L. Redfearn, *J. Heterocyclic Chem.*, 9, 447 (1972); (b) M. Ahmed and D. M. McKinnon, *Can. J. Chem.*, 48, 2142 (1970); (c) J. E. Oliver, S. C. Chang, R. T. Brown, J. B. Stokes and A. B. Borkovec, *J. Med. Chem.*, 15, 315 (1972).
635. (a) J. E. Oliver, *J. Org. Chem.*, 39, 2235 (1974); (b) J. E. Oliver and A. B. DeMilo, *J. Org. Chem.*, 39, 2225 (1974).
636. (a) M. B. Kolesova, L. I. Maksimova and A. V. El'tsov, *J. Org. Chem. USSR*, 7, 2315 (1971); (b) E. Fromm and H. Baumkauer, *Justus Liebigs Ann. Chem.*, 361, 319 (1908); (c) W. R. Diveley, *U.S. Pat.*, 3,166,564 (1965); *Chem. Abstr.*, 62, 9145 (1965).
637. J. Liebscher and H. Hartmann, *Justus Liebigs Ann. Chem.*, 1005 (1977).
638. H. Böhme and K. H. Ahrens, *Arch. Pharm. (Weinheim)*, 307, 828 (1974); *Chem. Abstr.*, 82, 72883 (1975).
639. (a) J. E. Oliver and A. B. DeMilo, *J. Heterocyclic Chem.*, 8, 1087 (1971); (b) J. E. Oliver, *J. Org. Chem.*, 36, 3465 (1971).
640. (a) L. Huestis and M. Brownell, *J. Heterocyclic Chem.*, 5, 427 (1968); (b) P. Hope and L. A. Wiles, *Chem. Ind. (London)*, 32 (1966); *J. Chem. Soc. C*, 642 (1967); (c) O. Maior and N. Arsenescu, *An. Univ. Bucuresti, Ser. Stiint. Nat.*, 14, 25 (1965); 16, 101 (1967); *Chem. Abstr.*, 67, 64319 (1967); 71, 3331 (1969).

641. N. Arsenescu and O. Maior, *Rev. Chim. (Bucharest)*, **17**, 172 (1966); *Chem. Abstr.*, **65**, 8895 (1966).
642. N. Arsenescu, *Rev. Roum. Chim.*, **12**, 427 (1967); *Chem. Abstr.*, **68**, 29660 (1968).
643. (a) L. Huestis, I. Emery and E. Steffensen, *J. Heterocyclic Chem.*, **3**, 518 (1966); (b) L. Huestis, M. L. Walsh and N. Hahn, *J. Org. Chem.*, **30**, 2763 (1965).
644. R. Neidlein, P. Leinberger, A. Gieren and B. Dederer, *Chem. Ber.*, **111**, 698 (1978).
645. J. P. Marino, in *Topics in Sulfur Chemistry* (Ed. A. Senning), Vol. 1, G. Thieme, Stuttgart, 1976, p. 86.
646. V. G. Kharchenko, S. N. Chalaya and T. M. Konovalova, *Chem. Heterocyclic. Compd.*, **11**, 125 (1975).
647. S. W. Schneller, *Adv. Heterocyclic Chem.*, **18**, 94 (1975).
648. R. Mayer, H. Hartmann, J. Fabian and A. Mehlhorn, *Z. Chem.*, **7**, 209 (1967).
649. (a) Z. Yoshida, H. Sugimoto and S. Yoneda, *Tetrahedron*, **28**, 5873 (1972); (b) S. Yoneda, T. Sugimoto and Z. Yoshida, *Tetrahedron*, **29**, 2009 (1973); (c) M. H. Palmer, R. H. Findlay, W. Moyes and A. J. Gaskell, *J. Chem. Soc. Perkin Trans. II*, 841 (1975); (d) T. E. Young and C. J. Ohnmacht, *J. Org. Chem.*, **32**, 444 (1967); (e) T. E. Young and C. R. Hamel, *J. Org. Chem.*, **35**, 816, 821 (1970); (f) Z. Yoshida, T. Sugimoto and S. Yoneda, *Tetrahedron Lett.*, 4259 (1971); (g) A. Mistr, M. Vavra, J. Skoupy and R. Zahradnik, *Coll. Czech. Chem. Commun.*, **37**, 1520 (1972); (h) A. Mistr and R. Zahradnik, *Coll. Czech. Chem. Commun.*, **38**, 1668 (1973).
650. (a) T. E. Young and C. J. Ohnmacht, *J. Org. Chem.*, **32**, 1558 (1967); (b) K. Dimroth, W. Kinzebach and M. Soyka, *Chem. Ber.*, **99**, 2351 (1966); (c) A. I. Tolmachev, L. M. Shulezhko and M. Y. Kornilov, *Ukr. Khim. Zh.*, 287 (1974); *Chem. Abstr.*, **81**, 3730 (1974); (d) I. Degani, R. Fochi and G. Spunta, *Ann. Chim. (Rome)*, **63**, 527 (1973); (e) F. C. Boccuzzi and R. Fochi, *Gazz. Chim. Ital.*, **104**, 671 (1974).
651. A. Rauk, J. D. Andose, W. G. Frick, R. Tang and K. Mislow, *J. Amer. Chem. Soc.*, **93**, 6507 (1971).
652. (a) G. A. Reynolds, *Synthesis*, 638 (1975); (b) G. A. Reynolds and J. A. Van Allan, *J. Heterocyclic Chem.*, **9**, 1105 (1972).
653. J. A. Van Allan, G. A. Reynolds and C. C. Petropoulos, *J. Heterocyclic Chem.*, **9**, 783 (1972).
654. C. C. Price, J. Follweiler, H. Pirelahi and M. Siskin, *J. Org. Chem.*, **36**, 791 (1971).
655. K. Kanai, M. Umehara, H. Kitano and K. Fukui, *Nippon Kagaku Zasshi*, **84**, 432 (1963); *Chem. Abstr.*, **59**, 13934 (1963).
656. G. V. Boyd, *J. Chem. Soc.*, 55 (1959).
657. (a) R. Wizinger and P. Ulrich, *Helv. Chim. Acta*, **39**, 207 (1956); (b) **39**, 217 (1956).
658. R. Wizinger and H. J. Angliker, *Helv. Chim. Acta*, **49**, 2046 (1966).
659. (a) V. G. Kharchenko, S. N. Chalaya and L. G. Chichenkova, *Chem. Heterocyclic. Compd.*, **11**, 561 (1975); (b) V. G. Kharchenko, S. K. Klimenko and M. N. Berezhnaya, *Chem. Heterocyclic Compd.*, **10**, 424 (1974); (c) V. G. Kharchenko, S. N. Chalaya, L. G. Chichenkova and N. I. Kozhevnikova, *J. Org. Chem. USSR*, **10**, 2433 (1974); (d) V. G. Kharchenko and A. A. Rassudova, *Chem. Heterocyclic Compd.*, **9**, 180 (1973); (e) V. G. Kharchenko, M. E. Stankevich, A. R. Yakoreva, A. A. Rassudova and N. M. Yartseva, *Chem. Heterocyclic. Compd.*, **8**, 832 (1972); (f) V. G. Kharchenko, M. E. Stankevich, N. M. Kupranets, A. R. Yakoreva, V. I. Kleimenova and S. K. Klimenko, *J. Org. Chem. USSR*, **8**, 197 (1972); (g) V. G. Kharchenko, V. I. Kleimenova and A. R. Yakoreva, *Khim. Geterotsikl. Soedin. Sb.* 3, 79 (1971); *Chem. Abstr.*, **78**, 71852 (1973); (h) V. G. Kharchenko and V. I. Kleimenova, *J. Org. Chem. USSR*, **7**, 618 (1971); (i) V. G. Kharchenko, V. I. Kleimenova and A. R. Yakoreva, *Chem. Heterocyclic Compd.*, **6**, 834 (1970); (j) V. G. Kharchenko, N. M. Kupranets, V. I. Kleimenova, A. A. Rassudova, M. E. Stankevich, N. M. Yartseva and A. R. Yakoreva, *J. Org. Chem. USSR*, **6**, 1123 (1970); V. G. Kharchenko, V. I. Kleimenova, N. M. Kupranets, N. V. Polikarpova and A. R. Yakoreva, *J. Org. Chem. USSR*, **4**, 1984 (1968); (l) V. G. Kharchenko, S. K. Klimenko, A. M. Plaksina and A. R. Yakoreva, *J. Org. Chem. USSR*, **2**, 1116 (1966).
660. S. K. Klimenko, T. V. Stolbova, M. N. Berezhnaya, N. S. Smirnova, I. Y. Evtushenko and V. G. Kharchenko, *J. Org. Chem. USSR*, **10**, 1952 (1974).
661. J. Liebscher and H. Hartmann, *Z. Chem.*, **13**, 342 (1973).

662. H. Hartmann, *J. Prakt. Chem.*, **313**, 1113 (1971).
663. J. Fabian and H. Hartmann, *Tetrahedron Lett.*, 239 (1969).
664. J. Bourdais, *Tetrahedron Lett.*, 2895 (1970).
665. H. Hartmann, *Tetrahedron Lett.*, 3977 (1972).
666. J. Liebscher and H. Hartmann, *Z. Chem.*, **15**, 16 (1975).
667. S. N. Baranov, R. O. Kochkanyan, G. I. Belova and A. N. Zaritovskii, *Proc. Acad. Sci. USSR*, **222**, 281 (1975).
668. G. Laban and R. Mayer, *Z. Chem.*, **7**, 227 (1967).
669. (a) V. N. Gogte, K. A. R. Sastry and B. D. Tilak, *Indian J. Chem.*, **12**, 1147 (1974); (b) S. L. Jindal and B. D. Tilak, *Indian J. Chem.*, **7**, 637 (1969); (c) B. D. Tilak and G. T. Panse, *Indian J. Chem.*, **7**, 191 (1969); (d) B. D. Tilak and S. L. Jindal, *Indian J. Chem.*, **7**, 737 (1969); (e) B. D. Tilak and S. K. Jain, *Indian J. Chem.*, **7**, 17 (1969); (f) B. D. Tilak, H. S. Desai, C. V. Deshpande, S. K. Jain and V. M. Vaidya, *Tetrahedron*, **22**, 7 (1966); (g) B. D. Tilak, R. B. Mitra and C. V. Deshpande, *Tetrahedron Lett.*, 3569 (1965).
670. (a) N. Engelhard and A. Kolb, *Angew. Chem. Int. Ed. Engl.*, **3**, 143 (1964); (b) N. Engelhard and A. Kolb, *Angew. Chem.*, **73**, 218 (1961).
671. (a) J. Ashby, M. Ayad and O. Meth-Cohn, *J. Chem. Soc. Perkin Trans. I*, 1104 (1973); (b) M. Ahmed, J. Ashby and O. Meth-Cohn, *J. Chem. Soc. Chem. Commun.*, 1094 (1970).
672. T. E. Young and P. H. Scott, *J. Org. Chem.*, **30**, 3613 (1965).
673. T. E. Young and C. J. Ohnmacht, *J. Org. Chem.*, **33**, 1306 (1968).
674. Z. Yoshida, S. Yoneda, T. Sugimoto and O. Kikukawa, *Tetrahedron Lett.*, 3999 (1971).
675. I. Degani, R. Fochi and G. Spunta, *Ann. Chim. (Rome)*, **58**, 263 (1968).
676. A. G. Hortmann, R. L. Harris and J. A. Miles, *J. Amer. Chem. Soc.*, **96**, 6119 (1974).
677. (a) I. Degani, R. Fochi and G. Spunta, *Boll. Soc. Fac. Chim. Ind. Bologna*, **24**, 75 (1966); *Chem. Abstr.*, **66**, 46292 (1967); (b) I. Degani and C. Vincenzi, *Boll. Soc. Fac. Chim. Ind. Bologna*, **23**, 245 (1965); *Chem. Abstr.*, **63**, 13197 (1965).
678. I. Degani, R. Fochi and C. Vincenzi, *Gazz. Chim. Ital.*, **94**, 451 (1964).
679. G. Canalini, I. Degani, R. Fochi and G. Spunta, *Ann. Chim. (Rome)*, **61**, 504 (1971).
680. W. Bonthorne and D. H. Reid, *Chem. Ind. (London)*, 1192 (1960).
681. A. Lüttringhaus and A. Kolb, *Z. Naturforsch.*, **16B**, 762 (1961).
682. S. Baklien, P. Groth and K. Undheim, *J. Chem. Soc. Perkin Trans. I*, 2099 (1975).
683. (a) V. G. Kharchenko, S. N. Chalaya, T. V. Stolbova and S. K. Klimenko, *J. Org. Chem. USSR*, **11**, 2510 (1975); (b) S. K. Klimenko, M. N. Berezhnaya and V. G. Kharchenko, *J. Org. Chem. USSR*, **10**, 2437 (1974); (c) V. G. Kharchenko, M. E. Stankevich, A. R. Yakoreva and E. G. Lillenfel'd, *Chem. Heterocyclic Compd.*, **7**, 391 (1971); (d) V. G. Kharchenko, N. M. Kupranets, S. K. Klimenko and M. N. Berezhnaya, *J. Org. Chem. USSR*, **8**, 390 (1972); (e) C. C. Price, M. Siskin and C. Miao, *J. Org. Chem.*, **36**, 794 (1971); (f) V. G. Kharchenko, N. M. Yartseva and M. N. Berezhnaya, *J. Org. Chem. USSR*, **8**, 2673 (1972); (g) V. G. Kharchenko, Z. K. Klimenko, T. V. Stolbova and N. S. Smirnova, *J. Org. Chem. USSR*, **9**, 2457 (1973); (h) V. G. Kharchenko, T. I. Krupina, S. K. Klimenko and A. A. Rassudova, *Chem. Heterocyclic Compd.*, **8**, 1081 (1972); (i) V. G. Kharchenko, N. M. Yartseva and A. A. Rassudova, *J. Org. Chem. USSR*, **6**, 1525 (1970); (j) B. D. Tilak, R. B. Mitra and Z. Muljiani, *Tetrahedron*, **25**, 1939 (1969).
684. A. Tadino, L. Christiaens, M. Renson and P. Cagniant, *Bull. Soc. Chim. Belges*, **81**, 595 (1972).
685. C. C. Price, M. Hori, T. Parasaran and M. Polk, *J. Amer. Chem. Soc.*, **85**, 2278 (1963).
686. (a) V. G. Kharchenko and N. M. Kupranets, *J. Org. Chem. USSR*, **6**, 192 (1970); (b) B. D. Tilak and S. L. Jindal, *Indian J. Chem.*, **7**, 948 (1969).
687. J. P. Pradere and H. Quiniou, *Ann. Chim. (Rome)*, **63**, 563 (1973).
688. B. Eistert and T. J. Arackal, *Chem. Ber.*, **108**, 2397 (1975).
689. J. A. Van Allan, G. A. Reynolds and C. H. Chen, *J. Heterocyclic Chem.*, **14**, 1399 (1977).
690. I. Degani, R. Fochi and C. Vincenzi, *Tetrahedron Lett.*, 1167 (1963).
691. S. V. Krivun, A. I. Buryak, S. N. Baranov, *Chem. Heterocyclic Compd.*, **9**, 1191 (1973).

692. D. M. McKinnon, *Can. J. Chem.*, **48**, 3388 (1970).
693. G. Suld and C. C. Price, *J. Amer. Chem. Soc.*, **84**, 2090 (1962).
694. C. C. Price and D. H. Follweiler, *J. Org. Chem.*, **34**, 3202 (1969).
695. A. Lüttringhaus and N. Engelhard, *Chem. Ber.*, **93**, 1525 (1960).
696. A. Lüttringhaus, N. Engelhard and A. Kolb, *Justus Liebigs Ann. Chem.*, **654**, 189 (1962).
697. A. Lüttringhaus and N. Engelhard, *Angew. Chem.*, **73**, 218 (1961).
698. S. A. Evans, H. Shine, E. T. Strom and A. L. Ternay, in *Organic Sulphur Chemistry*, (Ed. C. J. M. Stirling), Butterworths, London, 1975, p. 351.
699. K. K. Andersen, M. Cinquini and N. E. Papanikolaou, *J. Org. Chem.*, **35**, 706 (1970).
700. V. G. Kharchenko and A. A. Rassudova, *J. Org. Chem. USSR*, **9**, 2190 (1973).
701. G. A. Reynolds and J. A. Van Allan, *J. Heterocyclic Chem.*, **6**, 623 (1969).
702. (a) D. J. Sandman, A. J. Epstein, T. J. Holmes and A. P. Fisher, *J. Chem. Soc. Chem. Commun.*, 177 (1977); (b) N. K. Son, R. Pinel and Y. Mollier, *Bull. Soc. Chim. Fr.*, 1356 (1974); (c) J. L. Charlton, S. M. Loosmore and D. M. McKinnon, *Can. J. Chem.*, **52**, 3021 (1974); (d) S. Hünig and G. Ruider, *Justus Liebigs Ann. Chem.*, 1415 (1974); (e) K. Gewald, M. Buchwalder and M. Peukert, *J. Prakt. Chem.*, **315**, 679 (1973); (f) H. Behringer and A. Grimm, *Justus Liebigs Ann. Chem.*, **682**, 188 (1965); (g) H. J. Teague and W. P. Tucker, *J. Org. Chem.*, **32**, 3144 (1967); (h) J. Faust, G. Speier and R. Mayer, *J. Prakt. Chem.*, **311**, 61 (1969).
703. (a) R. C. Duty and R. E. Hallstein, *J. Org. Chem.*, **35**, 4226 (1970); (b) J. Faust, *Z. Chem.*, **8**, 171 (1968); (c) P. L. Pauson, G. R. Proctor and W. J. Rodger, *J. Chem. Soc.*, 3037 (1965); (d) B. Föhlisch and D. Krockenberger, *Chem. Ber.*, **101**, 3990 (1968).
704. (a) A. I. Tolmachev and M. A. Kudinova, *Chem. Heterocyclic Compd.*, **10**, 41 (1974); (b) A. G. Anderson, Jr. and W. F. Harrison, *J. Amer. Chem. Soc.*, **86**, 708 (1964).
705. R. Pettit, *Tetrahedron Lett.*, No. 23, 11 (1960).
706. V. Hanus and V. Cermak, *Coll. Czech. Chem. Commun.*, **24**, 1602 (1959).
707. S. Yoneda, T. Sugimoto, O. Tanaka, Y. Moriya and Z. Yoshida, *Tetrahedron*, **31**, 2669 (1975).
708. S. V. Krivun, A. I. Buryak, S. V. Sayapina, O. F. Voziyanova and S. N. Baranov, *Chem. Heterocyclic Compd.*, **9**, 926 (1973).
709. I. Degani, R. Fochi and C. Vincenzi, *Gazz. Chim. Ital.*, **97**, 397 (1967).
710. D. J. Harris, G. Y. P. Kan, V. Snieckus and E. Klingsberg, *Can. J. Chem.*, **52**, 2798 (1974).
711. Z. Yoshida, S. Yoneda, H. Sugimoto and T. Sugimoto, *Tetrahedron*, **27**, 6083 (1971).
712. I. Degani, R. Fochi and C. Vincenzi, *Boll. Sci. Fac. Chim. Ind. Bologna*, **23**, 241 (1965); *Chem. Abstr.*, **63**, 13197 (1965).
713. R. H. Nealey and J. S. Driscoll, *J. Heterocyclic Chem.*, **3**, 228 (1966).
714. S. K. Klimenko, M. N. Berezhnaya, T. V. Stolbova, I. Y. Evtushenko and V. G. Kharchenko, *J. Org. Chem. USSR*, **11**, 2207 (1975).
715. B. Tilak and G. T. Panse, *Indian J. Chem.*, **7**, 315 (1969).
716. (a) I. Degani, R. Fochi and G. Spunta, *Gazz. Chim. Ital.*, **97**, 388 (1967); (b) I. Degani, R. Fochi and G. Spunta, *Boll. Sci. Fac. Chim. Ind. Bologna*, **23**, 151 (1965); *Chem. Abstr.*, **63**, 13198 (1965).
717. F. Krollpfeiffer and A. Wissner, *Justus Liebigs Ann. Chem.*, **572**, 195 (1951).
718. J. P. Le Roux, J. C. Cherton and P. L. Desbene, *C. R. Acad. Sci., Ser. C*, **280**, 37 (1975).
719. S. V. Krivun, A. I. Buryak and S. N. Baranov, *Dopov. Akad. Nauk Ukr. RSR, Ser. B*, **34**, 931 (1972); *Chem. Abstr.*, **78**, 159364 (1973).
720. Z. Yoshida, H. Sugimoto, T. Sugimoto and S. Yoneda, *J. Org. Chem.*, **38**, 3990 (1973).
721. K. Akiba, K. Ishikawa and N. Inamoto, *Synthesis*, 862 (1977).
722. Y. A. Zhdanov, S. V. Krivun and V. A. Polenov, *Chem. Heterocyclic Compd.*, **5**, 279 (1969).
723. V. G. Kharchenko, A. A. Rassudova, T. I. Krupina, S. K. Klimenko and T. P. Chepurnenkova, *Chem. Heterocyclic Compd.*, **6**, 315 (1970).
724. U. Eisner and T. Krishnamurthy, *J. Org. Chem.*, **37**, 150 (1972).

518   Donald C. Dittmer and Bhalchandra H. Patwardhan

725. M. Hori, T. Kataoka, H. Shimizu, H. Hori and S. Sugai, *Chem. Pharm. Bull.*, **22**, 2754 (1974).
726. K. Dimroth, K. Wolf and H. Kroke, *Justus Liebigs Ann. Chem.*, **678**, 183 (1964).
727. (a) M. Hori, T. Kataoka, Y. Asahi and E. Mizuta, *Chem. Pharm. Bull.*, **21**, 1415 (1973); (b) M. Hori, T. Kataoka, K. Ohno and T. Toyoda, *Chem. Pharm. Bull.*, **21**, 1272, 1282 (1973); (c) M. Hori, T. Kataoka, H. Shimizu, S. Ohno and K. Narita, *Tetrahedron Lett.*, 251 (1978).
728. A. I. Tolmachev, M. A. Kudinova and L. M. Shulezhko, *Chem. Heterocyclic Compd.*, **13**, 142 (1977).
729. S. V. Krivun, *USSR Pat.*, 410,016 (1974); *Chem. Abstr.*, **80**, 120 907 (1974).
730. Y. Tamura, K. Sumoto and M. Ikeda, *Chem. Ind. (London)*, 498 (1972).
731. G. A. Reynolds and J. A. Van Allan, *J. Heterocyclic Chem.*, **8**, 301, 803 (1971).
732. H. W. Whitlock, *Tetrahedron Lett.*, 593 (1961).
733. A. I. Tolmachev and V. P. Sribnaya, *J. Gen. Chem. USSR*, **33**, 3802 (1963).
734. B. D. Tilak and G. T. Panse, *Indian J. Chem.*, **7**, 311 (1969).
735. S. V. Krivun, S. N. Baranov and A. I. Buryak, *Chem. Heterocyclic Compd.*, **7**, 1233 (1972).
736. D. Sullivan and R. Pettit, *Tetrahedron Lett.*, 401 (1963).
737. J. Daneke and H.-W. Wanzlick, *Justus Liebigs Ann. Chem.*, **740**, 52 (1970).
738. (a) G. Suld and C. C. Price, *J. Amer. Chem. Soc.*, **84**, 2094 (1962); (b) G. Suld and C. C. Price, *J. Amer. Chem. Soc.*, **83**, 1770 (1961).
739. (a) B. E. Maryanoff, J. Stackhouse, G. H. Senkler, Jr. and K. Mislow, *J. Amer. Chem. Soc.*, **97**, 2718 (1975); (b) G. H. Senkler, Jr., J. Stackhouse, B. E. Maryanoff and K. Mislow, *J. Amer. Chem. Soc.*, **96**, 5648 (1974); (c) J. Stackhouse, B. E. Maryanoff, G. H. Senkler, Jr. and K. Mislow, *J. Amer. Chem. Soc.*, **96**, 5650 (1974); (d) B. E. Maryanoff, G. H. Senkler, Jr., J. Stackhouse and K. Mislow, *J. Amer. Chem. Soc.*, **96**, 5651 (1974); (e) G. H. Senkler, Jr., B. E. Maryanoff, J. Stackhouse, J. D. Andose and K. Mislow, in *Organic Sulphur Chemistry*, (Ed. C. J. M. Stirling), Butterworths, London, 1975, p. 157.
740. (a) A. G. Hortmann and R. L. Harris, *J. Amer. Chem. Soc.*, **93**, 2471 (1971); (b) A. G. Hortmann and R. L. Harris, *J. Amer. Chem. Soc.*, **92**, 1803 (1970); (c) A. G. Hortmann, *J. Amer. Chem. Soc.*, **87**, 4972 (1965).
741. F. Ogura, W. D. Hounshell, C. A. Maryanoff, W. J. Richter and K. Mislow, *J. Amer. Chem. Soc.*, **98**, 3615 (1976).
742. M. Hori, T. Kataoka, H. Shimizu and H. Aoki, *Heterocycles*, **5**, 413 (1976).
743. M. Hori, T. Kataoka, H. Shimizu, K. Narita, S. Ohno and H. Aoki, *Chem. Lett.*, 1101 (1974).
744. H. Pirelahi, Y. Abdoh and M. Tavassoli, *J. Heterocyclic Chem.*, **14**, 199 (1977).
745. (a) C. C. Price, M. Hori, T. Parasaran and M. Polk, *J. Amer. Chem. Soc.*, **85**, 2278 (1963); (b) M. Polk, M. Siskin and C. C. Price, *J. Amer. Chem. Soc.*, **91**, 1206 (1969).
746. C. C. Price and H. Pirelahi, *J. Org. Chem.*, **37**, 1718 (1972).
747. (a) M. Hori, T. Kataoka, H. Shimizu and C. Hsu, *Chem. Lett.*, 391 (1973); (b) M. Hori, T. Kataoka, H. Shimizu, H. Hori and S. Sugai, *Chem. Pharm. Bull.*, **22**, 2754 (1974); (c) M. Hori, T. Kataoka, H. Shimizu and S. Ohno, *Tetrahedron Lett.*, 255 (1978); (d) M. Hori, T. Kataoka and H. Shimizu, *Chem. Lett.*, 1117 (1974).
748. M. Hori, T. Kataoka, Y. Asahi and E. Mizuta, *Chem. Pharm. Bull.*, **21**, 1692 (1973).
749. (a) Y. Tamura, H. Taniguchi, T. Miyamoto, M. Tsunekawa and M. Ikeda, *J. Org. Chem.*, **39**, 3519 (1974); (b) Y. Tamura, M. Tsunekawa, T. Miyamoto and M. Ikeda, *J. Org. Chem.*, **42**, 602 (1977).
750. Y. Kishida and J. Ide, *Chem. Pharm. Bull.*, **15**, 360 (1967).
751. T. Fujiwara, T. Hombo, K. Tomita, Y. Tamura and M. Ikeda, *J. Chem. Soc. Chem. Commun.*, 197 (1978).
752. S. Baklien, P. Groth and K. Undheim, *Acta Chem. Scand.*, **B30**, 24 (1976).
753. J. P. Sauve and N. Lozac'h, *Bull. Soc. Chim. Fr.*, 1196 (1974).
754. (a) R. Neidlein and I. Körber, *Arch. Pharm.*, **311**, 256 (1978); (b) R. Neidlein and I. Körber, *Arch. Pharm.*, **311**, 236 (1978).
755. J. A. Van Allan and G. A. Reynolds, *J. Heterocyclic Chem.*, **14**, 119 (1977).

756. V. G. Kharchenko, T. I. Krupina, S. K. Klimenko, N. M. Yartseva, M. N. Berezhnaya, V. I. Milovanova and N. I. Kozhevnikova, *Chem. Heterocyclic Compd.*, **10**, 56 (1974).
757. J. A. Van Allan and G. A. Reynolds, *Tetrahedron Lett.*, 2047 (1969).
758. N. F. Haley, *J. Heterocyclic Chem.*, **14**, 1245 (1977).
759. I. Degani and R. Fochi, *Ann. Chim. (Rome)*, **58**, 251 (1968).
760. F. Boccuzzi, I. Degani and R. Fochi, *Ann. Chim. (Rome)*, **62**, 528 (1972).
761. Z. Yoshida, T. Sugimoto and S. Yoneda, *J. Chem. Soc. Chem. Commun.*, 60 (1972).
762. S. Hünig, B. J. Garner, G. Ruider and W. Schenk, *Justus Liebigs Ann. Chem.*, 1036 (1973).
763. V. A. Izmail'skii, G. E. Ivanov and Y. A. Davidovskaya, *J. Gen. Chem. USSR*, **43**, 2488 (1973).
764. (a) T. Tamamura, M. Yokoyama, S. Kusabayashi and H. Mikawa, *Bull. Chem. Soc. Jap.*, **47**, 442 (1974); (b) T. Tamamura, H. Yasuba, K. Okamoto, T. Imai, S. Kusabayashi and H. Mikawa, *Bull. Chem. Soc. Jap.*, **47**, 448 (1974).
765. (a) Z. Yoshida, S. Yoneda and T. Sugimoto, *Chem. Lett.*, 17 (1972); (b) Z. Yoshida, T. Sugimoto and S. Yoneda, *Bull. Chem. Soc. Jap.*, **48**, 1519 (1975).
766. H. Yasuba, T. Imai, K. Okamoto, S. Kusabayashi and H. Mikawa, *Bull. Chem. Soc. Jap.*, **43**, 3101 (1970).
767. K. Kanai, T. Hashimoto, H. Kitano and K. Fukui, *Nippon Kagaku Zasshi*, **86**, 534 (1965); *Chem. Abstr.*, **63**, 6586 (1965).
768. H. D. Hoeltje, *Arch. Pharm.*, **311**, 311 (1978); *Chem. Abstr.*, **89**, 42136 (1978).
769. E. L. Eliel and R. L. Willer, *J. Amer. Chem. Soc.*, **99**, 1936 (1977).
770. G. Barbarella, P. Dembech, A. Garbesi and A. Fava, *Org. Magn. Reson.*, **8**, 469 (1976).
771. G. Barbarella, P. Dembech, A. Garbesi and A. Fava, *Tetrahedron*, **32**, 1045 (1976).
772. (a) E. L. Eliel, R. L. Willer, A. T. McPhail and K. D. Onan, *J. Amer. Chem. Soc.*, **96**, 3021 (1974); (b) G. Barbarella, P. Dembeck and A. Garbesi, *Tetrahedron Lett.*, 2109 (1980).
773. P. J. Halfpenny, P. J. Johnson, M. J. T. Robinson and M. G. Ward, *Tetrahedron*, **32**, 1873 (1976).
774. (a) J. B. Lambert, C. E. Mixan and D. H. Johnson, *Tetrahedron Lett.*, 4335 (1972); (b) J. B. Lambert, R. G. Keske and D. K. Weary, *J. Amer. Chem. Soc.*, **89**, 5921 (1967).
775. D. M. Roush and C. H. Heathcock, *J. Amer. Chem. Soc.*, **99**, 2337 (1977).
776. J. P. Schaefer and L. L. Reed, *J. Amer. Chem. Soc.*, **94**, 908 (1972).
777. N. P. Volynskii, *J. Gen. Chem. USSR*, **35**, 168 (1965).
778. D. K. Herron, *Tetrahedron Lett.*, 2145 (1975).
779. Y. Yano, M. Ishihara, W. Tagaki and S. Oae, *Int. J. Sulfur. Chem. A*, **2**, 169 (1972).
780. W. F. Cockburn and A. F. McKay, *J. Amer. Chem. Soc.*, **77**, 397 (1955).
781. F. Miyoshi, K. Tokuno, Y. Arata, S. Hiroki and T. Ohashi, *Yakugaku Zasshi*, **96**, 1440 (1976); *Chem. Abstr.*, **86**, 189680 (1977).
782. D. A. Tomalia and J. N. Paige, *J. Org. Chem.*, **38**, 3949 (1973).
783. N. N. Novitskaya, R. V. Kunakova, G. K. Samirkhanova and G. A. Tolstikov, *J. Org. Chem. USSR*, **9**, 650 (1973).
784. C. R. Johnson and J. J. Rigau, *J. Amer. Chem. Soc.*, **91**, 5398 (1969).
785. D. Swern, I. Ikeda and G. F. Whitfield, *Tetrahedron Lett.*, 2635 (1972).
786. J. B. Lambert, D. H. Johnson, R. G. Keske and C. E. Mixan, *J. Amer. Chem. Soc.*, **94**, 8172 (1972).
787. C. R. Johnson, J. J. Rigau, M. Haake, D. McCants, J. E. Keiser and A. Gertsema, *Tetrahedron Lett.*, 3719 (1968).
788. N. Furukawa, F. Takahashi, T. Akasaka and S. Oae, *Chem. Lett.*, 143 (1977).
789. D. C. Appleton, D. C. Bull, J. McKenna, J. M. McKenna and A. R. Walley, *J. Chem. Soc. Chem. Commun.*, 140 (1974).
790. R. Pellicciari, M. Curini and P. Ceccherelli, *J. Chem. Soc. Perkin Trans. I*, 1155 (1977).
791. M. Yoshimoto, S. Ishihara, E. Nakayama and N. Soma, *Tetrahedron Lett.*, 2923 (1972).
792. W. E. Parham and R. Koncos, *J. Amer. Chem. Soc.*, **83**, 4034 (1961).
793. W. J. Middleton, E. L. Buhle, J. G. McNally, Jr. and M. Zanger, *J. Org. Chem.*, **30**, 2384 (1965).
794. W. Schroth, G. Dill, N. T. K. Dung, N. T. M. Khoi, P. T. Binh, H. J. Waskiewicz and A. Hildebrandt, *Z. Chem.*, **14**, 52 (1974).

795. (a) N. J. Leonard and C. R. Johnson, *J. Amer. Chem. Soc.*, **84**, 3701 (1962); (b) N. J. Leonard and W. L. Rippie, *J. Org. Chem.*, **28**, 1957 (1963).
796. R. H. Eastman and G. Kritchevsky, *J. Org. Chem.*, **24**, 1428 (1959).
797. E. L. Eliel, *Angew. Chem. Int. Ed. Engl.*, **11**, 739 (1972).
798. H. Dorn, *Angew. Chem. Int. Ed. Engl.*, **6**, 371 (1967).
799. W. v. E. Doering and A. K. Hoffmann, *J. Amer. Chem. Soc.*, **77**, 521 (1955).
800. (a) K. W. Ratts, *J. Org. Chem.*, **37**, 848 (1972); (b) D. M. Roush, E. M. Price, L. K. Templeton, D. H. Templeton and C. H. Heathcock, *J. Amer. Chem. Soc.*, **101**, 2971 (1979); A. Garbesi, *Tetrahedron Lett.*, 547 (1980).
801. E. Vedejs, M. J. Arco and J. M. Renga, *Tetrahedron Lett.*, 523 (1978).
802. H. Kobayashi, N. Furukawa, T. Aida, K. Tsujihara and S. Oae, *Tetrahedron Lett.*, 3109 (1971).
803. (a) M. M. Campbell, G. Johnson, A. F. Cameron and I. R. Cameron, *J. Chem. Soc. Chem. Commun.*, 868, 974 (1974); (b) M. M. Campbell, G. Johnson, A. F. Cameron and I. R. Cameron, *J. Chem. Soc. Perkin Trans. I*, 1208 (1975); (c) M. M. Campbell and G. Johnson, *J. Chem. Soc. Perkin Trans. I*, 1212 (1975); (d) Y. Tamura, S. M. Bayomi, C. Mukai, M. Ikeda, M. Murose and M. Kise, *Tetrahedron Lett.*, 533 (1980).
804. C. Kissel, R. J. Holland and M. C. Caserio, *J. Org. Chem.*, **37**, 2720 (1972).
805. C. R. Johnson and D. McCants, *J. Amer. Chem. Soc.*, **87**, 5404 (1965).
806. C. R. Johnson, C. C. Bacon and J. J. Rigau, *J. Org. Chem.*, **37**, 919 (1972).
807. A. L. Maycock and G. A. Berchtold, *J. Org. Chem.*, **35**, 2532 (1970).
808. Y. Maki and T. Hiramitsu, *Chem. Pharm. Bull.*, **25**, 292 (1977).
809. K. T. Potts, R. Ehlinger and W. M. Nichols, *J. Org. Chem.*, **40**, 2596 (1975).
810. S. Wolfe, P. Chamberlain and T. F. Garrard, *Can. J. Chem.*, **54**, 2847 (1976).
811. I. Stahl and J. Gosselck, *Tetrahedron*, **30**, 3519 (1974).
812. J. B. Lambert, E. Vulgaris, S. I. Featherman and M. Majchrzak, *J. Amer. Chem. Soc.*, **100**, 3269 (1978).
813. S. Kato, H. Ishihara, M. Mizuta and Y. Hirabayashi, *Bull. Chem. Soc. Jap.*, **44**, 2469 (1971).
814. E. Hunt and B. Lythgoe, *J. Chem. Soc. Chem. Commun.*, 757 (1972).
815. I. Stahl and J. Gosselck, *Tetrahedron Lett.*, 989 (1972).
816. T. E. Young and R. A. Lazarus, *J. Org. Chem.*, **33**, 3770 (1968).
817. I. Stahl and J. Gosselck, *Tetrahedron*, **29**, 2323 (1973).
818. I. Stahl and J. Gosselck, *Tetrahedron*, **30**, 3519 (1974).
819. W. Schroth, M. Hassfeld and A. Zschunke, *Z. Chem.*, **10**, 296 (1970).
820. T. Oishi, H. Takechi and Y. Ban, *Tetrahedron Lett.*, 3757 (1974).
821. W. E. Parham and P. L. Stright, *J. Amer. Chem. Soc.*, **78**, 4783 (1956).
822. (a) N. E. Hester and G. K. Helmkamp, *J. Org. Chem.*, **38**, 461 (1973); (b) N. E. Hester, G. K. Helmkamp and G. I. Alford, *Int. J. Sulfur Chem. A*, **1**, 65 (1971); *Chem. Abstr.*, **75**, 110301 (1971).
823. F. Feher and K. Vogelbruch, *Chem. Ber.*, **91**, 996 (1958).
824. E. J. Corey and S. W. Walinsky, *J. Amer. Chem. Soc.*, **94**, 8932 (1972).
825. D. L. Coffen and M. L. Lee, *J. Org. Chem.*, **35**, 2077 (1970).
826. (a) J. A. Marshall and J. L. Belletire, *Tetrahedron Lett.*, 871 (1971); (b) F. A. Carey and J. R. Neergaard, *J. Org. Chem.*, **36**, 2731 (1971).
827. B. M. Trost and K. Hiroi, *J. Amer. Chem. Soc.*, **98**, 4313 (1976).
828. R. S. Glass, W. J. Britt, W. N. Miller and G. S. Wilson, *J. Amer. Chem. Soc.*, **95**, 2375 (1973).
829. W. Schroth, R. Borsdorf, R. Herzschuh and J. Seidler, *Z. Chem.*, **10**, 147 (1979).
830. T. Weiss and G. Klar, *Justus Liebigs Ann. Chem.*, 785 (1978).
831. K. D. Gundermann, W. Honig, M. Berrada, H. Giesecke and H. G. Paul, *Justus Liebigs Ann. Chem.*, 809 (1974).
832. T. E. Young and A. R. Oyler, *J. Org. Chem.*, **41**, 2753 (1976).
833. E. Deutsch, *J. Org. Chem.*, **37**, 3481 (1972).
834. P. G. Gassman and D. R. Amick, *Tetrahedron Lett.*, 3463 (1974).
835. C. P. Lillya, E. F. Miller and R. A. Sahatjian, *Int. J. Sulfur Chem. A*, **1**, 79 (1972); *Chem. Abstr.*, **75**, 151307 (1971).

836. S. R. Wilson and R. S. Myers, *Tetrahedron Lett.*, 3413 (1976).
837. N. H. Andersen, Y. Yamamoto and A. D. Denniston, *Tetrahedron Lett.*, 4547 (1975).
838. L. Field and D. L. Tuleen, in *The Chemistry of Heterocyclic Compounds*, (Ed. A. Rosowsky), Vol. 26, Wiley–Interscience, New York, 1972, p. 573.
839. V. J. Traynelis, in *The Chemistry of Heterocyclic Compounds*, (Ed. A. Rosowsky), Vol. 26, Wiley–Interscience, New York, 1972, p. 667.
840. F. Vögtle and P. Neumann, *Synthesis*, 85 (1973).
841. W. E. Truce and D. D. Emrick, *J. Amer. Chem. Soc.*, **78**, 6130 (1956).
842. J. v. Braun, *Chem. Ber.*, **43**, 3220 (1910).
843. N. J. Leonard and J. Figueras, Jr., *J. Amer. Chem. Soc.*, **74**, 917 (1952).
844. V. J. Traynelis and J. C. Sih, cited in reference 839, p. 685.
845. R. R. Fraser and F. J. Schuber, *Can. J. Chem.*, **48**, 633 (1970).
846. (a) R. B. DuVernet, O. Wennerstrom, J. Lawson, T. Otsubo and V. Boekelheide, *J. Amer. Chem. Soc.*, **100**, 2457 (1978); (b) V. Boekelheide, P. H. Anderson and T. A. Hylton, *J. Amer. Chem. Soc.*, **96**, 1558 (1974); (c) R. H. Mitchell and V. Boekelheide, *J. Amer. Chem. Soc.*, **96**, 1547 (1974); (d) M. Haenel and H. A. Staab, *Chem. Ber.*, **106**, 2203 (1973); (e) V. Boekelheide and C. H. Tsai, *J. Org. Chem.*, **38**, 3931 (1973); (f) V. Boekelheide and R. A. Hollins, *J. Amer. Chem. Soc.*, **95**, 3201 (1973); (g) N. Kannen, T. Umemoto, T. Otsubo and S. Misumi, *Tetrahedron Lett.*, 4537 (1973); (h) J. R. Davy and J. A. Reiss, *Tetrahedron Lett.*, 3639 (1972); (i) V. Boekelheide and R. A. Hollins, *J. Amer. Chem. Soc.*, **92**, 3512 (1970); (j) R. H. Mitchell and V. Boekelheide, *J. Amer. Chem. Soc.*, **92**, 3510 (1970); (k) R. H. Mitchell and V. Boekelheide, *Tetrahedron Lett.*, 1197 (1970); (l) V. Boekelheide and P. H. Anderson, *Tetrahedron Lett.*, 1207 (1970).
847. (a) V. Boekelheide, K. Galuszko and K. S. Szeto, *J. Amer. Chem. Soc.*, **96**, 1578 (1974); (b) J. Lawson, R. B. DuVernet and V. Boekelheide, *J. Amer. Chem. Soc.*, **95**, 956 (1973); (c) V. Boekelheide and J. A. Lawson, *J. Chem. Soc. Chem. Commun.*, 1558 (1970).
848. (a) T. Otsubo, R. Gray and V. Boekelheide, *J. Amer. Chem. Soc.*, **100**, 2449 (1978); (b) R. Danieli, A. Ricci and J. H. Ridd, *J. Chem. Soc., Perkin Trans. II*, 290 (1976).
849. T. Otsubo and V. Boekelheide, *Tetrahedron Lett.*, 3881 (1975).
850. K. Mislow, M. A. W. Glass, H. B. Hopps, E. Simon and G. H. Wahl, Jr., *J. Amer. Chem. Soc.*, **86**, 1710 (1964).
851. G. M. Bennett and E. G. Turner, *J. Chem. Soc.*, 813 (1938).
852. F. Bickelhaupt, K. Stach and M. Thiel, *Chem. Ber.*, **98**, 685 (1965).
853. V. J. Traynelis and J. C. Sih, cited in reference 839, p. 684.
854. (a) V. J. Traynelis, J. C. Sih, Y. Yoshikawa, R. F. Love and D. M. Borgnaes, *J. Org. Chem.*, **38**, 2623 (1973); (b) V. J. Traynelis, J. C. Sih and D. M. Borgnaes, *J. Org. Chem.*, **38**, 2629 (1973); (c) V. J. Traynelis and J. C. Sih, cited in reference 839, p. 696.
855. A. Ricci, R. Danieli, A. Boicelli and J. H. Ridd, *J. Heterocyclic Chem.*, **14**, 257 (1977).
856. V. J. Traynelis and J. C. Sih, cited in reference 839, p. 685.
857. C. R. Johnson and M. P. Jones, *J. Org. Chem.*, **32**, 2014 (1967).
858. T. Akasaka, T. Yoshimura, N. Furukawa and S. Oae, *Phosphorus Sulphur*, **4**, 211 (1978).
859. R. Neidlein and B. Stackebrandt, *Justus Liebigs Ann. Chem.*, 914 (1977).
860. A. Muller, E. Funder-Fritzsche, W. Konar and E. Rintersbacher-Wlasak, *Monatsh. Chem.*, **84**, 1206 (1953).
861. H. Hofmann and A. Molnar, *Abstracts of Papers, Eighth International Symposium on Organic Sulphur Chemistry*, Portoroz, Yugoslavia, June 18–23, 1978, p. 113.

The Chemistry of the Sulphonium Group
Edited by C. J. M. Stirling and S. Patai
© 1981 John Wiley & Sons Ltd

CHAPTER **14**

Organosulphur cation radicals

HENRY J. SHINE

Department of Chemistry, Texas Tech University, Lubbock, Texas 79409, USA

I. HISTORICAL DEVELOPMENTS

A. Phenothiazine, Thianthrene, and Phenoxathiin

1. Phenothiazine cation radicals

Following the discovery of electron-spin resonance (e.s.r.)[1], much interest developed in characterizing ion radicals by their e.s.r. spectra. Among the first cation radicals to be thus characterized were some which contained sulphur. Most of the early e.s.r. characterizations were made, furthermore, with solutions of the organosulphur compound in concentrated sulphuric acid, since it soon became evident that oxidation of electron-rich molecules in this acid often gave paramagnetic products. Lagercrantz showed in 1961 that phenothiazine in sulphuric acid gave a four-line e.s.r. spectrum, while in D_2SO_4 the spectrum was a triplet[2]. The four-line spectrum was deduced to arise from equivalent splitting by the nitrogen and its attached hydrogen atom in the phenothiazine cation radical **1**, the splitting due to hydrogen disappearing when hydrogen was replaced with deuterium in D_2SO_4. Analogously, 10-methylphenothiazine in both H_2SO_4 and D_2SO_4 gave a six-line e.s.r. spectrum which was attributed to equal coupling of the electron spin with the nitrogen atom and the three protons of the methyl group **2**. E.s.r. spectra obtained from solutions

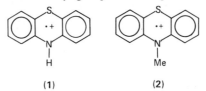

(1)   (2)

of 10-ethylphenothiazine and a number of phenothiazine-based drugs (promethezine·HCl, and chlorpromazine·HCl) in sulphuric acid were attributed to phenothiazine-type cation radicals, also.

The four-line e.s.r. spectrum of **1** was obtained also from the anodic oxidation of phenothiazine in acetonitrile solution[3], and this e.s.r. evidence confirmed electrochemical evidence for one-electron oxidation obtained with controlled-potential oxidation[4].

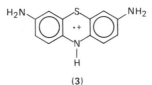

(3)

Improvements in the resolution of these early four-line spectra were not long in being made. Concentrated solutions of the leuco (i.e. reduced) form of a number of thiazine dyes in sulphuric acid also gave four-line e.s.r. spectra with nitrogen splitting (a^N) of about 0.66 mT (ref. 5) or 0.7 mT (ref. 6). However, dilute solutions of some leuco forms in sulphuric acid gave many hyperfine lines in the major quartet. In these cases, coupling with the ring protons was thus detected for the first time, e.g. with **3**[5].

Hyperfine splitting by the ring protons of **1** itself was first obtained from sulphuric acid solutions of phenothiazine in the same year[7]. Many-line e.s.r. spectra were reported later for both **1** and **2** obtained by anodic oxidation of the parent compounds in acetonitrile[8], and attempts were made to fit theoretical spectra to the experimental spectra with computed values of the hyperfine splitting constants for all ring protons[9]. These early e.s.r. findings established the nature of phenothiazine cation radicals clearly. It was not until 1966, however, that acceptable spectra and explanations of all coupling constants were obtained for **1** (see Table 1)[10,11].

The interest shown in cation radicals of the phenothiazine type had arisen from the presence of the phenothiazine nucleus in many well known dyestuffs and pharmacologically important drugs. As early as 1958 e.s.r. had been used to show that a radical product could be isolated from patients who had ingested chloropromazine, and that the same product could be obtained from the ultraviolet irradiation of chloropromazine in dilute aqueous solution[12]. Oxidation of chloropromazine and some other drugs (promazine, trifluoropromazine, and trifluoroperazine) by iron(III) or cerium(IV) salts in hydrochloric acid, by concentrated sulphuric acid, and anodically at a controlled potential in aqueous solution gave e.s.r. signals attributed to the cation radicals. Simultaneous e.s.r. and absorption spectrosocopy in a flow system was also carried out[13]. At the same time, Borg and Cotzias demonstrated with oxidimetric titrations and absorption and e.s.r. spectroscopy that chloropromazine and a number of its congeners were oxidized to the cation radicals by metal ions (e.g. Co^{3+}, Fe^{3+}, Mn^{3+}) in acid solution[14]. In the same year also, Billon showed that like phenothiazine itself and its 10-methyl and 2,7-dimethyl derivatives, numerous phenothiazine drugs could be oxidized anodically in acidic acetonitrile solution, and their comparative one-electron oxidation potentials could be measured[15]. A partial analysis of the e.s.r. spectra of promazine and chloropromazine was given in 1969[16].

It is noteworthy that the characterization of the cation radicals of phenothiazine and its derivatives by e.s.r. spectroscopy only confirmed proposals for their formation which had been made earlier, before the days of e.s.r. That is, Michaelis and coworkers carried out the potentiometric reductive titration of Lauth's violet (thionine) and methylene blue in strongly acidic solution with titanium(III) chloride. They were able to show clearly that a one-electron reduction occurred to give what was then called a semiquinone and, further, that the semiquinone **5** existed in equilibrium with dye **4** and the fully reduced leuco form **6** (Scheme 1)[17]. Analogous reductive titrations of other phenothiazine dicationic derivatives (3-amino, 3-hydroxy-, 3-amino-7-hydroxy-, and 3,7-diphenylamino-) were also carried out[18]. These reductive titrations from the dication side led eventually to oxidative titrations from the leuco side. For example, phenothiazine itself and N-methyl- and 3,7-dimethylphenothiazine were oxidized potentiometrically with bromine in aqueous acetic acid, while pyridinium derivatives were oxidized by potassium dichromate in 18.9 N sulphuric acid[19]. Here it was shown that the semiquinone (i.e. the cation radical) corresponded with the semiquinone postulated even earlier by Kehrmann et al.[20]. As early as 1902 Kehrmann had reported on the oxidation of phenothiazine with iron(III) chloride and had isolated a so-called crystalline iron chloride double salt[21]. However, it is remarkable that Kehrmann et al. deduced in 1914 that all direct oxidations of phenothiazine with, for example, bromine, iron(III) chloride, and cold concentrated sulphuric acid had been until then misunderstood in that they gave an incompletely oxidized product (called part- or semiquinoid) rather than what had been described as completely oxidized (called holoquinoid)[20]. Kehrmann et al. recognized that phenothiazine and its derivatives formed salts of two oxidation states, and that the lower (partial or semiquinoid) could be further oxidized to the higher oxidation state (the holoquinoid). These, of course, we now call the cation radical and dication.

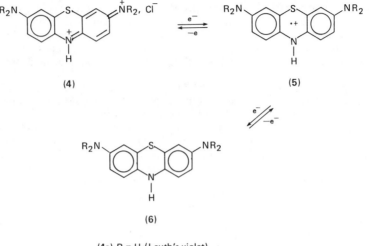

(4)   (5)

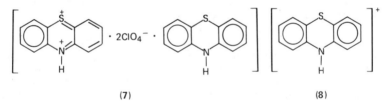

(6)

(4a) R = H (Lauth's violet)

(4b) R = Me (methylene blue)

SCHEME 1

Analogous deductions were made at the same time by Pummerer and co-workers[22,23]. This group began with phenothiazine 5-oxide and 3-phenothiazone, however, about which more is said later. Pummerer and coworkers also recognized that two classes of salts could be formed, one being called meri- and the other haloquinoid, but it is apparent that in formulating the meriquinoid as a double salt (e.g. 7) Pummerer *et al.*[23], in contrast to Kehrmann *et al.* (who designated the semiquinone correctly as 8), did not completely understand its nature.

(7)   (8)

2. Thianthrene cation radicals

The first class of organosulphur cation radicals to be identified and characterized directly by e.s.r. spectroscopy, the phenothiazine cation radicals, was thus known for a long time, but not in structural terms. A similar view can be held about the thianthrene cation radicals, whose history goes back even further.

The thianthrene cation radical **9a**, (R = R^1 = H) was first correctly identified by e.s.r. spectroscopy in the period 1961–62. Thianthrene dissolves in concentrated sulphuric acid, giving a purple solution; sulphur dioxide is evolved in the process. This reaction was first observed by Stenhouse in 1868 when he described the magnificent purple colour of the solution[24]. However, it was not until over 90 years later that, independently, the work in four laboratories explained correctly, and in

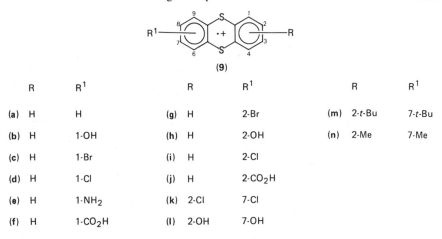

(9)

	R	R¹		R	R¹		R	R¹
(a)	H	H	(g)	H	2-Br	(m)	2-t-Bu	7-t-Bu
(b)	H	1-OH	(h)	H	2-OH	(n)	2-Me	7-Me
(c)	H	1-Br	(i)	H	2-Cl			
(d)	H	1-Cl	(j)	H	2-CO_2H			
(e)	H	1-NH_2	(k)	2-Cl	7-Cl			
(f)	H	1-CO_2H	(l)	2-OH	7-OH			

modern terms what was causing the purple colour. In each case a five-line e.s.r. spectrum was obtained from a purple solution of thianthrene in sulphuric acid. It was deduced that the e.s.r. spectrum arose from coupling of the electron spin with one of the two sets of four-equivalent protons in **9a**[25–29]. One-electron oxidation of thianthrene (as with phenothiazine, above) had occurred (equation 1).

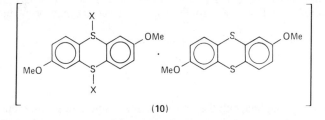

$$2 \quad + 3H_2SO_4 \longrightarrow 2 \quad + 2HSO_4^- + SO_2 + 2H_2O \quad (1)$$

The five-line e.s.r. spectrum of these purple solutions had been recorded much earlier[30,31], but curiously, in spite of the work of Michaelis with the analogues phenothiazine and phenoxazine, the correct structure of **9a** was not deduced. This is all the more curious since Fries and Engelbertz concluded in 1915 that thianthrene and 2,7-dimethoxythianthrene gave semiquinoid salts when oxidized by sulphuric acid, iron(III) chloride, and bromine[32]. These salts were represented as in **10**, which

$$
\left[
\begin{array}{c}
\text{(structure with X, OMe, MeO, S, S)} \quad \cdot \quad \text{(structure with S, OMe, MeO, S)}
\end{array}
\right]
$$

(10)

(a) X = OSO_3H;   (b) X = Cl:$FeCl_3$

of course is correct for 2 mol of cation-radical bisulphate or tetrachloroferrate. Fries and Engelbertz showed, furthermore, that the salts could be reduced by potassium iodide, tin(II) chloride, and hydroquinone to the parent heterocycles. Reduction of **10a** with potassium iodide gave, for example, 85% of the theoretical iodine. The low result was attributed correctly to competing hydrolysis of **10a** to 2,7-dimethoxythianthrene and its 5-oxide, for it was shown that reaction of **10a** with water gave equal amounts of these two compounds. Analogous results were obtained with the thianthrene salt. This reaction, in modern terms the reaction of the thianthrene cation radical with water, has had a major role in settling the mechanisms of reaction of organosulphur cation radicals with nucleophiles. Curiously, though, it remained unnoticed until re-discovered by later workers[25,28]. Most interestingly, Wizinger proposed in 1929 the correct structure of the thianthrene cation radical but this report also seems to have been overlooked by almost all later workers[33]. (Wizinger proposed the correct structures for the phenoxathiin cation radical, also, and its selenium and tellurium analogues, but he was slightly incorrect with phenothiazine and phenoxazine dyestuffs.)

The first e.s.r. spectra could not distinguish which set of four-equivalent protons in **9a** was responsible for the five lines. This was settled with the e.s.r. spectra of appropriately substituted thianthrene cation radicals. A series of 1-substituted thianthrenes gave, in concentrated sulphuric acid, a five-line e.s.r. spectrum **9b–f** like that of thianthrene itself, while a series of 2-substituted thianthrenes analogously gave distorted three- or four-line spectra **9g–j**[34]. These spectra indicated that coupling with the set of protons in the 2-, 3-, 7-, and 8-positions was being detected. This was clearly shown with the three-line spectra of some 2,7-derivatives **9k–m** and the nine-line spectrum of 2,7-dimethylthianthrene cation radical **9n**[25,34,35], in which equal coupling by the six methyl and two ring protons occurs. By resorting to low temperatures and a different medium (nitromethane–aluminium chloride) complete resolution of the e.s.r. spectrum of **9a** was achieved later[36]. The total spectrum showed not only the expected 25 proton lines but also, for the first time in a cation radical, coupling with naturally abundant ^{33}S (Table 2). This was the first complete resolution of the e.s.r. spectrum of an organosulphur cation radical.

Entwined in the discovery of the thianthrene cation radical are the behaviours of thiophenol and diphenyl disulphide in concentrated sulphuric acid. Each of these compounds dissolves in sulphuric acid to give small amounts of the thianthrene cation radical.

E.s.r. spectra of solutions of thiophenol and diphenyl disulphide in sulphuric acid were obtained in early years[30,37], but, again, not clearly understood. Most interestingly, when thiophenol dissolved in concentrated sulphuric acid, two e.s.r. spectra were observed[30], one of five lines (due to the thianthrene cation radical) and another of three lines and higher g-value. The nature of the second radical has never been discovered. Needler[25] has proposed that this radical may be of the type **11a**

(11)

(R = H) in which coupling with the protons in the 2- and 3-positions would lead to a three-line e.s.r. spectrum. Analogously, *p*-toluenethiol gave a five-line e.s.r. spectrum attributed to **11b** (R = Me). As for the high g-value (of the order 2.016 compared with 2.008 for thianthrene-type cation radicals), this was attributed to considerable localization of the unpaired electron on sulphur atoms in the cation radicals of type **11**.

3. Phenoxathiin cation radicals

Phenoxathiin dissolves in concentrated sulphuric acid giving, as with its analogues, a blue–purple solution and sulphur dioxide[38]. The phenoxathiin cation radical **12** is formed, the correct structure for which was proposed, as noted above, by Wizinger[33]. Drew, noting in 1928 that decomposition of the coloured solution with ice gave equal amounts of phenoxathiin and its 5-oxide, understood the oxidation state involved and deduced that one-electron transfer from phenoxathiin to sulphuric acid had occurred. The structure of the product was represented incorrectly, though, as the covalent dimer di-bisulphate **13**[39].

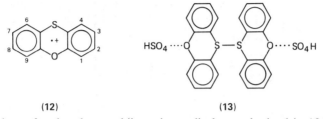

(12)   (13)

E.s.r. evidence for the phenoxathiin cation radical was obtained in 1962–63[40,41], while the e.s.r. of some substituted phenoxathiin cation radicals (e.g. 3,7-dimethyl-) gave some indication of the relative extents of coupling with ring protons[42]. Complete analyses of the e.s.r. spectrum were provided later[43,44], including one with a ^{33}S hfs of 1.19 mT (Table 3)[45].

4. General comments on e.s.r. data

These three sulphur heterocycles have been very important in the development of knowledge of organosulphur cation radicals. E.s.r. characteristics for the parent cation radicals and some of their derivatives are collected in Tables 1–3. It is interesting to note the effect of the nitrogen atom in the phenothiazines on the coupling constants, and hence spin density, at the ring positions. For convenience these data are given again on the structures **1**, **9a**, and **12** and compared with data for analogous cation radicals of dibenzodioxin **14**[43], phenoxazine **15**[11], and phenoselenazine **16**[47]. In the non-nitrogen-containing heterocycles, the largest

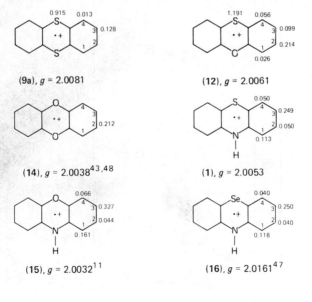

(9a), g = 2.0081   (12), g = 2.0061

(14), g = 2.0038[43,48]   (1), g = 2.0053

(15), g = 2.0032[11]   (16), g = 2.0161[47]

TABLE 1. E.s.r. data for some phenothiazine cation radicals (a in mT)

R	Medium	a^S	a^N	a_R^H	$a_{1,9}^H$	$a_{2,8}^H$	$a_{3,7}^H$	$a_{4,6}^H$	g	Reference
H	a	—	0.652	0.736	0.123	0.046	0.258	0.046	—	10
H	b	—	0.652	0.732	0.129	0.046	0.258	0.040	2.0050	11
H	b	—	0.654	0.696	0.122[k]	0.082	0.245	0.040	2.0051	64
H	c	—	0.650	0.750	0.123	0.046	0.259	0.046	2.0053	47
H	d	0.73	0.634	0.729	0.113	0.050	0.249	0.050	2.0053	47
H	f	—	0.641	0.741	0.114	0.049	0.250	0.049	—	46
Me	a	—	0.760	0.720	—	—	—	—	—	11
Me	b	—	0.734	0.734	0.105[k]	0.070	0.210	0.035	2.0053	64
Me	f	—	0.750	0.723	0.100	0.076	0.220	0.029	—	46
Et	b	—	0.769	0.384	0.104[k]	(0.069)	0.209	(0.035)	2.0053	64
C_6H_5	e	—	0.695	—	0.090	0.090	0.215	0.022	—	65
$(CH_2)_3NMe_2$	i	—	0.695	0.358	0.105	(0.050)	0.200	(0.050)	—	16
$(CH_2)_3NMe_2$	i	—	0.700	0.350	0.100	(0.050)[j]	0.190	(0.050)	—	16

[a] MeCN.
[b] Conc. H_2SO_4.
[c] MeCN–H_2SO_4.
[d] MeNO$_2$–AlCl$_3$.
[e] MeNO$_2$–AlCl$_3$ or Tl(OAc)$_3$ in MeNO$_2$–AcOH.
[f] MeNO$_2$–H_2SO_4, –40°C.
[g] MeNO$_2$–AlCl$_3$, low temperature.
[h] Conc. H_2SO_4 or MeNO$_2$–SbCl$_5$.
[i] Benzene, reaction with Bz_2O_2.
[j] 2-Cl (chlorpromazine).
[k] These assignments were not made by the original author, but have been made in the table to fit other assignments.

TABLE 2. E.s.r. data for some thianthrene cation radicals

Medium*	R	R^1	a^S	a_1^H	$a_2^{H(R)}$	a_3^H	a_4^H	g	Reference
b	H	H	—	0.012	0.130	0.130	0.012	2.0081	43
b	H	H	—	—	0.126^k	0.126	—	2.0078	64
g	H	H	0.915	0.013_5	0.128	0.128	0.013_5	—	36
b	Me	H	—	—	0.140	0.140	—	—	35
b	Me	H	—	—	0.165	0.165	—	—	34
g	Me	H	0.873	—	0.164	0.164	—	—	45
b	Cl	H	—	—	—	0.130	—	—	35
b	Cl	H	—	—	—	0.149	—	—	34
g	Cl	H	0.892	—	—	0.145	—	—	45
g	Br	H	0.894	—	—	0.139	—	—	45
b	OH	H	—	—	—	0.190	—	—	34
g	OH	H	0.810	0.00	0.036	0.205	0.011	—	45
b	t-Bu	H	—	—	—	0.160	—	—	34
g	t-Bu	H	0.872	—	—	0.152	—	—	45
g	MeO	MeO	0.792	—	—	—	—	—	45

*See footnotes to Table 1.

TABLE 3. E.s.r. data for some phenoxathiin cation radicals

Medium*	R	R^1	a^S	a_1^H	a_2^H	a_3^H	a_4^H	g	Reference
a	H	H	—	0.026	0.211	0.097	0.054	—	44
b	H	H	—	0.030	0.240	0.120	0.060	—	40
b	H	H	—·	0.029^k	0.260	0.120	0.058	2.0066	64
b	H	H	—	0.026	0.208	0.104	0.052	2.0061	43
g	H	H	1.19	0.026	0.214	0.099	0.056	—	45
h	H	H	—	—	0.208	0.104	—	2.0061	66
h	2-F	8-F	—	—	0.207	0.415^l	—	2.0064	66

*See footnotes to Table 1.
$^l a^F$.

coupling (a^H) constants are in the 2- and 3-positions. Introduction of nitrogen into the ring causes a marked increase in coupling constants (and thus spin densities) at the 1- and 3-positions, these now becoming the dominant positions *ortho* and *para* to nitrogen, while the 2-position (*meta* to nitrogen) undergoes a large decrease in spin density.

Another feature that stands out in the e.s.r. data is the size of the g-values for the cation radicals containing sulphur (and selenium). These relatively large g-factors reflect the extent of spin density at these heteroatoms and also the relatively large spin–orbit coupling parameters (151, 382, and 1688 cm^{-1}, respectively, for O, S, and Se).

Detection of hyperfine splitting by naturally occurring ^{33}S (0.75% abundance) is also shown for some of the cation radicals in Tables 1–3. The coupling constants (a^S) are relatively large, which again reflects the relatively large spin density at sulphur. Since ^{33}S has a nuclear spin of 3/2, the e.s.r. spectrum due to one ^{33}S per molecule should be a quartet. This is not seen in its entirety since the two inner lines are usually covered by the proton e.s.r. lines. The end lines of the quartet are seen and usually split to show the major proton coupling of the radical. Thus, for **9a** the ^{33}S lines are quintets[36], and for **9m** they are triplets[45].

B. Formation of Organosulphur Cation Radicals from Sulphoxides

One of the curious reactions which pervades the literature of these compounds and is entangled with cation radical chemistry is the formation of organosulphur cation radicals from sulphoxides in acid solution. For example, if a solution of thianthrene 5-oxide **17** in concentrated sulphuric acid is allowed to stand, the cation radical **9a** is formed (equation 2)[28]. The same reaction occurs with phenoxathiin 5-oxide **18**[49]

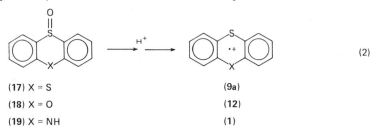

$$\tag{2}$$

(17) X = S	(9a)
(18) X = O	(12)
(19) X = NH	(1)

and some diaryl sulphoxides[50,51] in concentrated sulphuric acid, and with phenothiazine 5-oxide **19** in a variety of less concentrated acids[52]. These conversions have been expressed for many years as in equation 3[42,53,54]. Equation 3

$$\tag{3}$$

calls, in effect for the reduction of a sulphoxide, from 0 to −1 state, in acid solution, and implies that hydrogen peroxide might be formed. Many years ago Fries and Volk found that when the blue solution of 2,7-dimethylthianthrene 5-oxide in concentrated sulphuric acid was poured into water, the mother liquor liberated iodine from iodide ion[55a]. They concluded that a peroxide may have been formed, and this conclusion has persisted in sulphoxide chemistry[28,42,52], although, in fact, Fries felt later that his earlier conclusion was probably not right. That is, he and Vogt[55b] found that when a solution of **17** in sulphuric acid was poured on to ice and the solution was treated with iodide ion, not as much iodine was liberated as would have been expected if hydrogen peroxide had been formed. Fries concluded, therefore, that although the fate of the sulphoxide's oxygen atom was not known, it no longer appeared to have become part of hydrogen peroxide.

There is no doubt that the early workers knew to a considerable extent what was formed from sulphoxides in sulphuric acid. Fries and Engelbertz, for example, reported that 2,7-dimethoxythianthrene 5-oxide gave the holoquinoid salt (our dication) in concentrated sulphuric acid, and that when the solution took up water on standing it gave the semiquinoid salt (our cation radical)[32]. It is also clear that

Kehrmann recognized the analogous behaviour in the phenothiazine series. Solutions of phenothiazine 5-oxide in 80–100% sulphuric acid were characterized spectroscopically as containing the holoquinoid salts, while corresponding solutions in 10% sulphuric acid were characterized as containing the semiquinoid salts[20]. Kehrmann was able to clarify in this way the work of others[22,56] but did not know how the semiquinone was formed.

Some bearing on these reactions may perhaps be found in the work of Hanson and Norman with the phenothiazinium ion **20** and the cation radical **1** in buffered solutions[57]. They prepared crystalline **20**(ClO$_4$) and studied its reactions spectroscopically in acetonitrile solution with water and with aqueous buffers of various pH. Reaction with unbuffered water gave the cation radical **1**, whose concentration reached a limiting value and then fell. During this fall the dimer cation **27** appeared and increased in concentration. Phenothiazine was also observed with increasing concentration. When a weakly acidic buffer (pH 4) was used, similar changes occurred except that the neutral radical **24** was observed instead of **1** (Scheme 2).

The key to these reactions (Scheme 2) is the reduction of **20** and the proposal[57] that the required reducing agent is initially 3-hydroxyphenothiazine **22** or its tautomer **21**. Once reduction (equation 5) has got underway it can be continued not only with **21** or **22** but also with the dimer **26** or its precursor **25**. The overall proposal is shown in Scheme 2. The important feature is the proposed participation of **21** (or **22**), showing that *some* 3-oxy derivative of phenothiazine *must* be formed. Hanson and Norman found that with a weakly acid buffer (pH 4), 78% of **20** was converted into the dimer cation **27** and about 10% into **23**. Thus, the major reducing agent in these circumstances was the dimer **25** (or **26**). It is apparent that by adjusting the pH on the more acidic side the dimerization of **24** (equation 8) might be hindered and the pathway of reduction of **20** be diverted more toward the 3-phenothiazone **23** side.

The question that now follows is whether reactions of this kind can account for the formation of the phenothiazine cation radical **1** from the oxide **19** in acidic solutions. The oxide is easily converted into phenothiazinium ion **20** in acidic solution. We would expect that conversion into **1** would follow and that one or more of the products of Scheme 2 would also be formed besides **1** when **19** stands in moderately strong acidic solution, but that certainly **22** or one of its derivatives must be formed.

Shine and Mach[52] found that stable solutions of the dication **28**, i.e. protonated **20**, were obtained from **19** in concentrated sulphuric acid (equation 10)*. In contrast, **19** in aqueous sulphuric acid gave the cation radical **1**. However, this was not the only cation radical formed; eventually the hydroxy cation radical **29** was also detectable by e.s.r., although it appeared that **29** was formed at a later stage than **1**. Shine and Mach proposed that these cation radicals were formed by homolysis of the protonated sulphoxides **30** and **31** (Scheme 3). It is possible, however, that the sequence of reactions leading to **1** and **29** is, in fact, similar to that of Scheme 2. In that case, the cation radical **29** represents the necessary derivative of **22**. In order to account for the apparent unequal timing of the formation of the two cation radicals, Scheme 4 is proposed in which the hydroxy cation **32** is formed first, along with **1**, and is converted into **29** in subsequent redox steps such as in equation 11*.

Whether or not Scheme 4 is valid for explaining the conversion of **19** into **1** cannot be said yet. In 15% sulphuric acid the conversion is characterized spectroscopically with clean isosbestic points, indicating a direct connection between **19** and **1** with

*See page 535.

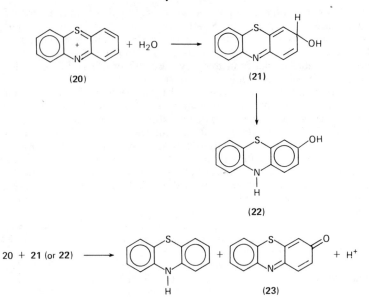

(4)

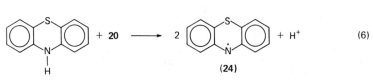

20 + 21 (or 22) ⟶   +   (23)   + H⁺   (5)

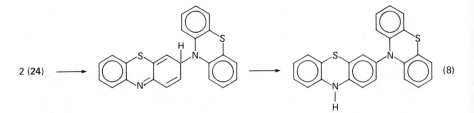

           + 20 ⟶ 2   + H⁺   (6)

               (24)

24 + H⁺ ⇌ 1   (7)

2 (24) ⟶   ⟶   (8)

         (25)   (26)

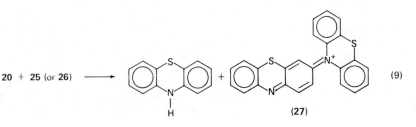

20 + 25 (or 26) ⟶   +   (9)

                          (27)

SCHEME 2

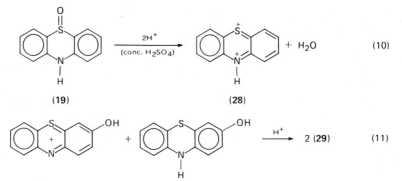

(10)

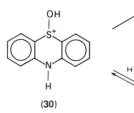

(11)

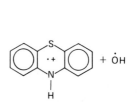

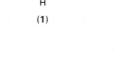

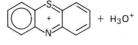

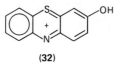

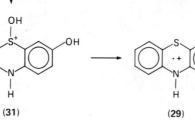

SCHEME 3

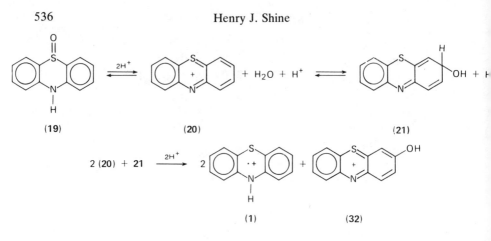

(19)   (20)   (21)

(1)   (32)

SCHEME 4

intermediates of short life and very low concentration. Kinetic analysis of these features is now needed.

Analogous questions may be asked about the reactions of thianthrene 5-oxide **17** and phenoxathiin 5-oxide **18**. Each of these oxides dissolves in concentrated sulphuric acid (96%) and rapidly gives the corresponding de-oxygenated dication. The dication next is slowly converted into the cation radical. Again, however, a hydroxylated heterocycle is also obtained: a hydroxythianthrene 5-oxide **40a** was

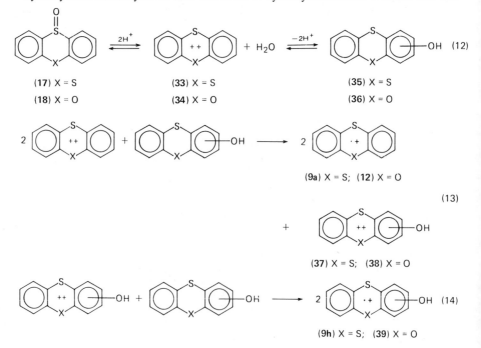

(17) X = S   (33) X = S   (35) X = S
(18) X = O   (34) X = O   (36) X = O

(9a) X = S; (12) X = O

(37) X = S; (38) X = O

(9h) X = S; (39) X = O

SCHEME 5

isolated from reaction of 17[28,58] and a hydroxyphenoxathiin 5-oxide **40b** from 18[49]. In the latter case, e.s.r. spectroscopy showed also that whereas cation radical 12 was obtained early in reaction a mixture of 12 and the hydroxy cation radical 39 was obtained later. By analogy with Schemes 2 and 4, these findings would be fitted into Scheme 5.

The conversion of 33 into **9a** was followed spectroscopically by Shine and Piette and, again, the spectra of the two ions were linked with clean isosbestic points[28]. Therefore, if Scheme 5 is to be valid, the intermediates (e.g. 35) must have a very short life. The spectrum of the hydroxy dication 37 was also observed, linked initially by isosbestic point to the bands of 33 and **9a**; but eventually departure from this isosbestic convergence occurred, as would be expected if 37 were to disappear by a subsequent redox process.

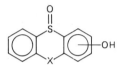

(40a) X = S; (40b) X = O

The foregoing comments on early work with the heterocyclic 5-oxides is made in a speculative way from the vantage point of latter-day overview. Further, our representation of water as a nucleophile in 96% sulphuric acid solutions (e.g. equation 12) may be an oversimplification. Possibly the nucleophile is the bisulphate ion, but in that case the corresponding product, the acid sulphate ester, would presumably undergo rapid protonolysis, as is seen in the analogous Wallach rearrangement[59].

Diaryl sulphoxides carrying electron-donating substituents also give cation radicals in acid solutions[50,51]. For example, di-p-tolyl sulphoxide gives the same e.s.r. spectrum in concentrated sulphuric acid as is obtained with di-p-tolyl sulphide[50]. Oae and coworkers have carried out studies of racemization and oxygen exchange with aryl sulphoxides in acidic solutions and have implicated cation radicals as intermediates[60–62]. For example, (+)-phenyl p-tolyl [^{18}O]sulphoxide 41 was found to undergo racemization and exchange with equal rates in 95.5% sulphuric acid, and a plot of log $k_{rac.}$ against $-H_0$ in the range of 91.8–96.3% sulphuric acid had a slope of 1.07. Further, the rates of exchange and racemization of a series of aryl p-tolyl sulphoxides were not strongly affected by substituents in the aryl group.

Since diaryl sulphoxides are probably completely protonated in concentrated sulphuric acid[63], a route involving a monoprotonated sulphoxide is improbable. Oae and Kunieda prefer the homolytic (equation 15) rather than heterolytic scission (equation 16) of a diprotonated sulphoxide because of the small effect of substituents

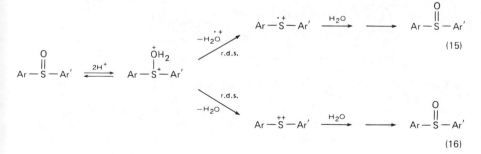

on rates of reaction. The rationale is that in the rate-determing step (r.d.s.) of homolysis there is little change in charge character at sulphur, whereas in the heterolytic cleavage the charge character is increased. Therefore, the electron-donating/-attracting properties of substituents would be expected to be more noticeable in heterolytic than homolytic cleavage. The conclusions[60] are plausible, but some further points are worth noting. It has now been found that monoprotonation of sulphoxides is relatable to the H_A rather than the H_0 function, the sulphoxides functioning somewhat like amides[63]. Also, the effect of substituents on the monoprotonation of some aryl methyl sulphoxides was small ($\rho = +0.85$)[63]. It is probable, therefore, that addition of a second proton, not only to aryl methyl but also to diaryl sulphoxides, would also not be much affected by substituents. If, furthermore, in the racemization and exchange reactions, the transfer of the second proton was rate determining, and followed by or coincident with rapid scission, the rationale favouring homolysis over heterolysis would not be necessary. Thus, even though e.s.r. shows that aryl sulphoxides in strong acids give aryl sulphide cation radicals, and the rate data implicate homolytic scission, the latter is not entirely certain.

Oae and coworkers have obtained equal rates of exchange and racemization for **41** in 85% phosphoric acid and trichloroacetic acid at $120°C$[62]. Again, the effect of substituents on rates was small and reaction was attributed to homolysis, but of the monoprotonated sulphoxide.

In summary, it seems fair to say, from what is presently known, that the evidence for formation of sulphide cation radicals from sulphoxides in acidic solutions by homolyses of protonated forms is unreliable. Between the homolyses which have been proposed, that involving the diprotonated sulphoxide is better suited to kinetic data than that involving the monoprotonated form. An alternative pathway (at least with the heterocyclic sulphoxides) involving reduction of the dication is very attractive. Supporting kinetic evidence would be highly desirable. The reducing agent is another product of reaction rather than a water molecule, for in that regard, Hanson and Norman have calculated from redox potential data that reduction of the phenothiazinium ion by water is improbable[57].

II. METHODS OF MAKING ORGANOSULPHUR CATION RADICALS

A. Use of Sulphuric and Other Brønsted Acids

We have already discussed in detail the historical use of concentrated sulphuric acid. The mechanism by which oxidation with this acid occurs is not known. Proposals have been made for the oxidation of aromatic hydrocarbons (ArH), as summarized in equations 17–19[67], which, if valid, would presumably apply to organosulphur

$$\text{ArH} + \text{H}_2\text{SO}_4 \rightleftharpoons \text{ArH}_2^+ + \text{HSO}_4^- \tag{17}$$

$$\text{ArH} + \text{ArH}_2^+ \rightleftharpoons \text{ArH}^{\cdot+} + \text{ArH}_2^{\cdot} \tag{18}$$

$$\text{ArH}_2^{\cdot} + 2\text{H}_2\text{SO}_4 \rightleftharpoons \text{ArH}^{\cdot+} + \text{HSO}_4^- + \text{SO}_2 + 2\text{H}_2\text{O} \tag{19}$$

compounds also. The steps shown allow for the expected easier oxidation of the free radical (shown as $\text{ArH}_2^{\cdot}$) by sulphuric acid as compared with the parent compound[68]. A major disadvantage to the use of concentrated sulphuric acid is its sulphonating ability. If a compound is not too readily oxidized it may instead undergo sulphonation, as is the case with diphenyl sulphide which is converted into its 4,4'-disulphonic acid[50]. Also, the use of low temperatures, at which e.s.r. spectra may become better resolved by the lessening of exchange phenomena, is made difficult by the high freezing point of sulphuric acid.

Frequently, a mixture of nitromethane and sulphuric acid is used for making cation radicals for e.s.r. characterization. A mixture of acetonitrile and sulphuric acid may also be used. These mixtures have the advantage of moderating sulphonation possibilities and allowing the use of low temperatures. Whether or not the organic solvent participates in the electron transfer (as an acceptor) is not known. Examples are given in the tables.

In this connection, thianthrene has been oxidized to its cation radical by methanesulphonic acid in nitrobenzene[69]. This acid is not an oxidizing agent. Therefore, the nature of the electron transfer is not known, particularly as to whether sequences such as those in equations 17–19 are involved, and whether the nitrobenzene or atmospheric oxygen might be the acceptor. Solutions of thianthrene in hydrofluoric acid are colourless and not paramagnetic[70], indicating that an electron acceptor, which is absent in this acid alone, is needed for electron transfer in non-oxidizing acids.

Perchloric acid is useful for some oxidations. It is usually used as the commercial 70% solution in conjunction with acetic anhydride to remove water. Thus, thianthrene in carbon tetrachloride[28] and phenoxathiin in benzene[71] give their crystalline cation radical perchlorates when treated in this way. Again, the mechanism of oxidation is not known.

B. Use of Persulphuric Acid

Oxidation of a number of substituted phenoxathiins with potassium persulphate in concentrated sulphuric acid gave the corresponding cation radicals[42]. The same cation radicals were obtained from the corresponding 5-oxides in sulphuric acid and the g-values were in the range expected for phenoxathiin cation radicals.

On the other hand, the thianthrene cation radical was found to be short-lived in solutions of potassium persulphate–sulphuric acid. Depending on the amount of potassium persulphate used the cation radical was converted into the tri- or tetra-S-oxide[72].

Oxidation of some diaryl sulphides with potassium persulphate–concentrated sulfuric acid also gave the cation radicals[73]. In contrast, other investigations found that persulphate in sulphuric acid oxidized not only di-p-tolyl sulphide but also some sulphoxides, sulphones, and sulphonic acids to radicals with g-values far too low to be attributable to ordinary organosulphur cation radicals (Table 4)[74]. Whether or not these persulphate solutions contained neutral rather than charged radicals was never established. It is apparent from the g-values that the radicals have little spin density on sulphur. Further, the data suggest that a sulphide and its sulphoxide give the same radical, and this is different from that obtained from the corresponding sulphone. Persulphate was thought, therefore, to convert sulphides into radicals of hydroxylated sulphoxides, and sulphones and sulphonic acids also into hydroxylated radicals[74]. The situation is reminiscent of the persulphate oxidation of p-xylene into the cation radical of the corresponding quinone rather than that of p-xylene[75].

C. Use of Lewis Acids

Electron transfer from aromatics and thioaromatics to Lewis acids has been known for a long time[68].

TABLE 4. g-Values of organosulphur radicals obtained in H_2SO_4 and $K_2S_2O_8$–H_2SO_4 solutions[74]

	H_2SO_4		$K_2S_2O_8$–H_2SO_4	
Substrate	Lines	g	Lines	g
p-Tolyl sulphide	15[a]	2.00737	13	2.00336
p-Tolyl sulphoxide	15	2.00738	13	2.00336
p-Tolyl sulphone	—		[b]	2.00354
Methyl p-tolyl sulphoxide	13	2.00818	7	2.00351
Methyl p-tolyl sulphone	—		17	2.00375
Phenyl sulphone	—		3–6	2.0036
2-Hydroxyphenyl phenyl sulphone	—		3–6	2.0035
2,5-Dihydroxyphenyl phenyl sulphone	—		5	2.0036
2-Hydroxy-4,4'-dimethylphenyl sulphone	—		21	2.0035
Hydroquinone-2,5-disulphonic acid	—		3	2.0035
Biphenyl-4,4'-disulphonic acid	—		3–5	2.0037

[a]Split further into overlapping multiplets.
[b]Complex.

Of early used Lewis acids, only $SbCl_5$ and $AlCl_3$ are now commonly used. Kinoshita and coworkers used reactions of thianthrene with $SbCl_5$[26,27] and $AlCl_3$[26] in one of the first correct deductions (e.s.r.) of the structure of the thianthrene cation radical. Other cation radicals were obtained from reaction of $SbCl_5$ with dibenzothiophene, and 2-naphthalene disulphide[26], and with phenothiazine and 10-methylphenothiazine[76]. A cation radical salt was also obtained from 1,2,4,5-tetra(methylthio)benzene and $SbCl_5$[77].

Far more versatile has been the use of $AlCl_3$ in nitromethane solution. Although this reagent had been used by others earlier[78], its use at low temperature to give high-resolution e.s.r. spectra (of dialkoxyaromatic cation radicals) was first reported by Forbes and Sullivan in 1966[79]. Examples of its use with organosulphur compounds soon followed, some of which are given here.

Conformational changes in di(alkylthio)benzene cation radicals at $-38°C$ were detected from e.s.r. line widths. Thus, equilbria in cis- and trans-42 and cis- and trans-43 were reported by Forbes and Sullivan[80]. An e.s.r. spectrum of nine lines with splitting of 0.117 mT was obtained for the cation radical 44 and was interpreted as

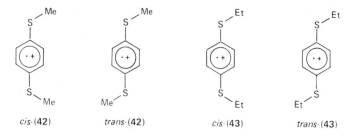

cis-(42)   trans-(42)   cis-(43)   trans-(43)

showing delocalization of the odd electron over both rings (from coupling with the eight methylene protons of the ethyl groups)[81].

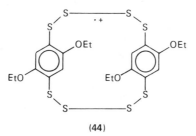

(44)

Other examples with heterocyclic sulphur compounds are listed in Table 5, while Table 6 gives e.s.r. data for a number of diaryl sulphides. The data for these sulphides have been viewed by Sullivan and Norman as showing that there is no detectable barrier to rotation about the aryl−S bonds[82].

The use of $AlCl_3$-CH_2Cl_2 was the preferred method of oxidizing some tetrakis(alkylthio)ethylenes to the cation radicals[83]. Here, also, the use of variable temperatures, made possible by the choice of the medium, demonstrated the existence of conformational isomers.

What is the electron acceptor in these facile oxidations? When $SbCl_5$ is used there is little doubt that it, itself, is the electron acceptor. The fate of the $SbCl_5$ has been discussed in detail[68], and seems to be best represented as in equations 20 and 21. It is possible, therefore, that where

$$SbCl_5 + e^- \longrightarrow \overset{\cdot -}{SbCl_5} \qquad (20)$$

$$2\overset{\cdot -}{SbCl_5} \longrightarrow \overline{SbCl_6} + \overline{SbCl_4} \qquad (21)$$

a salt analyses as $Ar_2S \cdot SbCl_5$ it may be a mixture of the tetrachloro- and hexachlorantimonates.

Analogous definitions of oxidations by $MeNO_2$−$AlCl_3$ cannot be given. Aluminate salts are not customarily isolated, so that analytical data are not available (except for those of the early 1900s[87]). It was thought originally that the nitromethane was the electron acceptor in solutions of aluminium chloride in that solvent[78]. This may be correct; no evidence for or against has been presented. E.s.r. spectra of aromatic hydrocarbons have been obtained with aluminium chloride in other solvents, e.g. chloroform, carbon disulphide, and benzene[88], and of thianthrene in chloroform and benzene[28]. These solvents can, of course, also function as electron acceptors, but if they (and nitromethane) do, when used with aluminium chloride, no one has yet shown what happens to the electron.

D. Use of Halogens

The oxidation of an organosulphur compound by halogens can be expressed in principal by equation 22. (In this, and many of the equations that follow, the symbol ⁄S⧵ is used to represent either a diaryl sulphide or one of the heterocyclic

$$2 \underset{\diagup}{S}_{\diagdown} + X_2 \rightleftharpoons 2 \underset{\diagup}{\overset{\cdot +}{S}}_{\diagdown} + 2X^- \qquad (22)$$

sulphides, such as thianthrene.) The use of fluorine is virtually out of the question because of its fluorinating and extensive oxidizing abilities. Chlorine is not often used for similar reasons. Fries and Engelbertz[32] found, for example, that oxidation of thianthrene with chlorine went to the so-called holoquinoid state, which would be

TABLE 5. E.s.r. data for cation radicals of dithiins and analogous compounds in $MeNO_2$–$AlCl_3$ at low temperatures

Compound	a^S	a_1^H	a_2^H	a_3^H	a_4^H	a_5^H	a_6^H	a_{Me}^H	Reference
	0.984	0.282	0.282	—	—	—	—	—	45
	0.935	0.0201	0.1056	0.1056	0.0201	—	0.332	—	45
	—	—	0.450	0.092	0.537	0.537	0.092	—	84
	0.716	—	0.456	0.096	0.552	0.552	0.096	—	84a
	—	—	—	0.098	0.038	0.183	0.010	—	84

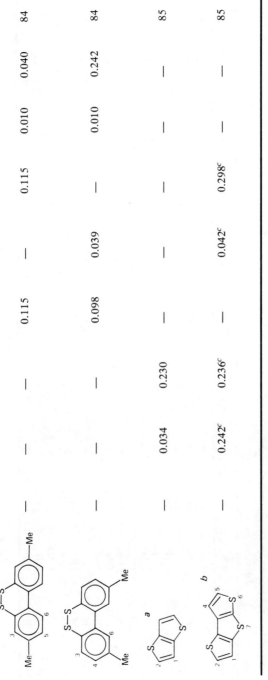

—	—	—	0.115	—	0.115	0.010	0.040	84
—	—	—	0.098	0.039	—	0.010	0.242	84
—	0.034	0.230	—	—	—	—	—	85
—	0.242[c]	0.236[c]	—	0.042[c]	0.298[c]	—	—	85

[a] Correct name and numbering is thieno[3,2-b]thiophene; numbering in the table is for convenience only.
[b] Correct name and numbering is dithiene[2,3-b,2',3'-d]thiophene; numbering in the table is for convenience only.
[c] Assignments made intuitively by the author, and may be incorrect.

TABLE 6. E.s.r. data for some diaryl sulphide cation radicals in MeNO$_2$–AlCl$_3$

| Substituents | | | | | a^H | | | | | | |
2,2'	3,3'	4,4'	5,5'	6,6'	2,2'	3,3'	4,4'	5,5'	6,6'	g	Reference
H	H	OH	H	H	0.161	0.0115	0.102	0.0115	0.161	2.00687	a
H	H	OMe	H	H	0.155	0.016	0.110	0.016	0.155	2.00686	a
H	Me	OH	Me	H	0.145	0.025	0.115	0.025	0.145	2.00645	a
H	Me	OH	Me	H	0.145	0.0275	0.115	0.0275	0.145	2.0070	b
Me	H	OH	H	Me	0.103	0.0153	0.103	0.0153	0.103	2.00672	a
Me	H	OMe	H	Me	0.10	—	0.10	—	0.10	2.00669	a
Me	Me	OH	Me	Me	0.095	0.046	0.095	0.046	0.095	2.00647	a
Me	Me	OMe	Me	Me	0.117	0.029	0.117	0.029	0.117	2.00654	a
H	t-Bu	OH	t-Bu	H	0.145	—	0.076	—	0.145	2.0070	b
Me	H	OH	Me	H	0.152	0.017	0.100	0.067	0.100	2.00646	a
Me	H	OMe	Me	H	0.130	0.075	0.113	0.019	0.113	2.00644	a
H	Me	OH	t-Bu	H	0.145	0.0275	0.112	—	0.112	2.00657	a
H	Me	OH	t-Bu	H	0.145	0.0275	0.112	—	0.112	2.0070	b
H	H	OMe	t-Bu	H	0.162	0.020	0.116	—	0.141	2.00645	b
Me	H	OH	t-Bu	H	0.147	0.025	0.098	—	0.122	2.00649	a
Me	H	OMe	t-Bu	H	0.160	0.020	0.105	—	0.135	2.00658	a

^{a}Reference 82, low temperature.
^{b}Reference 86, room temperature.

represented now as in **45** or **46**. On the other hand, they found that 2,7-dimethoxythianthrene gave the cation radical chloride represented, though as **10c** (X = Cl) (Section IA2).

(45)   (46)

It is probable that careful control of the amount of chlorine used would allow for oxidations to the cation radical stage. This is the case with tetrathiofulvalene **47**, which is oxidized in carbon tetrachloride solution to the cation radical chloride by the stoichiometric amount, but to the dication dichloride with excess of chlorine (equation 23)[89]. Oxidation of tetrathiotetracene **48** by chlorine (in

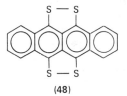

$$47^{\cdot+} \cdot Cl^- \xrightarrow{Cl_2} 47^{2+} \cdot 2Cl^- \qquad (23)$$

(47)

trichlorobenzene) also gave the dication, but interestingly **48** was oxidized by hydrogen peroxide (or PbO_2 or MnO_2) in hydrochloric acid to $48^{\cdot+} \cdot Cl^-$

S — S

(48)

hydrate[90]. Here, it seems that controlled oxidation of chloride ion to chlorine may have been the basis of the successful one-electron oxidation of **48**.

Reference has already been made to the potentiometric bromine titrations of phenothiazines by Michaelis *et al.*[19]. Oxidation of phenothiazine with a 1–2% excess of bromine in diethyl ether gave the solid cation-radical bromide, while use of excess of bromine gave the so-called holoquinoid perbromide[91]. The cation radical bromides of phenothiazine[92,93], and dibenzophenothiazine[94] were more recently easily made by this method for studies of optical and electrical (conducting) properties of the salts.

Oxidations of phenothiazines with bromine are successful when carried out in diethyl ether, probably because the salt precipitates. Reaction with 1 equivalent of bromine in acetic acid at room temperature gave phenothiazinium perbromide[95]. When 10-alkylphenothiazines were similarly treated monobromination occurred, and when 2 equivalents of bromine were used 3,7-dibromo derivatives **49** were formed. Further treatment of **49** in acetic acid with bromine gave the corresponding cation radicals $49^{\cdot+}$ (equation 24).

Oxidations with iodine are used very conveniently in the presence of a silver salt to prepare cation radical salts. These salts may coprecipitate with silver iodide or may remain in solution to be used as such or to be recovered after filtration of the silver iodide. Phenothiazine[96], 10-methylphenothiazine[97], 10-phenylphenothiazine[98], and 10-methyl-3,4:6,7-dibenzophenothiazine[97] can be converted into their cation radical perchlorates in this way. 10-Methylphenothiazine and dibenzophenothiazine have also given other salts (of $\overline{BF}_4$ and $\overline{SbF}_6$) in good yields[93]. In contrast, oxidations of

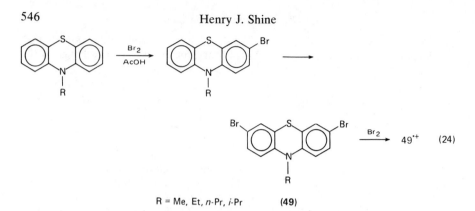

$$R = Me, Et, n\text{-}Pr, i\text{-}Pr \qquad (49)$$

thianthrene and phenoxathiin by this method have not been so successful. Oxidation occurred, but only slowly, and isolation of the cation radicals (as their perchlorates) in high purity failed[99].

E. Disproportionation of Sulphides and Sulphoxides in Acidic Solutions

A useful method of preparing some organosulphur cation radicals involves the reaction of the sulphide *and* sulphoxide in an acidic solution (equation 25). The reaction appears to be that of the dication with sulphide (equations 26 and 27) although no one has yet shown this to be the case.

$$\backslash S + \backslash S{=}O + 2HX \longrightarrow 2 \backslash S^{\cdot+}X^- + H_2O \qquad (25)$$

$$\backslash S{=}O + 2HX \longrightarrow \backslash S^{2+} 2X^- + H_2O \qquad (26)$$

$$\backslash S + \backslash S^{2+} \longrightarrow 2 \backslash S^{\cdot+} \qquad (27)$$

By this technique thianthrene cation radical perchlorate and tetrafluoroborate were prepared, the anhydrous HX being added to thianthrene and its 5-oxide in nitromethane[100]. The perchlorates of phenothiazine[96], 10-methylphenothiazine[98], dibenzophenothiazine **50**[94], and both benzo[a]- and benzo[c]phenothiazine[94] cation radicals have been made, using 70% perchloric acid. Interestingly, **50**$^{\cdot+} \cdot Cl^- \cdot H_2O$ and **50**$^{\cdot+} \cdot I^-$ were obtained when **50** and its 5-oxide were ground together and moistened with concentrated hydrochloric and hydriodic acid[94].

F. By Anodic Oxidation

Oxidation of organic compounds at an anode is an extremely valuable method of forming cation radicals. A detailed discussion of the method and its various techniques, however, would be out of place in this review of organosulphur cation radicals. Recent reviews of a general type are available[68,101–103]. It is not customary to use anodic oxidation for preparative-scale formation of cation radicals[68]. This can, of course, be done in particular cases, but the scale is rather small. For preparative-scale oxidation, all or most of the organosulphur compound would be oxidized in a stirred solution at a fixed applied potential. More commonly, anodic oxidation for diagnostic or mechanistic information is carried out in unstirred solutions, in which only an extremely small amount of the compound is oxidized in the vicinity of the anode, and the applied potential is varied continuously over an appropriate range. Oxidations of this kind, carried out either polarographically or by cyclic voltammetry, can provide a measure of the relative ease of oxidizing the organic compound (expressed frequently as the first one-electron half-wave oxidation potential, $E^1_{1/2}$), and the stability (i.e. lifetime) of the cation radical in the oxidizing medium. Frequently it is possible to measure the oxidation potential not only for cation-radical formation but also for oxidizing the cation radical to the dicationic state ($E^2_{1/2}$). When these oxidations are carried out under reversible conditions (i.e. when neither the cation radical nor dication react with the medium during the time of measurement) the oxidation potentials are extremely informative. Values of $E^1_{1/2}$ alone indicate the relative ease of forming a cation radical, while values of $E^1_{1/2}$ and $E^2_{1/2}$ indicate either the formation constant or, expressed in the reverse way, the disproportionation (equation 30) constant (equation 31) for the cation radical[104].

$$D - e^- \rightleftharpoons D^{\cdot+} \tag{28}$$

$$D^{\cdot+} - e^- \rightleftharpoons D^{2+} \tag{29}$$

$$2D^{\cdot+} \rightleftharpoons D + D^{2+} \tag{30}$$

$$\log K_d = -\frac{F}{2.303\,RT} (E^2_{1/2} - E^1_{1/2}) \tag{31}$$

Data of these kinds will be used during discussions of the reactions of organosulphur cation radicals (Section IV), but some recent examples of collected data are given now in illustration.

In general, electron-donating substituents make oxidation of an organic compound easier[101], unless the substituent introduces particular, unfavourable, steric or solvation factors into the cation radical. The effects of thiomethyl substituents on oxidation potentials in aryl methyl thioethers are shown in Table 7[105-107]. The comparative values of $E_{1/2}$ (in volts) for the parent aromatics are benzene (2.30), naphthalene (1.54), anthracene (1.09), biphenyl (1.78), and pyrene (1.16). The data in Table 7 show that methylthio groups lead (for the most part) to lower oxidation potentials, and the oxidation potential is the lower when the number of groups increases. The relative resonance-stabilizing effects of groups *ortho*, *para*, and *meta* to each other are also to be seen, for example, in the higher oxidation potentials with 1,3- and 1,3,5-substituents as compared with 1,2- and 1,4-substituents in benzene. Oxidation potentials of aryl methyl ethers are also given in Table 7 to illustrate that it is generally easier to make the cation radical of a thioether.

Eberson and Nyberg[101] have pointed out that it is logical to try to correlate $E_{1/2}$ data with the energy of the HOMO of the compounds being oxidized and that this has been done successfully in a number of cases. In this connection the data for the aryl methyl thioethers have been fitted reasonably well by Zweig and coworkers[106,107] to the calculated energies of the corresponding HOMOs after adjustment of appropriate parameters for α_s and β_{cs} terms. Furthermore, theory requires that an electron donor will raise the energy of the HOMO if the substituent is attached to a position of high electron density in the HOMO. In such a position, therefore, the substituent will cause a lowering in $E_{1/2}$. This is seen (Table 7) in the data for 1- (0.425) *vs.* 2-napthyl (0.263) and for 4,4'- (0.440) *vs.* 3,3'- (0.220) and

Henry J. Shine

TABLE 7. Half-wave oxidation potentials of some aryl methyl ethers and thioethers[105–107]

Parent aromatic: positions of substituents and $E_{1/2}(V)^a$

	1-	1,2-	1,3-	1,4-	1,3,5-	1,2,4,5-
MeO	1.76	1.45	—	1.34	1.49	0.81
MeS	1.565	1.35	1.45	1.19	1.43	1.08

	1-	2-	1,4-	1,5-	1,8-	2,3-	2,6-	2,7-
MeO	1.38	1.52	1.10	1.28	1.17	1.39	1.33	1.47
MeS	1.32	1.365	1.07	1.265	1.09	1.355	1.10	1.33

	9,10-	2,2'-	3,3'-	4,4'-	1,6-
MeO	0.98	1.51	1.60	1.30	0.82
MeS	1.11	1.39	1.475	1.255	0.96

^{a}Measured polarographically in MeCN *vs.* S.C.E. at a rotating Pt electrode, and in 0.1 M tetrapropylammonium perchlorate.

2,2'-biphenyl (0.311) methyl thioethers, where the numbers in parentheses are the coefficients of charge density at the positions indicated in the HOMO.

Analogous results have been obtained by Bock and Brähler[108], who have shown that the ionization potentials of 1,4- and 2,6-bis(methylthio)naphthalenes are 7.58 and 7.59 eV, respectively, compared with 8.15 eV for naphthalene. Thus, in raising the energy of the HOMO the substituents make it easier for one-electron removal. Interestingly, also, the 1,4- and 2,6-isomers (as well as 9,10-bis(methylthio)anthracene) are among the few sulphur compounds from which both cation and anion radicals can be made. It is evident from the g-values [e.g., 2.0075 for the 1,4-($\cdot+$) and 2.0025 for the 1,4-($\cdot-$)] that there is much spin density on sulphur for the former but not for the latter. This is also noticeable from the methyl coupling constants [0.365 mT for 1,4-($\cdot+$) and 0.021 mT for 1,4-($\cdot-$)], and is in accord with the expected effect of the substituents on the HOMO and LUMO of naphthalene[108].

The use of $E^1_{1/2}$ and $E^2_{1/2}$ data for calculating disproportionation constants is illustrated in Table 8 for the well known heterocyclic cation radicals 1, 9a, and 12. It is seen that K_d is very small indeed for these cation radicals, and that it varies (for 9a) considerably with the medium. The small sizes of K_d show how stable these organosulphur cation radicals are toward disproportionation, and have been used, in the case of 9a, particularly, in demonstrating that the dication is unlikely to participate as a reactant when 9a reacts with nucleophiles (Section IVB).

TABLE 8. Half-wave oxidation potentials (*vs.* S.C.E.) and disproportionation constants for phenothiazines, thianthrene, and phenoxathiin

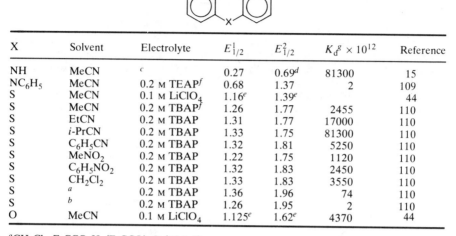

X	Solvent	Electrolyte	$E^1_{1/2}$	$E^2_{1/2}$	$K_d^g \times 10^{12}$	Reference
NH	MeCN	[c]	0.27	0.69[d]	81300	15
NC$_6$H$_5$	MeCN	0.2 M TEAP[f]	0.68	1.37	2	109
S	MeCN	0.1 M LiClO$_4$	1.16[e]	1.39[e]		44
S	MeCN	0.2 M TBAP[f]	1.26	1.77	2455	110
S	EtCN	0.2 M TBAP	1.31	1.77	17000	110
S	i-PrCN	0.2 M TBAP	1.33	1.75	81300	110
S	C$_6$H$_5$CN	0.2 M TBAP	1.32	1.81	5250	110
S	MeNO$_2$	0.2 M TBAP	1.22	1.75	1120	110
S	C$_6$H$_5$NO$_2$	0.2 M TBAP	1.32	1.83	2450	110
S	CH$_2$Cl$_2$	0.2 M TBAP	1.33	1.83	3550	110
S	[a]	0.2 M TBAP	1.36	1.96	74	110
S	[b]	0.2 M TBAP	1.26	1.95	2	110
O	MeCN	0.1 M LiClO$_4$	1.125[e]	1.62[e]	4370	44

[a]$CH_2Cl_2–F_3CCO_2H–(F_3CCO)_2O$ (45:1:5).
[b]$F_3CO_2H–(F_3CCO)_2O$ (9:1).
[c]$MeCN–0.1$ M LiClO$_4$–0.1 M HClO$_4$.
[d]Dependent on pH and amount of water in medium.
[e]Measured against Ag/Ag$^+$ (10^{-2} M) and adjusted to correspond with S.C.E. by adding 0.3 V.
[f]TEAP and TBAP = tetraethyl- and tetrabutylammonium percholate, respectively.
[g]Assumed at 25°C.

III. SOME NEWER TYPES OF ORGANOSULPHUR CATION RADICALS

A. Dimer-Sulphide and Disulphide Cation Radicals

When dimethyl sulphide was subjected to pulse radiolysis in aqueous solution a dimeric cation radical **51** was formed (equation 32)[111]. The same cation radical was

$$2\,Me_2S \longrightarrow \longrightarrow (Me_2S{-}SMe_2)^{\cdot+} \qquad (32)$$

(51)

also obtained when dimethyl sulphide was oxidized in a flow system by hydroxyl radicals generated with acidic solutions of either Ti(III)-hydrogen peroxide or Ti(III)–sodium persulphate. Analogous oxidations were achieved with diethyl sulphide, tetrahydrothiophene, and β-hydroxyethyl methyl sulphide[112], and the dimeric cation radicals were identified by e.s.r. In some cases of the flow oxidations a second radical was also identified, resulting from the loss of an H atom from the sulphide; e.g. dimethyl sulphide gave ·CH$_2$SMe.

The suggestion was made[112] that the dimeric cation radical **52** arises from a first-formed R$_2$S$^{\cdot+}$ transient cation radical (equation 33). As for the other (neutral) radical **53**, this also, presumably, arises from the R$_2$S$^{\cdot+}$ by proton loss, since the relative amount of a neutral radical increased with increase in pH.

$$R_2S \xrightarrow{\cdot OH} R_2S^{\cdot+} \xrightarrow{R_2S} R_2S-SR_2^{\cdot+}$$

$$\textbf{(52)} \qquad\qquad\qquad (33)$$

$$\xrightarrow{(-H^+)} R_2S(-H)\cdot$$

$$\textbf{(53)}$$

In fact, these straightforward reactions may not be the source of the radicals. By generating hydroxyl radicals with pulse radiolysis and analysing the formation and decay of products conductimetrically and spectroscopically, Asmus and coworkers have found the reactions to be more complex. It is proposed that the first reaction is the diffusion controlled addition of ·OH to the sulphide, and it is the adduct **54** which leads to the other radicals (Scheme 6)[113]. High concentrations of R_2S favour formation of **52** while low concentrations favour that of **53**.

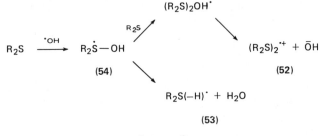

$$\text{SCHEME 6}$$

The dimeric cation radical has a relatively long life and is thought to decay essentially by dissociation into $R_2S^{\cdot+}$ which next reacts with the medium (equations 34–36).

$$R_2S^{\cdot+} + \bar{O}H \longrightarrow R_2S(-H)^{\cdot} + H_2O \qquad (35)$$

$$R_2S-SR_2^{\cdot+} \rightleftharpoons R_2S^{\cdot+} + R_2S \qquad (34)$$

$$R_2S^{\cdot+} + H_2O \longrightarrow R_2S(-H)^{\cdot} + H_3O^+ \qquad (36)$$

In related work, Norman and coworkers found that reaction of β-hydroxyalkyl sulphides with hydroxyl radical caused, in part, fragmentation in which it is thought the cation radical is the precursor (e.g., equations 37 and 38)[114]. Whether these reactions should be re-interpreted as involving an adduct analogous to **54** will presumably be decided by kinetic analysis.

$$HOCH_2CH_2SR \xrightarrow{\cdot OH} H-O-CH_2-CH_2\overset{\cdot+}{S}R \longrightarrow H^+ + HCHO + \dot{C}H_2SR \qquad (37)$$

$$\underset{\underset{OH}{|}}{R'CHCH_2SR} \xrightarrow{\cdot OH} \underset{\underset{OH}{|}}{R'\overset{\overset{H}{|}}{C}-CH_2\overset{\cdot+}{S}R} \longrightarrow \underset{\underset{O}{\|}}{R'CCH_3} + H^+ + RS^{\cdot} \qquad (38)$$

Application of pulse radiolysis oxidation to 1,4-dithian led to analogous conclusions. A monomeric 'internal' $R_2S-SR_2^{\cdot+}$-type cation radical **56** was formed at low and a dimeric type **57** at high dithian concentrations. These also are thought to arise, however, from a first-formed complex radical **55**[115].

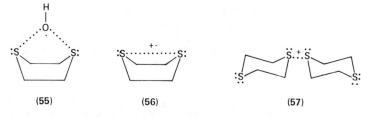

(55)   (56)   (57)

Pulse radiolysis studies with dialkyl disulphides have shown that similar reactions occur, but here the attack of ·OH on a disulphide appears to lead directly both to the cations radical **58** and an adduct **59** equations 39 and 40[116]. The two reactions occur essentially at diffusion-controlled rates, and the portion of the hydroxyl radicals reacting according to equation 39 was 55–56% for R = Me, Et, and i-Pr, and 42% for R = t-Bu.

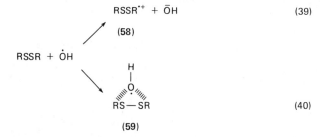

$$RSSR^{\cdot +} + \bar{O}H \qquad (39)$$

(58)

RSSR + ·OH

(40)

(59)

Although these disulphide cation radicals have been characterized by absorption spectroscopy (λ_{max} 420 nm for R = Me, Et; 410 nm for R = i-PR, t-Bu)[116], they have eluded e.s.r. characterization[117]. Reaction with Ti(III)–H_2O_2 led only to the e.s.r. detection of secondary-product radicals, such as $MeSO_2\cdot$ from MeSSMe and $HO_2CCH_2CH_2SO\cdot$ from $(HO_2CCH_2CH_2S)_2$. These radicals appear to come from fragmentation (and further oxidation of fragments) of adducts **59**[116,117]. The cation radicals **58** were found to decay via a second-order process (equation 41) and it is thought that the dication then reacts with the medium (Section IV)[116].

$$2RSSR^{\cdot +} \longrightarrow RSSR + RSSR^{2+} \qquad (41)$$

Direct one-electron oxidation of dimethyl and diethyl disulphide by Ag^{2+}, $Ag(OH)^+$, Tl^{2+}, $SO_4^{\cdot -}$, and even the 1,3,5-trimethoxybenzene cation radical occurs quantitatively. One-electron oxidation also occurs with $R_2S^{\cdot +}$ (which is obtained by scission of $R_2S-SR_2^{\cdot +}$, equation 34). These oxidations are very fast, but not within the range of diffusion control, i.e., k is of the order of 10^8-10^9 M^{-1}S^{-1} (reference 118).

Dimeric disulphide cation radicals of the type **52** have also been made from mesocyclic dithioethers. Thus, oxidation of 1,5-dithiacyclooctane **60** with NOBF$_4$ or Cu(MeCN)$_4$(BF$_4$)$_2$ in acetonitrile gave the cation radical **60**$^{\cdot +}$. This is so stable (isolable as the tetrafluoroborate), in contrast with other thioether cation radicals, that Musker and coworkers have proposed the structure shown in which stabilization

(60$^{\cdot +}$)

of the cation radical by the second sulphur occurs. Nevertheless, e.s.r. data indicate that the unpaired spin is localized on one sulphur atom and the neighbouring four hydrogen atoms, the spectrum consisting in a triplet of triplets ($a^H = 0.152$ and 0.104 mT). This is in contrast with, say, $Me_2SSMe_2^{.+}$, for which the 13-line spectrum shows complete spin delocalization. 1,5-Dithiacyclononane also gave a stable cation radical, but the seven- and ten-membered analogues did not[119]. Further oxidation gave **60**$^{2+}$, which undergoes reaction with **60** to give **60**$^{.+}$.

Another interesting development in radicals of the type $R_2S-SR_2^{.+}$ **52** comes from the irradiation of thiodiglycolic acid ($HO_2CCH_2SCH_2CO_2H$) with X-rays at 4.2°C. A radical is obtained with high g-factor (value depending on crystal orientation) and an e.s.r. spectrum attributable to coupling with four sets of two equivalent protons, rather than two sets of four equivalent protons expected of a dimer (**52**, R = HO_2CCH_2-). On the basis of the e.s.r. data (obtained from ENDOR measurements), Box and Budzinski designate the radical as $R_2S^{.+}$, with the four extra proton couplings coming from hyperfine interaction with neighbouring molecules. That is, it is concluded that **52** is not formed in this case[120].

B. Tetrathiafulvalene and Related Cation Radicals

Tetrathiafulvalene is the common name given to compound **47**. The systematic name of **47** is 2,2'-bis-1,3-dithiole, but it has become such an important compound

(47)

TABLE 9. Oxidation potentials (*vs.* S.C.E.) and e.s.r. data for some tetrathiofulvalenes and cation radicals

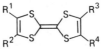

Note	R^1	R^2	R^3	R^4	$E^1_{1/2}$	$E^2_{1/2}$	a^H	a^H_R	g	Reference
a	H	H	H	H	0.33	0.70	—	—	—	121
b	H	H	H	H	0.320	0.680	—	—	—	122
c	H	H	H	H	—	—	0.126^e	—	2.00838	123
d	H	H	H	H	—	—	0.122	—	—	124
b	Me	H	Me	H	0.290	0.705	—	—	—	122
c	Me	Me	H	H	—	—	0.118	0.082	2.0078	123
c	Me	H	H	Me	—	—	0.128	0.064	2.0079	123
c	Me	Me	Me	Me	—	—	—	0.074	2.0077	123
b	Me	Me	Me	Me	0.245	0.615	—	—	—	122
b	Et	Et	Et	Et	0.230	0.750	—	—	—	122
b	Me	Et	Me	Et	0.240	0.610	—	—	—	122
a	MeS	MeS	MeS	MeS	0.470	0.710	—	0.024	2.00764	125

^{a}Cyclic voltammetry, MeCN, 0.1 M TEAP.
^{b}Polarography, MeCN, 0.1 M TBAP.
^{c}Fluoroborate salt in solvent propylene carbonate.
^{d}Oxidation with Br_2 in CH_3CO_2H.
$^e a^s$ is 0.42 mT (reference 84a).

that its common name and abbreviation (TTF) are now used almost without exception. TTF is very easily oxidized anodically to the cation radical and dication. In reaction with a variety of electron acceptors TTF forms, however, solid salts with extraordinary conducting properties. Therefore, much attention has been given to TTF, and a variety of homologues of TTF have also been synthesized and characterized for studying their capabilities of forming conducting salts. The oxidation potentials of some of these compounds are given in Table 9, together with some e.s.r. data for the cation radicals.

The remarkable electrical properties of salts of tetrathiafluvalenes have spurred interest in analogous sulphur compounds. The aryl-linked dimer **61** has $E_{1/2}^1$ and

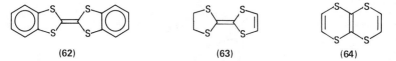

(61)

$E_{1/2}^2$ of 0.43 and 0.84 V in benzonitrile *vs.* S.C.E.[126]. The cation radical obtained by oxidation with hydrogen peroxide in sulpholanacetonitrile–concentrated HBF$_4$ has a quartet e.s.r. spectrum, $a^H = 0.125$ mT, $g = 2.0078$. These data indicate that only one of the TTF units is involved in the oxidation.

The dibenzo analogue **62** has been characterized also, polarographically, with oxidation potentials of 0.72 and 1.06 V (*vs.* Ag/AgCl–CH$_3$CN) in acetonitrile, and $a_1^H = 0.0164$ mT, $a_2^H = 0.0492$ mT in acetic acid[124,127]. Compounds **47** and **62** are among the class of violenes investigated extensively by Hünig (Section IIIC).

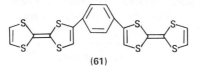

(62)   **(63)**   **(64)**

Dihydrotetrahiafulvalene **63**, $E_{1/2}^1$ and $E_{1/2}^2$ 0.405 and 0.89 V (*vs.* S.C.E.)[121] and so-called tetrathiotetralin **64** (systematic name 1,4-dithiino[2,3-*b*]-1,4-dithiin), $E_{1/2}^1$ and $E_{1/2}^2$ 0.561 and 0.965 V (*vs.* Ag/Ag$^+$–CH$_3$CN)[128] have also been made. The relatively high oxidation potentials of **64**, however, do not make it attractive for making highly conducting salts.

The characterization of cation radicals of these compounds spectroscopically is normally carried out after anodic oxidation, because chemical oxidation may lead to precipitates of complex salts. TTF is oxidized to TTF$^{\cdot+}$Cl$^-$ by a controlled amount of chlorine[89] (Section IID). In contrast, oxidation of TTF with H$_2$O$_2$–HBF$_4$ gave the salt (TTF)$_3$(BF$_4$)$_2$. Exchange of anions between this and ordinary inorganic salts gave conducting salts with complex formulae, e.g. (TTF)$_{15}$(NCS)$_8$, (TTF)$_{11}$I$_8$, (TTF)$_{24}$I$_{63}$ (reference 129).

Oxidation of the saturated analogue **65** [2,2'-bi(1,3-dithiolanylidene)] and its homologue **66** [2,2'-bi(1,3-dithianylidene)] has also been characterized[121,130–132].

(65)   **(66)**

Most curiously the cation radical **65**$^{\cdot+}$ is obtained also by the oxidative fragmentation of the orthothiooxalate **67** when it is subjected to cyclic voltammetry[130]. It is suggested that the cation radical **67**$^{\cdot+}$ decomposes into **65**$^{\cdot+}$

(equation 42), which is detected by its e.s.r. spectrum and also its electrochemical behaviour.

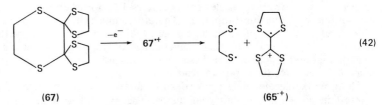

$$-e^{-} \qquad 67^{\cdot +} \longrightarrow \qquad + \qquad \qquad (42)$$

(67)   (65$^{\cdot +}$)

Oxidative rearrangements leading to the saturated systems **65**$^{\cdot +}$ and **66**$^{\cdot +}$ are also obtained with tetrathiaoctalin **68** {systematic name 2,5,7,10-tetrathiabicyclo-[4.4.0]dec-1(6)-ene}, and its homologue **69** {systematic name 2,6,8,12-tetrathia-bicyclo[5.5.0]dodec-1(7)-ene}, as illustrated with equation 43. Of course, **65**$^{\cdot +}$ is obtained and detected by repeated cyclic voltammetry[132].

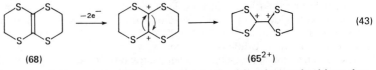

$$\xrightarrow{-2e^{-}} \qquad \longrightarrow \qquad \qquad (43)$$

(68)   (65^{2+})

The same system of rearrangements is also obtained with other orthothiooxalate fragmentations. For example, **70a** and **70b** undergo oxidative conversion into **69**, and this on being subjected to repeated cyclic voltammetry eventually leads to **66**$^{\cdot +}$ (Scheme 7)[133].

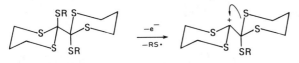

(70)

(a) R = Me; (b) R = Et

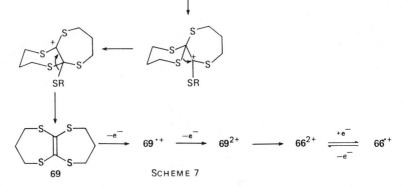

$$69^{\cdot +} \xrightarrow{-e^{-}} 69^{2+} \longrightarrow 66^{2+} \underset{-e^{-}}{\overset{+e^{-}}{\rightleftharpoons}} 66^{\cdot +}$$

69   SCHEME 7

C. Violenes

This name has been given by Hünig[134] to a class of cation radicals of the general type **71**$^{\cdot +}$. In the formulation **71**, X may be one of a variety of organic groupings, particularly heterocyclic, while n may be 0, 1, etc. The name violene is derived from the violet colour of $Ph_2N-NPh_2^+$, the prototype of one family of violenes, in which X is Ph_2N and n is 0. A large number of violenes (of many colours besides violet) have been prepared. They have been characterized mostly by their formation constants (K_f, equation 45) calculated from their one- and two-electron oxidation

$$\ddot{X}-(CH=CH)_n-\ddot{X} \underset{}{\overset{-e^-}{\rightleftharpoons}} \ddot{X}-(CH=CH)_n-\overset{\cdot+}{X} \underset{}{\overset{-e^-}{\rightleftharpoons}} \overset{+}{X}=(CH-CH)_n=\overset{+}{X} \quad (44)$$

$$(71) \qquad\qquad (71^{\cdot+}) \qquad\qquad (71^{2+})$$

$$K_f = (71^{\cdot+})^2/(71)(71^{2+}) \tag{45}$$

potentials, and sometimes by e.s.r. spectroscopy. Among the violenes are several types containing sulphur, some of which have already been described, e.g. $47^{\cdot+}$ and its homologues (Section IIIB). Most of the violenes are very stable toward disproportionation; i.e. K_f is very high, as is seen with those from compounds $72-74^{124,135}$. An exception is seen with $75^{\cdot+}$.[136]

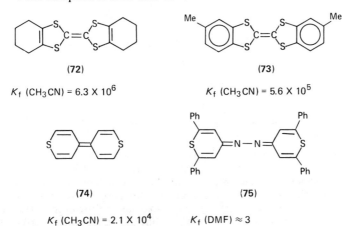

(72)

K_f (CH$_3$CN) = 6.3 X 10^6

(73)

K_f (CH$_3$CN) = 5.6 X 10^5

(74)

K_f (CH$_3$CN) = 2.1 X 10^4

(75)

K_f (DMF) $\approx$ 3

It is interesting that violenes $47^{\cdot+}$ (i.e. TTF$^{\cdot+}$) and $72^{\cdot+}$ were made for e.s.r. characterization by oxidation with bromine in acetic acid, and $62^{\cdot+}$ and $73^{\cdot+}$ similarly in methylene chloride–trifluoroacetic acid[124]; cf. Section IID.

IV. REACTIONS OF ORGANOSULPHUR CATION RADICALS

A. General Comments

For the most part the reactions of organosulphur cation radicals which have been studied have been with nucleophiles. Therefore, the known chemistry is mostly of the cationic nature of the radical ions. Furthermore, most of the reactions have been those of the heterocyclic cation radicals, that is, of phenothiazine 1, and its derivatives, thianthrene 9a and phenoxathiin 12 cation radicals. Reactions of these cation radicals with nucleophiles occur usually at the sulphur atom, and sometimes at a ring position. The overall stoichiometry of these reactions is shown by equation 46, in which an aromatic or heterocyclic organosulphur cation radical is represented by S and the nucleophile as charged Nu$^-$. Equal amounts of product and parent sulphide are formed. The equation is a simplification, however, since

$$2 \quad \underset{/}{\overset{\backslash}{S}}{}^{\bullet+} + Nu^- \longrightarrow \underset{/}{\overset{\backslash+}{S}}-Nu + \underset{/}{\overset{\backslash}{S}} \qquad (46)$$

one or two protons may also be released in reaction if either the nucleophile is uncharged (e.g., equation 47) or if substitution occurs at a ring position in the cation radical. Examples are given in the following sections.

$$2 \quad \underset{/}{\overset{\backslash}{S}}{}^{\bullet+} + H_2O \longrightarrow \underset{/}{\overset{\backslash}{S}}{=}O + \underset{/}{\overset{\backslash}{S}} + 2H^+ \qquad (47)$$

Various reactions with nucleophiles have been studied, some of them kinetically. In general, two mechanisms have been considered, one involving prior disproportionation (equations 48 and 49), and the other the so-called half-regeneration sequence (equations 50 and 51). It has been found that the latter sequence describes all cases thus far studied.

$$2 \quad \underset{/}{\overset{\backslash}{S}}{}^{\bullet+} \rightleftharpoons \underset{/}{\overset{\backslash}{S}} + \underset{/}{\overset{\backslash}{S}}{}^{2+} \qquad (48)$$

$$\underset{/}{\overset{\backslash}{S}}{}^{2+} + Nu^- \longrightarrow \underset{/}{\overset{\backslash+}{S}}-Nu \qquad (49)$$

$$\underset{/}{\overset{\backslash}{S}}{}^{\bullet+} + Nu^- \rightleftharpoons \underset{/}{\overset{\backslash\bullet}{S}}-Nu \qquad (50)$$

$$\underset{/}{\overset{\backslash}{S}}-Nu^{\bullet} + \underset{/}{\overset{\backslash}{S}}{}^{\bullet+} \longrightarrow \underset{/}{\overset{\backslash+}{S}}-Nu + \underset{/}{\overset{\backslash}{S}} \qquad (51)$$

Apart from the direct reaction (equation 50), there is another which has yet to be studied in detail. This is one of electron exchange, which may or may not be followed by further reaction. Electron exchange occurs in reactions with some nucleophiles which are easily oxidized but is also particularly important in some halide ion reactions (equation 52). Exchange may also be involved in reactions with some other anions. Exchange is a complicating feature because further reaction with X_2 may occur and in overall *product* terms be indistinguishable from the direct-reaction pathway. Particular cases will be discussed.

$$\underset{/}{\overset{\backslash}{S}}{}^{\bullet+} + X^- \rightleftharpoons \underset{/}{\overset{\backslash}{S}} + \tfrac{1}{2} X_2 \qquad (52)$$

$$\underset{/}{\overset{\backslash}{S}} + X_2 \longrightarrow product \qquad (53)$$

B. Reactions with Water

This was the first reaction to be studied in detail. It was already known in the older German work that the so-called semiquinones of phenothiazine and thianthrene reacted with water according to equation 47. The reaction was rediscovered by later workers, particularly with thianthrene[25,28] and phenoxathiin[44,49], after the nature of the cation radicals was settled (see Sections IA2 and IA3).

The mechanism of the water reaction turned out to be complex. In the early years it was found by Murata and Shine[137] that the reaction with **9a** was second order in **9a**, and the disproportionation pathway was proposed without further kinetic analysis. Parker and Eberson found, however, from reactions at various anode potentials that reaction had to be directly between the cation radical and water, rather than between the dication and water as required by disproportionation[138]. The complete and complex nature of the reaction was elucidated by Evans and Blount[139], who found that the reaction was not only second order in **9a** but also third order in water and inverse first order in acid. The overall scheme is given in equations 54–57

$$Th^{\cdot +} + H_2O \underset{}{\overset{K_1}{\rightleftharpoons}} Th(OH_2)^{\cdot +} \tag{54}$$

$$Th(OH_2)^{\cdot +} + H_2O \underset{}{\overset{K_2}{\rightleftharpoons}} Th(OH)^{\cdot} + H_3O^+ \tag{55}$$

$$Th(OH_2)^{\cdot +} + Th(OH)^{\cdot} \overset{k}{\longrightarrow} Th + Th(OH)^+ + H_2O \tag{56}$$

$$Th(OH)^+ + H_2O \overset{fast}{\longrightarrow} ThO + H_3O^+ \tag{57}$$

in which $Th^{\cdot +}$ represents **9a**, Th represents thianthrene, and ThO its 5-oxide. The intermediates, e.g., $Th(OH_2)^{\cdot +}$, represent bonding of the nucleophile at sulphur. From this sequence of reactions the rate expression (equation 58) was derived.

$$-d[Th^{\cdot +}]/dt = 2kK_1K_2[Th^{\cdot +}]^2[H_2O]^3/[H_3O^+] \tag{58}$$

Evans and Blount found also that the reaction with water was catalysed by small amounts of pyridine (see also Section IVC), and proposed the following reactions (equations 59–62, showing an S atom of $Th^{\cdot +}$):

$$Th^{\cdot +} + Py \rightleftharpoons {\Large >}S-Py^{\cdot +} \tag{59}$$

$${\Large >}S-Py^{\cdot +} + Th^{\cdot +} \longrightarrow {\Large >}S-Py^{2+} + Th \tag{60}$$

$${\Large >}S-Py^{2+} + H_2O \overset{fast}{\longrightarrow} Th(OH)^+ + PyH^+ \tag{61}$$

$$Th(OH)^+ + Py \overset{fast}{\longrightarrow} ThO + PyH^+ \tag{62}$$

How complex these reactions may be is seen further in the reaction of the chloropromazine (CPZ) cation radical with water. This reaction in the presence of phosphate and citrate buffers, has the usual stoichiometry (equation 63). McCreery

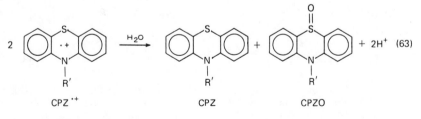

$$R' = (CH_2)_3NMe_2$$

et al.[140] found that the reaction, however, is second order in cation radical ($CPZ^{\cdot +}$), first order in buffer, and inverse order in both acid and CPZ. The following pathway is proposed (equations 64–66), and the rate expression of equation 67. In

$$CPZ^{\cdot+} + RCO_2^- + H_2O \;\xrightleftharpoons{k_1}\; [CPZ(RCO_2)(OH)]^{\cdot-} + H^+ \tag{64}$$

$$[CPZ(RCO_2)(OH)]^{\cdot-} + CPZ^{\cdot+} \;\xrightleftharpoons[k_{-2}]{k_2}\; CPZ + [CPZ(RCO_2)(OH)] \tag{65}$$

$$[CPZ(RCO_2)(OH)] \;\xrightarrow{k_3}\; CPZO + RCO_2H \tag{66}$$

$$-d[CPZ^{\cdot+}]/dt \;=\; \frac{2K_1 k_3 (k_2/k_{-2})[CPZ^{\cdot+}]^2 [RCO_2^-][H_2O]}{[H^+]([CPZ] + k_3/k_{-2})} \tag{67}$$

this proposal the intermediates are again bonded at sulphur. Thus, $[CPZ(RCO_2)(OH)]$ is thought to be as in **76**. It is noteworthy that an intermediate of analogous nature is probably involved in the pyridine-catalysed hydrolysis of thianthrene cation radical (equation 61).

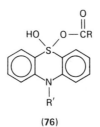

(76)

Reaction of the phenothiazine cation radical **1** in water at several pHs has also been studied but not with detailed kinetics. The reactions are linked to those of the phenothiazinium ion (**20**, Section IB). In basic solution **1** is rapidly deprotonated and the neutral phenothiazinyl radical **24** for the most part dimerizes. In weakly acidic solution deprotonation is not as extensive. Both **1** and **24** are present in solution, therefore, and undergo disproportionation (equation 68) giving **20** and

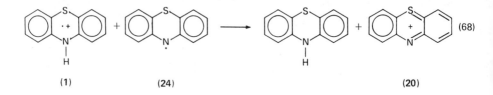

(1)   (24)   (20)

phenothiazine. Some of the radicals **24** also dimerize, and the dimer is oxidized by ion **20**. At the same time ion **20** may also react with water to give either the sulphoxide **19** or 3-hydroxyphenothiazine or its tautomer. From the last route 3-phenothiazone is eventually formed. Most of these reactions are found in Scheme 2 (Section IB).

That one may obtain reaction at both sulphur and ring carbon in the case of **1** as compared with, say **9a**, reflects the ease with which deoxygenation of phenothiazine

5-oxide **19** occurs in moderately acidic solution, thus allowing for facile competition between reactions at the two sites (Scheme 4).

C. Reactions with Ammonia and Amines

Reaction of an organosulphur cation radical with ammonia should, in principle, give a sulphilimine (an iminosulphurane, **77**), equation 69. In practice, this reaction

$$2 \; \diagup\hspace{-0.3em}\underset{\diagup}{S}{}^{\cdot +} + 3NH_3 \; \longrightarrow \; \diagup\hspace{-0.3em}\underset{\diagup}{S}{=}NH + \diagup\hspace{-0.3em}\underset{\diagup}{S} + 2NH_4{}^+ \tag{69}$$

(77)

has been achieved only with phenoxathiin cation radical **12**[71]. Reaction of 10-methyl- **2** and 10-phenylphenothiazine **78**[98] and thianthrene cation radical **9a**[141] with ammonia gave instead the dimeric cations **79**, isolated as their perchlorates, equation 70. The formation of compounds **79** no doubt involves the

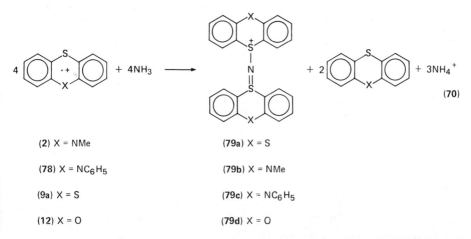

$$4 \quad [\text{ring}] \cdot{}^+ + 4NH_3 \longrightarrow [\text{ring}] + 2 [\text{ring}] + 3NH_4{}^+ \tag{70}$$

(2) X = NMe	(79a) X = S
(78) X = NC₆H₅	(79b) X = NMe
(9a) X = S	(79c) X = NC₆H₅
(12) X = O	(79d) X = O

formation and further reaction of the sulphilimine. In fact, Mani and Shine[71] found that **12** could give either **79d** or the sulphilimine **80d** (X = O) depending on the speed with which ammonia was bubbled into a solution of **12**. Isolation of the analogues of **80d** was never achieved, except that **80a** (X = S) was obtained by careful hydrolysis of **79a**[141].

Reaction of these heterocyclic cation radicals with alkyl- and dialkylamines also occurs at sulphur to give protonated *N*-alkylsulphilimines **81** and *NN*-dialkylaminosulphonium ions, **82**, equation 71, in which $\diagup\hspace{-0.3em}S{}^{\cdot +}$ is used

(80)

symbolically[98,142–144]. Treatment of **81** with strong base led readily to the N-alkylsulphilimines.

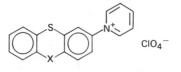

$$2 \quad \overset{\backslash}{\underset{/}{S}}\!\cdot^{+} + 2R^1R^2NH \quad \longrightarrow \quad \overset{\backslash}{\underset{/}{S}}\!\!-\!\!N\overset{R^1}{\underset{R^2}{\diagdown}} + \overset{\backslash}{\underset{/}{S}} + R^1R^2\overset{+}{N}H_2 \qquad (71)$$

(81) R^1 = H, R^2 = alkyl

(82) R^1 = R^2 = alkyl

In contrast with these reactions at sulphur, reaction with pyridine led to the ring-substituted compounds **83**. Shine et al.[96] commented on this distinction, and

ClO$_4^-$

(83a) X = S;   (83b) X = NH[96];   (83c) X = NC$_6$H$_5$[145]

pointed out that reaction of a nucleophile at sulphur can succeed only when the intermediate dication (shown symbolically as **84**, equation 72) formed during

$$\overset{\backslash}{\underset{/}{S}}\!\cdot^{+} + NuH \quad \longrightarrow \quad \longrightarrow \quad \overset{\backslash}{\underset{/}{S}}\!\!-\!\!\overset{+}{N}uH \quad \longrightarrow \quad \overset{\backslash}{\underset{/}{S}}\!\!-\!\!Nu + H^+ \qquad (72)$$

(84)

reaction can lose one or more protons (e.g. when the nucleophile is H$_2$O, NH$_3$, RNH$_2$, etc). In the case of pyridine as nucleophile it was proposed that the intermediate **85** could not be stabilized and therefore its formation was reversible, giving place instead to reaction at a ring position **86**, from which position a proton

(85a) X = S;   (85b) X = NC$_6$H$_5$ (86)

is removable. Evans et al.[109] found by kinetic analysis of the reaction of **78** with pyridine that **85b** was indeed formed and was, in fact, the direct precursor to ring pyridination. That is, the S-pyridinium group was displaced by attack of pyridine at the 3-position of **85**, the overall reaction being second order in pyridine (as well as second order in **78** and inverse first order in 10-phenylphenothiazine)[109]. It has already been noted (Section IVB) that **85a** is also, in fact, an intermediate in the pyridine-catalysed reaction of **9a** with water, reaction of a second molecule of nucleophile (H$_2$O) now occurring at sulphur, too.

Reactions of aromatic amines with organosulphur cation radicals have not been investigated much, probably because electron exchange occurs so easily. NN-Dimethylaniline and acetanilide react with **9a**, however (Section IVD).

D. Reactions with Aromatics

Organosulphur cation radicals behave as weak electrophiles towards aromatics. Reaction has been studied mostly with **9a** and proceeds as in equation 73[146,147]. The

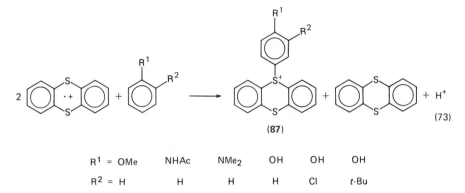

(73)

(87)

$$R^1 = OMe \qquad NHAc \qquad NMe_2 \qquad OH \qquad OH \qquad OH$$

$$R^2 = H \qquad H \qquad H \qquad H \qquad Cl \qquad t\text{-}Bu$$

aromatic must carry one or more electron-donating substituents, and reaction is too slow to be useful with benzene, chlorobenzene, and nitrobenzene. Reaction with aniline led only to oxidation of the aniline, but the corresponding product (87; $R^1 = NH_2 \; R^2 = H$) has been obtained by hydrolysis of the acetyl derivative (87; $R^1 = NHAc, R^2 = H$)[147].

Studies with phenol, anisole, and acetanilide have shown[148–150] that these reactions also follow, in principle, the half-regeneration mechanism (equations 74–76) rather than, as earlier proposed[147], the disproportionation mechanism, but particular variations in mechanism are seen with some aromatics.

$$Th^{\cdot +} + C_6H_5X \rightleftharpoons (Th-C_6H_5X)^{\cdot +} \tag{74}$$

$$(Th-C_6H_5X)^{\cdot +} + Th^{\cdot +} \rightleftharpoons (Th-C_6H_5X)^{2+} + Th \tag{75}$$

$$(Th-C_6H_5X)^{2+} \longrightarrow (Th-C_6H_4X)^+ + H^+ \tag{76}$$

That is, at relatively high concentrations of $Th^{\cdot +}$ (i.e. 9a) reaction with anisole is second order in $Th^{\cdot +}$ and follows the half-regeneration sequence. At relatively low concentration the reaction becomes first order in $Th^{\cdot +}$ and leads to thianthrene rather than 87 ($R^1 = OMe$, $R^2 = H$). Apparently there is in this case insufficient $Th^{\cdot +}$ competitively to oxidize (equation 75) the first-formed complex (equation 74). The complex therefore undergoes internal electron transfer to give thianthrene and, it is presumed, the anisole cation radical[148].

In the case of phenol, the fate of the first-formed complex depends on whether the solution is initially acidic or neutral[149]. In acidic solution the half-regeneration sequence prevails. In neutral solution, however, the reaction again becomes first order in $Th^{\cdot +}$, the rate-limiting process now being deprotonation of the complex (equation 77) rather than its oxidation by a second molecule of $Th^{\cdot +}$ (equation 75).

$$(Th-C_6H_5OH)^{\cdot +} \rightleftharpoons (Th-C_6H_4OH)^{\cdot} + H^+ \tag{77}$$

$$(Th-C_6H_4OH)^{\cdot} + Th^{\cdot +} \rightleftharpoons (Th-C_6H_4OH)^+ + Th \tag{78}$$

The competition between electron exchange and substitution noted with anisole[148] is also seen in reaction with NN-dimethylaniline. Not only is 87 ($R^1 = NMe_2$, $R^2 = H$) obtained but also tetramethylbenzidine (as its cation radical)[147].

Analogous substitution reactions have been obtained with anodic oxidations of phenyl sulphides (C_6H_5SR, R = alkyl and phenyl)[151]. The products were sulphonium perchlorates 88, obtained presumably from reaction of the cation radical ($C_6H_5SR^{\cdot +}$) with the sulphide.

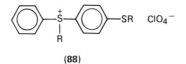

(88)

E. Reactions with Organometallics

This is a relatively newly discovered reaction of the heterocyclic cation radicals (equation 79)[152]. It works best between thianthrene **9a** and phenoxathiin cation

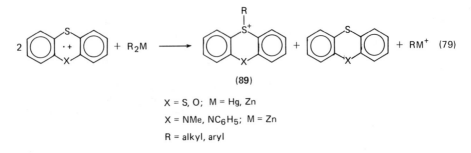

(89)

X = S, O; M = Hg, Zn

X = NMe, NC_6H_5; M = Zn

R = alkyl, aryl

radical **12** and organomercurials, e.g. R_2Hg in which R is Me, Et, C_6H_5, o-MeC_6H_4, p-ClC_6H_4, etc. The reaction is very convenient and wider in scope than reaction with aromatics (Section IVD) because not only can mildly deactivated aryl groups be used (e.g., R = C_6H_5, C_6H_4Cl) but also substitution in the aryl ring can now be achieved at the *ortho*- and *meta*- as well as the *para*-position. Reaction with highly deactivated mercurials, e.g. $(O_2NC_6H_4)_2Hg$, does not take place. The reaction allows alkyl groups to be placed on sulphur, too, although not much exploration of the scope in this case has been carried out.

The phenothiazine cation radicals (X = N-Me, N-C_6H_5) appear to be inert to reaction with organomercurials. However, reaction occurs with organozincs (e.g., R_2Zn, R = Et, C_6H_5).

The yields of product **89** are good in most cases. Thus for X = S the yields were for R = Me 83%, Et 39%, E_6H_5 95%, p-MeC_6H_4 86%, o-MeC_6H_4 76%, p-ClC_6H_4 84%, and m-ClC_6H_4 82%. Comparable yields were obtained for X = O. On the other hand with X = N-C_6H_5 and use of diethylzinc only 31% of product was obtained.

Nothing is yet known about the mechanism of these organometal reactions. It is evident that displacement of R occurs such that bonding to sulphur is at the carbon atom previously bonded to mercury. Therefore, the reactions may be direct displacements such as has been observed in protonolyses of some R_2Hg compounds with retention of configuration[153]. Some reactions of dialkylmercurials with electron acceptors (e.g. $IrCl_6^{2-}$)[154] involve electron exchange first. Whether or not the cation radical reactions may follow such a path, particularly when R is alkyl, needs to be considered. The low yield of **89** (X = S, R = Et) suggests that in this case, especially, a route to reaction may be involved in which competing reactions are severe, and this may be symptomatic of radical reactions following electron exchange.

F. Reactions with Ketones

This reaction was discovered by accident when thianthrene cation radical was allowed to come into contact with acetone. The reaction can be expressed in general terms by equation 80 in which $\diagup S$ represents one of the heterocyclic compounds. To our knowledge simpler organosulphur cation radicals have not been used. The

$$2 \diagup{S}^{\cdot +} + R_2CHCOR' \longrightarrow \underset{\underset{ClO_4^-}{\diagup S^+ \diagdown}}{\overset{R_2CCOR'}{|}} + \diagup{S}\diagdown + HClO_4 \qquad (80)$$

ketone must possess an α-H atom. Among ketones which have been used are acetone and its homologues, cyclopentanone and cyclohexanone, 1-indanone and 1-tetralone, and some diketones such as dimedone and dibenzoylmethane[155–157]. Following the stoichiometry of equation 80 the products are β-ketoalkylsulphonium perchlorates isolable in good yields. In some cases, particularly with diones the product is an ylide, e.g. **90** from reaction of pentane-1,3-dione with **12**[156].

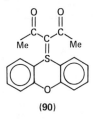

(90)

The β-ketoalkylsulphonium salts are themselves easily converted into ylides by treatment with base, and also serve as a source for preparing α-substituted ketones by reaction with nucleophiles[156,157]. The heterocycles are excellent leaving groups, shown schematically in equation 81.

$$\underset{\diagup S^+ \diagdown}{\overset{X^- \frown CH_2COR}{|}} \longrightarrow XCH_2COR + \diagup{S}\diagdown \qquad (81)$$

The mechanism of reaction of these organosulphur cation radicals with ketones has not beeen explored. It has been suggested, though, that reaction may occur with the enol (equation 82) and in conformity with this possibility reaction occurs at the more substituted α-C atom of an unsymmetrical ketone[157]. Whether or not deprotonation of the adduct occurs before reaction with the second $\diagup S^{\cdot +}$ is also a matter of speculation.

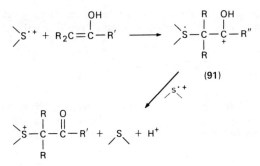

G. Reactions with Alkenes and Alkynes

Equation 82 shows the addition of an organosulphur cation radical to a C−C double bond. Firm evidence that such additions do take place has been obtained in reactions of **9a** and **12** with a number of alkenes, but the stoichiometry of addition follows equation 83 (X = S, O). The reaction is thought to occur in stages as shown

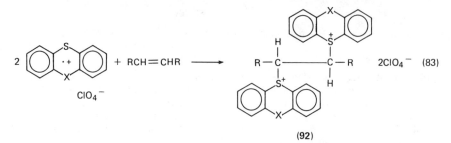

(92)

schematically in equation 84, and it appears that adduct **93** differs from **91** in combining with another cation radical than losing a proton. However, the

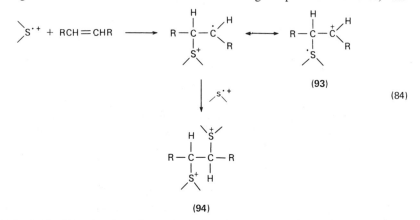

(93)

(84)

(94)

mechanism of this reaction (equation 83), like that of the ketone reaction (equation 80), has not been explored.

Among alkenes which react according to equation 83 are oct-1-ene, but-2-ene, cyclopentene, cyclohexene, and cycloheptene, each giving isolable adducts **92** with either **9a** or **12**[158].

Most interestingly also, analogous addition reactions occur with alkynes (equation 85). Again, diperchlorate adducts have been isolated and characterized by

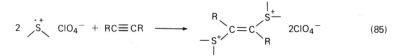

n.m.r. and elemental analysis[158]. The latter has been troublesome but successfully managed†, since the adducts are sensitive to heating. Adducts have been obtained from acetylene, propyne, but-2-yne, phenylacetylene, and diphenylacetylene.

†We thank Schwarzkopf Laboratories, Woodside, N.Y., for their patient work.

The scope of these addition reactions with alkenes and alkynes has not been explored, although it appears that electron exchange competes with addition (e.g. to *cis*- and *trans*-stilbene), and that electron-poor alkynes are inert to addition (e.g. propargyl chloride). The phenothiazine cation radicals **2** and **78** also appear to be unreactive.

H. Reactions with Inorganic Anions (Halide, Nitrite, Nitrate)

The simplest reaction to be expected with halide and nitrite ions is ring substitution expressed in general terms in equation 86. Whether or not direct substitution occurs is not yet known, because reaction is complicated by the possibility of electron exchange (equation 87) followed by electrophilic substitution (equation 88). The overall stoichiometry is the same for the two possibilities.

$$2(ArH)_2S^{\cdot+} + X^- \longrightarrow (ArX)S(ArH) + (ArH)_2S + H^+ \qquad (86)$$

$$2(ArH)_2S^{\cdot+} + 2X^- \rightleftharpoons 2(ArH)_2S + X_2 \qquad (87)$$

$$(ArH)_2S + X_2 \longrightarrow (ArX)S(ArH) + H^+ + X^- \qquad (88)$$

Electron exchange is often complete with excess of iodide ion, and is used in fact in iodimetric assay of some organosulphur cation radicals, **9a** and its analogues. The exchange is written as a reversible reaction since, as described in Section IID, oxidation by halogens, even iodine, can be used to prepare some cation radicals. Exchange with fluoride ion is in principle not possible because of the high oxidation potential of fluoride ion. Therefore, we would anticipate that nucleophilic fluorination should occur (equation 86). However, to our present knowledge no report of a successful reaction of an organosulphur cation radical with fluoride ion has ever been made. Reaction of phenothiazine cation radical **1** with fluoride ion gave only phenothiazine, its 3,10'-dimer, and the green dimer cation[96]. Apparently, fluoride served only as a base, and a sequence of reactions followed like those in Scheme 2 (Section IB). Reaction of phenoxathiin cation radical **12** also failed to give a fluorophenoxathiin[159].

Reaction of **1** with chloride and bromide ion gave phenothiazine (PT, about 45% conversion), 3-halogenophenothiazine **95** (about 35%), and some 3,7-dihalogenophenothiazine **96** (about 5%)[96]. The formation of the last product is particularly perplexing and leads to the feeling that electron exchange (equation 87) occurred and that electrophilic halogenation was responsible for the mono- and dihalogeno products. It is not known, though, that successive half-regeneration reactions (equations 89–93) can be ruled out since it is probable that the oxidation potentials of **95** and phenothiazine would not be too far apart.

$$1 + X^- \rightleftharpoons (1-X)^{\cdot} \qquad (89)$$

$$(1-X)^{\cdot} + 1 \longrightarrow PT + 95 + H^+ \qquad (90)$$

$$95 + 1 \rightleftharpoons 95^{\cdot+} + PT \qquad (91)$$

$$95^{\cdot+} + X^- \rightleftharpoons (95-X)^{\cdot} \qquad (92)$$

$$(95-X)^{\cdot} + 1 \longrightarrow 96 + PT + H^+ \qquad (93)$$

Reaction of chloride ion with **9a** resulted in part in electron exchange since some chlorine (9%) was removed by nitrogen flow, but only a very small amount (0.3%) of 2-chlorothianthrene was obtained[137]. In the reaction of **12** with both chloride and bromide ion free halogen was detectable by smell and starch–iodide test, and a monohalogenophenoxathiin was obtained[159]. These results show that more study is needed if a complete understanding of the halide ion reactions is to be obtained.

Reactions with nitrite ion can lead to both nitration and oxygen atom transfer. Thus, **1** gave 3-nitrophenothiazine according to the usual stoichiometry for nucleophilic reactions (equation 86)[96]. In contrast, **9a** and **12** gave total conversion into the 5-oxides[96,159]. This reaction has been interpreted as in the sequence of equation 94, but we do not really know if, perhaps, electron exchange does not occur first here too; that is if reaction is not between the parent compounds and NO_2 (equation 95). A more complicated reaction has been found with

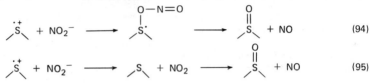

N-phenylphenothiazine cation radical **78**. In the absence of air reaction goes mainly according to equation 94, but in the presence of air continued 'nitration' occurs, giving a mixture of the 5-oxide and its 3-nitro- and two dinitro derivatives. Apparently the 5-oxide reacts further with NO_2 obtained from reaction of NO (equation 94) with air[160].

Reactions of **9a** and **12** with nitrate ion also gave the 5-oxides[96,159]. These reactions have been interpreted as in equation 96, but although NO_2 was formed, further confirmatory mechanistic studies have not been made[161].

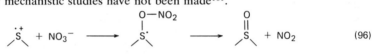

V. ACKNOWLEDGEMENTS

The writer is indebted to numerous co-workers, some of whom are named in references to his work, for their collaborative contributions to the chemistry described herein. He acknowledges also generous support of the Robert A. Welch Foundation (Grant No. D-028), the National Science Foundation, Texas Tech University Institute for Research, and in the early years the Air Force Office of Scientific Research.

VI. REFERENCES

1. E. J. Zavoiskii, *J. Phys. (USSR)*, **9**, 211 (1945).
2. C. Lagercrantz, *Acta Chem. Scand.*, **15**, 1545 (1961).
3. J.-P. Billon, G. Cauquis, J. Combrisson and A.-M. Li, *Bull. Soc. Chim. Fr.*, 2062 (1961).
4. J.-P. Billon, *Bull. Soc. Chim. Fr.*, 1784 (1960).
5. F. W. Heineken, M. Bruin and F. Bruin, *J. Chem. Phys.*, **37**, 1479 (1962).
6. L. D. Tuck and D. W. Schiesser, *J. Phys. Chem.*, **66**, 937 (1962).
7. D. Gagnaire, H. Lemaire, A. Rassat and P. Servoz-Gavin, *C.R. Acad. Sci.*, **255**, 1441 (1962).
8. J.-P. Billon, G. Cauquis and J. Combrisson, *J. Chim. Phys.*, **61**, 374 (1964).
9. S. Odiot and F. Tonnard, *J. Chim. Phys.*, **61**, 382 (1964).
10. B. C. Gilbert, P. Hanson, R. O. C. Norman and B. T. Sutcliffe, *J. Chem. Soc. Chem. Commun.*, 161 (1966).
11. J. M. Lhoste and F. Tonnard, *J. Chim. Phys.*, **63**, 678 (1966).
12. I. S. Forest, F. M. Forest and M. Berger, *Biochim. Biophys. Acta*, **29**, 441 (1958).
13. L. H. Piette and I. S. Forest, *Biochim. Biophys. Acta*, **57**, 419 (1962).
14. D. C. Borg and G. C. Cotzias, *Proc. Natl. Acad. Sci. USA*, **48**, 617, 623, 643 (1962).
15. J.-P. Billon, *Ann. Chim.*, **7**, 183 (1962).

16. H. Fenner and H. Möckel, *Tetrahedron Lett.*, 2815 (1969).
17. L. Michaelis, M. P. Schubert and S. Granick, *J. Amer. Chem. Soc.*, **62**, 204 (1940).
18. S. Granick, L. Michaelis and M. P. Schubert, *J. Amer. Chem. Soc.*, **62**, 1802 (1940).
19. L. Michaelis, S. Granick and M. P. Schubert, *J. Amer. Chem. Soc.*, **63**, 351 (1941).
20. F. Kehrmann, J. Speitel and E. Grandmougin, *Chem. Ber.*, **47**, 2796 (1914).
21. F. Kehrmann, *Justus Liebigs Ann. Chem.*, **322**, 1 (1902).
22. R. Pummerer and S. Gassner, *Chem. Ber.*, **46**, 2310 (1913).
23. R. Pummerer, F. Eckert and S. Gassner, *Chem. Ber.*, **47**, 1494 (1914).
24. J. Stenhouse, *Proc. R. Soc. (London)*, **17**, 62 (1868); *Justus Liebigs Ann. Chem.*, **149**, 247 (1869).
25. W. C. Needler, *PhD Dissertation*, University of Minnesota, 1961, *Diss. Abstr.*, **22**, 3873 (1962).
26. M. Kinoshita and H. Akamatu, *Bull, Chem. Soc. Jap.*, **35**, 1040 (1962).
27. M. Kinoshita, *Bull. Chem. Soc. Jap.*, **35**, 1137 (1962).
28. H. J. Shine and L. Piette, *J. Amer. Chem. Soc.*, **84**, 4798 (1962).
29. E. A. C. Lucken, *J. Chem. Soc.*, 4963 (1962).
30. J. E. Wertz and J. L. Vivo, *J. Chem. Phys.*, **23**, 2193 (1955).
31. A. Fava, P. B. Sogo and M. Calvin, *J. Amer. Chem. Soc.*, **79**, 1078 (1957).
32. K. Fries and E. Engelbertz, *Justus Liebigs Ann. Chem.*, **407**, 194 (1915).
33. R. Wizinger, *Z. Angew. Chem.*, **42**, 668 (1929).
34. H. J. Shine, C. F. Dais and R. J. Small, *J. Org. Chem.*, **29**, 21 (1964).
35. E. A. C. Lucken, *Theor. Chim. Acta*, **1**, 397 (1963).
36. H. J. Shine and P. D. Sullivan, *J. Phys. Chem.*, **72**, 1390 (1968).
37. J. M. Hirshon, D. M. Gardner and C. R. Fraenkel, *J. Amer. Chem. Soc.*, **75**, 4115 (1953).
38. T. P. Hilditch and S. Smiles, *J. Chem. Soc., Trans.*, **49**, 408 (1911).
39. H. D. K. Drew, *J. Chem. Soc.*, 511 (1928).
40. B. Lamotte, A. Rassat and P. Servoz-Gavin, *C.R. Acad. Sci.*, **255**, 1508 (1962).
41. M. Tomita, S. Ueda, Y. Nakai, Y. Deguchi, and H. Takaki, *Tetrahedron Lett.*, 1189 (1963).
42. U. Schmidt, K. Kabitzke and K. Markau, *Chem. Ber.*, **97**, 498 (1964).
43. B. Lamotte and G. Berthier, *J. Chim. Phys.*, **63**, 369 (1966).
44. G. Barry, G. Cauquis and M. Maurey, *Bull. Soc. Chim. Fr.*, 2510 (1966).
45. P. D. Sullivan, *J. Amer. Chem. Soc.*, **90**, 3618 (1968).
46. P. D. Sullivan and J. Bolton, *J. Magn. Reson.*, **1**, 356 (1969).
47. M. F. Chiu, B. C. Gilbert and P. Hanson, *J. Chem. Soc. B*, 1700 (1970).
48. G. C. Yang and A. E. Pohland, *J. Phys. Chem.*, **76**, 1504 (1972).
49. H. J. Shine and R. J. Small, *J. Org. Chem.*, **30**, 2140 (1965).
50. H. J. Shine, M. Rahman, H. Seeger and G.-S. Wu, *J. Org. Chem.*, **32**, 1901 (1967).
51. U. Schmidt, *Angew. Chem.*, **76**, 629 (1964).
52. H. J. Shine and E. E. Mach, *J. Org. Chem.*, **30**, 2130 (1965).
53. H. J. Shine, in *Organosulfur Chemistry* (Ed., M. J. Janssen), Interscience, New York, 1967, pp. 93–115.
54. H. J. Shine, *Mech. Reac. Sulfur Compd.*, **3**, 155 (1968).
55. (a) K. Fries and W. Volk, *Chem. Ber.*, **42**, 1170 (1909).
    (b) K. Fries and W. Vogt, *Justus Liebigs Ann. Chem.*, **381**, 312 (1911).
56. E. de B. Barnett and S. Smiles, *J. Chem. Soc.*, **95**, 1253 (1909); **97**, 186 (1910).
57. P. Hanson and R. O. C. Norman, *J. Chem. Soc. Perkin Trans. II*, 264 (1973).
58. H. J. Shine and T. A. Robinson, *J. Org. Chem.*, **28**, 2828 (1963).
59. E. J. Buncel and W. M. J. Strachan, *Can. J. Chem.*, **47**, 911 (1969). See also R. A. Cox and E. Buncel, in *Chemistry of the Hydrazo, Azo and Azoxy Groups, Part 2* (Ed. S. Patai), Wiley, New York, 1975, p. 814.
60. S. Oae, and N. Kunieda, *Bull. Chem. Soc. Jap.*, **41**, 696 (1968).
61. N. Kunieda and S. Oae, *Bull. Chem. Soc. Jap.*, **42**, 1324 (1969).
62. S. Oae, in *Organic Chemistry of Sulfur* (Ed. S. Oae), Plenum Press, New York, 1977, pp. 413–419, for a summary of racemization and exchange in sulphoxides.
63. D. Landini, G. Modena, G. Scorrano and F. Taddei, *J. Amer. Chem. Soc.*, **91**, 6703 (1969).

64. T. N. Tozer, *PhD Dissertation*, University of California, San Francisco, 1964.
65. D. Clarke, B. C. Gilbert and P. Hanson, *J. Chem. Soc. Perkin Trans II*, 1078 (1975); 114 (1976).
66. I. Baciu, M. Hillebrand and V. E. Sahini, *J. Chem. Soc. Perkin Trans II*, 986 (1974).
67. A. Carrington, F. Dravnieks and M. C. R. Symons., *J. Chem. Soc.*, 947 (1959).
68. A. J. Bard, A. Ledwith and H. J. Shine, *Adv. Phys. Org. Chem.*, 13, 155 (1976).
69. P. A. Malachesky, L. S. Marcoux and R. N. Adams, *J. Phys. Chem.*, 70, 2064 (1966).
70. C. MacLean and J. H. van der Waals, *J. Chem. Phys.*, 27, 287 (1957).
71. S. R. Mani and H. J. Shine, *J. Org. Chem.*, 40, 2756 (1975).
72. H. J. Shine and T. A. Robinson, unpublished work, quoted in ref. 74.
73. U. Schmidt, K. Kabitzke, K. Markau and A. Müller, *Justus Liebigs Ann. Chem.*, 672, 78 (1964).
74. H. J. Shine, M. Rahman, H. Nicholson and R. K. Gupta, *Tetrahedron Lett.*, 5255 (1968).
75. J. A. Brivati, R. Hulme and M. C. R. Symons, *Proc. Chem. Soc.*, 384 (1961); J. R. Bolton and A. Carrington, *Proc. Chem. Soc.*, 385 (1961).
76. Y. Sato, M. Kinoshita, M. Sano and H. Akamatu, *Bull. Chem. Soc. Jap.*, 40, 2539 (1967); 42, 548 (1969).
77. A. Zweig and W. G. Hodgson, *Proc. Chem. Soc.*, 417 (1964).
78. H. M. Buck, W. Bloemhoff and L. J. Oosterhoff, *Tetradedron Lett.*, 5 (1960).
79. W. Forbes and P. D. Sullivan, *J. Amer. Chem. Soc.*, 88, 2862 (1966).
80. W. F. Forbes and P. D. Sullivan, *Can. J. Chem.*, 46, 317 (1968).
81. P. D. Sullivan, *Int. J. Sulfur Chem. (A)*, 2, 149 (1972).
82. P. D. Sullivan and L. J. Norman, *J. Magn. Reson.*, 23, 395 (1976).
83. D. H. Geske and M. V. Merritt, *J. Amer. Chem. Soc.*, 91, 6921 (1969).
84. G. F. Pedulli, V. Vivarelli, P. Dembech, A. Ricci and G. Seconi, *Int. J. Sulfur Chem.*, 8, 255 (1973).
   (a) F. B. Bramwell, R. C. Haddon, F. Wudl, M. L. Kaplan and J. H. Marshall, *J. Amer. Chem. Soc.*, 100, 4612 (1978).
85. L. Lunazzi, G. Placucci and M. Tieco, *Tetrahedron Lett.*, 3847 (1972).
86. G. Brunton, B. C. Gilbert and R. J. Mawby, *J. Chem. Soc. Perkin Trans. II*, 1267 (1976).
87. I. Boeseken, *Rec. Trav. Chim. Pays Bas*, 24, 209 (1905).
88. J. J. Rooney and R. C. Pink, *Proc. Chem. Soc.*, 142 (1961).
89. F. Wudl, G. M. Smith and E. J. Hufnagel, *J. Chem. Soc. Chem. Commun.*, 1453 (1970).
90. C. Marschalk, *Bull. Soc. Chim. Fr.*, 147 (1952).
91. F. Kehrmann and L. Diserens, *Chem. Ber.*, 48, 318 (1915).
92. Y. Matsunaga and K. Shono, *Bull. Chem. Soc. Jap.*, 43, 2007 (1970).
93. Y. Iida, *Bull. Chem. Soc. Jap.*, 44, 663 (1971).
94. Y. Matsunaga and Y. Suzuki, *Bull. Chem. Soc. Jap.*, 46, 719 (1973).
95. H.-S. Chiou, P. C. Reeves, and E. R. Biehl, *J. Heterocycl. Chem.*, 13, 77 (1976).
96. H. J. Shine, J. J. Silber, R. J. Bussey, and T. Okuyama, *J. Org. Chem.*, 37, 2691 (1972).
97. M. H. Litt and J. Radovic, *J. Phys. Chem.*, 78, 1750 (1974).
98. B. K. Bandlish, A. G. Padilla, and H. J. Shine, *J. Org. Chem.*, 40, 2590 (1975).
99. H. J. Shine and coworkers, unpublished work.
100. W. Rundel and K. Scheffler, *Tetrahedron Lett.*, 993 (1963).
101. L. Eberson and K. Nyberg, *Adv. Phys. Org. Chem.*, 12, 1 (1976).
102. L. Eberson and K. Nyberg, *Tetrahedron*, 32, 2185 (1976).
103. D. H. Evans, *Acc. Chem. Res.*, 10, 313 (1977).
104. K. J. Vetter, *Electrochemical Kinetics*, Academic Press, New York, 1967, pp. 10–22.
105. A. Zweig, W. G. Hodgson and W. H. Jura, *J. Amer. Chem. Soc.*, 86, 4124 (1964).
106. A. Zweig and J. E. Lehnson, *J. Amer. Chem. Soc.*, 87, 2647 (1965).
107. A. Zweig, A. H. Maurer and B. G. Roberts, *J. Org. Chem.*, 32, 1322 (1967).
108. H. Bock and G. Brähler, *Angew. Chem. Int. Ed. Engl.*, 16, 855 (1977).
109. J. F. Evans, J. R. Lenhard and H. N. Blount, *J. Org. Chem.*, 42, 983 (1977).
110. O. Hammerich and V. D. Parker, *Electrochim. Acta*, 18, 537 (1973).
111. G. Meissner, A. Henglein and G. Beck, *Z. Naturforsch.*, 22b, 13 (1967).
112. B. C. Gilbert, D. K. C. Hodgeman and R. O. C. Norman, *J. Chem. Soc. Perkin Trans. II*, 1748 (1973).

113. M. Bonifacic, H. Möckel, D. Bahnemann and K.-D. Asmus, *J. Chem. Soc. Perkin Trans. II*, 675 (1975).
114. B. C. Gilbert, J. P. Larkin and R. O. C. Norman, *J. Chem. Soc. Perkin, Trans. II*, 272 (1973).
115. D. Bahnemann and K.-D. Asmus, *J. Chem. Soc. Chem. Commun.*, 238 (1975).
116. M. Bonifacic, K. Schäfer, H. Möckel, and K.-D. Asmus, *J. Phys. Chem.*, **79**, 1496 (1975).
117. B. C. Gilbert, H. A. H. Laue, R. O. C. Norman and R. C. Sealy, *J. Chem. Soc. Perkin, Trans. II*, 892 (1975).
118. M. Bonifacic and K.-D. Asmus, *J. Phys. Chem.*, **80**, 2426 (1976).
119. W. K. Musker and T. L. Wolford, *J. Amer. Chem. Soc.*, **98**, 3055 (1976).
     W. K. Musker and P. B. Roush, *J. Amer. Chem. Soc.*, **98**, 6745 (1976).
120. H. C. Box and E. E. Budzinski, *J. Chem. Soc. Perkin Trans. II*, 553 (1976).
121. D. L. Coffen, J. Q. Chambers, D. R. Williams, P. E. Garrett and N. D. Canfield, *J. Amer. Chem. Soc.*, **93**, 2258 (1971).
122. A. Mas, J.-M. Fabre, E. Torreilles, L. Giral, and G. Brun, *Tetrahedron Lett.*, 2579 (1977).
123. F. Wudl, A. A. Kruger, M. L. Kaplan and R. S. Hutton, *J. Org. Chem.*, **42**, 768 (1977).
124. S. Hünig, G. Kiesslich, H. Quast, and D. Scheutzow, *Justus Liebigs Ann. Chem.*, 310 (1973).
125. P. R. Moses and J. Q. Chambers, *J. Amer. Chem. Soc.*, **96**, 945 (1974).
126. M. L. Kaplan, R. C. Haydon and F. Wudl, *J. Chem. Soc. Chem. Commun.*, 388 (1977).
127. S. Hünig, H. Schlaf, G. Kiesslich and D. Scheutzow, *Tetrahedron Lett.*, 2271 (1969).
128. M. Mizuno, M. P. Cava and A. F. Garito, *J. Org. Chem.*, **41**, 1484 (1976).
129. F. Wudl, *J. Amer. Chem. Soc.*, **97**, 1962 (1975).
130. N. D. Canfield, J. Q. Chambers and D. L. Coffen, *J. Electroanal. Chem.*, **24**, A7 (1970).
131. J. Q. Chambers, N. D. Canfield, D. R. Williams and D. L. Coffen, *Mol. Phys.*, **19**, 581 (1970).
132. R. M. Harnden, P. R. Moses and J. Q. Chambers, *J. Chem. Soc. Chem. Commun.*, 11, (1977).
133. P. R. Moses, R. M. Harnden and J. Q. Chambers, *J. Electroanal. Chem.*, **84**, 187 (1977).
134. S. Hünig, *Pure Appl. Chem.*, **15**, 109 (1967).
135. S. Hünig, B. J. Garner, G. Ruider and W. Schenk, *Justus Liebigs Ann. Chem.*, 1036 (1973).
136. S. Hünig and G. Ruider, *Justus Liebigs Ann. Chem.*, 1415 (1974).
137. Y. Murata and H. J. Shine, *J. Org. Chem.*, **34**, 3368 (1969).
138. V. D. Parker and L. Eberson, *J. Amer. Chem. Soc.*, **92**, 7488 (1970).
139. J. Evans and H. N. Blount, *J. Org. Chem.*, **42**, 976 (1976).
140. H. Y. Cheng, P. H. Sackett and R. L. McCreery, *J. Amer. Chem. Soc.*, **100**, 962 (1978).
141. H. J. Shine and J. J. Silber, *J. Amer. Chem. Soc.*, **94**, 1026 (1972).
142. H. J. Shine and K. Kim, *Tetrahedron Lett.*, 99 (1974).
143. K. Kim and H. J. Shine, *J. Org. Chem.*, **39**, 2537 (1974).
144. B. K. Bandlish, S. R. Mani and H. J. Shine, *J. Org. Chem.*, **42**, 1538 (1972).
145. H. J. Shine, B. K. Bandlish and A. G. Padilla, unpublished work.
146. J. J. Silber and H. J. Shine, *J. Org. Chem.*, **36**, 2923 (1971).
147. K. Kim, V. J. Hull and H. J. Shine, *J. Org. Chem.*, **39**, 2534 (1974).
148. U. Svanholm, O. Hammerich and V. D. Parker, *J. Amer. Chem. Soc.*, **97**, 101 (1975).
149. U. Svanholm and V. D. Parker, *J. Amer. Chem. Soc.*, **98**, 997 (1976).
150. U. Svanholm and V. D. Parker, *J. Chem. Soc. Perkin Trans. II*, 1567 (1976).
151. S. Torii, Y. Matsuyama, K. Kawasaki and K. Uneyama, *Bull. Chem. Soc. Jap.*, **46**, 2912 (1973).
152. B. K. Bandlish, W. R. Porter, Jr. and H. J. Shine, *J. Phys. Chem.*, **82**, 1168 (1978).
153. F. R. Jensen and B. Rickborn, *Electrophilic Substitution of Organomercurials*, McGraw-Hill, New York, 1968.
154. J. Y. Chen, H. C. Gardner and J. K. Kochi, *J. Amer. Chem. Soc.*, **98**, 6150 (1976).
155. K. Kim and H. J. Shine, *Tetrahedron Lett.*, 4413 (1974).
156. K. Kim, S. R. Mani and H. J. Shine, *J. Org. Chem.*, **40**, 3857 (1975).
157. A. G. Padilla, B. K. Bandlish and H. J. Shine, *J. Org. Chem.*, **42**, 1833 (1977).

158. H. J. Shine, B. K. Bandlish, S. R. Mani and A. G. Padilla, *J. Org. Chem.*, **44**, 915 (1979).
159. S. R. Mani, *PhD Dissertation*, Texas Tech. University, 1976.
160. B. K. Bandlish, K. Kim and H. J. Shine, *J. Heterocycl. Chem.*, **14**, 209 (1977).
161. The kinetics and mechanisms of the reaction of thianthrene cation radical with nitrate ion have now been studied. The reaction is second order in $Th^{\cdot+}$ and first order in nitrate ion, and it is proposed that a complex of these reactants leads to two molecules of ThO and one of NO^+. The NO^+ is thought to be converted into N_2O_4 by reaction with NO_3^-. J. E. Pemberton, G. L. McIntire, H. N. Blount, and J. F. Evans, *J. Phys. Chem.*, **83**, 2696 (1979).

The Chemistry of the Sulphonium Group
Edited by C. J. M. Stirling and S. Patai
© 1981 John Wiley & Sons Ltd

CHAPTER **15**

Heterosulphonium salts

SHIGERU OAE, TATSUO NUMATA, and TOSHIAKI YOSHIMURA

Department of Chemistry, University of Tsukuba, Sakura-mura, Ibaraki, 305 Japan

I. INTRODUCTION

This chapter describes the chemistry of heterosulphonium salts, each of which possesses at least one C–S bond and a discrete positive charge on the trivalent sulphur atom. Triheterosulphonium salts, such as triaminosulphonium salts, and sulphonium-like compounds, such as sulphide – metal complexes, are not included.

II. PREPARATION OF STABLE HETEROSULPHONIUM SALTS

A. Oxysulphonium Salts

Among oxysulphonium salts. alkoxysulphonium salts have been the most widely synthesized, while other types, such as acyloxy-, aryloxy-, and hydroxysulphonium salts, have not been synthesized successfully except in a few cases.

Alkylation of sulphoxide oxygen is used widely for the syntheses of alkoxysulphonium salts. Since Meerwein found 40 years ago that trialkyloxonium tetrafluoroborates (Meerwein reagents) can alkylate sulphoxide oxygen to yield the corresponding alkoxysulphonium tetrafluoroborates[1], the Meerwein reagent has been used to alkylate a wide variety of sulphoxides to afford the corresponding alkoxysulphonium salts generally in quantitative yields[2,3], (equations 1 and 2). Typical examples are described in later sections (Sections IIE and IIIA2b).

$$R-SO-R' + Me_3O^+ \ \bar{B}F_4 \longrightarrow R-\overset{+}{\underset{\underset{\displaystyle OMe}{|}}{S}}-R' \ \bar{B}F_4 \tag{1}$$

$$R-SO-R' + Et_3O^+ \ \bar{B}F_4 \longrightarrow R-\overset{+}{\underset{\underset{\displaystyle OEt}{|}}{S}}-R' \ \bar{B}F_4 \tag{2}$$

Dialkoxycarbonium ions are also powerful alkylating agents. For example, diphenyl sulphoxide was readily ethylated with diethoxycarbonium ion in 92% yield[4] (equation 3).

$$Ph-SO-Ph + EtO-\overset{+}{\underset{\underset{\displaystyle \bar{S}bCl_6}{}}{C}}H-OEt \longrightarrow Ph-\overset{+}{\underset{\underset{\displaystyle OEt}{|}}{S}}-Ph \ \bar{S}bCl_6 \tag{3}$$

Alkyl halides, alkyl tosylates, and alkyl sulphates are also versatile alkylating reagents (equation 4). When alkyl and benzyl tosylates are heated in dimethyl

$$R-SO-R + R'-Y \longrightarrow R-\overset{+}{\underset{\underset{\displaystyle OR'}{|}}{S}}-R \ Y^- \ \overset{x^-}{\longrightarrow} \ R-\overset{+}{\underset{\underset{\displaystyle OR'}{|}}{S}}-R \ X^- \tag{4}$$

$$R'-Y = MeI, \ R''OSO_2Tol\text{-}p, \ R''OSO_2CF_3, \ R''OSO_2F, \ \text{etc.}$$

sulphoxide alkylation takes place, and several dimethylalkoxysulphonium salts have been obtained in 50–99% yields[5]. However, on prolonged heating of dimethyl sulphoxide with benzyl tosylate the S-alkylated product 1 resulted[5]. Apparently the S-alkylated product is the thermodynamic product, whereas the O-alkylated one is the kinetic product (equation 5). Isomerization of the O-alkylated to the S-alkylated

$$Me-SO-Me + ROSO_2Tol\text{-}p \longrightarrow Me-\overset{+}{\underset{\underset{\displaystyle OR}{|}}{S}}-Me \ TsO^- \ \overset{\Delta}{\longrightarrow} \ Me-\overset{+}{\underset{\underset{\displaystyle R}{|}}{S}}O-Me \ TsO^- \tag{5}$$

$$(1)$$

product was found[5] to be accelerated when the nucleophilic character of the counter anion increased, in the order $TsO^- < NO_3^- < I^-$. Dimethyl sulphoxide afforded only the S-methylated product on heating with methyl iodide[6,7], whereas O-methylation of dimethyl sulphoxide with methyl iodide was achieved only in the presence of a silver salt such as silver perchlorate[8].

Other alkyl sulphoxides, such as methyl phenyl, ethyl phenyl, dibenzyl and dimethyl sulphoxides, were methylated with a 2-fold excess of methyl iodide in the presence of an equimolar amount of silver tetrafluoroborate at room temperature to give the corresponding methoxysulphonium salts in 40–70% yelds[3,9] (equation 6).

$$R-SO-R' + MeI + AgBF_4 \xrightarrow{\text{r.t.}} R-\overset{+}{\underset{|}{S}}-R' \ \ \bar{B}F_4 \qquad (6)$$
$$\quad\quad\quad\quad\quad\quad\quad\quad\quad\quad OMe$$

O-Ethylation of sulphoxides is generally carried out successfully by treatment with ethyl halide in the presence of $AgBF_4$[10] or $AgClO_4$[11]. However, when ethyl phenyl sulphoxide was heated with methyl iodide in the presence of HgI_2, the S-methylated product 2 was produced in 18% yield. The partially reserved oxosulphonium salt 2 can be prepared by either resolution of the racemic salt[12], or by S-methylation of the optically active (R)-ethyl phenyl sulphoxide[13] (equation 7).

$$Ph-SO-Et + MeI + HgI_2 \longrightarrow R-\overset{+}{\underset{|}{S}}O-Et \qquad (7)$$
$$\quad(R)\ \ (+)\quad\quad\quad\quad\quad\quad\quad\quad\quad Me$$
$$\quad\quad\quad\quad\quad\quad\quad\quad\quad\quad\quad (R)\ \ (+)$$
$$\quad\quad\quad\quad\quad\quad\quad\quad\quad\quad\quad (2)$$

When dimethyl sulphoxide and other sulphoxides were treated with 'magic methyl' (FSO_3CH_3)[14] or ethyl triflate $(CF_3SO_3C_2H_5)$[15,16] in non-polar media or with dimethyl sulphate in the presence of $NaBPh_4$ in methanol[17], alkylation was found to occur exothermically to give the corresponding alkoxysulphonium salts in good yields.

Interesting examples of the alkylation of dimethyl sulphoxide as shown in equations 8[11] and 9[18].

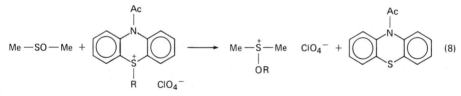

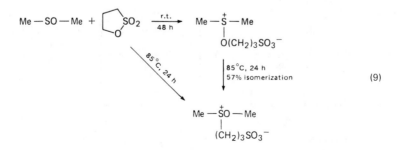

Reaction of various epoxides with dimethyl sulphoxide in the presence of trinitrobenzenesulphonic acid usually gives the corresponding β-hydroxy-alkoxysulphonium salts in 40–80% yields and the reaction was found to take place both regioselectively and stereoselectively[19,20] (equation 10). Styrene oxide

$$Ph-CH-CH_2 + Me-SO-Me \xrightarrow[\text{r.t.}]{ArSO_3H} Me-\overset{+}{\underset{\underset{Ph\ \ OH}{O-CH-CH_2}}{S}}-Me \quad ArSO_3^- \xrightarrow{^-OH}$$

$$Ph-\underset{\underset{OH}{|}}{CH}-\underset{\underset{OH}{|}}{CH_2} + Me-SO-Me$$

(10)

$$ArSO_3^- = O_2N-\underset{\underset{NO_2}{\bigcirc}}{\overset{NO_2}{}}-SO_3^- \quad (TNBS^-)$$

derivatives were reported to afford β-hydroxy-α-phenylalkoxysulphonium salts in this reaction through preferential attack of the sulphinyl oxygen on the benzylic carbon of the epoxide. The β-hydroxyalkoxysulphonium salts thus formed are know to undergo alkaline hydrolysis to give the corresponding 1,2-diols generally in good yields via attack of hydroxide on the sulphonium sulfur through S_N2 process on the sulphur atom. Thus the *cis*-epoxide is transformed to the *threo*-glycol while the *trans*-epoxide gives the *erythro*-glycol in the overall process[20]. Other strong acids such as CF_3COOH, H_2SO_4, HNO_3, and BF_3 are also useful for the ring opening of epoxides with dimethyl sulphoxide[21].

Alkoxysulphonium salts can be prepared by treatment of dimethyl sulphoxide with alkyl chloroformates[17] (equation 11) or with alkyl chlorosulphinates at $-78°C$[22] (equation 12).

$$Me-SO-Me + i\text{-BuOCOCl} \xrightarrow[\text{MeOH/H}_2O]{NaBPh_4} \left[Me-\overset{+}{\underset{\underset{OCOOBu\text{-}i}{|}}{S}}-Me \quad Cl^- \right] \xrightarrow{-CO_2} Me-\overset{+}{\underset{\underset{OBu\text{-}i}{|}}{S}}-Me \quad \bar{B}Ph_4 \quad (11)$$

$$R-SO-R + R'O-\underset{\underset{O}{\|}}{S}-Cl \xrightarrow[-78°C, CH_2Cl_2]{SbCl_5, -SO_2} R-\overset{+}{\underset{\underset{OR'}{|}}{S}}-R \quad \bar{S}bCl_6 \quad (12)$$

R = Me, Ph

R = Me, Et, i-Pr

$$Me-SO-Me + MeO-\underset{\underset{O}{\|}}{S}-Cl \xrightarrow{-78°C} \left[Me-\overset{+}{\underset{\underset{O-\underset{\underset{O}{\|}}{S}-OMe}{|}}{S}}-Me \quad Cl^- \right] \xrightarrow[(2)\ SbCl_5]{(1)\ NuH} Me-\overset{+}{\underset{\underset{Nu}{|}}{S}}-Me \quad \bar{S}bCl_6 \quad (13)$$

NuH (%) = EtOH (72), t-BuOH (46)

EtSH (77), PhSSPh (80)

Et$_2$NH (35)

When external nucleophiles such as alcohols, thiols, disulphides, and secondary amines were added to the mixture of methyl chlorosulphinate and dimethyl sulphoxide at $-78°C$, the alkoxy-, thia-, and azasulphonium salts, respectively, were obtained in 35–80% yield[22] (equation 13).

Another useful synthetic method for alkoxysulphonium salts is the alcoholysis of the chlorosulphonium salt. Dimethyl chlorosulphonium hexachloroantimonate was shown to afford the alkoxysulphonium salts in 62–100% yields upon treatment with alcohols or diols[23,24] (equation 14).

$$\overset{+}{\underset{Cl}{Me-S-Me}} \quad \bar{S}bCl_6 + ROH \longrightarrow \overset{+}{\underset{OR}{Me-S-Me}} \quad \bar{S}bCl_6 \qquad (14)$$

ROH = EtOH, PrOH, i-PrOH, $HO(CH_2)_2OH$, $HO(CH_2)_2OH$, $HO(CH_2)_6OH$

'One-pot' synthesis of alkoxysulphonium salts was successfully carried out by alcoholysis of halosulphonium salts prepared in situ. Treatment of a sulphide with alkyl hypochlorite usually gives the alkoxysulfonium salt in a 'one-pot' procedure via alcoholysis of the chlorosulphonium salt formed initially. Johnson and Jones[25] reported syntheses of a variety of t-butoxysulphonium salts in 71–82% yields by the treatment of sulphides with t-butyl hypochlorite at $-78°C$, followed by addition of $SbCl_5$. However, in the treatment with isopropyl hypochlorite the yields were found to be substantially lower (23–66%)[25]. During the reaction a sulphurane-type adduct **3** was considered, from n.m.r. measurements ($-45°C$), to be formed from the mixture of methyl phenyl sulphide and t-butyl hypochlorite, in the absence of $SbCl_5$[26] (equation 15).

$$R-SO-R + t\text{-}BuOCl \xrightarrow{-78°C} \left[\overset{Cl}{\underset{OBu\text{-}t}{R-S-R}} \right] \xrightarrow{SbCl_5} \overset{+}{\underset{OBu\text{-}t}{R-S-R}} \quad \bar{S}bCl_6 \qquad (15)$$

$$(3)$$

When the reaction of a thiane with t-butyl hypochlorite was carried out in ethanol in the presence of $HgCl_2$, the ethoxysulphonium salt was successfully isolated[26]. Reaction of 4-t-butylthiane with t-BuOCl or NBS in ethanol, followed by addition of $AgBR_4$, was found to afford the axially oriented ethoxysulphonium salt[27] (equation 16).

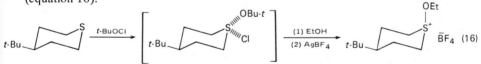

Reaction of methyl phenyl sulphide with N-chlorobenzotriazole also gave the adduct **4**, which was then attacked by added alcohol on the sulphur atom to afford the alkoxysulphonium salt in 30–42% yields[28] (equation 17).

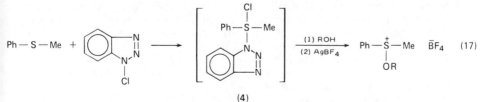

A large number of stable tetravalent sulphur compounds, i.e. sulphuranes, have been prepared by Martin and coworkers[29]. Their fundamental synthetic route involves the alkoxysulphonium salt as an intermediate which is formed by the reaction of the sulphide with the alkyl hypochlorite[30] (equation 18).

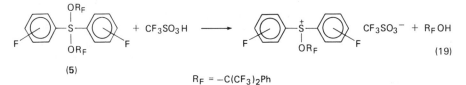

$$ (18) $$

$$ Ar = \underset{\text{COOC(CF}_3)_2\text{Ph}}{ \bigcirc } $$

When the sulphuranes **5** are treated with trifluoromethanesulphonic acid, the reverse reaction usually takes place and the alkoxysulphonium salts are obtained in 74–83% yields[31] (equation 19).

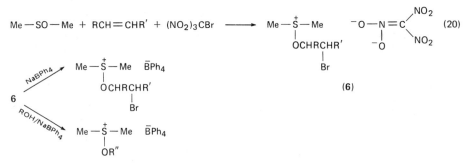

$$ (19) $$

$$ R_F = -C(CF_3)_2Ph $$

Torssell reported that β-bromoalkoxysulphonium salts **6** are formed in 42–67% yields by treatment of DMSO with olefins and bromotrinitromethane in the presence of NaBPh₄[17]. The stable salts **6** are then converted into other simple alkoxysulphonium salts by treatment with alcohols[17] (equation 20).

$$
Me-SO-Me + RCH=CHR' + (NO_2)_3CBr \longrightarrow
\underset{\underset{Br}{\overset{|}{\underset{OCHRCHR'}{|}}}}{Me-\overset{+}{S}-Me}
\quad
\underset{-O}{\overset{NO_2}{\underset{|}{-O-N=C}}}\overset{NO_2}{}
\qquad (20)
$$

(6)

$$
6 \quad
\begin{array}{c}
\overset{NaBPh_4}{\nearrow} \quad Me-\overset{+}{\underset{\underset{Br}{\overset{|}{\underset{OCHRCHR'}{|}}}}{S}}-Me \quad \bar{B}Ph_4 \\[3em]
\underset{\searrow}{ROH/NaBPh_4} \quad Me-\overset{+}{\underset{\underset{OR''}{|}}{S}}-Me \quad \bar{B}Ph_4
\end{array}
$$

A hydroxysulphonium salt was reported to be formed by protonation of a sulphoxide with a suitable counter anion in the reaction mixture. Thus, when dimethyl sulphoxide was treated with the phosphorus chloride compound **7** in dry chloroform, the dimethyl hydroxysulphonium salt **8** with the phosphate as counter ion was obtained quantitatively, together with chloromethyl methyl sulphide as the Pummerer reaction product[32,33] (equation 21).

Cyanuric acid **9** also gives the hydroxysulphonium salt[32] (equation 22).

An adduct formed from dimethyl sulphoxide and an equimolar amount of trinitrobenzenesulphonic acid is also known[20].

Another example is the reaction of the adduct of DMSO and SbCl₅ with hydrogen chloride in CH₂Cl₂ at −78°C to yield dimethyl hydroxysulphonium

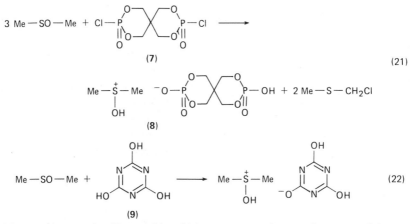

(21)

(22)

hexachloroantimonate in 90% yield, which reverts to the starting meaterials on warming the solution to 20°C. However, the crystalline salt isolated at −78°C was found to be stable[24] (equation 23).

$$Me-SO-Me \cdot SbCl_5 + HCl \underset{20°C}{\overset{-78°C}{\rightleftharpoons}} \overset{+}{Me-S-Me} \quad \bar{S}bCl_6 \qquad (23)$$
$$\qquad\qquad\qquad\qquad\qquad\quad OH$$

There is one example of a stable acyloxysulphonium salt. Sharma and Swern reported that treatment of dimethyl sulphoxide with trifluoroacetic anhydride in methylene chloride at −78°C affords dimethyl trifluoroacetoxysulphonium trifluoroacetate as white crystals, although no attempt at isolation was carried out. This salt apparently undergoes the Pummerer reaction on warming to room temeprature[34] (equation 24).

$$Me-SO-Me + (CF_3CO)_2O \xrightarrow{-78°C} \overset{+}{Me-S-Me} \quad CF_3COO^- \xrightarrow{r.t.} Me-S-CH_2OCOCF_3$$
$$\qquad\qquad\qquad\qquad\qquad\qquad\qquad\quad OCOCF_3 \qquad\qquad\qquad\qquad (24)$$

A sulphonoxysulphonium salt has been successfully isolated. Hendrickson and Schwarzman reported that treatment of dimethyl sulphoxide with trifluoromethanesulphonic anhydride at −78°C afforded the sulphonoxysulphonium salt **10**, which decomposed at room temperature or exposure to air or water[35] (equation 25).

$$Me-SO-Me + (CF_3SO_2)_2O \longrightarrow \overset{+}{Me-S-Me} \quad CF_3SO_3^- \qquad (25)$$
$$\qquad\qquad\qquad\qquad\qquad\qquad\quad OSO_2CF_3$$

$$\text{(10)}$$

Dialkoxysulphonium salts also comprise a family of alkoxysulphonium salts and can be prepared. When sulphinates are alkylated with alkyl triflates the products are usually unstable and can be detected only in solution by n.m.r. spectroscopy in many cases[15,36]. However, a successful preparation of a dialkoxysulphonium salt has been reported by Warthmann and Schmidt in the reaction of a dichlorosulphonium salt with an alcohol[24] (equation 26).

Other diheterosulphonium salts, the alkoxyaminosulphonium salts, were also prepared in two different ways by Minato and co-workers. One involves treatment of a sulphenamide with N-chlorobenzotriazole, followed by alcoholysis[37] (equation 27), and the other involves alkylation of a sulphinamide[15,38] (equation 28).

$$\overset{+}{Me-S}-Cl \quad \bar{S}bCl_6 \; + \; ROH \; \longrightarrow \; Me-\overset{+}{S}-OR \quad \bar{S}bCl_6 \qquad (26)$$
$$\overset{|}{Cl} \qquad\qquad\qquad\qquad\qquad \overset{|}{OR}$$

ROH = MeOH (77%), EtOH (16%)

$$(27)$$

R = i-Pr, p-Tol; (i) ;(ii) R′OH = MeOH, PhOH, (−)-menthol; (iii) NaBPh₄

$$(28)$$

B. Azasulphonium Salts

1. Unsubstituted azasulphonium salts

Many stable azasulphonium salts have been synthesized in connection with the synthesis of sulphimides, which are usually prepared by deprotonation of azasulphonium salts bearing a hydrogen atom on nitogen.

A useful method is direct amination of sulphides with hydroxylamine O-sulphate derivatives. In 1959, Appel and Büchner reported that diethyl sulphide was aminated with hydroxylamine O-sulphonic acid at $-20°C$ to afford the solid diethyl aminosulphonium salt in 50–60% yields[39,40] (equation 29).

$$2\;Et-S-Et \; + \; 2\;H_2NOSO_3H \; \xrightarrow[-20°C]{NaOMe/MeOH} \; \left(Et-\overset{+}{\underset{|}{S}}-Et \atop NH_2 \right)_2 SO_4{}^{2-} \qquad (29)$$

Many other aminosulphonium salts were prepared successfully in good yields (57–100%) by the reaction of dialkyl, alkyl aryl and diaryl sulphides with hydroxylamine O-mesitylenesulphonate[41–43], a well known and versatile aminating agent[44] (equation 30). Cyclic sulphides such as phenothiazine, thianthrene, and

$$R-S-R' \; + \; H_2NOMes \; \longrightarrow \; R-\overset{+}{\underset{|}{S}}-R' \quad MesO^- \qquad (30)$$
$$NH_2$$

phenoxathiin were aminated with the same reagent in 50–80% yields[45,46]. With 2-dimethylaminoethyl phenyl sulphide, however, treatment with the same reagent gave the corresponding N-aminated product in 72% yield, indicating that there is a limitation to the use of the reagent for S-amination if other groups which are susceptible to amination with hydroxylamine O-mesitylenesulphonate are present in the molecule[43]. Other substituted hydroxylamines with leaving groups on oxygen, such as hydroxylamine O-2,4,6-triisopropylbenzenesulphonate, O-(2,4-dinitrophenyl)hydroxylamine, and O-picrylhydroxylamine, can also aminate diphenyl sulphide in 61–90% yields[47].

In connection with the use of these hydroxylamines in amination, it was found that the reaction of O-(2,4-dinitrophenyl)hydroxylamine with uncharged nucleophiles such as pyridines, phosphines, and sulphides afforded the corresponding aminated products. The reaction was found to proceed by way of a typical S_N2 process on sp^3 nitrogen, as concluded from kinetic observations[48,49] (equation 31). One of the interesting features of the reaction is that it is not markedly influenced by the steric bulk of the nucleophile, unlike the S_N2 reaction at sp^3 carbon[48,49].

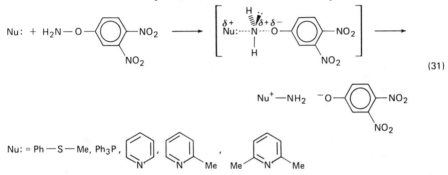

(31)

Aminosulphonium salts here found by Appel and Büchner to be formed by treatment of sulfides with chloramine[40]. Thus, dimethyl sulphide was transformed into dimethyl aminosulphonium salt in 2.3% yield by treatment of dimethyl sulphide with a large excess of chloramine in diethyl ether at $-78°C$. Later they reported that the use of acetonitrile instead of diethyl ether as solvent increased the yield of the salt to 31%[50] (equation 32). The reaction is considered to proceed through a nucleophilic displacement reaction on the nitrogen atom of chloramine.

$$Me-S-Me + NH_2Cl \longrightarrow Me-\overset{+}{\underset{\underset{NH_2}{|}}{S}}-Me \quad Cl^- \qquad (32)$$

A US patent[51] describes the reaction of methyl n-octyl and methyl n-dodecyl sulphide with chloramine–ammonia in tetrahydrofuran solution. In the presence of CO_2, which prevents formation of the sulphimides, this affords the corresponding aminosulphonium salts in good yields. Dialkyl sulphides with one ethylphenylamino group at the ω-position were also shown to give the corresponding aminosulphonium salts by treatment with chloramine in diethyl ether[52].

In the reaction of alkyl sulphides with chloramine further amination was found to take place, eventually yielding the N-aminoazasulphonium salt hydrochloride **11**, which dimerized to the sulphurane-type compound **12** on removal of hydrogen chloride[53]. Compound **11** was also reported to be formed by treatment of free diethyl sulphimide with chloramine[53] (equation 33). However, Appel et al. claimed that the monomeric sulphone diimide **13** was formed by treatment of alkyl sulphides with chloramine in isopropanol and then with base, while its structure was confirmed by mass spectral analysis[50] (equation 34). A Raman spectral study also showed that the

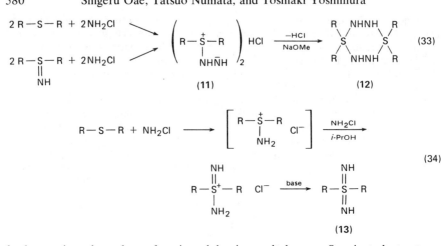

$$2\,R-S-R + 2NH_2Cl$$

$$2\,R-\underset{\underset{NH}{\|}}{S}-R + 2NH_2Cl$$

$$\left(R-\overset{+}{\underset{\underset{NH\bar{N}H}{|}}{S}}-R\right)_{\!2} HCl \xrightarrow[NaOMe]{-HCl} \underset{R}{\overset{R}{>}}S\underset{NHNH}{\overset{NHNH}{<}}S\underset{R}{\overset{R}{<}} \qquad (33)$$

$$(11) \qquad\qquad (12)$$

$$R-S-R + NH_2Cl \longrightarrow \left[R-\overset{+}{\underset{\underset{NH_2}{|}}{S}}-R \;\; Cl^-\right] \xrightarrow[i\text{-PrOH}]{NH_2Cl}$$

$$(34)$$

$$R-\overset{\overset{\displaystyle NH}{\|}}{\underset{\underset{NH_2}{|}}{S^+}}-R \;\; Cl^- \xrightarrow{base} R-\overset{\overset{\displaystyle NH}{\|}}{\underset{\underset{NH}{\|}}{S}}-R$$

$$(13)$$

further aminated product of aminosulphonium salt has an *S*-aminated structure, sulphone diimide **13**, and not an *N*-aminated structure such as **11**[54].

When the stable di(*p*-methoxyphenyl) sulphide–chlorine complex was treated with ammonia, the corresponding aminosulphonium salt was isolated in 45% yield[55] (equation 35).

$$MeO-\!\!\!\bigcirc\!\!\!-\overset{\overset{\displaystyle Cl}{|}}{\underset{\underset{Cl}{|}}{S}}-\!\!\!\bigcirc\!\!\!-OMe + NH_3 \longrightarrow MeO-\!\!\!\bigcirc\!\!\!-\overset{+}{\underset{\underset{NH_2}{|}}{S}}-\!\!\!\bigcirc\!\!\!-OMe \;\; Cl^- \qquad (35)$$

Reaction of a dialkyl sulphide with *N*-chlorosuccinimide (NCS) or *N*-bromosuccinimide (NBS) and ammonia usually affords the corresponding aminosulphonium salt, accompanied by the sulphone diimide as a minor side-product (equation 36). In the reaction with *t*-butyl hypochlorite, another halogenating reagent, the sulphone diimide becomes the sole product in good yield[56] (equation 37).

$$R-S-R + NCS + NH_3 \longrightarrow R-\overset{+}{\underset{\underset{NH_2}{|}}{S}}-R \;\; Cl^- + R-\overset{\overset{\displaystyle NH}{\|}}{\underset{\underset{NH}{\|}}{S}}-R \qquad (36)$$

$$\text{minor}$$

$$R-S-R + t\text{-BuOCl} + NH_3 \longrightarrow R-\overset{\overset{\displaystyle NH}{\|}}{\underset{\underset{NH}{\|}}{S}}-R \qquad (37)$$

Incidentally, a convenient synthesis for *N*-tosyl sulphone diimides consists simply in treatment of free sulphimides with chloramine-T in a polar aprotic solvent[57].

Although there is a limitation in synthetic applications, *N*-chlorosulphimides were

$$Ph-\overset{\overset{\displaystyle \|}{S}}{\underset{\underset{\underset{Cl}{\diagdown}}{N}}{}}-Ph \xrightarrow{air} Ph-\overset{+}{\underset{\underset{NH_2}{|}}{S}}-Ph \;\; Cl^- \qquad (38)$$

also found to be converted into the *N*-unsubstituted azasulphonium salts. Thus, when *N*-chlorodiphenyl sulphimide was in contact with air in the crystalline state for 20 days at room temperature the aminosulphonium chloride was obtained in 75% yield[58] (equation 38).

2. N-Substituted azasulphonium salts

Numerous *N*-substituted azasulphonium salts have been prepared simply by treatment of sulphides with *N*-haloamides or *N*-haloimides. The best known is succinimidodimethylsulphonium chloride **14**, prepared quantitatively by the reaction of dimethyl sulphide with NCS[59] (equation 39). The cation of **14** becomes

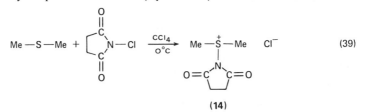

(14)

more stable by replacement of chloride with tetrafluoroborate as counter anion[60]. This azasulphonium salt **14** is useful for the oxidation of alcohols (see Section IIIE), and for syntheses of other azasulphonium salts (see below).

In 1947, Likhosherstov reported that *N*-chloroacetamide reacted with dimethyl sulphide in acetone–CCl$_4$ solution to form crude dimethyl *N*-acetylazasulphonium chloride (90%), which decomposed partially during recrystallization[61]. Recently, several pure *N*-acetylazasulphonium bromides were prepared in 38–84% yields by treatment of sulphides with *N*-bromoacetamide[62,63] (equation 40). *N*-Carboethoxyazasulphonium salts were synthesized in 50–60% yields by treatment of alkyl sulphides with *N*-chlorocarbamates in an aprotic non-polar solvent[64,65] (equation 41).

$$R-S-R' + MeCONHBr \longrightarrow R-\overset{+}{\underset{|}{S}}-R' \quad Br^- \qquad (40)$$
$$NHCOMe$$

$$R-S-R' + ClNHCOOEt \xrightarrow[-20 \text{ to } 0°C]{CHCl_3} R-\overset{+}{\underset{|}{S}}-R' \quad Cl^- \qquad (41)$$
$$NHCOOEt$$

N-Chlorourea was also found to be useful[66] (equation 42).

$$Ar-S-Ar' + ClNHCONH_2 \xrightarrow[40-70\%]{CH_3CN} Ar-\overset{+}{\underset{|}{S}}-Ar' \quad Cl^- \qquad (42)$$
$$NHCONH_2$$

Dimethyl sulphide and other alkyl sulphides were successfully converted into *N*-substituted azasulphonium salts by treatment with other *N*-chloroamides such as *N*-chlorobenzamidine[67,68], *N*-chlorobenzimidates[69,70], *N*-chlorocarbamate[71], and *N*-chloroguanidine[72].

Gassman *et al.* reported that *N*-arylazasulphonium salts were obtained in 59–80% yields on treatment of dimethyl sulphide with *N*-*t*-butyl-*N*-chloroanilines, prepared by chlorination of the anilines with calcium hypochlorite[73] (equation 43).

Another type of azasulphonium compound **15** was prepared by alkylation of the alkylidene sulphenamide[74] (equation 44).

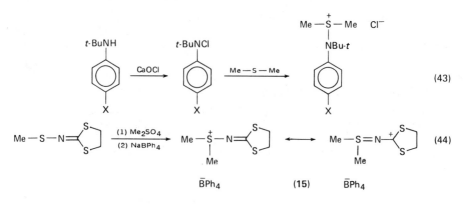

Azasulphonium salts can be prepared by nucleophilic substitution by amines at the sulphur of sulphonium salts, e.g. the reaction of stable chlorosulphonium salts with amines (equations 45[75] and 46[76]).

$$Me\overset{+}{\underset{Cl}{-S}}-R \quad \bar{S}bCl_6 + i\text{-}PrNH_2 \xrightarrow[55-83\%]{} Me\overset{+}{\underset{NHPr\text{-}i}{-S}}-R \quad \bar{S}bCl_6 \qquad (45)$$

$$Me\overset{+}{\underset{Cl}{-S}}-Me \quad \bar{S}bCl_6 + RNHCOR' \longrightarrow Me\overset{+}{\underset{NRCOR'}{-S}}-Me \quad \bar{S}bCl_6 \qquad (46)$$

Johnson *et al.* reported that alkyl aryl sulphides are converted into the azasulphonium salts in 40–84% yields by initial treatment with a halogenating agent such as N-chlorobenzotriazole, followed by addition of a primary or secondary amine such as 1-phenylethylamine[28] (equation 47).

$$Ar-S-R + \left[\underset{\underset{Cl}{N}}{\overset{N}{\bigodot}}\right] \xrightarrow{-78°C} \left[\begin{array}{c} Cl \\ | \\ Ar-S-R \\ | \\ \underset{N}{\overset{N}{\bigodot}}\end{array}\right] \xrightarrow[\text{(2) AgBF}_4]{\text{(1) R'NH}_2} Ar\overset{+}{\underset{NHR'}{-S}}-R \quad \bar{B}F_4 \qquad (47)$$

Treatment of dimethyl sulphide with NCS, followed by addition of an amine or an amide, was found by Vilsmaier and Sprügel to give dimethylazasulphonium salts in 49–91% yields[59] (equation 48). N-Aryl- (77–91%), N-α-naphthyl- (75%),

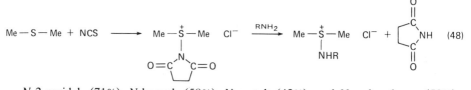

N-2-pyridyl- (71%), N-benzyl- (59%), N-acetyl- (42%), and N-carboethoxy- (59%) azasulphonium chlorides have been prepared by this reaction, which is effectively an amine exchange reaction.

Dowson and Swern also synthesized about twenty N-substituted azasulphonium salts in 42–96% yields by treatment of dimethyl sulphide with either NCS or N-chlorobenzotriazole and then with amines or amides[77]. They reported that tetrahydrothiophene gave the corresponding azasulphonium salt similarly but di-*t*-butyl and diphenyl sulphides and thiophene did not form any azasulphonium salt by the same treatment.

Indole could not be used for this type of amine exchange reaction. When dialkyl and alkylaryl azasulphonium salts were treated with indole, 3-sulphonioindoles were obtained and eventually gave 3-alkylthio-substituted indoles with or without heating[78] (equations 49 and 50).

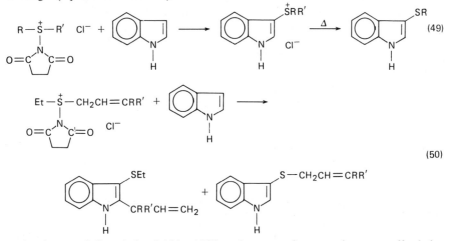

A mixture of dimethyl sulphide, NCS and an enamine was shown to afford the sulphonium salt **16** in 73% yield, which was converted into the ylide **17** in 51% yield on treatment with hydrogen chloride and then alkali. In some cases the ylide demethylated on distillation[79] (equation 51).

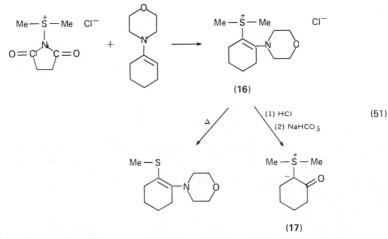

Treatment of dialkyl sulphides with NCS, *t*-BuOCl, or SO_2Cl_2 and subsequent treatment of the reaction mixture with primary aromatic amines or sodium salts of amines or amides affords the corresponding sulphimides[80,81] (equation 52).

Reaction of a reactive sulphurane with secondary alkylamines was shown to afford

$$R-S-R + X-Cl + R'NH_2 \longrightarrow \left[\begin{array}{c} R-\overset{+}{\underset{NHR'}{S}}-R \quad X^- \end{array} \right] \xrightarrow{-HX} R-\overset{}{\underset{NR'}{S}}-R \quad (52)$$

the azasulphonium salts in 54–74% yields. These undergo further reaction to yield the imine and diphenyl sulphide in the case of benzylamines, while reaction with ammonia, primary amines, or amides gives the corresponding sulphimides in good yields[82] (equation 53). N-Alyl or N-acyl sulphimides are also known to be formed by treatment of other sulphuranes with amines or amides[83,84].

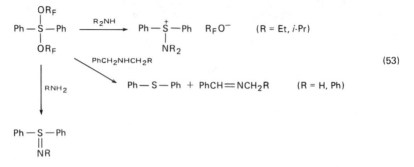

(53)

Several cyclic sulphimides were prepared by Claus et al. in good yields from N-aryl sulphimides via an azasulphonium intermediate[85] (equation 54).

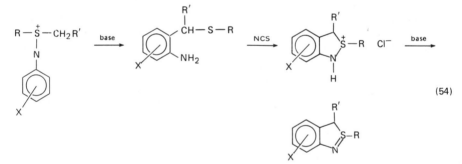

(54)

Sharma and coworkers found that trifluoroacetoxysulphonium salts, prepared from dimethyl sulphoxide and trifluoroacetic anhydride, readily reacted with aromatic amines, amides, and sulphonamides to give the corresponding azasulphonium salts as intermediates, which on further treatment with a base in situ yielded sulphimides in 40–90% yields[34,86] (equation 55).

$$Me-SO-Me + (CF_3CO)_2O \xrightarrow[CH_2Cl_2]{-60°C} Me-\overset{+}{\underset{OCOCF_3}{S}}-Me \quad CF_3COO^- \xrightarrow{RNH_2}$$

(55)

$$Me-\overset{+}{\underset{NHR}{S}}-Me \quad CF_3COO^- \xrightarrow{base} Me-\overset{}{\underset{NR}{S}}-Me$$

Analogously, dimethyl sulphoxide, activated by electrophiles such as SO_3, P_4O_{10},

BF$_3$, Ac$_2$O, DCC, POCl$_3$, and CH$_3$SO$_2$Cl, was found to react with amines, amides and sulphonamides in the presence of a base to yield the sulphimides.[87-92]. Reaction of dimethyl sulphoxide with POCl$_3$ and a tertiary amine such as dimethylaniline gave p-dimethylsulphoniodimethylaniline[93].

There is one interesting case of intramolecular cyclization to form an azasulphonium salt. When methionine was oxidized with an equimolar amount of iodine in alkaline media, the cyclized azasulphonium salt was obtained in good yield[94] (equation 56). This reaction is a useful method for the quantitative

$$Me-S-CH_2CH_2\underset{\underset{NH_2}{|}}{CH}COOH \xrightarrow[\text{NaOMe/MeOH}]{I_2} \overset{+}{Me-S}\cdots \qquad (56)$$

determination of methionine. Another interesting example is the reaction of dimethyl sulphoxide to give the azasulphonium salt in 46% yield by treatment with an N-sulphinyldimethylimmonium salt[95] (equation 57).

$$Me-SO-Me + \underset{Me}{\overset{Me}{>}}\overset{+}{N}=S=O \quad \bar{B}F_4 \xrightarrow{-SO_2} Me-\overset{+}{\underset{NMe_2}{S}}-Me \quad \bar{B}F_4 \qquad (57)$$

When the alkoxysulfonium salt **18**, prepared by treatment of phenoxathiin S-oxide with methyl iodide in the presence of AgClO$_4$, was treated with alkylamines, the corresponding azasulphonium salts were obtained, while on reaction with carbanions the sulphonium ylides **19** were formed[11] (equation 58).

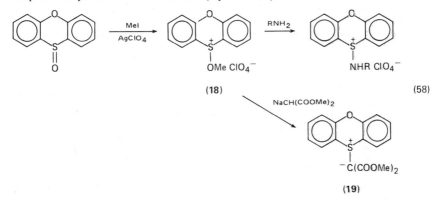

(**18**)   (**58**)

(**19**)

Reaction of the lactam oxime **20** with dimethyl sulphide gave the azasulfonium salt **21**, either via a nitrenium-type cation or by direct nucleophilic substitution (S$_N$2) at nitrogen by dimethyl sulphide[96] (equation 59).

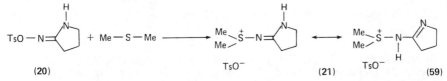

(**20**)   (**21**)   (**59**)

Sulphimides have a basic nitrogen atom. The pK$_{BH^+}$ values of N-arenesulphonyl aryl methyl sulphimides are in the range −1.81 to −3.0[97]. and those of N-unsubstituted ('free') diaryl sulphimides are in the range 7.30–8.79[98]. Thus, sulphimides can be readily protonated to form azasulphonium salts, the conjugated acids of the sulphimides, although this reaction has no synthetic utility, since sulphimides are generally prepared by deprotonation of the azasulphonium salts.

N-Acetyl sulphimides are protonated by hydrogen chloride at 0°C to yield the corresponding azasulphonium chloride in 79–87% yields[63] (equation 60).

$$R-\underset{\underset{NCOMe}{\|}}{S}-R' \xrightarrow[0°C]{HCl} R-\underset{\underset{NHCOMe}{|}}{\overset{+}{S}}-R' \quad Cl^- \tag{60}$$

When N-tosyl diphenyl sulphimide was dissolved in concentrated sulphuric acid at 5°C and the resultant purple solution was poured into ice–water, the diphenyl aminosulphonium salt **22** was obtained quantitatively. Treatment with base afforded the 'free' sulphimide **23**, which can be reconverted into the aminosulphonium salt **22** by protonation with p-toluenesulphonic acid[99,100] (equation 61).

$$Ph-\underset{\underset{NTs}{\|}}{S}-Ph \xrightarrow{\text{conc. }H_2SO_4} Ph-\underset{\underset{NH_2}{|}}{\overset{+}{S}}-Ph \quad TsO^- \underset{TsOH}{\overset{\text{base}}{\rightleftharpoons}} Ph-\underset{\underset{NH}{\|}}{S}-Ph \tag{61}$$

$$\qquad\qquad\qquad\qquad\qquad\qquad\text{(22)}\qquad\qquad\qquad\qquad\qquad\text{(23)}$$

For characterization, liquid or unstable free sulphimides can be converted into their picrates, which are also aminosulphonium salts[100,101]. N-Aryl sulphimides can also be characterized as the picrates[80].

N-Alkyl azasulphonium salts can be obtained by alkylation of sulphimides with alkylating agents such as the Meerwein reagent[102,103], dimethyl sulphate[104], methyl triflate[103], and alkyl iodides[82,105], generally in quantitative yields. N-Alkyl sulphimides are acylated by acid chlorides at −20–40°C, yielding the N-acyl azasulphonium salts, which are easily hydrolysed[82] (equation 62).

$$Ph-\underset{\underset{NR}{\|}}{S}-Ph + R'COCl \longrightarrow Ph-\underset{\underset{NRCOR'}{|}}{\overset{+}{S}}-Ph \quad Cl^- \tag{62}$$

Free dialkyl sulphimides are known to be carbonated by carbon dioxide to yield the N-carboxy azasulphonium salts **24**, which upon heating undergo disproportionation to afford the salt **25**[106] (equation 63).

$$R-\underset{\underset{NH}{\|}}{S}-R + CO_2 \longrightarrow R-\underset{\underset{NHCOO^-}{|}}{\overset{+}{S}}-R \xrightarrow{\Delta} R-\underset{\underset{NH_2}{|}}{\overset{+}{S}}-R \cdot R-\underset{\underset{NCOO^-}{\|}}{S}-R \tag{63}$$

$$R = Me, Et \qquad\qquad\qquad \text{(24)} \qquad\qquad\qquad \text{(25)}$$

An interesting derivative of the azasulphonium salt **26** was synthesized in good yield by reaction of dimethyl sulphoxide with either $(NSCl)_3$[107], chlorocyanate, or bromocyanate[108] (equation 64). A more general synthetic method for compounds of

$$Me-SO-Me + (NSCl)_3 \xrightarrow{-SO_2} \underset{Me}{\overset{Me}{>}}\overset{+}{S}-\bar{N}-\overset{+}{S}\underset{Me}{\overset{Me}{<}} \quad Cl^- \tag{64}$$

$$\text{(26)}$$

this type is reaction of the N-chloro- or the N-bromosulphimide **27** with sulphides to give **28** (equation 65), while **27** can give **29**, **30**, or **31** by treatment with triphenylphosphines, tertiary amines or sulphoxides, respectively[58] (equations 66, 67, and 68). Further, a benzene solution of the N-chlorosulphimide **27**, on standing at room temperature, gave an azasulphonium compound **32** in 74% yield[58,109] (equation 69).

$$\text{Ph}-\underset{\underset{\text{Cl}}{\overset{\|}{\text{N}}}}{\overset{}{\text{S}}}-\text{Ph} + \text{Ph}-\text{S}-\text{Me} \longrightarrow \underset{\text{Ph}}{\overset{\text{Ph}}{>}}\overset{+}{\text{S}}-\overset{-}{\text{N}}-\overset{+}{\text{S}}\underset{\text{Me}}{\overset{\text{Ph}}{<}} \quad \text{Cl}^- \tag{65}$$

(27)   (28)

$$27 + \text{Ph}_3\text{P} \longrightarrow \underset{\text{Ph}}{\overset{\text{Ph}}{>}}\overset{+}{\text{S}}-\overset{-}{\text{N}}-\overset{+}{\text{P}}\text{Ph}_3 \quad \text{Cl}^- \tag{66}$$

(29)

$$27 + \text{R}_3\text{N} \longrightarrow \underset{\text{Ph}}{\overset{\text{Ph}}{>}}\overset{+}{\text{S}}-\overset{-}{\text{N}}-\overset{+}{\text{N}}\text{R}_3 \quad \text{Cl}^- \tag{67}$$

$$\text{R}_3\text{N} = \text{Et}_3\text{N}, \text{DABCO} \qquad \text{(30)}$$

$$27 + \text{R}-\text{SO}-\text{Me} \longrightarrow \underset{\text{Ph}}{\overset{\text{Ph}}{>}}\overset{+}{\text{S}}-\overset{-}{\text{N}}=\underset{\overset{|}{\text{R}}}{\overset{\overset{\text{O}}{\|}}{\text{S}}}-\text{Me} \quad \text{Cl}^- \tag{68}$$

$$\text{R} = \text{Me, Ph, } p\text{-Tol} \qquad \text{(31)}$$

$$27 \xrightarrow[\text{r.t.}]{\text{C}_6\text{H}_6} \underset{\text{Ph}}{\overset{\text{Ph}}{>}}\overset{+}{\text{S}}-\overset{-}{\text{N}}-\overset{+}{\text{S}}\underset{\text{Ph}}{\overset{\text{Ph}}{<}} \quad \text{Cl}^- \tag{69}$$

(32)

Compound **32** can also be prepared by treatment of diphenyl sulphide with ammonium trichloride[110] (equation 70) or by treatment of the free sulphimide with a reactive sulphurane[82] (equation 71). Compound **33** was similarly prepared by treatment of the *N*-halosulphoximide with dimethyl sulphide[50] (equation 72).

$$\text{Ph}-\text{S}-\text{Ph} + \text{NCl}_3 \longrightarrow \underset{\text{Ph}}{\overset{\text{Ph}}{>}}\overset{+}{\text{S}}-\overset{-}{\text{N}}-\overset{+}{\text{S}}\underset{\text{Ph}}{\overset{\text{Ph}}{<}} \quad \text{Cl}_3^- \tag{70}$$

(32)

$$\text{Ph}-\underset{\overset{|}{\text{NH}}}{\overset{\|}{\text{S}}}-\text{Ph} + \text{Ph}-\underset{\overset{|}{\text{OR}_F}}{\overset{\overset{\text{OR}_F}{|}}{\text{S}}}-\text{Ph} \xrightarrow{55\%} \underset{\text{Ph}}{\overset{\text{Ph}}{>}}\overset{+}{\text{S}}-\overset{-}{\text{N}}-\overset{+}{\text{S}}\underset{\text{Ph}}{\overset{\text{Ph}}{<}} \quad \text{R}_F\text{O}^- \cdot \text{R}_F\text{OH} \tag{71}$$

$$\text{Me}-\text{S}-\text{Me} + \text{Me}-\underset{\overset{|}{\text{N}}\diagdown_{\text{Br}}}{\overset{\overset{\text{O}}{\|}}{\text{S}}}-\text{Me} \longrightarrow \underset{\text{Me}}{\overset{\text{Me}}{>}}\overset{+}{\text{S}}-\text{N}=\underset{\overset{|}{\text{Me}}}{\overset{\overset{\text{O}}{\|}}{\text{S}}}-\text{Me} \quad \text{Br}^- \tag{72}$$

(33)

Some interesting azasulphonium salts have been synthesized by Shine and coworkers. When thianthrene cation radical was treated with anhydrous ammonia in nitromethane, an azasulphonium salt **34** was obtained in 49% yield[111] (equation 73). The reaction presumably involves a sulphur dication intermediate formed by disproportionation of the cation radical. Similar results were obtained with both phenothiazine and phenoxathiin cation radicals[112,113].

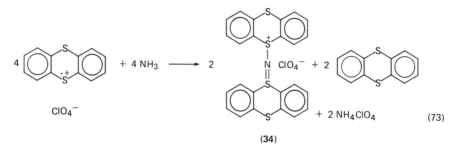

(34)

A mixture of phenoxathiin sulphimide and thianthrene or phenoxathiin cation radical was also found to give an azasulphonium salt **35**[113] (equation 74).

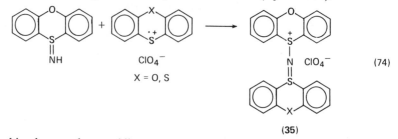

X = O, S

(74)

(35)

When thianthrene, phenoxathiin, and phenothiazine cation radicals were allowed to react with primary and secondary amines, the corresponding azasulphonium salts were obtained in good yields[105,112] (equation 75).

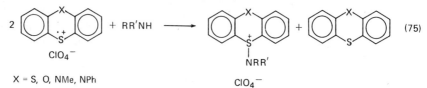

(75)

X = S, O, NMe, NPh

3. Diaminosulphonium salts

Richards and Tarbell reported the formation of the water-stable diaminosulphonium salt **36** by ethylation of dimorpholinosulphide with the Meerwin reagent[114] (equation 76). Unfortunately, this method has only limited utility for the

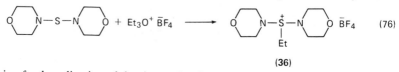

(76)

(36)

synthesis of other diaminosulphonium salts since reaction of other aminosulphides with the Meerwein reagent or other alkylating agents yields no isolable products[114]. Heimer and Field reported that methylation of alkanesulphenyl morpholide with methyl iodide gave not the aminosulphonium salt but dialkyl disulphide and iodine, presumably via an initial N-methylation[115].

The diaminosulphonium salt **37**, stable above its melting point, was obtained by Haake and Benack in good yields (81–90%) by treatment of the sulphenamide with

NCS in dichloromethane[116]. Exchange of the amino group in **37** did not occur on treatment with primary and secondary amines but gave a ring-opened product **38** in 67–89% yields[116] (equation 77).

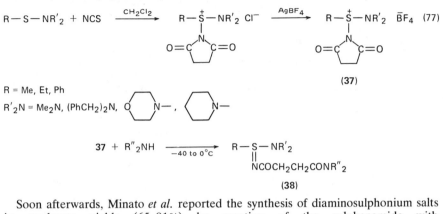

R = Me, Et, Ph

R'_2N = Me$_2$N, (PhCH$_2$)$_2$N,

$$37 + R''_2NH \xrightarrow[-40\ to\ 0°C]{} R-S-NR'_2$$
$$\underset{NCOCH_2CH_2CONR''_2}{\overset{\parallel}{}}$$

(38)

Soon afterwards, Minato *et al.* reported the synthesis of diaminosulphonium salts in moderate yields (65–91%) by reaction of the sulphenamide with *N*-chlorobenzotriazole in CH$_2$Cl$_2$ at −80°C, followed by addition of secondary amines at −80°C and subsequent addition of sodium tetraphenylborate at room temperature[37] (equation 78). The reaction with NCS instead of

N-chlorobenzotriazole also gave the diaminosulphonium salt, although in lower yields. Reaction of the sulphenamide with primary amines was unsuccessful[37].

The diaminosulphonium salt **39** was also synthesized in 62% yield by methylation of the iminosulphinamide **40**, prepared from the sulphenamide and chloramine-T[37] (equation 79).

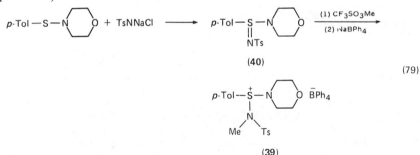

When a sulphinamide was treated with an *N*-sulphinyl dimethylimmonium salt, the diaminosulphonium salt was obtained in 67% yield[95] (equation 80).

$$Me-SO-NMe_2 + \underset{Me}{\overset{Me}{\diagdown}}\overset{+}{N}=S=O\ \bar{B}F_4 \xrightarrow{-SO_2} Me-\overset{+}{S}-NMe_2\ \bar{B}F_4$$
$$\underset{NMe_2}{\overset{\mid}{}}$$
(80)

Chlorosulphonium salts gave the corresponding aminosulphonium salts on treatment with trimethylsilylamines[117] (equation 81).

$$[Me_{3-n}SCl_n]^+ \ \bar{Sb}Cl_6 + Me_2NSiMe_3 \xrightarrow{CH_2Cl_2} [Me_{3-n}S(NMe_2)_n]^+ \ \bar{Sb}Cl_6 \qquad (81)$$

$n = 1, 2, 3$

C. Thiasulphonium Salts

The stable thiasulphonium salt **41** is obtained in 90% yield by treatment of dimethyl disulphide with trimethyloxonium trinitrobenzenesulphonate in nitromethane solution[118] (equation 82). Treatment of dimethyl sulphide with

$$Me-S-S-Me + Me_3O^+ \ TNBS^- \longrightarrow Me-\overset{+}{\underset{Me}{S}}-S-Me \ TNBS^- \qquad (82)$$

(41)

methanesulphenyl bromide and silver trinitrobenzenesulphonate also gives the related salt[119] (equation 83). Other alkyl disulphides such as diethyl disulphide and

$$Me-S-Me + MeSBr + AgTNBS^- \xrightarrow{50\%} Me-\overset{+}{\underset{Me}{S}}-S-Me \ TNBS^- \qquad (83)$$

1,2-dithiane were also found to be alkylated with trimethyloxonium or triethyloxonium trinitrobenzenesulphonate in 67–90% yields[120,121]. However, diaryl disulphides such as p-methoxyphenyl disulphide afforded a disproportionation product, dimethyl p-methoxyphenyl sulphonium salt, by the same treatment[120]. Dimethyl methylthiasulphonium salt was also prepared by treatment of dimethyl disulphide with the Meerwein reagent, trimethyloxonium tetrafluoroborate[23].

When 1,2-dithian was alkylated with methyl iodide in the presence of AgTNBS in refluxing methylene chloride, the thiasulphonium salt **42** was obtained in 58% yield[121] (equation 84).

$$(84)$$

(42)

Relative rates of the S-methylation of various disulphides with methyl fluorosulphonate were measured and found to be dependent on the CSSC dihedral angle. Lipoic acid, a five-membered cyclic disulphide, was about 11 times more reactive than 1,2-dithian, a six-membered cyclic disulphide, which was 33 times more reactive than an open-chain alkyl disulphide such as diethyl disulphide. Thus, the rate was found to decrease with increase in CSSC dihedral angle, and this relationship was rationalized[122].

Reaction of some alkyl sulphides with methane- and ethanesulfenyl chlorides in the presence of Lewis acid was also found to afford the corresponding thiasulphonium salts in 70–90% yields[123] (equation 85). Similarly the reaction of

$$R-S-R + MeSCl \xrightarrow{SbCl_5} R-\overset{+}{\underset{R}{S}}-S-Me \ \bar{Sb}Cl_6 \qquad (85)$$

dimethyl sulphide with p-toluenesulphenyl chloride in the presence of silver perchlorate in acetonitrile at 0°C gave crystalline dimethyl p-tolylthiasulphonium perchlorate[124].

The reaction in equation 86 also gave the thiasulphonium salt[123].

$$Me-S-\overset{+}{\underset{\underset{Me}{|}}{S}}-S-Me \ \ \bar{S}bCl_6 \ + \ R-S-R \ \longrightarrow \ R-\overset{+}{\underset{\underset{R}{|}}{S}}-S-Me \ \ \bar{S}bCl_6 \ + \tag{86}$$

(43)

$$Me-S-S-Me$$

Dimethyl methylthiasulphonium salt was formed in 63% yield by treatment of dimethyl chlorosulphonium salt with dimethyl disulphide, or by keeping the chlorosulphonium salt in acetic acid[23] (equations 87 and 88). The latter reaction gave as a by-product a small amount of dimethyl methanethiamethyl sulphonium salt.

$$Me-\overset{+}{\underset{\underset{Cl}{|}}{S}}-Me \ \ \bar{S}bCl_6 \ + \ Me-S-S-Me \ \xrightarrow{5^\circ C} \ Me-\overset{+}{\underset{\underset{Me}{|}}{S}}-S-Me \ \ \bar{S}bCl_6 \ + \ MeSCl \tag{87}$$

$$Me-\overset{+}{\underset{\underset{Cl}{|}}{S}}-Me \ \ \bar{S}bCl_6 \ \xrightarrow{AcOH} \ Me-\overset{+}{\underset{\underset{Me}{|}}{S}}-S-Me \ \ \bar{S}bCl_6 \ + \ Me-\overset{+}{\underset{\underset{CH_2SMe}{|}}{S}}-Me \ \ \bar{S}bCl_6 \tag{88}$$

$$(3 \ : \ 1)$$

Similar thiasulphonium salts were synthesized in 25–84% yields by treatment of chlorosulphonium salts with alkane- and arenethiols[76] (equations 89 and 90).

$$\underset{\bar{S}bCl_6}{\overset{+}{S}}-Cl \ + \ PhCH_2SH \ \xrightarrow{51\%} \ \overset{+}{S}-S-CH_2Ph \ \ \bar{S}bCl_6 \tag{89}$$

$$Me-\overset{+}{\underset{\underset{Cl}{|}}{S}}-Me \ \ \bar{S}bCl_6 \ + \ RSH \ \xrightarrow{25-84\%} \ Me-\overset{+}{\underset{\underset{Me}{|}}{S}}-S-R \ \ \bar{S}bCl_6 \tag{90}$$

$$RSH = PhSH, \ p\text{-}ClC_6H_4SH, \ HSCH_2COOH, \ HSCH_2CH_2OH$$

The dithiasulphonium salt **43** can be prepared in 95% yield by treatment of dimethyl disulphide with methanesulphenyl chloride in CH_2Cl_2 in the presence of a Lewis acid such as $SbCl_5$[125] (equation 91). An n.m.r. study showed that such a

$$Me-S-S-Me \ + \ MeSCl \ \xrightarrow[CH_2Cl_2]{SbCl_5} \ Me-S-\overset{+}{\underset{\underset{Me}{|}}{S}}-S-Me \ \ \bar{S}bCl_6 \tag{91}$$

(43)

dithiasulphonium salt was formed in liquid SO_2 solutions of dialkyl disulphides and alkanesulphenyl chlorides in the presence of a Lewis acid or a strong mineral acid[125].

When dimethyl disulphide was treated with $SbCl_5$, the salt **43** was obtained in 92% yield, presumably involving *in situ* formation of methanesulphenyl chloride from the complex of the disulphide and $SbCl_5$[126] (equation 92).

$$3 \ Me-S-S-Me \ + \ 3 \ SbCl_5 \ \xrightarrow{CH_2Cl_2} \ 2 \ Me-S-\overset{+}{\underset{\underset{Me}{|}}{S}}-S-Me \ \ \bar{S}bCl_6 \ + \ SbCl_3 \tag{92}$$

Alkyl trisulphides similarly can be methylated with trimethyloxonium tetrafluoroborate to give the corresponding dithiasulphonium salts[127].

The interesting thiasulphonium-type compounds **44** and **45** were prepared by Musker and Roush[128]. When the cyclic dithioethers **46** and **47** were oxidized with nitrosyl compounds, the stable dications **44** and **47** were oxidized with nitrosyl compounds, the stable dications **44** and **45** were obtained. The cation radical **44a** was detected by e.s.r. spectrometry when one equivalent of the oxidant was used[129]. The dications were found to behave either as electrophiles or as two-electron oxidizing agents[128] (equation 93).

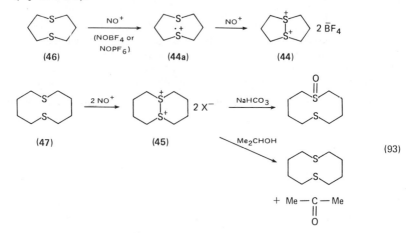

When a mixture of methyl *p*-toluenethiolsulphonate and methyl triflate was allowed to stand at room temperature for 10 days, the colourless crystalline deposit was identified as a sulphonyl-substituted sulphonium salt **48**[130] (equation 94).

$$p\text{-TolSO}_2-\text{S}-\text{Me} + \text{CF}_3\text{SO}_3\text{Me} \longrightarrow p\text{-TolSO}_2-\overset{+}{\underset{\underset{\text{Me}}{|}}{\text{S}}}-\text{Me} \ \ \text{CF}_3\text{SO}_3^- \quad (94)$$

(48)

D. Halosulphonium Salts

Sulphide–halogen complexes formed by treatment of sulphides with molecular halogens had already appeared in the literature in the last century. These complexes are known to be generally unstable upon exposure to air or moisture. For easy handling, the gegen anion usually halide, has been exchanged for other less nucleophilic anions to form stable halosulphonium salts.

Meerwein *et al.* prepared the stable salt dimethyl chlorosulphonium hexachloroantimonate in high yield by reaction of the dimethyl sulphide–SbCl adduct with chlorine at low temperature[23] (equation 95). Instead of chlorine, other

$$\text{Me}-\text{S}-\text{Me}\cdot\text{SbCl}_5 + \text{Cl}_2 \longrightarrow \text{Me}-\overset{+}{\underset{\underset{\text{Cl}}{|}}{\text{S}}}-\text{Me} \ \ \overset{-}{\text{SbCl}}_6 \quad (95)$$

chlorinating reagents such as SO_2Cl_2, $SOCl_2$, $COCl_2$, and acetyl chloride were used successfully for the preparation of the chlorosulphonium salt[23]. They also prepared the hygroscopic dimethyl chlorosulphonium tetrafluoroborate by treatment of dimethyl sulphide with sulphuryl chloride in the presence of HBF_4[23].

The dimethyl chlorosulphonium salt was also prepared by reaction of dimethyl sulphide with chlorine in the presence of boron trichloride in liquid hydrogen chloride at low temperatures[131]. When dimethyl or ethyl methyl sulphide was treated with a 2–4 molar excesses of $SbCl_5$ alone, the chlorosulphonium hexachloroantimonates were obtained in 68–88% yields[23,75] (equation 96).

$$\text{Me}-\text{S}-\text{R} + 2\,\text{SbCl}_5 \longrightarrow \text{Me}-\overset{+}{\underset{\underset{\text{Cl}}{|}}{\text{S}}}-\text{R}\ \ \bar{\text{S}}\text{bCl}_6 \tag{96}$$

R = Me, Et

The same dimethyl chlorosulphonium salt was also prepared by the reaction between DMSO–SbCl$_5$ adduct and thionyl chloride[23] (equation 97).

$$\text{Me}-\text{SO}-\text{Me}\cdot\text{SbCl}_5 + \text{SOCl}_2 \xrightarrow{\ -\text{SO}_2\ } \text{Me}-\overset{+}{\underset{\underset{\text{Cl}}{|}}{\text{S}}}-\text{Me}\ \ \bar{\text{S}}\text{bCl}_6 \tag{97}$$

Bromosulphonium salts are known to be stable compared with the corresponding chlorosulphonium salts, and several bromosulphonium bromides have been isolated simply by treatment of sulphides with bromine without any exchange of gegen anion. In 1909 Zincke and Frohneberg isolated the dark-red bisbromosulphonium salt **49**, consisting of two separable isomers, by either treatment of the sulphide with bromine or of the sulphoxide with hydrogen bromide[132] (equation 98).

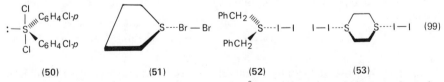

Several other stable bromosulphonium bromides, such as dimethyl[133], diphenyl[134–136], and other alkyl aryl compounds[137,138] have been obtained by treatment of the corresponding sulphides with bromine in non-polar aprotic media below 0°C.

By replacement of the gegen anion of the bromide by other less nucleophilic anions, the resultant bromosulphonium salts are converted into more stable salts. Thus, dimethyl bromosulphonium bromide, on treatment with sodium perchlorate in methanol, is converted into the perchlorate, which is stable for 12–18 days at room temperature if free from methanol[139].

A few stable sulphide–halogen complexes have been subjected to X-ray crystallographic analysis. Thus, the sulphide–chlorine complex **50** was found to have a covalent sulphurane structure[140], while the sulphide–bromine complex **51**[141] and sulphide–iodine complexes such as **52**[142] and **53**[143] were found to have charge-transfer structures in the solid state (equation 99).

$$\underset{(50)}{\overset{\text{Cl}}{\underset{\underset{\text{Cl}}{|}}{:-\overset{|}{\underset{|}{S}}}}\underset{\text{C}_6\text{H}_4\text{Cl-}p}{\overset{\text{C}_6\text{H}_4\text{Cl-}p}{}}} \qquad \underset{(51)}{\langle\ \rangle\text{S}\cdots\text{Br}-\text{Br}} \qquad \underset{(52)}{\overset{\text{PhCH}_2}{\underset{\text{PhCH}_2}{S\cdots\text{I}-\text{I}}}} \qquad \underset{(53)}{\text{I}-\text{I}\cdots\text{S}\langle\ \rangle\text{S}\cdots\text{I}-\text{I}} \tag{99}$$

Both S—Cl bond lengths (2.26 and 2.32 Å) in **50** are much longer than the covalent radius sum (2.03 Å), in keeping with the apical bonding arrangements of the two S—Cl linkages in the hypervalent trigonal bipyramidal sulphur structure[140]. Wilson and Chang, however, have claimed, based on ^{19}F n.m.r. measurements, that the sulphide and chlorine complexes are in rapid equilibrium between the sulphurane **54** and the sulphonium salt **55** in solution[144,145] (equation 100).

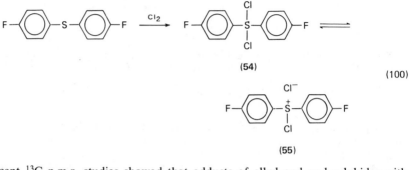

(54)

(100)

(55)

Recent ^{13}C n.m.r. studies showed that adducts of alkyl and aryl sulphides with bromine or iodine were simple molecular complexes[146,147]. It is also known that the stability of the charge-transfer complexes formed between substituted diphenyl sulphides and iodine is substantially influenced by substituents on the benzene ring[148].

Sulphide–halogen complexes would generally exist as sulfonium salts in solution, but the equilibrium between the sulphonium salt and the sulphurane is markedly influenced by the solvent. The azasulphonium bromide **56** exists as a covalent sulphurane **57** in non-polar solvents such as chloroform, while in polar solvents such as water or dimethyl sulphoxide the sulphonium structure **56** was found by n.m.r. spectral measurements to predominate[149] (equation 101).

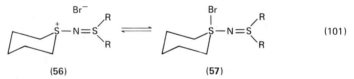

(101)

(56)   (57)

A low-temperature n.m.r. study suggested that the counter anion plays an important role in the equilibrium between the sulphonium salt and the sulphurane; methyl phenyl *t*-butoxysulphonium tetrafluoroborate exists as the sulphonium salt at $-46°C$, while the same salt with chloride as counter anion, prepared by reaction of the sulphide with *t*-BuOCl, has the sulphurane structure[26].

E. Optically Active Heterosulphonium Salts

Several optically active alkoxysulphonium salts have been prepared by alkylation of the corresponding optically pure sulphoxides with the Meerwein reagent (Table 1) (equation 102).

$$p\text{-Tol} - \overset{*}{S}O - R + R'_3O^+ \ \bar{B}F_4 \longrightarrow p\text{-Tol} - \overset{*+}{\underset{OR'}{S}} - R \ \ \bar{B}F_4 \qquad (102)$$

An optically active alkoxysulphonium salt has also been prepared by separation of diastereomeric alkoxysulphonium salts. (+)-Benzyl *p*-tolyl menthoxysulphonium salt with $[\alpha]_D +137°$ was obtained on recrystallization from benzene–hexane (2:1) of the diastereomeric mixture of the alkoxysulphonium salts with $[\alpha]_D +58.4°$ from reaction of benzyl *p*-tolyl sulphide with (−)-menthol in the presence of *N*-chlorobenzotriazole[28] (equation 103). This salt was obtained in at least 87% enantiomeric excess with the *R*-configuration at sulphur since, when the salt was

TABLE 1. Optically active alkoxysulphonium salts, $p\text{-Tol}-\overset{+}{\underset{|}{S}}-R^1 \quad \bar{B}F_4$
$$\underset{OR^2}{}$$

R^1	R^2	Absolute configuration	$[\alpha]_D$ (solvent)	Reference
CH_2Ph	Et	R	$+203°(CHCl_3)$	200
			$+202.6°(CHCl_3)$	209
			$+205°(CHCl_3)$	356
CH_2Ph	Et	S	$-202°$	200
CH_2Ph	Me	R	$+189°(CH_2Cl_2)$	3
$n\text{-Bu}$	Et	R	$+188°(acetone)$	206
Me	Me	R	$+135.8°(CH_2Cl_2)$	211
			$+186°(acetone)$	205
Me	Et	R	$+149°(acetone)$	356
α-Naphthyl	Et	S	$-410°(acetone)$	355
Me	1-Adamantyl[a]	—	$+68.4°(acetone)$	207

[a]Counter anion is $\overline{SbCl_6}$ for $\bar{B}F_4$.

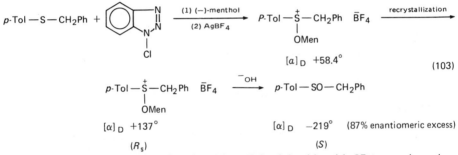

$$(103)$$

subjected to alkaline hydrolysis, S-benzyl p-tolyl sulphoxide with 87% enantiomeric excess was obtained. The azasulphonium salt **59** with $[\alpha]_D +4.2°$ was prepared by protonation of the optically active N-acyl sulphimide **58**[103,150], (equation 104).

$$\underset{\underset{NCOMe}{\overset{\|}{}}}{Et-S-Me} + P\text{-TolSO}_3H \xrightarrow[\text{acetone}]{0°C} \underset{\underset{NHCOMe}{|}}{Et-\overset{+}{S}-Me} \quad TsO^- \qquad (104)$$

$[\alpha]_D +83°$ $\qquad\qquad\qquad\qquad\qquad\qquad$ $[\alpha]_D +4.2°$

$\qquad$ **(58)** $\qquad\qquad\qquad\qquad\qquad\qquad\qquad\qquad$ **(59)**

Menson and Darwish carried out a kinetic study of the thermal racemization caused by pyramidal inversion of both this azasulphonium salt **59** and of the sulphimide **58** in acetonitrile[150]. Racemization of **59** was found to be 32–46 times faster than that of **58** in the range 70–90°C. The ready racemization of the azasulphonium salt is due to the low activation enthalpy (29.2 kcal/mol) relative to those of sulphimides **58** (34.1 kcal/mol), $[p\text{-ClC}_6H_4(Me)S{=}NTs]$ (27.9 kcal/mol)[151], and $[Ph(o\text{-MeOC}_6H_4)S{=}NCOPh]$ (30.0 kcal/mol)[152] [cf. those of alkyl aryl and diaryl sulphoxides (36–43 kcal/mol)[153]]. The activation enthalpy for racemization of the azasulphonium salt **59** was, however, higher than those of the sulphonium ylide $[Me(Et)S{-}CHCOPh]$ (23.3 kcal/mol)[154] and sulphonium salts[155–158], e.g. adamantylethylmethylsulphonium perchlorate ($\Delta H^{\ddagger} = 26$ kcal/mol)[155].

Darwish and Datta also observed that another optically active azasulphonium salt **61** with $[\alpha]_{546}$ $-17°$, prepared by alkylation of the sulphimide **60** with methyl triflate, undergoes ready thermal racemization[103] (equation 105) 40 times faster than

$$(-)\text{-}p\text{-Tol} \overset{\|}{\underset{\text{NTs}}{-S}}-\text{Me} + \text{CF}_3\text{SO}_3\text{Me} \longrightarrow (-)\text{-}p\text{-Tol}\overset{+}{\underset{\underset{\text{Me}\diagdown\text{Ts}}{N}}{-S}}-\text{Me}\quad \text{CF}_3\text{SO}_3{}^- \qquad (105)$$

(60)   (61)

60 in ethyl acetate at 90°C. Activation parameters were obtained both for **61** ($\Delta H^{\ddagger} = 25.8$ kcal/mol, $\Delta S^{\ddagger} = -1.08$ e.u.) and for **60** ($\Delta H^{\ddagger} = 28.7$ kcal/mol, $\Delta S^{\ddagger} = -1.08$ e.u.)[103].

Methylation of **60** with trimethyloxonium salt was also successful[103] (equation 106).

$$(-)\text{-}p\text{-Tol} \overset{\|}{\underset{\text{NTs}}{-S}}-\text{Me} + \text{Me}_3\text{O}^+ \text{X}^- \longrightarrow (-)\text{-}p\text{-Tol}\overset{+}{\underset{\underset{\text{Me}\diagdown\text{Ts}}{N}}{-S}}-\text{Me}\quad \text{X}^- \qquad (106)$$

(60)

$[\alpha]_{546}$ $-5°$ for X = $\bar{\text{B}}\text{F}_4$

$[\alpha]_{546}$ $-8.3°$ for X = TNBS$^-$

(61)

Thermal racemization of alkoxysulphonium salts has not been explored, but they are probably more stable thermally than azasulphonium salts as the planar sp^2 transition state of the former should have a higher energy than that of the latter, as in the difference of barrier to stereomutation for the sulphoxides and sulphimides[152,159].

The crystal structures of the azasulphonium salts **62**[160] and **63**[94] have been determined (equation 107). Both are pyramidal at sulphur and have shorter S–N

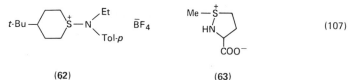

(62)   (63)

bonds (1.64 Å for **62**; 1.68 Å for **63**) than that of the sum of the covalent radii of sulphur and nitrogen (1.74 Å). The configuration at nitrogen in **62** is sp^2, indicating that there is a significant pπ–dπ bonding between sulphur and nitrogen while that of **63** is sp^3 indicating the absence of pπ–dπ bonding.

X-ray crystallogrpahy of salts **64** and **65** showed that the symmetrical one **64** has the structure **64b** as a major contributor in which two lone pairs on nitrogen and one on each sulphur atom are not localized, because both S—N bond lengths (1.63, 1.64 Å) and both C—N—S angles at sulphur are nearly the same[161] (equation 108). In the unsymmetrical salt **65**, the positive charge is localized predominantly on the sulphur atom of methyl phenyl sulphide (structure **65c**), since one S—N bond length

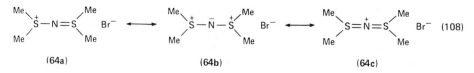

(64a)   (64b)   (64c)

(1.67 Å) is nearly the same as that of the azasulphonium salt and the other (1.60 Å) as that of the sulphimide[162] (equation 109).

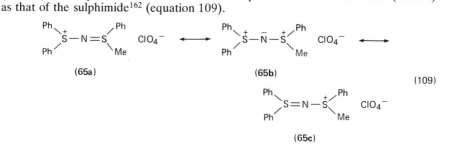

(65a)   (65b)   (65c)   (109)

III. REACTIONS

A. Nucleophilic Substitution at Sulphonium Sulphur

1. Introduction

Nucleophilic substitution at trivalent sulphur is considered to proceed by way of a bipyramidal sulphurane or transition state intermediate. Pseudorotation, although slower than that for analogous phosphorus intermediates[163], must also be considered. The stereochemistry depends on the steric arrangement and the electronic properties of both incoming and outgoing groups. In the hypervalent trigonal bipyramid **66** (equation 110), apical positions are considered to be occupied by the more

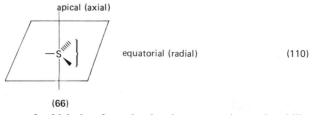

(66)

(110)

electronegative groups, one of which is often the leaving group in nucleophilic substitution. Since the apical bonds are longer than equatorical bonds, most nucleophilic substitutions proceed with both the entering and leaving groups at apical positions, much as in S_N2 reactions at carbon. The result is inversion of configuration, and many examples are cited in this section.

When the entering and leaving groups both assume equatorical positions, the situation is less favourite, since both groups are electronegative and tend to occupy apical positions, but the stereochemical result of the reaction is similarly net inversion. Cram and coworkers suggested that the formation of methyl p-tolyl N-p-tosylsulphimide from the corresponding sulphoxide, which is second order in diimide and proceeds with 98% inversion, may be an example of such a process[164-166] (equation 111).

If incoming and outgoing groups can assume apical and radial positions, i.e. the perpendicular arrangement, reaction will lead to net retention via one pseudoratation without violating the hypervalency concept. One of the earliest examples is the oxygen exchange reaction between methyl p-tolyl sulphoxide and dimethyl sulphoxide[167] (equation 112). By changing the medium from pyridine to a apolar

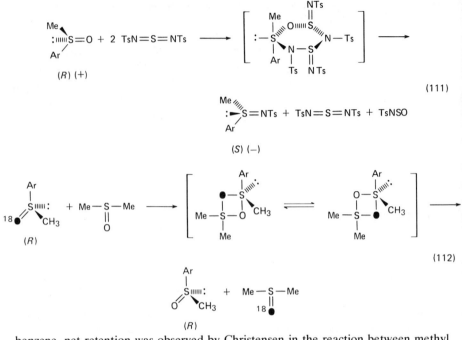

(111)

(112)

benzene, net retention was observed by Christensen in the reaction between methyl
p-tolyl sulphoxide and NN'-bis(p-tosyl)sulphur diimide[168]. The same author had
earlier observed retention in the reaction between N-phthaloylmethionine sulphoxide
and N-sulphinyl tosylamide[169]

The following example also results in retention of configuration and involves a
type of thiasulphonium salt as intermediate[170] (equation 113). Some reactions

between diaryl menthoxysulphonium salts and acylamides were found to give the
corresponding sulphimide with net retention of configuration[171].

2. General reactions

Since the heterosulphonium salt has a positive charge on the sulphonium sulphur,
nucleophilic substitution on the trivalent sulphur of the heterosulphonium salt occurs
easily.

a. Halosulphonium salts. Since 1909, when it was reported that halosulphonium
salts were easily hydrolysed by water to give sulphoxides[132], this reaction has been
widely used in syntheses of sulphoxides. When ^{18}O-enriched water is used,
^{18}O-labelled sulphoxides are obtained. The synthesis of ^{18}O-labelled sulphoxides is
easily achieved by treatment of a sulphide with either bromine–DABCO or
bromine–pyridine complex, with hydrolysis *in situ* of the bromosulphonium salt[172]
(equation 114).

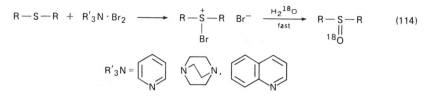

$$R-S-R + R'_3N \cdot Br_2 \longrightarrow R-\overset{+}{\underset{\underset{Br}{|}}{S}}-R \ Br^- \xrightarrow[\text{fast}]{H_2{}^{18}O} R-\overset{\underset{18O}{\|}}{S}-R \qquad (114)$$

Alternatively, an acetic acid solution of any dialkyl, alkyl aryl or diaryl sulphide, a 3–10 molar excess of H_2O and one equivalent of pyridine is treated with an acetic acid solution of one equivalent of bromine.

In a kinetic investigation of oxidation of alkyl aryl sulphides with bromine in aqueous methanol, formation of the respective bromosulphonium bromide was suggested to be rate-determining, based on the observation of the large steric effect of the alkyl group and the large ρ-value (-3.2)[173]. When alkyl aryl sulphides were treated with bromine in 95% acetic acid, bromination was found to take place predominantly at the *para*-position on the benzene ring. The rate decreased with increase in the size of the alkyl group, also supporting the intial formation of the bromosulphonium salt during the reaction[174].

Various sulphoxides have been synthesized by treatment of sulphides with many other halogenating agents, followed by *in situ* hydrolysis or alcoholysis. Examples are NBS[175–177], NCS[176,178], chloramine-B[179], N-chlorobenzotriazole[180], N-chloronylon 66[181], N-chloro-ε-caprolactam[182], iodine[183,184], iodobenzene dichloride[185,186], iodobenzene diacetate[187,188], tribromocresol[189], SO_2Cl_2[190], and t-BuOCl[26,191–193]. Asymmetric oxidation of sulphides to sulphoxides by halogenating agents in the presence of optically active alcohols[194–196] or dicarboxylic acids[197] has also been reported.

b. Oxysulphonium salts. Smith and Winstein observed that a dimethyl alkoxysulphonium salt was easily hydrolysed by water[5]. Leonard and Johnson reported that when the eight-membered cyclic sulphoxide **67** was treated with 70% perchloric acid, followed by acetylation, the alkoxysulphonium salt **68** was obtained and was easily hydrolysed with $H_2^{18}O$ at room temperature to afford the ^{18}O-labelled initial sulphoxide, indicating that water attacked the sulphonium sulphur atom of the sulphonium salt **68**[8] exclusively (equation 115). A similar transannular sulphoxide ketone interaction was observed in a seven-membered cyclic γ-keto sulphoxide[198].

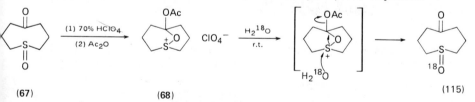

$$(67) \qquad (68) \qquad (115)$$

A simple alkoxysulphonium salt such as dimethyl methoxysulphonium perchlorate is hydrolysed with $H_2^{18}O$ to ^{18}O-labelled dimethyl sulphoxide[8] (equation 116).

$$Me-\overset{+}{\underset{\underset{OMe}{|}}{S}}-Me \ ClO_4{}^- \xrightarrow[\text{r.t.}]{H_2{}^{18}O} Me-\overset{\underset{18O}{\|}}{S}-Me \qquad (116)$$

The stereochemical course of the alkaline hydrolysis of alkoxysulphonium salts was established by Johnson[199]. *cis-p*-Chlorophenylthiane S-oxide **69** was transformed into the *trans*-sulphoxide **71** by alkaline hydrolysis of the intermediate ethoxysulphonium salt **70**, and the *trans*-sulphoxide **71** was converted into the *cis*-isomer **69** in the same way (equation 117). Alkaline hydrolysis of the ethoxysulphonium salts **70** and **72** was found to proceed via with over 96% inversion of configuration.

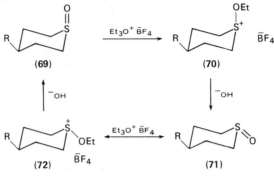

(117)

$$R = p\text{-}ClC_6H_4$$

The same stereochemical cycle was completed by Johnson and McCants for benzyl *p*-tolyl sulphoxide. In this case the alkaline hydrolysis of the alkoxysulphonium salt with 0.1 N NaOH proceeded with more than 98% inversion of configuration[200] (equation 118).

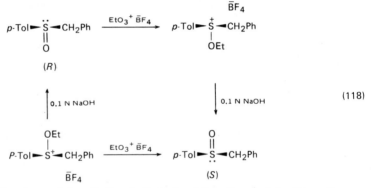

(118)

The stereochemistry of the alkaline hydrolysis of the alkoxysulphonium salts was demonstrated in the examples shown in equations 119[201], 120[163], 121[202], and 122[203],

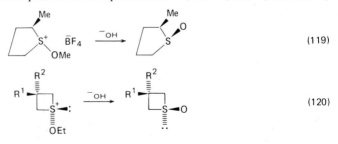

(119)

(120)

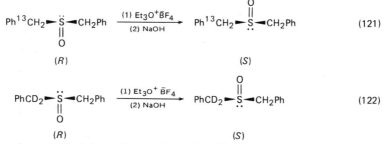

$$(121)$$

$$(122)$$

(R)   (S)

and was found to proceed through inversion of configuration at sulphur with high stereospecificity in every case.

[18]O tracer experiments revealed that in alkaline hydrolysis of the steroidal alkoxysulphonium salts **73** (*anti*) and **74** (*syn*) in *N*-methylpyrrolidone solution, the site of attack of hydroxide ion changed from sulphur to the α-carbon of the alkoxy group owing to steric crowding in the molecule[204] (equation 123).

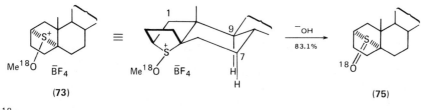

(73)

[18]O: 1.16 excess atom-%

1.18 excess atom-%

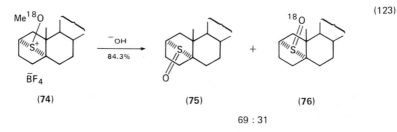

$$(123)$$

(74)   (75)   +   (76)

69 : 31

[18]O: 1.14 excess atom-%   0.06 excess atom-%   1.18 excess atom-%

[18]O-labelled *syn* salt **74** afforded both the *anti*-sulphoxide **75**, which retained little [18]O (4.8%), and the *syn*-sulphoxide **76**, with complete retention of ^{18}O, in the ratio 69:31, indicating that the *anti*-sulphoxide **75** is produced by attack of hydroxide at sulphur, and the *syn*-fulphoxide **76** by attack on alkoxy α-carbon. The ^{18}O-labelled *anti*-alkoxysulphonium salt **73** afforded only *anti*-sulphoxide **75** with completely retained [18]O. Attack of hydroxide occurs exclusively on the alkoxy α-carbon, undoubtedly due to severe steric crowding of the rear side of the *anti*-salt **73** by the 1α-, 7α-, and 9α-hydrogens of the adjacent cyclohexane ring.

Alkoxysulphonium salts are known to undergo facile alkoxy exchange in the presence of alkoxides[2.3,205], whereas under neutral conditions the reaction with alcohol is surprisingly slow[2.5,26]. Johnson and Rigau found that alkoxy exchange in the ethoxysulphonium salt **77** took place very slowly in neutral ^{14}C-labelled ethanol, but that on addition of catalytic amounts of hydrogen chloride or tetrabutylammonium chloride, akoxy exchange took place rapidly at room temperature, suggesting that chloride ion is catalytic[26] (equation 124).

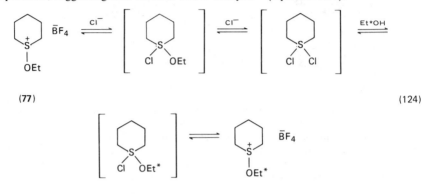

(77) (124)

*[14]C label

When the primary and secondary β-hydroxyalkoxysulphonium salts **78** were treated with methanol, nucleophilic attack of alcohol at the sulphonium sulphur took place slowly to give the glycol and dimethyl methoxysulphonium salt[19,20] (equation 125). When the benzylic β-hydroxyalkoxysulphonium salt **79** was allowed to react with methanol, the four products obtained indicated that the major reaction is nucleophilic attack at the benzylic carbon atom of the alkoxy group and the S_N1 reaction on sulphur is the minor process[20] (equation 126).

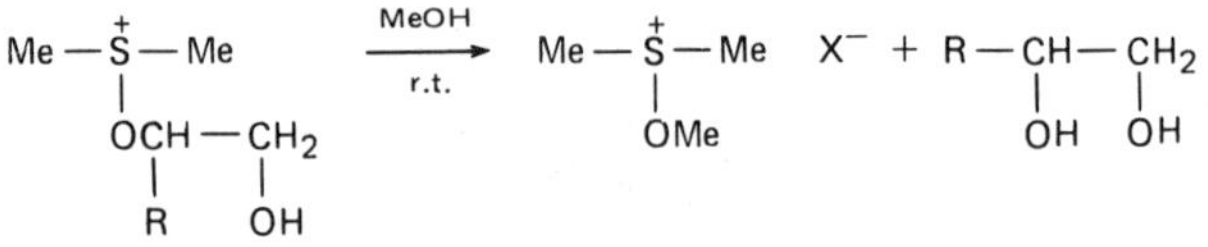

$$ Me-\overset{+}{S}-Me \underset{OCH-CH_2}{\overset{}{|}} \xrightarrow[\text{r.t.}]{\text{MeOH}} Me-\overset{+}{S}-Me \quad X^- + R-CH-CH_2 $$

(125)

(78)

R = H, alkyl X⁻ = TNBS⁻

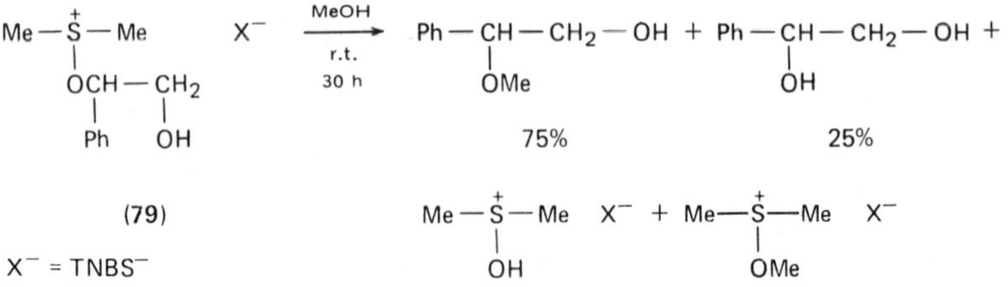

$$ Me-\overset{+}{S}-Me \quad X^- \xrightarrow[\substack{\text{r.t.} \\ 30 \text{ h}}]{\text{MeOH}} Ph-CH-CH_2-OH + Ph-CH-CH_2-OH + $$

75%   25%

(126)

(79) $Me-\overset{+}{S}-Me \quad X^- + Me-\overset{+}{S}-Me \quad X^-$

X⁻ = TNBS⁻   OH   OMe

25%

Attack of alcohol at benzylic carbon probably proceeds through an S_N1 process, since the salt **79** when kept in DMSO-d₆ was hexadeuterated on recovery **80**[20] (equation 127).

Andersen *et al.* reported that when the *cis*- and *trans*-methoxysulphonium salts **81** were treated with methylmagnesium bromide at −78°C or with dimethylcadmium at room temperature, the sulphonium salt **82** was obtained with more than 85%

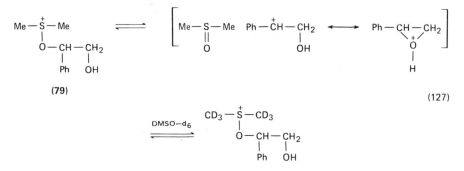

(127)

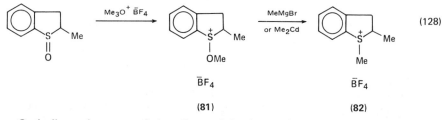

(80)

inversion. They concluded that loss of stereospecificity may occur by isomerization of the starting methoxysulphonium salt under the reaction conditions[205] (equation 128).

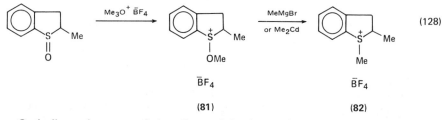

(128)

Optically active open-chain alkoxysulphonium salts, such as n-alkyl p-tolyl alkoxysulphonium salts, react with an alkyl Grignard reagent or alkylcadmium to give optically active dialkyl aryl sulphonium salts. Diaryl alkoxysulphonium salts afforded racemic alkyl diaryl sulphonium salts, and dialkyl alkoxysulphonium salts gave no sulphonium salt on the same treatment[206,207]. Treatment of diaryl alkoxysulphonium salts with aryl Grignard reagents afforded racemic triaryl sulfonium salts[208,209].

Diphenyl ethoxysulphonium ion was attacked by p-toluenethiolate at sulphonium sulphur to give diphenyl sulphide and di(p-tolyl) disulphide[210] while in the reaction of aryl methyl methoxysulphonium ion with thiophenoxide, nucleophilic attack was at sulphur and the α-carbon of the alkoxy group in nearly the same ratio[211].

Johnson and Phillips claimed, on the basis of a deuterium tracer experiment, that in the reaction of methyl phenyl methoxysulphonium ion with sodium borohydride, the methoxy group is displaced from the sulphur by $\overline{B}H_4$ and the protonated sulphide is formed[212] (equation 129).

$$Ph-\overset{+}{\underset{OMe}{S}}-Me \quad \overline{B}F_4 \xrightarrow{\overline{B}H_4} Ph-\overset{H}{\underset{}{\overset{|}{S^+}}}-Me \xrightarrow{-H^+} Ph-S-Me \qquad (129)$$

Durst *et al.* have developed a mild and selective method for reduction of sulphoxides to sulphides via alkoxysulphonium salts. The salts are easily reduced at $0°C$ in 77–91% yields with $NaBH_3CN$ in methanol or in CH_2Cl_2 in the presence of a catalytic amount of a crown ether[14].

Reaction of sulphoxides in acidic media may involve a protonated sulphoxide, i.e.

hydroxysulphonium salt, which is easily attacked on sulphur by any nucleophile present. Sulphoxides have a basic oxygen atom (pK_{BH^+} -5 to 0)[213–219]. Landini et al. have claimed that sulphoxides are non-Hammett bases and in reactions of sulphoxides in acidic media the acidity function which should be chosen is H_A and not H_0[216]. Olah et al. reported ^1H n.m.r. studies[220] showing that sulphoxides are protonated on sulphur in super-acids, while a ^{13}C n.m.r. study carried out by Gatti et al.[221] and reinvestigation by Olah et al.[222] using ^{13}C n.m.r. and MINDO/3 calculations showed that dimethyl sulphoxide is protonated on oxygen in strong acids. Protonation of sulphoxides now is generally considered to take place on oxygen[223,224].

Diphenyl sulphoxide protonates (equation 130) in concentrated sulphuric acid to afford a deep blue–green solution which gives a Van't Hoff i-factor of 2 in cryoscopic measurements[225–227] (equation 130).

$$Ph-\underset{\underset{O}{\|}}{S}-Ph + H_2SO_4 \longrightarrow Ph-\underset{\underset{OH}{|}}{\overset{+}{S}}-Ph + HSO_4^- \qquad (130)$$

When the sulphoxide was dissolved in a large excess of concentrated ^{18}O-enriched sulphuric acid, ^{18}O-labelled diphenyl sulphoxide was recovered on dilution with a large excess of ordinary water, and found to incorporate roughly the same content of ^{18}O as that of ^{18}O-enriched sulphuric acid. $(-)$-p-Aminophenyl p-tolyl sulphoxide was racemized under these conditions[228].

A careful kinetic study with optically active and ^{18}O-labelled phenyl ^{18}O-labelled phenyl p-tolyl sulphoxide and substituted diphenyl sulphoxides revealed that the rate of oxygen exchange in 95.5% sulphuric acid was identical with that of racemization ($k_{ex}/k_{rac} = 0.97$) and correlated with the acidity function H_A (or H_0). The effect of polar substituents was very small and a large positive activation entropy ($\Delta S^{\ddagger} = 9.86$ e.u.) was obtained. A clear e.s.r. signal of a cation radical species was also observed. Based on these observations, the following A-1 mechanism involving S—O bond fission was suggested[229,230] (equation 131).

$$Ar-\underset{\underset{^{18}\bullet H}{|}}{\overset{*+}{S}}-Ar' \underset{}{\overset{H^+}{\rightleftharpoons}} Ar-\underset{\underset{^+\bullet H_2}{|}}{\overset{+}{S}}-Ar' \xrightarrow{-H_2\bullet} \left[\begin{array}{c} Ar-\overset{2+}{S}-Ar' \\ \text{or} \\ Ar-\overset{+.}{S}-Ar' \end{array} \right] \xrightarrow[H_2O]{fast} Ar-\underset{\underset{OH}{|}}{\overset{+}{S}}-Ar' \quad (131)$$

$$\text{racemic}$$

Methyl p-tolyl and n-butyl methyl sulphoxide were similarly found to undergo concurrent oxygen exchange and racemization through an A-1 route in both 90% and 60% sulphuric acid. The rates showed linear correlation with H_A, and positive or small negative activation entropies and a k_{ex}/k_{rac} value of unity were observed[230,231].

In less concentrated sulphuric acid (e.g. below 60%), concurrent oxygen exchange and racemization reactions were also found to occur, but the rate of racemization no longer correlated with H_A and k_{ex}/k_{rac} became approximately 0.5[230,232]. An A-2 mechanism, (nearly S_N2 on sulphur) was suggested with the rate-determining step involving nucleophilic attack of water on the conjugated acid and every oxygen exchange resulting in inversion at sulphur. Apparently there is a gradual change of mechanism from A-1 to A-2 with decrease in sulphuric acid concentration. Associated with this change the activation entropy also became negative and k_{ex}/k_{rac} also gradually changed from 1 to 0.5[230,232] (equation 132).

Similar concurrent oxygen exchange and racemization reactions were found to take place with acids, such as p-toluenesulphonic acid[233], orthophosphoric acid[233], acetic acid, and monochloro-, dichloro- and trichloroacetic acid[234]. In all these cases,

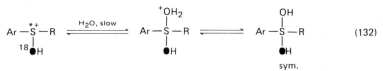

$$ (132) $$

the reactions were found to be acid-catalysed, the effect of polar substituents was small, and k_{ex} was identical with k_{rac}. These reactions are suggested to proceed via homolytic S—O bond cleavage in the rate-determining step.

The concurrent oxygen exchange and racemization reaction of diaryl sulfoxides in N_2O_4 is different in that the polar effect of substituents on the rate is large and correlates with the σ^+ values ($\rho = -1.30$), although the rate of oxygen exchange is identical with that of racemization. The reaction was suggested to involve the rate-determining heterolysis of the S—O linkage, via incipient formation of a sulphinyl oxygen–N_2O_4 adduct[235], which is known to be partly ionized[236].

t-Butyl phenyl sulphoxide undergoes racemization and cleavage in aqueous perchloric acid[237]. Experiments with ^{18}O-labelled sulphoxide and (R_S, R_C)-1-phenylethyl phenyl sulphoxide showed that racemization at sulphur did not proceed via the usual acid-catalysed oxygen exchange and was accompanied by partial racemisation of the carbon chiral centre, suggesting that the reaction proceeds through acid-catalysed heterolysis and recombination of an alkyl cation–sulphenic acid ion molecular pair[237] (equation 133). Similar cleavage was also observed in reactions of optically active alkyl t-butyl sulphoxides in perchloric acid[238].

$$ (133) $$

Hydrochloric acid[239–241] and hydrobromic acid[242] were found to effect similar concurrent oxygen exchange and racemization reactions in which the rate of oxygen exchange is identical with that of racemization[239,240]. The concurrent oxygen exchange reaction with hydrochloric acid or chloride ion in strong acids such as $HClO_4$ or H_2SO_4 was found to be markedly retarded by bulky substituents around sulphur[239,240,243,244], while a linear dependence between k_{rac} and the acidity function was observed for various sulphoxides[240,243–245]. On the basis of these observations, a mechanism involving the formation of a halogen–sulphide complex was suggested. The identity of the rates for oxygen exchange and racemization is considered to be due to the rapid interconversions between **83** and **84** in polar media (equation 134).

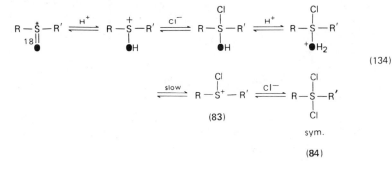

$$ (134) $$

Kwart and Omura have suggested that the hydrochloric acid-catalysed stereomutation of sulphoxides can occur through an SS-dihydroxysulphide intermediate via slow formation of S-chloro-S-hydroxysulphide[246] (equation 135).

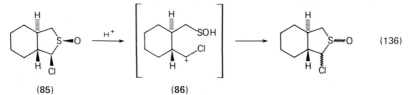

$$(135)$$

The α-chlorosulphoxide **85** was shown to isomerize on treatment with hydrogen chloride in dioxane at $0\,^{\circ}C$ to a mixture of isomers. In this reaction, epimerization could have taken place either at sulphur or at carbon bearing the chlorine, or at both. Since no deuterium or bromine was incorporated when the reaction was carried out with DCl or HBr, epimerization was suggested to proceed through C—S bond breaking and reformation presumably via the intermediate **86**[247] (equation 136).

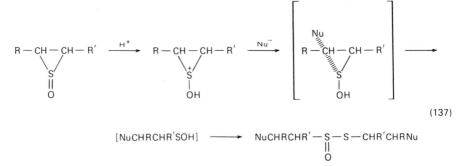

$$(136)$$

The three-membered cyclic sulphoxide, ethylene episulphoxide, undergoes acid-catalysed ring opening in dilute hydrochloric acid to give the disulphide and thiolsulphonate, presumably via a sulphenic acid intermediate[248]. An A-2 mechanism was proposed for the reaction based on a kinetic[249] and stereochemical study of the reaction (inversion of configuration around carbon)[250], and also by analyses of the product with the change of episulphoxide and solvent used[250,251] (equation 137).

$$
R-CH-CH-R' \xrightarrow{H^+} R-CH-CH-R' \xrightarrow{Nu^-} \left[R-\overset{Nu}{\underset{}{CH}}-CH-R' \right] \longrightarrow
$$

$$(137)$$

$$
[\text{NuCHRCHR'SOH}] \longrightarrow \text{NuCHRCHR'}-\underset{O}{\overset{\parallel}{S}}-S-\text{CHR'CHRNu}
$$

The steroidal thiolsulphinate shown in equation 138 was prepared from the episulphoxide in ethanol containing a small amount of sulphuric acid, and the resulting diastereomeric mixture was separated[252].

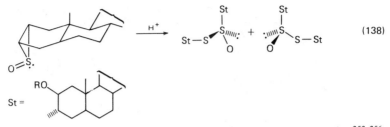

$$\tag{138}$$

Treatment of sulphoxides with hydrogen iodide causes reduction[253-256] (equation 139). Landini and coworkers found[243,244,257] that plots of log k for the

$$R-\overset{\underset{\displaystyle O}{\|}}{S}-R' + 2\,HI \longrightarrow R-S-R' + H_2O + I_2 \tag{139}$$

reduction of several sulphoxides with iodide ion in perchloric acid were linear with H_A (slope 1.5) or H_0. The large slopes were attributed to the involvement of two protons in the rate-determining step. This was supported by the high Bunnett–Olsen ϕ values and by Kruger's observation of a second-order dependence on proton concentration for the reduction of dimethyl sulphoxide with iodide ion in aqueous dimethyl sulphoxide[258]. These observations are consistent with a mechanism involving rate-determining formation of an iodosulphonium salt (equation 140).

$$R-\overset{+}{\underset{\underset{\displaystyle OH}{|}}{S}}-R' + I^- \rightleftharpoons R-\overset{|}{\underset{\underset{\displaystyle OH}{|}}{S}}-R' \overset{H^+}{\rightleftharpoons} R-\overset{|}{\underset{\underset{\displaystyle {}^+OH_2}{|}}{S}}-R' \rightleftharpoons R-\overset{|}{\underset{|}{\overset{+}{S}}}-R' \overset{I^-}{\rightleftharpoons}$$

$$R-S-R' + I_2 \tag{140}$$

In some cases general acid-catalysed reduction may occur rather than specific acid-catalysed reaction[259].

Since the reduction of sulphoxides is stoichiometric, reaction with iodide ion either in acetic acid[260] or in acetic acid–acetyl chloride[261] has been suggested as a convenient method for the determination.

Ring-size effects on these nucleophilic hydrogen halide-catalysed racemizations and reductions of sulphoxides have been studied by several groups. Sagrainora et al. reported that the rate of hydrogen chloride-catalysed stereomutation of thiolane S-oxide 87 was 390 times larger than that of thiane S-oxide 88[262]. A similar order of the relative rates of reduction of sulphoxides with hydrogen iodide was found by Tamagaki et al.[263] (equation 141). The high reactivity of thiolane S-oxide 87 was

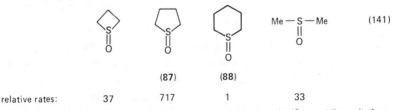

$$\tag{141}$$

	(87)	(88)	
relative rates:	37	717	1	33

attributed to its much more favourable entropy of activation ($\Delta S^\ddagger = -15$ e.u.) than that for dimethyl sulphoxide. The slow rate of reduction of thiane S-oxide 88 is believed to be due to steric hindrance by β-axial hydrogens to nucleophilic attack of iodide on sulphur, while such a steric repulsion can be minimized by twisting of the ring into a half-chair form 89. Evidence in support of this view comes from the extremely low reactivity of bicyclo[3.3.1]nonane 9-sulphoxide, which is so rigid that it cannot bend to minimize steric effects[263] (equation 142).

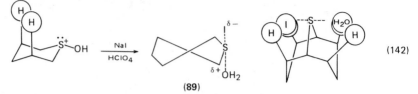

$$(142)$$

(89)

Curci et al. observed a similar tendency in the reduction of the cyclic sulphoxides with NaI–HClO$_4$. trans-4-t-Butylthiane S-oxide was reduced ten times faster than the cis-isomer[264] due mainly to preferential axial orientation (by 1.6 kcal/mol) of the hydroxy group of the protonated thiane S-oxide[218].

Several studies on the reduction of sulphoxides with hydrogen chloride[188,265] and hydrogen bromide[266–268] have been made. Interestingly, dialkyl and alkyl aryl sulphoxides are reduced catalytically with HBr–Br$_2$[268].

Reduction of sulphoxides with sulphydryl compounds under acidic conditions is considered to proceed through a nucleophilic substitution on sulphur in the conjugate acid.

Mikołajczyk and Para[269] and Oae et al.[270] reported independently that several sulphoxides are quantitatively reduced in an exothermic reaction at room temperature by treatment with dithioacetic acid 90 (equation 143).

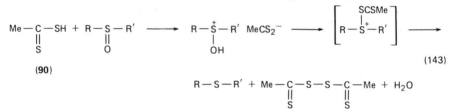

$$(143)$$

(90)

Mikołajczyk also showed that dimethyl sulphoxide can be reduced with dithiophosphoric acids 91 and other highly acidic sulphydryl compounds[271,272]. Independently, Nakanishi and Oae found that dialkyl, alkyl aryl and diaryl sulphoxides were quantitatively reduced to sulphides in exothermic reactions with other dithiophosphoric acid 91a and 91b[273,274] (equation 144). Reduction with

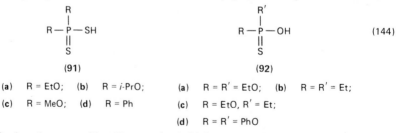

$$(144)$$

(91)   (92)

(a) R = EtO; (b) R = i-PrO;   (a) R = R' = EtO; (b) R = R' = Et;

(c) R = MeO; (d) R = Ph   (c) R = EtO, R' = Et;

                                  (d) R = R' = PhO

monothiophosphorus acids 92 requires high temperatures and addition of electrophilic catalysts such as p-toluenesulphonic acid or BF$_3$·Et$_2$O[269,271,272]. Optically active monothiophosphorus acids have been used in the asymmetric reduction of racemic sulphoxides[275].

Senning[276] and Numata et al.[277] reported a further method for the reduction of sulfoxides involving an oxygen transfer reaction (equation 145). It is not clear

$$R—S—R' + R''—S—Cl \longrightarrow R—S—R' + R''—SO_2Cl \qquad (145)$$
$$\overset{\displaystyle \|}{O} \qquad\quad \overset{\displaystyle \|}{O}$$

R'' = Me, p-NO$_2$C$_6$H$_4$

whether the reduction proceeds in one step, via formation of a covalent intermediate **93**, or through the consecutive formation of two sulfonium salts **94** and **95** (equation 146).

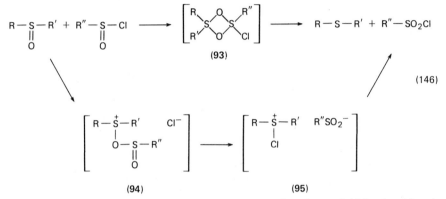

(93)

(94)   (95)

(146)

Volynskii *et al.* reported that sulphoxides were reduced to sulphides by thionyl chloride[278] and thiolane *S*-oxide gives the sulphide in 25% yield in reaction with thionyl chloride together with a small amount of the 2-chlorosulphide[279].

Sulfoxides are readily reduced to the corresponding sulphides by iodide[280], H$_2$S[281], and Me$_2$S[282] after activation with trifluoroacetic anhydride. Yields are nearly quantitative and the reaction conditions are mild (0°C with iodide ion and Me$_2$S, and −60°C with H$_2$S) (equation 147).

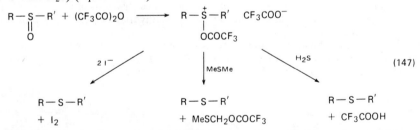

(147)

Acetyl chloride is also a good reducing agent for sulfoxides. A mechanism involving an acyloxysulphonium intermediate is sugested[283] (equation 148). Similarly, sulphimides were also reduced to sulphides by acetyl chloride[283]. Dimethyl sulphoxide affords chloromethyl methyl sulphide quantitatively as the Pummerer

$$R—S—R' + CH_3COCl \longrightarrow \left[\begin{array}{cc} R—\overset{+}{S}—R' & Cl^- \\ | \\ OCOMe \end{array} \rightleftharpoons \begin{array}{cc} R—\overset{+}{S}—R' & MeCOO^- \\ | \\ Cl \end{array} \right] \xrightarrow{MeCOCl}$$
$$\overset{\displaystyle \|}{O}$$

(148)

$$R—S—R' + Cl_2$$

product on treatment with acetyl chloride[284]. The reaction with optically active sulphoxides produced the first examples of optically active sulphuranes (equations 149[285] and 150[286]).

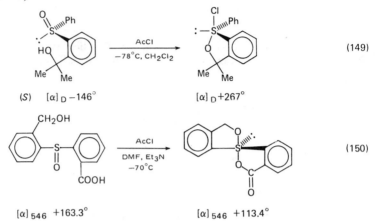

$$(S) \quad [\alpha]_D -146°$$
$$[\alpha]_D +267°$$

$$[\alpha]_{546} +163.3°$$
$$[\alpha]_{546} +113.4°$$

Diaryl sulphoxides are known to undergo concurrent oxygen exchange and racemization reactions in acetic anhydride, the rate of racemization being twice that of oxygen exchange[287,288]. Reaction clearly proceeds through an S_N2 process at sulphur (equation 151).

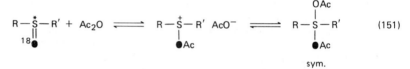

$$R-\overset{*}{\underset{18\bullet}{S}}-R' + Ac_2O \rightleftharpoons R-\overset{+}{\underset{\bullet Ac}{S}}-R' \ AcO^- \rightleftharpoons R-\overset{\overset{\displaystyle OAc}{|}}{\underset{\bullet Ac}{S}}-R' \quad (151)$$

$$\text{sym.}$$

The addition of a Brönsted or Lewis acid markedly accelerates the reaction[289–291]; racemization of methyl p--tolyl sulphoxide takes place rapidly even at room temperature on treatment with trifluoroacetic anhydride alone, again probably by oxygen exchange[292].

Both alkane- and arenesulphinates undergo concurrent oxygen exchange and racemization without cleavage of the S—OR' bond on treatment with trichloroacetic anhydride[292] (equation 152). The rate of racemization of (−)-menthyl

$$R-\overset{*}{\underset{18\bullet}{S}}-OR' + (Cl_3CCO)_2O \overset{slower}{\rightleftharpoons} R-\overset{+}{\underset{\bullet COCCl_3}{S}}-OR' \ Cl_3CCO_2^- \overset{slow}{\rightleftharpoons} R-\overset{\overset{\displaystyle OCOCCl_3}{|}}{\underset{\bullet COCCl_3}{S}}-OR' \quad (152)$$

$$\text{sym.}$$

(−)-p-toluenesulphinate was twice that of oxygen exchange, suggesting that the reaction involves Walden inversion at sulphur. When the reaction was carried out in benzene, the rate of racemization was found to be first order in both sulphinate and trichloroacetic anhydride. The value of ρ is large (−1.53), suggesting that the initial acylation is rate determining. Racemization of the sulphinate was 10^3 times slower than that of the sulphoxide under the same conditions, presumaly owing to the lower basicity of the sulphinate oxygen[292].

　　　c. *Azasulphonium salts.* Azasulphonium salts are hydrolysed by aqueous hydroxide to sulphoxides with inversion of configuration at sulphur. However, nucleophilic attack at the α-carbon of the nitrogen substituent competes, and the ratio of the competitive reactions was highly dependent on the structure of the azasulphonium salt[102] (equation 153).

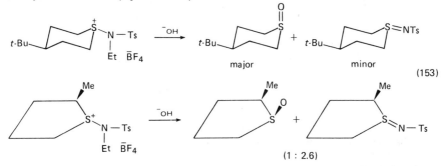

(153)

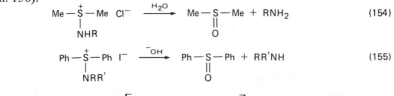

(1 : 2.6)

Several N-acyl and N-aryl dimethyl azasulphonium salts are hydrolysed to dimethyl sulphoxide and the corresponding amines[59] (equation 154). NN-Dialkyl diphenyl azasulphonium salts are hydrolysed quantitatively by aqueous 15% KOH at 95°C[82] (equation 155), while the NN-diethyl azasulphonium salt **96** underwent oxidative cleavage via the ylide intermediate **97** on similar treatment[28] (equation. 156).

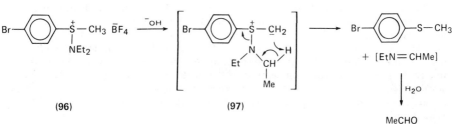

(154)

(155)

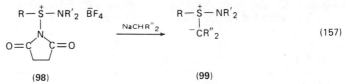

(96)   (97)

(156)

An optically active sulphinamide has been converted into its enantiomer by O-methylation with CF_3SO_3Me and subsequent alkaline hydrolysis of the resulting azasulphonium salt[293].

When the diaminosulphonium salts **98** were treated with carbanions, the sulphonium ylides **99** were obtained in 30–70% yields as products of a typical nucleophilic substitution on sulphonium sulphur[294] (equation 157). Treatment of the

$$R-\overset{+}{\underset{\underset{O=C}{\overset{|}{N}}\diagdown_{C=O}}{S}}-NR'_2 \ \bar{B}F_4 \quad \xrightarrow{NaCHR''_2} \quad R-\overset{+}{\underset{\underset{CR''_2}{\overset{|}{S}}}{S}}-NR'_2$$

(157)

(98)   (99)

azasulphonium salt **100** with equimolar cyclopentadienylthallium in CH_2Cl_2 at -20 to $-70°C$ gave bis-sulphonium compounds in 20–25% yields[295] (equation 158).

$$R-\overset{+}{\underset{\underset{NR'_2}{|}}{S}}-R\ Cl^- + \ \bigcirc\!\!\!\!-\overset{+}{Tl} \ \xrightarrow{LiClO_4}\ \overset{R}{\underset{R}{\diagdown}}\!\!\overset{+}{S}\!\!-\!\!\bigcirc\!\!\!\!-\!\!\overset{+}{\underset{\diagdown R}{S}}\!\!\overset{R}{\diagup}\ ClO_4^- \qquad (158)$$

(100)

$$R = Me,\ RR = -(CH_2)_5-$$

$$NR'_2 = \bigcirc\!\!\!\!\overset{\displaystyle N}{\underset{N}{\diagup\!\!\!\diagdown}}\!\!\overset{\|}{\underset{N}{}},\quad O=C\overset{\displaystyle N}{\underset{\diagup\!\!\!\diagdown}{}}\!\!\!C=O$$

Displacement of *N*-methyltoluene-*p*-sulphonamide from an azasulphonium salt with phenylmagnesium bromide gave the triphenylsulphonium salt in 90% yield[104] (equation 159).

$$Ph-\overset{+}{\underset{\underset{Me}{\overset{|}{\diagup}N\diagdown}{Ts}}{S}}-Ph\ ClO_4^- + PhMgBr \longrightarrow Ph-\overset{+}{\underset{\underset{Ph}{|}}{S}}-Ph\ ClO_4^- + TsNHMe \qquad (159)$$

The first work on the acid hydrolysis of sulphimides was reported by Tarbell and Weaver[296]. *N*-Tosyl diphenyl sulphimide in warm hydrochloric acid gave diphenyl sulphoxide and tosylamide, undoubtedly by way of the azasulphonium salt, the conjugate acid of the sulphimide[296]. *N*-Carbamoyl sulphimides are also hydrolysed in acid conditions.[66].

Kresze and Wustrow reported that optically active *N*-tosyl *m*-carboxyphenyl methyl sulphimide was hydrolysed stereospecifically with 12 N hydrochloric acid[297], but Cram *et al.* found that acid hydrolysis of optically active *N*-tosyl methyl *p*-tolyl sulphimide with 12 N sulphuric acid at 100°C or with 12 N hydrochloric acid at 25°C afforded the racemic sulphoxide. Presumably the sulphoxide underwent racemization under these conditions[166].

A kinetic study of the acid hydrolysis of *N*-tosyl sulphimides in strong acids was carried out by Kapovits *et al.* and the rate-determining step was found to be nucleophilic attack of water on sulphur of the protonated sulphimide. The observed ϕ values (0.94–1.5) suggest that a water molecule participates in the reaction both as a nucleophile and as a proton transfer agent at the rate-determining step[298] (equation 160).

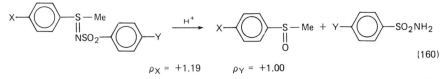

$$\rho_X = +1.19 \qquad \rho_Y = +1.00$$

A thianthrene sulphimide (equation 161) was hydrolysed to the sulphoxide together with the sulphide while treatment with aqueous hydrochloric acid in acetonitrile afforded predominantly the sulphide[299].

When *N*-alkyl diphenyl sulphimides were treated with hydrogen chloride in chloroform and the mixture then diluted with diethyl ether, predominant formation of diphenyl sulphoxide (73–83%) was observed for *N*-methyl and *N*-*t*-butyl sulphimides, but in the case of *N*-isopropyl derivative diphenyl sulphide (41%) and diphenyl sulphoxide (23%) were obtained[82].

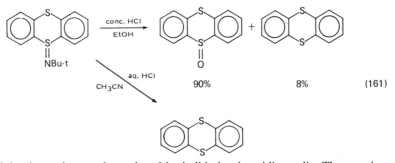

Azasulphonium salts are also reduced by iodide ion in acidic media. The reaction is third order, being first order in azasulphonium salt, in iodide, and in acid. A negative activation entropy ($\Delta S^{\ddagger} = -19$ e.u.) and a large solvent isotope effect ($k_{H_2O}/k_{D_2O} = 0.27$) were observed, suggesting that nucleophilic substitution at sulphur by iodide is rate-determining[300,301] (equation 162).

$$
\text{Me}-\overset{+}{\underset{\underset{NH_2}{|}}{S}}-\text{Me} \;\; ClO_4^{-} \;\; \underset{}{\overset{H^+}{\rightleftharpoons}} \;\; \text{Me}-\overset{+}{\underset{\underset{{}^+NH_3}{|}}{S}}-\text{Me} \;\; \xrightarrow{\text{slow, } I^-} \;\; \left[\; \overset{\delta -}{I}---\overset{\overset{\displaystyle Me}{|}\,\delta +}{\underset{\underset{Me}{|}}{S}}---\overset{\delta +}{N}H_3 \;\right] \longrightarrow
$$

$$
\text{Me}-\overset{\overset{\displaystyle I}{|}}{\underset{}{S^+}}-\text{Me} \;\; \xrightarrow{I^-} \;\; \text{Me}-\text{S}-\text{Me} + I_2 \quad (162)
$$

Other azasulphonium salts of phenothiazine derivatives are reduced quantitatively with hydrogen iodide to the sulphides[112].

N-Arenesulfonyl dialkyl sulphimides are reduced quantitatively with sodium iodide in aqueous perchloric acid. Again, third-order kinetics with first-order dependence in sulphimide, acid, and iodide, and a small substituent effect of the leaving group ($\rho_p = +0.33$) were observed, suggesting nucleophilic substitution at trivalent sulphur[302]. Reduction of sulphimides with thiophenol occurs similarly[303].

When benzyl phenyl N-tosyl sulphimide was treated with a mixture of triphenylphosphine, carboxylic acid, and alcohol (or amine) at 100°C, reduction of the sulphimide and concurrent reaction to afford benzyl phenyl sulphide and the corresponding ester (or amide) quantitatively occurred[304] (equation 163).

$$
\underset{\underset{NTs}{\overset{\|}{}}}{\text{Ph}-\text{S}-\text{CH}_2\text{Ph}} + \text{Ph}_3\text{P} + \text{RCOOH} + \underset{\underset{R'NH_2}{\text{or}}}{R'OH} \longrightarrow \text{Ph}-\text{S}-\text{CH}_2\text{Ph} + \underset{\underset{RCONHR'}{\text{or}}}{\text{RCOOR}'} + \text{Ph}_3\text{PONH}_2\text{Ts}
$$

$$\tag{163}$$

When N-tosyl dialkyl, alkyl, aryl, and diaryl sulphimides were treated with cold concentrated sulphuric acid, the azasulphonium salts **101** were obtained quantitatively and further treatment of the salts with base gave the 'free' sulphimides **102**. Under highly acidic conditions, unimolecular cleavage of the S—N bond in the sulphonamide group appears to take place, presumably on account of the extremely low activity of water in concentrated sulphuric acid[99,100] (equation 164).

The most common synthetic procedure for sulphimides is reaction of a sulphide with sodium N-p-tolenenesulphonyl chloramine (chloramine-T)[305,306] (equation 165).

N-Acyl sulphimides are similarly prepared by reaction of sulphides with chloramine derivatives of carboxylic amides[307].

$$R-\underset{\underset{NHTs}{|}}{\overset{+}{S}}-R' \quad HSO_4^- \longrightarrow R-\underset{\underset{NH_2}{|}}{\overset{+}{S}}-R' \quad TsO^- \overset{base}{\longrightarrow} R-\underset{\underset{NH}{||}}{S}-R'$$

$$(101) \qquad\qquad (102)$$

$$(164)$$

$$R-S-R' + p\text{-TolSO}_2-N\overset{Na}{\underset{Cl}{\diagdown}} \longrightarrow \underset{\underset{NSO_2Tol\text{-}p}{||}}{R-S-R'} + NaCl \qquad (165)$$

The mechanism of the reaction of aryl methyl sulphides with chloramine-T has been investigated by Japanese and Hungarian groups[308–310]. The reaction was found to be much affected by pH and to be facilitated by electron-releasing groups in the benzene ring. The suggested mechanism is shown in equation 166 and involves an intermediate chlorosulphonium salt.

$$TsNClNa \overset{H^+}{\rightleftharpoons} TsNHCl$$

$$R-S-R' + TsNHCl \overset{slow}{\longrightarrow} \left[R-\underset{\underset{Cl}{|}}{\overset{+}{S}}-R' \quad TsNH^- \right] \overset{fast}{\longrightarrow} \left[R-\underset{\underset{Cl}{|}}{\overset{+}{S}}-R' \quad Cl^- \right] \longrightarrow$$

$$(166)$$

$$\underset{\underset{NTs}{||}}{R-S-R'} + HCl$$

The inductive effect of the alkyl group plays an important role in controlling the rate of the formation of dialkyl sulphimides, while the steric effect is apparently of only minor importance. The rate-determining step is S-chlorination[310].

d. Neighbouring group participation. Several examples of neighbouring group participation in the racemization and reduction of sulphoxides in acidic media have been reported by Allenmark and co-workers[311–315]. Reaction of (+)-3-benzylsulphinylbutyric acid with hydrogen halide is believed to proceed through a cyclic acyloxysulphonium salt **103**, which can subsequently undergo either reduction or racemization (equation 167).

$$\underset{\underset{\underset{O}{||}}{R-S}}{Me-CH}-\underset{\underset{COOH}{|}}{CH}-Me \overset{H^+}{\underset{-H_2O}{\longrightarrow}} \underset{\underset{\underset{O}{\diagdown}}{R-S^+}}{Me-CH}-\underset{C=O}{CH}-Me \overset{HX}{\longrightarrow} \underset{\underset{X}{|}}{R-S^+}\underset{COOH}{Me-CH}-CH-Me \longrightarrow$$

$$(103)$$

$$(167)$$

$$\underset{\underset{COOH}{|}}{Me-CH}-\underset{R-S}{CH}-Me \quad or \quad \underset{\underset{\underset{O}{||}}{R-S}}{Me-CH}-\underset{COOH}{CH}-Me$$

(reduction)   (racemization)

It was found by Oae and coworkers[316,317], using optically active and ^{18}O-labelled *o*-carboxyphenyl phenyl sulphoxides, that oxygen exchange reactions in 65.7% sulphuric acid proceed about 10^4 times faster than racemization. In the *meta*- and the *para*-carboxyl-substituted diphenyl sulphoxides these processes occur at similar rates.

The high rate of oxygen exchange is undoubtedly caused by neighbouring group participation by the *ortho*-carboxyl group forming a cyclic acyloxysulphonium salt **104**, and subsequent attack of water yields the original sulphoxide. Thus the whole process leads to net retention of configuration by double inversion (equation 168).

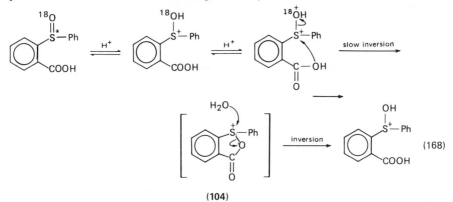

(168)

(104)

A similar anchimeric assistance (10^3–10^4 acceleration) by an *ortho*-carboxyl group was observed in the reduction or racemization of *o*-carboxyphenyl methyl sulphoxide with halide ion in acidic media[318,319]. Strangely, the reaction with chloride ion is markedly influenced by the steric effect of alkyl groups compared with that with bromide ion[320].

Anchimeric assistance of an *o*-carboxyl group was also observed in the oxidation of sulphides with iodine, and an intermediate similar to **104** was suggested[321].

o-Alkylthio groups display substantial neighbouring group participation in the heterolysis of the S—O bond. However, in this case the reaction was a net oxygen transfer[322] (equation 169).

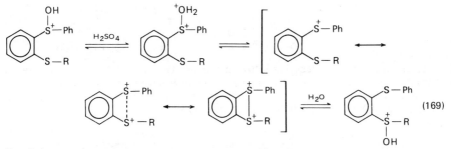

(169)

R = alkyl group

Reduction of the cyclic sulphoxide **105** by iodide ion in aqueous acid proceeds 10^6 times more rapidly than for simple sulphoxides. The rate was essentially independent of iodide ion concentration and showed second-order dependence on the concentration of perchloric acid. The accelerated rate of reduction was attributed to anchimeric assistance by the transannular thioether group which resulted in the formation of an intermediate dithioether dication **106**[323] (equation 170).

The nucleophilic nature of the sulphinyl group is seen in several examples of neighbouring group participation, resulting in formation of cyclic alkoxysulphonium intermediates.

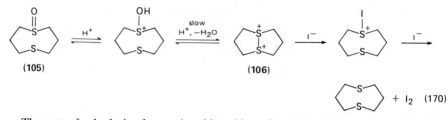

(105)   (106)

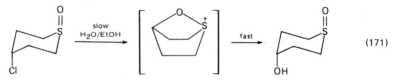

$+$ I_2 (170)

The rate of solvolysis of *trans*-4-*p*-chlorothiane S-oxide in 50% aqueous ethanol is 630 times larger than that of the *cis*-isomer, which underwent normal unassisted S_N1 solvolysis[324] (equation 171).

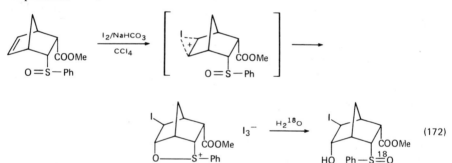

(171)

The ready iodohydrin formation in the *syn*-ester (equation 172) shows neighbouring group participation by the sulphinyl oxygen[325]. Similar sulphinyl oxygen participation was observed in iodohydrin formation with norbornene sulphoxides[326,327].

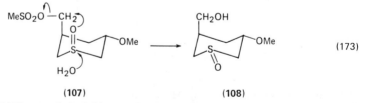

(172)

Hydrolysis of the mesylate **107** proceeds via intramolecular nucleophilic substitution and subsequent nucleophilic attack of water on the sulphur atom of the resulting alkoxysulphonium salt, giving the hydroxymethyl thiane S-oxide **108** with inverted configuration at sulphur[328] (equation 173).

MeSO$_2$O$-$CH$_2$ ... OMe $\longrightarrow$ CH$_2$OH ... OMe

H_2O

(173)

(107)   (108)

3β-Hydroxy-5(*R*)-α-methylsulphinylcholestane **109** reacts with methanesulphonyl chloride in pyridine to give the alkoxysulphonium salt **110** via an intramolecular nucleophilic substitution by sulphinyl oxygen[329] (equation 174).

Rates of solvolyses of ω-chloroalkyl sulphoxides, PhS(O)(CH$_2$)$_n$CMe$_2$Cl (n = 1–4) and PhS(O)CH$_2$CH$_2$CRR′Cl (R, R′ = H, H; Me, H; Me, Me; *t*-Bu, H), in

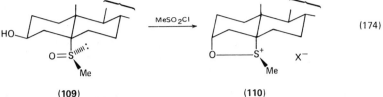

(174)

(109)   (110)

dimethylformamide or sulpholane fall in the order $\delta > \gamma > \beta > \epsilon \lesssim t\text{-Bu}$[330-332], similar to reactivity in the $PhCO(CH_2)_n Br$ series[333], indicating neighbouring group participation by δ- and γ-sulphinyl groups.

Solvolyses of benzyl chloride derivatives with a sulphinyl group at an appropriate position within the molecule have been reported to proceed rapidly owing to anchimeric assistance by sulphinyl oxygen[334].

Oxidation of 4-hydroxy-2,6-diphenylthiane 111 with t-BuOCl afforded the equatorial sulphoxide 112. In this case, the cyclic alkoxysulphonium salt 113 was the probable intermediate, while oxidation of cis-2,6-diphenylthiane also afforded the epimer[335] (equation 175).

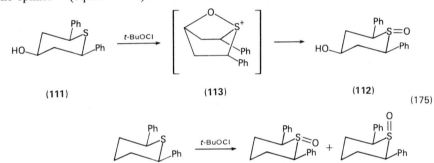

(111)   (113)   (112)
                                                        (175)

1.1 : 1

The optically active sulphimide shown in equation 176 was readily hydrolysed in sulphuric acid to the sulphoxide, in which the configuration at sulphur is the result of net retention of configuration by double inversion, involving neighbouring carboxyl group participation during the reaction[336] (equation 176).

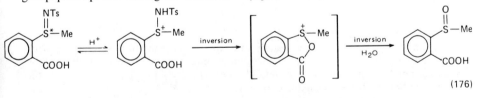

(176)

B. Nucleophilic Substitution on the Heteroatom

An n.m.r. study by Kice and Favstritsky revealed that the thiasulfonium salt 114 undergoes exchange of the sulphide extremely rapidly ($k \geq 10^5\,\mathrm{M}^{-1}\,\mathrm{s}^{-1}$), and that nucleophilic substitution at sulphur by Me_2S was at least 10^9–10^{10} times faster than displacement by Me_2S on any sp^3 carbon. Displacement of Me_2S is faster by a factor of *ca.* 10^5 than that of MeS⁻ Me—S—S—Me, presumably because the leaving dimethyl sulphide group in **114** is a much better one[337].

$$Me-\overset{+}{\underset{\underset{Me}{|}}{S}}-S-Me \ X^- + \underset{\underset{Me}{|}}{S}-Me \ \rightleftharpoons \ Me-\underset{\underset{Me}{|}}{S} \ + \ Me-S-\overset{+}{\underset{\underset{Me}{|}}{S}}-Me \ X^- \quad (177)$$

(**114**)

Smallcombe and Caserio also studied the equilibrium reaction shown in equation 178 and showed that reactivity of the S—S bond toward nucleophilic attack is enhanced by a factor in excess of 10^6 by methylation and methylsulphenylation of the disulphide[338].

$$Me-\overset{+}{\underset{\underset{Me}{|}}{S}}-S-Me \ X^- + \underset{\underset{Me}{|}}{S}-S-Me \ \underset{k_{-1}}{\overset{k_1}{\rightleftharpoons}} \ Me-\underset{\underset{Me}{|}}{S} \ + \ Me-S-\overset{+}{\underset{\underset{Me}{|}}{S}}-S-Me \ X^- \quad (178)$$

$K = k_1/k_{-1} = 7.4 \times 10^{-4}; k_1 = 1.06 \times 10^4 \ \mathrm{M^{-1}s^{-1}}; k_{-1} = 1400 \times 10^4 \ \mathrm{M^{-1}s^{-1}}$

Minato *et al.* measured the equilibrium constant of the reaction shown in equation 179 by n.m.r., and observed large electronic effects of substituents[339].

$$Me-\overset{+}{\underset{\underset{Me}{|}}{S}}-S-Me \ + \ \underset{\underset{R}{|}}{S}-R' \ \overset{K}{\rightleftharpoons} \ Me-\underset{\underset{Me}{|}}{S} \ + \ Me-S-\overset{+}{\underset{\underset{R}{|}}{S}}-R' \quad (179)$$

$R-S-R'$ (K) = MeSEt (1.0), MeSPr-*i* (2.6), EtSEt (4.5), PhSPh (0.0047)

A later study using D-labelled and optically active sulphides indicated that there is an irreversible process (involving a 1,2-shift of alkyl group within the intermediate) in the exchange reaction[340] (equation 180).

$$Me-\overset{+}{\underset{\underset{Me}{|}}{S}}-S-Me \ + \ \underset{\underset{CD_3}{|}}{S}-R^* \ \rightleftharpoons \ Me-\underset{\underset{Me}{|}}{S} \ + \ Me-S-\overset{+}{\underset{\underset{CD_3}{|}}{S}}-R^*$$

$$(180)$$

$$Me-\overset{+}{\underset{\underset{R}{|}}{S}}-Me \ + \ Me-S-S-CD_3 \ \underset{Me_2S}{\overset{slow}{\longleftarrow}} \ Me-\overset{R}{\underset{}{\overset{|}{\underset{}{S^+}}}}-S-CD_3$$

R* = optically active 1-phenylethyl group

The thiasulfonium salt **114** is readily attacked at the sulphenyl sulphur atom by other nucleophiles, such as potassium cyanide, phenylmagnesium bromide, sodium iodide, anisole, phenols, benzenethiol, and phenyl phosphite, to give the corresponding methylsulphenylated products[23,119,341]. Biologically active thiols are converted into the asymmetric disulphides by treatment with **114**[342]. When the thiasulfonium salt **114** was treated with alkyl- or dialkyl-substituted olefins, the adduct **115** was obtained generally in high yield via *trans* ring opening of the intermediate **116**[343,344]. Acetylenic compounds were similarly found to give the adduct (equation 181).

An intermediate such as **116** can be isolated under mild conditions. For example, treatment of the salt **114** with tetramethylethylene in liquid SO_2 at $-60°C$ afforded the thiiranium salt **116** in 78% yield[123] (equation 182).

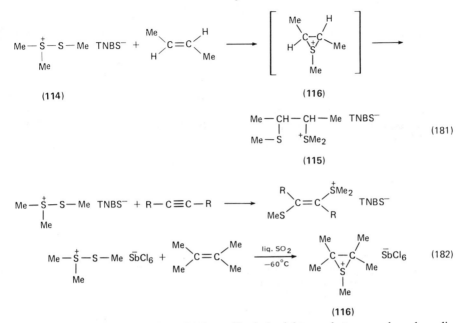

(181)

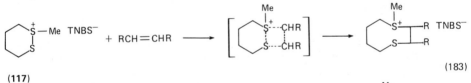

(182)

(116)

The thiasulphonium salt **117** also afforded eight- and ten-membered cyclic sulphonium salts in 51–90% yelds[120,345] (equation 183). Concerted [2 + 2] and

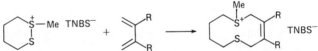

(183)

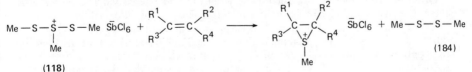

[2 + 4] addition mechanisms were suggested and n.m.r. study of the products revealed the total absence of a transannular interaction between the two sulphur atoms[120].

The dithiasulphonium salt **118** can also act as a sulphenylating agent. When the salt was treated with alkenes in methylene chloride at 0°C or in liquid SO_2 at −60°C, the thiiranium salts were produced stereospecifically in 85–95% yields. The S-methyl group is on the less crowded side[346] (equation 184).

$$Me-S-\overset{+}{\underset{\underset{Me}{|}}{S}}-S-Me \ \ \bar{S}bCl_6 \ + \ \underset{R^3}{\overset{R^1}{}}C=C\underset{R^4}{\overset{R^2}{}} \longrightarrow \underset{R^3}{\overset{R^1}{}}\overset{}{\underset{\underset{Me}{|}}{\underset{S}{C-C}}}\underset{R^4}{\overset{R^2}{}} \ \ \bar{S}bCl_6 \ + \ Me-S-S-Me$$

(184)

(118)

With di-*t*-butylacetylene, the dithiasulfonium salt in liquid SO_2 at low temperatures gave the stable thiirenium salt **119**[347,348] (equation 185).

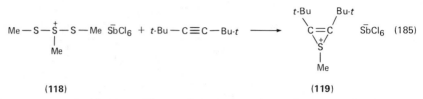

(118)   (119)

On treatment with tetrachlorocyclopropene, the salt **118** afforded the cyclopropenium cation **119** in 80% yield, which then gave the cyclopropene thione **120**[349] (equation 186).

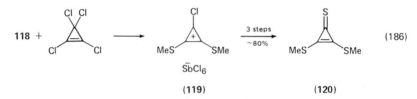

(119)   (120)

Reaction of 'free' sulphimide with either Z- or E-dibenzoylethylene in benzene at room temperature yielded the trans-aziridine **121** and the enamine **122** each in 50% yield, together with 100% of diphenyl sulphide[98,350] (equation 187).

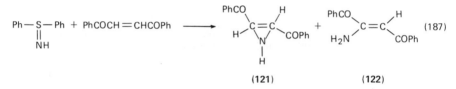

(121)   (122)

Aziridines with optical purities of nearly 30% were synthesized by treatment with optically active sulphimides[351] (equation 188). The reaction is believed to proceed by

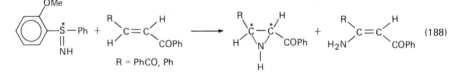

initial addition of the sulphimide to the alkene to give the azasulphonium salt **123**, followed by intramolecular attack of the carbanion centre on nitrogen. The enamine is believed to be formed via the intermediate **124**. Asymmetric induction is considered to arise during formation of the intermediate **123**, since the cyclization step appears to be stereospecific in view of the predominant formation of the *trans*-aziridine (equation 189).

Other electrophilic alkenes such as acrylonitrile and phenyl vinyl sulphone gave the N-(β-substituted-alkyl)-sulphimides in good yields[350] (equation 190).

N-Alkyl sulphimides also gave N-alkylaziridines and enamines on treatment with dibenzoylethylene[352]. N-Phthalimido dimethyl sulphimide afforded an aziridine on treatment with methyl acrylate at 80°C[353] but on reaction without alkene present the *cis*-phthaloyltetrazene **125** was a major product, suggesting that the aziridine may be formed by reaction of the alkene with a nitrene formed during the reaction[353] (equation 191).

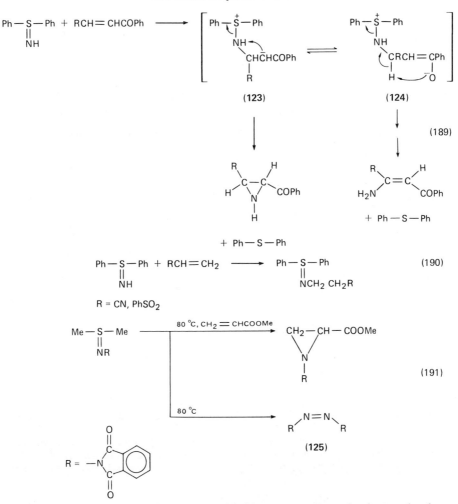

(123)   (124)

(189)

+ Ph—S—Ph

$$Ph—S—Ph + RCH=CH_2 \longrightarrow Ph—S—Ph \quad (190)$$
with NH and NCH₂CH₂R

R = CN, PhSO₂

(191)

(125)

R = phthalimido

When conjugatively stabilized sulphimides were heated, intramolecular displacement reactions occurred giving heterocyclic products (equations 192[354], 193[68], and 194[70]).

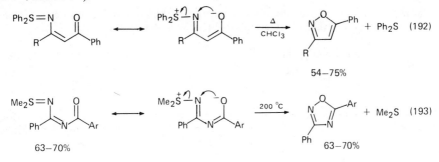

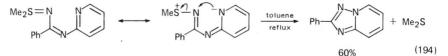

60%   (194)

C. Nucleophilic Substitution on the Carbon Atom Attached to the Heteroatom (Transalkylation)

In reactions of alkoxysulphonium salts with nucleophiles, attack occurs not only at sulphonium sulphur as mentioned in Section IIIA2b, but also at the carbon atom attached to the heteroatom, resulting in transalkylation. Generally, strong nucleophiles such as carbanions, hydroxide ion, and amines attack at sulphur, while weak nucleophiles such as halide anions, pyridine and sulphides attack the α-carbon.

One of the early examples is the reaction of dimethyl methoxysulphonium salt with sodium iodide to give dimethyl sulphoxide[17]. A study on the reaction of the alkoxysulphonium salt with halide ion was carried out by Annunziata et al.[355]. In the reaction of optically active and ^{18}O-labelled diaryl alkoxysulphonium salt with halide ion, they found that with a two-fold excess of chloride, bromide, or iodide, the sulphoxide produced completely retained both the ^{18}O label and the configuration. However, in the reaction with fluoride, the sulphoxide produced was found to have retained about 30% optical purity and only 10% of the original ^{18}O content (equation 195).

$$p\text{-Tol} - \overset{+}{\underset{\underset{\overset{|}{\bar{B}F_4}}{^{18}O - Et}}{S}} - C_{10}H_7\text{-}\alpha + X^- \longrightarrow p\text{-Tol} - \overset{\overset{..}{\underset{^{18}O}{\parallel}}}{S} - C_{10}H_7\text{-}\alpha + [EtX] \qquad (195)$$

(S) (S)

$$X = Cl^-, Br^-, I^-$$

Annunziata et al. also found transalkylation of alkoxysulphonium salts to occur on treatment with pyridine. When the optically active diaryl alkoxysulphonium salt was heated with pyridine in refluxing dichloromethane, the sulphoxide obtained was found to have retained over 93% of the original configuration, in yielding the N-ethylpyridinium salt[356] (equation 196). In reactions of alkyl aryl ethoxysulphonium

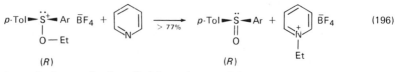

salts having α-hydrogen in the alkyl branch, α-pyridinosulphides, products of the Pummerer reaction, were obtained in 60–70% yields[356] (equation 197).

Dimethyl methoxysulphonium salts can act as methylating agents for β-ketoenolates. Potassium dibenzoylmethides gave the O-methylated products, while methyl iodide gave the C-methylated products[357].

The optically active methyl *p*-tolyl methoxysulphonium salt shown in equation 198 acts as a methylating agent. Benzylethylphenylamine was converted into the partially optically active ammonium salt and benzyl ethyl sulphide afforded the sulphonium salt with 60% enantiomeric excess on treatment with the salt[211] (equation 198).

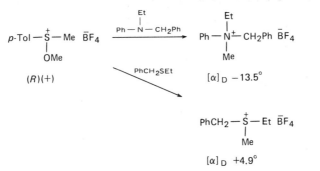

(198)

Transalkylation reactions were observed in the reaction of aminoalkoxysulphonium salts[15,38] (equation 199).

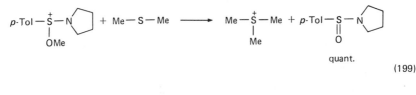

(199)

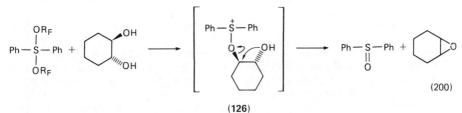

Reaction of *trans*-cyclohexane-1,2-diol with the sulphurane (equation 200) gave the epoxide, presumably via the initial formation of the alkoxysulphonium salt

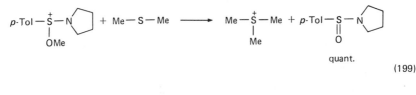

(200)

(126)

126[358]. Other 1,2-diols gave epoxides in excellent yields, while reaction with alkylidene-1,3-, 1,4-, 1,5-, and 1,6-diols afforded the cyclic ethers in only moderate yields since formation of the open-chain ether was competitive with ring closure[358]. 1,4-, 1,5-, and 1,6-diols are also dehydrated with dimethyl sulphoxide at high temperatures to give the corresponding cyclic ethers[359].

D. Ylide Formation

Heterosulphonium salts possess acidic hydrogens α- to the sulphonium sulphur atom, and hence treatment of the salts with bases readily gives an ylide **127** which undergoes further reactions (equation 201).

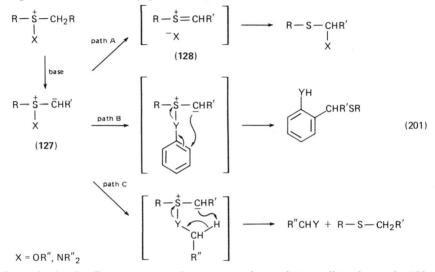

In path A, the Pummerer reaction occurs via an intermediate ion pair **128** produced by S—X bond cleavage in the ylide **127**. Path B is the Sommelet–Hauser rearrangement, which involves a nucleophilic attack of the carbanion at an *ortho*-position of the aromatic ring. Path C represents intramolecular oxidation–reduction of the heterosulphonium salt with formation of a carbonyl compound and sulphide. It is described further in Section IIIE.

1. Pummerer reaction

When sulphoxides with α-protons are treated with acetic anhydride, the Pummerer reaction takes place and the corresponding α-acetoxysulphides are obtained in good yields[360] (equation 202). The reaction[361] was named after the discoverer by

$$RSOCH_3 + (CH_3CO)_2O \longrightarrow R—S—CH_2OCOCH_3 + CH_3COOH \tag{202}$$

Horner[362] and the mechanism was studied using dimethyl sulphoxide and uniformly ^{18}O-labelled acetic anhydride in diethyl ether[363]. Based on the observation that all of the oxygen atoms were completely scrambled during the reaction, involvement of intermolecular nucleophilic attack of acetate at the α-carbon was suggested. Oae *et al.* pointed out possible involvement of the acyloxysulphonium ylide, but did not specify involvement of the sulphur-stabilized carbonium ion as an intermediate[363] (equation 203).

Although the initial formation of the sulphonium ylide **129** was also presumed, involvement of the carbonium ion **130** was first suggested by Johnson[364] (equation 204).

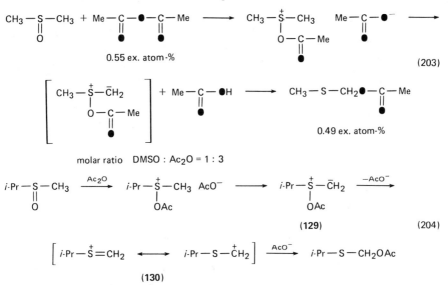

(203)

molar ratio   DMSO : Ac$_2$O = 1 : 3

(129)   (204)

(130)

When alkoxysulphonium salts are treated with base, two competing reactions are known to occur; one is the Pummerer reaction and the other is oxidative cleavage to give carbonyl compounds and sulphides, and in addition a rapid alkoxy exchange also takes place during the reaction. Benzylic alkoxysulphonium salt undergoes the former reaction preferentially, whereas for methyl phenyl methoxysulphonium salt, the latter reaction is preferred[2,3] (equation 205).

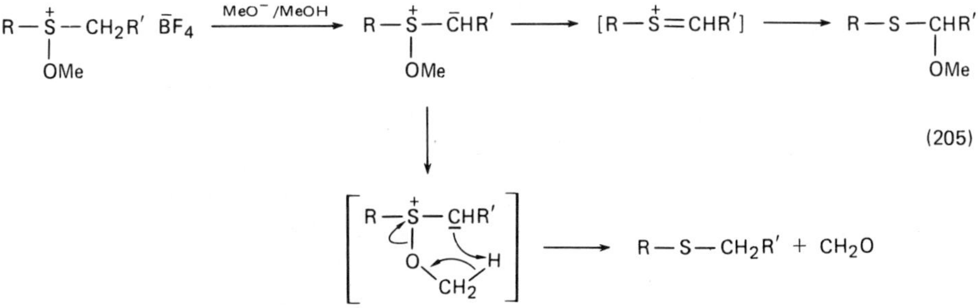

(205)

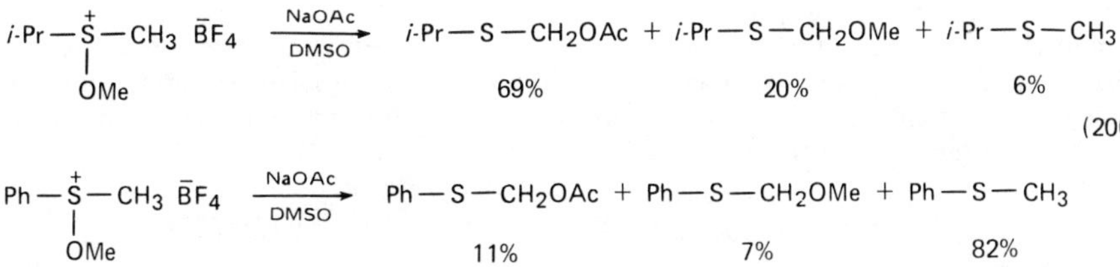

(206)

Two examples are in equation 206. Treatment with sodium salts of propionic, benzoic, and *p*-nitrobenzoic acids give nearly the same results, but, sodium salts of the less nucleophilic bromoacetic and trifluoroacetic acid do not yield the corresponding α-acyloxysulphides[364,365].

Because of the facile hydrolysis of α-acetoxymethyl sulphide, the Pummerer reaction was suggested as a possible pathway for enzymatic demethylation of methionine on the basis of the experiments shown in equation 207[366].

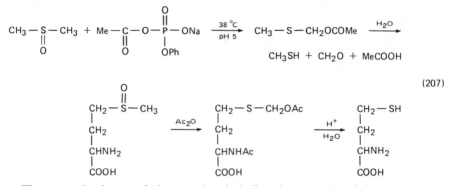

$$(207)$$

The general scheme of the reaction is believed to consist of four sequential reactions, the rate-determining step varying with change of both acylating agent and sulphoxide. Examples of the Pummerer reaction in which each of the four steps is rate determining are known, and the whole mechanistic spectrum of the reaction is now complete (equation 208).

$$\text{R—S—CH}_3 + \text{Ac}_2\text{O} \xrightarrow{\text{step 1}} \text{R—}\overset{+}{\text{S}}\text{—CH}_3 + \text{AcO}^-$$

(with O and OAc on the sulphur)

$$\xrightarrow[{-\text{H}^+}]{\text{step 2}} \left[\text{R—}\overset{+}{\text{S}}\text{—CH}_2 \longleftrightarrow \text{R—S=CH}_2 \right] + \text{AcOH}$$

$$(208)$$

$$\xrightarrow{\text{step 3}} \left[\text{R—}\overset{+}{\text{S}}\text{=CH}_2 \longleftrightarrow \text{R—S—}\overset{+}{\text{C}}\text{H}_2 \right]$$

$$\xrightarrow{\text{step 4}} \text{R—S—CH}_2\text{OAc}$$

The initial step of the Pummerer reaction of the sulphoxide is acylation of sulphinyl oxygen to form the acyloxysulphonium salt. In fact, on treatment of dimethyl sulphoxide with trifluoroacetic anhydride in methylene chloride at −60°C, Sharma and Swern isolated what appeared to be trifluoroacetoxysulphonium salt which, on warming to room temperature, underwent the Pummerer reaction to give the corresponding α-trifluoroacetoxysulphide[34] (equation 209).

$$\text{CH}_3\text{—S—CH}_3 + (\text{CF}_3\text{CO})_2\text{O} \xrightarrow[\text{CH}_2\text{Cl}_2]{-60\,°\text{C}} \text{CH}_3\text{—}\overset{+}{\text{S}}\text{—CH}_3\ \text{CF}_3\text{COO}^- \xrightarrow[\text{CH}_2\text{Cl}_2]{\text{r.t.}}$$

$$(209)$$

$$\text{CH}_3\text{—S—CH}_2\text{OCOCF}_3$$

A kinetic study of the Pummerer reaction of dimethyl sulphoxide with p-substituted benzoic anhydride gave a good Hammett correlation betwen the log k and σ values with a relatively large positive ρ value (1.40), and a small kinetic isotope effect (k_H/k_D = 1.21) with DMSO-d_6. This suggested the rate-determining step to be the initial acylation[367].

Another example is in equation 210[368,369].

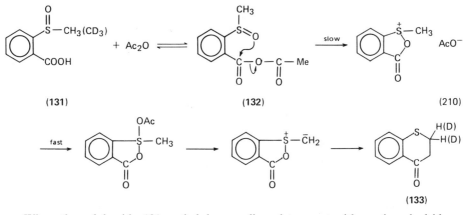

(131) (132) (210)

(133)

When the sulphoxide 131-methyl-d_3 was allowed to react with acetic anhydride to 50% completion, the recovered sulphoxide was found to retain all three deuterium atoms completely while the cyclized product 133 also was found to contain two deuterium atoms per molecule. The kinetic isotope effect was small, i.e. k_H/k_D = 1.07. The reactivity falls in the order of alkyl substituents i-Pr > n-Pr > Et > Me > CH$_2$Ph, which is the order of the basicities of these alkyl phenyl sulphoxides, clearly supporting the conclusion that the rate-determining step of this intramolecular Pummerer reaction is acylation of the sulphinyl oxygen[370].

The initial step of this reaction is believed to be the formation of the mixed acid anhydride 132, which acylates the sulphoxide intramolecularly in the rate-determining step. The rate of this reaction is 140 times larger than that for methyl phenyl sulphoxide. The formation of the mixed acid anhydride 132 was supported by the spectroscopic observation of a similar mixed acid anhydride in the treatment of o-isopropylsulphinylbenzoic acid with diphenyl ketene[370].

The second step of the Pummerer reaction is proton removal from the acyloxysulphonium salt to form an acyloxysulphonium ylide 136 which has neither been isolated nor confirmed spectroscopically. However, in view of the sizeable kinetic isotope effects in a number of cases, rate-determining proton removal by acylate ion is evident in many Pummerer reactions. A sizeable kinetic isotope effect was first observed by Oae and Kise in the reaction between aryl methyl sulphoxides and acetic anhydride, k_H/k_D = 2.9[371,372] (equation 211). The large Hammett ρ value of −1.6 obtained for p-substituents implies that the acylation equilibrium is also important in the energy profile of the reaction. Sulphoxides are known to undergo oxygen exchange with acetic anhydride[287–289], presumably via a sulphurane intermediate, and this oxygen exchange is known from the usual ^{18}O tracer experiments to be responsible for the racemization of optically active sulphoxides[287,288]. For aryl methyl sulphoxides, however, the Pummerer reaction was found to proceed about six times faster than oxygen exchange at 120°C (E_a for the Pummerer reaction is 21.2 kcal/mol, $\Delta S^{\ddagger}$ = −20.7 e.u.).

The kinetic isotope effect, k_H/k_D, remained about 3 even with p-nitrophenyl methyl sulphoxide, while a stable selenurane, tetracoordinated seleno-analogue of

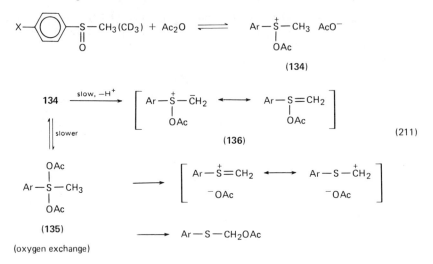

sulphurane, was isolated in the similar reaction of a selenide with benzoyl peroxide[373,374]. Thus, it was once suggested that a sulphurane-type intermediate **135** may be also formed in the Pummerer reaction. Proton removal takes place through a six-membered cyclic transition state with an angle close to 110–120°, as in the Ei reaction, via either a five- or six-membered cyclic transition state with isotope effects generally in the range 3–4[375] (equation 212). The much slower rate of oxygen exchange than in the Pummerer reaction, however, alone can rule out the scheme involving the sulphurane **135**.

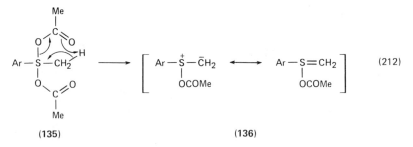

In some cases, e.g. the reaction of *o*-carboxyphenyl alkyl sulphoxides, reaction proceeds via a cyclic acyloxysulphonium ylide to form the cyclized product, and intervention of a sulphurane intermediate may be conceivable[370].

When proton removal takes place intermolecularly, a large kinetic isotope effect is observed. Examples are shown in equations 213[376] and 214[377].

$$R-\overset{+}{S}-CH_3(CD_3)\ Cl^- + R_3N \longrightarrow R-S-CH_2(D_2)-\overset{+}{N}R_3$$

or

$$R-S-CH_2\ (D_2)-N\overset{CO}{\underset{CO}{\diagdown}}$$

(213)

$$k_H/k_D = 9\text{--}10$$

$$Ph-CH_2-\underset{\underset{O}{\|}}{S}-CD_2-Ph + Ac_2O \longrightarrow Ph-\underset{\underset{OAc}{|}}{CH}-S-CD_2-Ph + Ph-CH_2-S-\underset{\underset{OAc}{|}}{CD}-Ph$$

$$(214)$$

$$k_H/k_D = 9$$

A large primary deuterium isotope effect (at least 4) was observed in the reaction of dimethyl sulphoxide (or DMSO-d_6) with acetyl chloride to form chloromethyl methyl sulphide[378].

The importance of proton removal is shown by the regioselectivity of the Pummerer reaction. The migration aptitude of the acyloxy group is usually determined by the relative acidities of the α-hydrogen atoms in the acyloxysulphonium salt. In other words, a more stable acyloxysulphonium ylide is formed preferentially to a less stable one upon removal of the α-proton prior to the acyloxy migration. This is in keeping with the mechanism involving the acyloxysulphonium ylide intermediate.

In the reactions of alkyl methyl sulphoxides with acetic anhydride, the acetoxy group migrates only to the methyl group[364] (equation 215), while with p-nitrobenzyl

$$RCH_2-\underset{\underset{O}{\|}}{S}-CH_3 + Ac_2O \longrightarrow RCH_2-S-CH_2OAc \qquad (215)$$

$$RCH_2 = n\text{-Pr}, i\text{-Pr}, n\text{-Bu}$$

benzyl sulphoxide, acetoxy migration occurs at the p-nitrobenzylic site selectively[379]. A clearer case is the Pummerer reaction of cyanomethyl benzyl sulphoxide, in which the acetoxy group migrates only to the cyanomethyl side[380].

In the earlier study on the Pummerer reaction of methionine sulphoxide, migration of the acetoxy group was found to take place nearly exclusively to the methyl side, giving homocysteine after hydrolysis[366]. Thus, the migration of the acetoxy group generally takes place at the α-carbon atom which bears the most acidic proton. Qualitatively, the acidity of the α-methylene protons of various substituted sulphoxides falls in the order CH_2COR, $CH_2CN > PhCH_2 > CH_3 > n$-alkyl $> sec$-alkyl $> -CH=CHR$.

There is a case in which regioselectivity is caused by steric strain, as shown by Jones et al.[381] in the reactions shown in equation 216.

The α-isomeric steroid with the methylsulphinyl group at position 6 **137** affords the 6-acetoxymethylthio derivative, the normal Pummerer product. In the β-isomer **138**, the acetoxysulphonium salt **139** is sterically crowded by axial hydrogens at positions 4 and 8 and the 19-methyl group. Deprotonation would result in the release of steric strain to form a nearly coplanar ylide **140**. Acetoxy migration is presumed to take place at the ring carbon, followed by subsequent elimination of acetic acid resulting in formation of an alkene **141**.

The third step of the Pummerer reaction is S—O bond cleavage which results in the formation of an ion pair between the sulphur-stabilized carbonium ion and acetate (equation 217). The sulphur-stabilized carbonium ion pair may or may not be very intimate depending upon the nature of the substituent, R'. The lifetime of the ion pair depends very much on the substituent R', and on other factors such as solvent and acylating agent. None of these intermediates has yet been detected. There are, however, reactions in which the third step is obviously rate determining. These are called 'ElcB' Pummerer reactions. In these reactions, proton removal is fast and reversible, and S—O bond cleavage is the slowest step. Usually, the R' group is an electron-withdrawing one such as cyano, carbonyl, alkynyl, cyclopropyl,

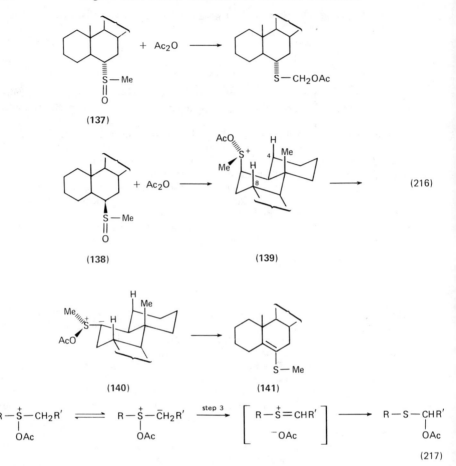

$$R-\overset{+}{\underset{\underset{OAc}{|}}{S}}-CH_2R' \;\rightleftharpoons\; R-\overset{+}{\underset{\underset{OAc}{|}}{S}}-\bar{C}H_2R' \;\xrightarrow{\text{step 3}}\; \left[R-\overset{+}{S}=CHR' \atop \;\;\;\;^-OAc \right] \;\longrightarrow\; R-S-\underset{\underset{OAc}{|}}{CHR'}$$

(217)

or phosphoryl, which facilitates proton removal, and the S—O bond cleavage is retarded by the a + I effect of R', as for alkene formation by the ElcB mechanism.

In the typical case of cyanomethyl phenyl sulphoxide, almost no kinetic isotope effect ($k_H/k_D = 1.02$) was observed in reaction with acetic anhydride[382], and the cyanomethyl-d_2 derivative was found to lose deuterium completely before half-completion of the reaction, demonstrating fast reversible proton removal. The rate-determining step is believed to be S—O bond cleavage.

The ElcB mechanism is more evident in the Pummerer reactions of sulphimides and sulphonium ylides. A well studied case is the reaction between the five-membered N-tosyl sulphimide **142** and alcoholic potassium hydroxide, which gives the α-alkoxysulphide[383] (equation 218). The kinetic isotope effect ($k_H/k_D = 1.09$) was of the magnitude of the secondary one due to hyperconjugation, which is found in solvolysis. The recovered sulphimide was found to have lost its original deuterium during the reaction, and the Hammett ρ value (+2.0) obtained with σ values was substantial. All of these observations fit the ElcB mechanism. Again, the original sulphimide has a poor leaving group, namely the arenesulphonamide anion. In the case of alkyl aryl N-tosyl sulphimides, treatment

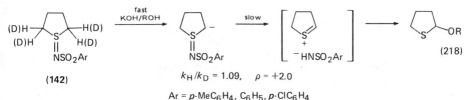

$k_H/k_D = 1.09, \quad \rho = +2.0$

Ar = p-MeC$_6$H$_4$, C$_6$H$_5$, p-ClC$_6$H$_4$

with alcoholic alkai metal hydroxide solution gives only partly the Pummerer product and mainly the sulphides, presumably formed by oxidative cleavage of alkoxysulphonium salt formed during the reaction[384]. When N-tosyl alkyl sulphimides having β-hydrogen were treated with t-BuOK in benzene, the corresponding vinyl sulphides were obtained in more than 50% yields[385].

Aryl methyl N-tosyl sulphimides undergo the Pummerer reaction with acetic anhydride[386]. The values of the activation parameters ($\Delta H^{\ddagger} = 15.2$ kcal/mol, $\Delta S^{\ddagger} = -41.2$ e.u.) are similar to those of the S$_N$2 oxygen exchange reactions of diaryl sulphoxides with acetic anhydride ($\Delta H^{\ddagger} = 13.1$ kcal/mol, $\Delta S^{\ddagger} = -45.1$ e.u.). Moreover, the corresponding sulphoxide was isolated when the reaction was stopped halfway. The observations fit a mechanism which involves rate-determining S$_N$2 acetoxy exchange (equation 219). The Hammett ρ value ($\rho_X = -0.71$, $\rho_Y = -0.57$) and the small kinetic isotope effect ($k_H/k_D = 1.57$) support the mechanism.

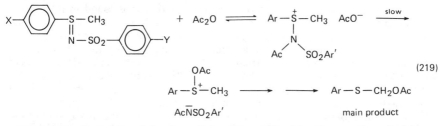

Treatment of a stable sulphonium ylide such as **143** with either acetic anhydride or benzoyl peroxide similarly gives the Pummerer product, α-acetoxymethyl aryl sulphide or α-benzoyloxymethyl aryl sulphide, quantitatively[387] (equation 220). The

X—⟨○⟩—S$^+$—CH$_3$ + Ac$_2$O ⟶ Ar—S—CH$_2$OAc + AcCH(COOMe)$_2$

   |
   $^-$C(COOMe)$_2$

(143) $\Delta H^{\ddagger} = 21.4$ kcal/mol

            $\Delta S^{\ddagger} = -22.2$ e.u.

(220)

   + BPO ⟶ Ar—S—CH$_2$OCOPh + PhCO$_2$CH(COOMe)$_2$

            $\Delta H^{\ddagger} = 17.3$ kcal/mol

            $\Delta S^{\ddagger} = -18.7$ e.u.

kinetic isotope effect (k_H/k_D) observed with the trideuterated methyl aryl sulphonium ylide **143** and acetic anhydride was 1.57 while that with benzoyl peroxide was 1.13. The ylide **143**, recovered after half-completion of the reaction, had lost over 80% of the original deuterium, revealing the reaction to proceed via ElcB mechanism.

Step 4 of the Pummerer reaction involves recombination of the acyloxy group with the sulphur-stabilized carbonium ion within the ion pair which is formed by S—O bond cleavage in the third step. The shift is believed to be fast, and the manner of the acyloxy shift differs from one sulphoxide to another. It also depends on the nature of the migrating group. There has therefore been some controversy about the nature of the migration. The most controversial point has been whether the migration is intra- or intermolecular. The intermolecular mechanism was first suggested for the acetoxy migration on the basis of [18]O tracer experiments by Oae *et al.*[363], and has prevailed for more than a decade. Recent stereochemical investigations, however, have revealed the occurrence of intramolecular acetoxy migration in various cases[382,388,389].

Treatment of the optically active cyanomethyl *p*-tolyl sulphoxide **144** with acetic anhydride gave α-acetoxy-α-cyanomethyl sulphide **145** with induced optical rotation at the α-carbon to the extent of 29% enantiomeric excess on measurement with europium shift reagent[382] (equation 221).

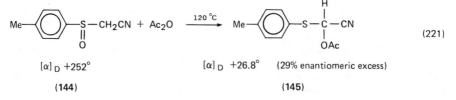

$$[\alpha]_D +252° \qquad\qquad [\alpha]_D +26.8° \quad (29\% \text{ enantiomeric excess})$$

(144)   (145)

[18]O-Labelled sulphoxide, recovered after 50% reaction, was found to retain 96% of the label while the product **145** was found to contain 85% of the label. When the loss of 4% of [18]O due to possible oxygen exchange is taken into the account, the reaction is considered to proceed via intramolecular acetoxy migration to the extent of at least 90%. Analysis of the distribution of [18]O in the ester showed that 63% of [18]O is at the carbonyl group and the remaining 37% is at the ethereal oxygen. This uneven distribution of [18]O suggests that the precursor of the ester is not a dissociated ion pair but is an intimate ion pair or even an undissociated ylide-like intermediate.

The rate-determining step of this reaction is the cleavage of the S—O bond of the acetoxysulphonium ylide **146** in the ElcB process, while the effect of *p*-substituents on the rate of the reaction gives a Hammett ρ value of −0.65. The [18]O tracer experimental results, the substantial asymmetric induction at α-carbon and other pertinent data suggest that the acyloxy migration proceeds via both five-membered cyclic and three-membered sliding transient intimate ion pairs. The former is the

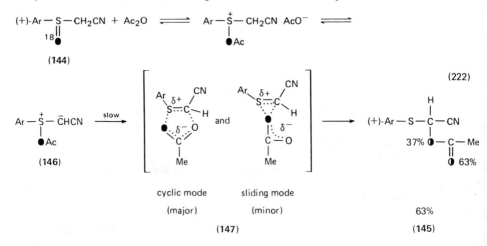

major route, probably due to the necessary anchimeric assistance of the carbonyl group to ease the S—O bond cleavage (equation 222).

The asymmetric induction cannot be caused by the stereoselective proton removal from the acyloxysulphonium salt, since the resulting α-cyanocarbanion, the ylide **146**, is known to be a planar resonance-stabilized sp^2 carbanion and proton removal is reversible. Asymmetric induction takes place when the acetoxy group shifts from the chiral sulphur atom to the α-carbon. Proposal of dual pathways for the acetoxy migration in which the five-membered cyclic path predominates over the three-membered sliding path is similar to those made for acyloxy migration in several rearrangements of t-amine N-oxides with acylating agents[390].

There are a few other examples of asymmetric induction in the Pummerer reaction. Optically active p-tolyl propargyl sulphoxide **148** gives the optically active α-acetoxysulphide[391], as does p-tolyl phenacyl sulphoxide **149**[392] (equation 223). In

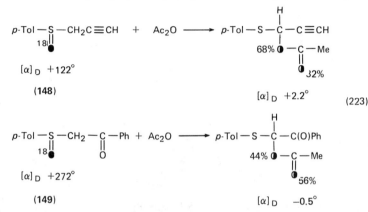

both cases, ^{18}O tracer experiments revealed the acetoxy migration to proceed at least 50% intramolecularly and the distribution of ^{18}O was found to be uneven. One obvious case of intramolecular Pummerer reaction proceeding with induction is that of (−)-o-benzylsulphinylbenzoic acid with acetic anhydride (equation 224). Asymmetric induction was 19.5% in enantiomeric excess at α-carbon. When the sulphoxide was treated with DCC (dicyclohexylcarbodiimide) in THF, the cyclic ester was formed with asymmetric induction to the extent of 29.8% enantiomeric excess[388,389]. With DCC present, an unusually high induction (70%) was achieved with β-carbonyl sulphoxides[392].

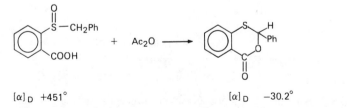

Another recent example of asymmetric induction in the Pummerer reaction is that of α-phosphoryl sulphoxide **150** with acetic anhydride[393] (equation 225).

In all cases in which asymmetric induction takes place, the intermediate acyloxysulphonium ylide is substantially stabilized either by electron-withdrawing substituents or conjugative resonance with substituents or both, and the heterolytic cleavage of the S—O bond generally leaves highly unstable, very short-lived

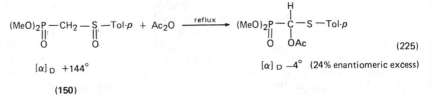

$[\alpha]_D$ +144° $[\alpha]_D$ −4° (24% enantiomeric excess)

(150)

carbonium ions which undergo very rapid recombination with the intimate counter anion.

Several other examples of Pummerer reactions give geometric isomers stereoselectively. One is the reaction of phenyl cyclopropyl sulphoxide with acetic anhydride. This proceeds with 69–76% stereoselectivity[394] (equation 226).

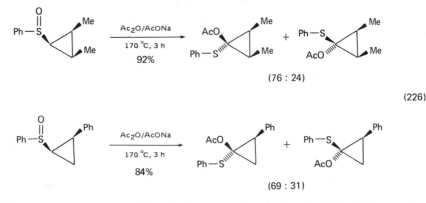

This reaction has a small kinetic isotope effect k_H/k_D = 1.13–1.49, and the Hammett plot (σ values) gave a U-shaped curve. The enthalpy and entropy of activation are unusually large ($\Delta H^{\ddagger}$ = 41.3 kcal/mol, $\Delta S^{\ddagger}$ = +10.4 e.u.) and deuterium in the starting sulphoxide **151** is gradually lost during the reaction[395] (equation 227). There is no product of ring opening in contrast to the acetolysis of

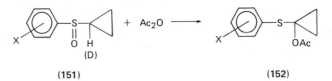

k_H/k_D = 1.13 for X = p-MeO

1.24 for X = H

1.49 for X = m-CF$_3$

1-chloro-1-phenylthiocyclopropane, which proceeds through a sulphur-stabilized carbonium ion intermediate to give chiefly the products of ring opening[395,396].

All of these observations suggest that the Pummerer reaction of cyclopropyl sulphoxides also proceeds via an ElcB type route with S—O bond cleavage in the rate-determining step. ^{18}O tracer experiments with the sulphoxide **151** showed that the rearranged acetoxy derivatives **152** retained roughly 22–30% of the ^{18}O label of the sulphoxide **151**, and the acetoxy group of the rearranged product **152** lies at the opposite side of the α-hydrogen which was removed.

Several other stereoselective Pummerer reactions have been observed: reaction of 2,2-dialkyl-1,3-oxathiolane-5-one *S*-oxide with acetic anhydride[397], reaction of 3-cephem *S*-oxides with ethyl chlorocarbonate[398,399], reaction of sulphoxide derivatives of a thiosugar with acetic anhydride[400,401] and pyrolysis of α-trimethylsilyl sulphoxides[402–405].

When azasulphonium salts were treated with triethylamine at 0°C, Pummerer reactions, which appeared to be intramolecular, took place as shown in equations 228[406], 229[407], 230[408], 231[409], and 232[410], although no mechanistic details were presented.

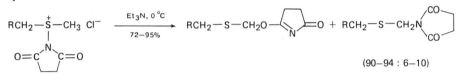

R = H, Me, Et, *i*-Pr   (228)

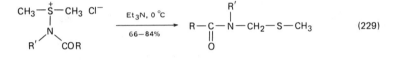

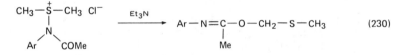

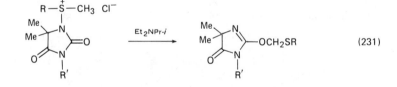

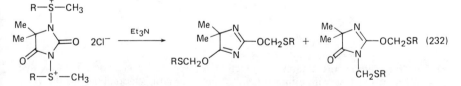

Sulphonium ylides were reported to undergo what appears to be an intramolecular Pummerer reaction on refluxing with water or ethanol[411,412].

[18]O tracer experiments on the Pummerer reaction of dimethyl sulphoxide with uniformly ^{18}O-labelled acetic anhydride in diethyl ether revealed that all of the oxygen atoms of the starting materials were completely scrambled in the resulting α-acetoxymethyl methyl sulphide[363]. Intermolecular attack of acetate at the α-carbon is thus involved in this rearrangement.

The discovery of several intramolecular Pummerer reactions with the aid of [18]O tracer experiments prompted reinvestigation of the intermolecular reaction using ^{18}O-labelled sulphoxides. In the reaction of benzyl methyl sulphoxide, only 5% of [18]O was found to migrate from the sulphoxide to the resulting ester, while with methyl phenyl sulphoxide the Pummerer product retained only 3% of [18]O from the starting material[413].

Earlier, Johnson and coworkers carried out a cross-over experiment between ^{14}C-labelled and unlabelled aryl methyl methoxysulphonium salts[364,365] (equation 233). This methoxy-sulphonium salt would, however, undergo rapid

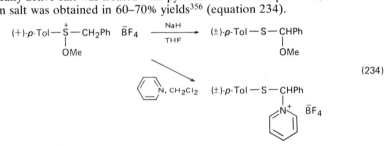

$$Ar - \overset{+}{\underset{\underset{OCH_3}{|}}{S}} - CH_2CH_3 \ \ \bar{B}F_4 \ + \ Ph - \overset{+}{\underset{\underset{O\overset{*}{C}H_3}{|}}{S}} - CH_2CH_3 \ \ \bar{B}F_4^- \ \xrightarrow[acetone]{base} \ Ar - S - \underset{\underset{O\overset{*}{C}H_3}{|}}{CHCH_3}$$

1389 cpm/mol   423 cpm/mol

$$+ \ \ Ph - S - \underset{\underset{O\overset{*}{C}H_3}{|}}{CHCH_3}$$ (233)

939 cpm/mol

base = 2,6-lutidine

* = [14]C label

methoxy exchange at sulphur. This is another example of intermolecular migration in the Pummerer reaction and this mechanism is supported by the observation that racemic α-methoxybenzyl p-tolyl sulphide is obtained by treatment of the optically active (+)-benzyl p-tolyl methoxysulphonium salt with NaH in THR[3]. When the same optically active salt was treated with pyridine at room temperature, the racemic pyridinium salt was obtained in 60–70% yields[356] (equation 234).

$$(+)\text{-}p\text{-Tol} - \overset{+}{\underset{\underset{OMe}{|}}{S}} - CH_2Ph \ \ \bar{B}F_4 \ \xrightarrow[THF]{NaH} \ (\pm)\text{-}p\text{-Tol} - \underset{\underset{OMe}{|}}{S} - CHPh$$

(234)

$$\xrightarrow{\text{N, CH}_2Cl_2} \ (\pm)\text{-}p\text{-Tol} - S - CHPh$$

In the late 1960s, the Pummerer reaction was generally considered to involve intermolecular reaction with either solvent or an external nucleophile. Thus the ylide **153** would react directly with solvent to afford the α-substituted sulphide or lose acyloxy ion to form the sulphur-stabilized carbonium ion **154**, which reacts immediately with solvent to give the final product (equation 235).

Accordingly, several attempts have been made to obtain clear evidence in support of carbonium ion formation. Unsuccessful examples are the Pummerer reactions of cyclopropyl phenyl, cyclopropylcarbinyl phenyl, and cyclobutylcarbinyl phenyl sulphoxides with acetic anhydride. If any dissociated ion pair is involved in reaction of these sulphoxides, all compounds should give, at least partially, products of ring

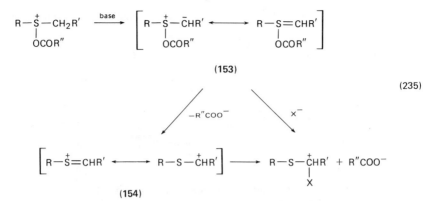

(235)

opening, which were not found. Instead, only normal Pummerer rearrangement products were obtained[414] (equation 236). Acetolyses of the corresponding

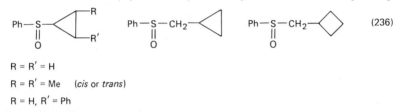

(236)

R = R' = H

R = R' = Me   (*cis* or *trans*)

R = H, R' = Ph

α-chlorocycloalkyl phenyl sulphides, on the other hand, give products of ring opening. Probably, the acyloxysulphonium ylide **153** undergoes heterolysis to form the sulphur-stabilized carbonium ion **154**, paired intimately with the acylate ion and the reaction therefore appears to involve intramolecular migration.

When the sulphur-stabilized carbonium ion pairs are partially dissociated in polar media, the carbonium ion **154** does appear to recombine with other nucleophiles. One such example of an intermolecular Pummerer reaction may be seen in aromatic electrophilic substitution by the Pummerer reaction intermediate[415] (equation 237).

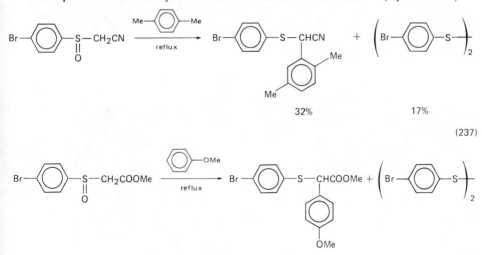

(237)

The reaction is presumed to involve the sulphur-stabilized carbonium ion **155** (equation 238).

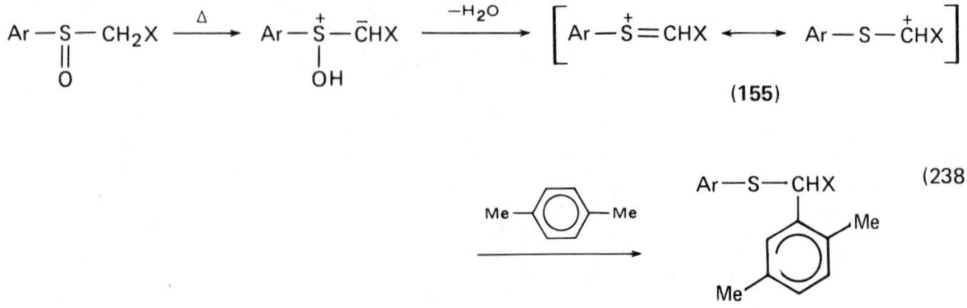

$$Ar-\underset{\underset{O}{\|}}{S}-CH_2X \xrightarrow{\Delta} Ar-\underset{\underset{OH}{|}}{\overset{+}{S}}-\bar{C}HX \xrightarrow{-H_2O} \left[Ar-\overset{+}{S}=CHX \longleftrightarrow Ar-S-\overset{+}{C}HX\right]$$

(155)

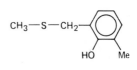

$$Ar-S-CHX$$ (238)

Phenol and *p*-substituted thiophenols are powerful trapping agents for carbonium ions and are reported to be substituted by the electrophile **155** formed as an intermediate in reaction of dimethyl sulphoxide with trifluoroacetic anhydride[416] (equation 239).

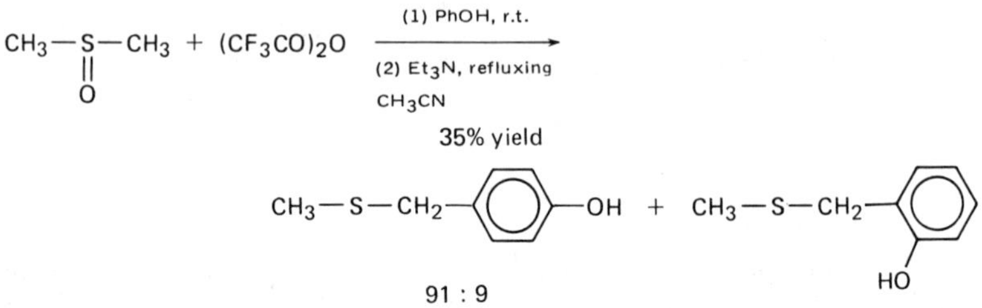

$$CH_3-\underset{\underset{O}{\|}}{S}-CH_3 + (CF_3CO)_2O \xrightarrow[\begin{subarray}{c}(2)\ Et_3N,\ refluxing\\ CH_3CN\end{subarray}]{(1)\ PhOH,\ r.t.}$$

35% yield

$$CH_3-S-CH_2-\langle\bigcirc\rangle-OH + CH_3-S-CH_2-\langle\bigcirc\rangle$$
HO

91 : 9

(239)

$$CH_3-\underset{\underset{O}{\|}}{S}-CH_3 + (CF_3CO)_2O \xrightarrow[r.t.,\ CH_3CN]{X-\langle\bigcirc\rangle-SH} CH_3-S-CH_2-S-\langle\bigcirc\rangle-X$$

55–59%

X = Cl, H, Me

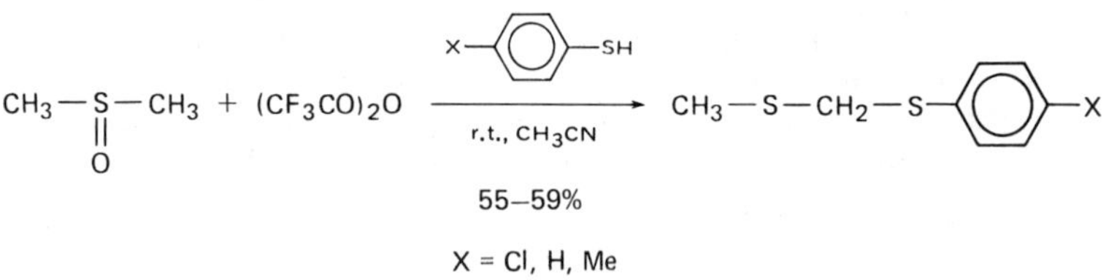

A similar reaction is seen in equation 240[417].

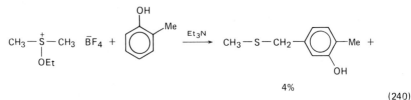

$$CH_3-\underset{\underset{OEt}{|}}{\overset{+}{S}}-CH_3\ \bar{B}F_4 + \underset{\text{Me}}{\overset{OH}{\langle\bigcirc\rangle}} \xrightarrow{Et_3N} CH_3-S-CH_2-\langle\bigcirc\rangle-Me +$$
OH

4%

(240)

$$CH_3-S-CH_2-\langle\bigcirc\rangle$$
HO   Me

2%

trans-1,4-Dithiane disulphoxide **156**, on treatment with acetic anhydride, gives the ring contracted product **158**, indicating that the carbonium ion intermediate **157**, stabilized by both α- and β-sulphur atoms, is involved[418] (equation 241).

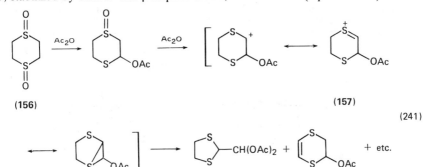

(156)   (157)

(241)

(158)

The β-ketosulphoxide in equation 242 undergoes the Pummerer reaction in dilute hydrochloric acid, presumably via a carbonium ion intermediate[419] (equation 242).

On the basis of an ^{18}O tracer study, the Pummerer reaction of a conformationally fixed six-membered sulphur heterocycle such as 4-*p*-chlorophenylthiane *S*-oxide (*cis* or *trans*) was found to proceed intermolecularly to yield the product ester with very little ^{18}O from the original sulphoxide. The recovered sulphoxide was found to have retained the label completely. The reaction of both *cis*- and *trans*-sulphoxides proceeded stereo-specifically to afford predominantly equatorial α-acetoxy-sulphide[420] (equation 243).

$$\text{Ar} \underset{\text{DCC}}{\overset{Ac_2O}{\longrightarrow}} \text{Ar} \underset{\text{DCC}}{\overset{Ac_2O}{\longleftarrow}} \text{Ar} \qquad (243)$$

Ar = *p*-ClC₆H₄⁻

$Ar = p\text{-ClC}_6H_4^-$

The allylic sulphoxide **159** gave the two γ-acetoxy sulphides in good yields in the ratio of 3:1, apparently intermolecularly via an intermediate **160**[421] (equation 244).

The base-catalysed rearrangement of the thioxanthene sulphimide **161** may be classified as an intermolecular Pummerer reaction (path a)[422] (equation 245). In this case, an intramolecular 1,4-sigmatropic rearrangement (path b) is less likely, although it cannot be ruled out completely. A partially stereoselective 1,4-sigmatropic rearrangement of a 10-aryl group is seen in the reaction of an optically active 10-aryl 10-thioxanthenium salt with base[423].

Sulphides are known to form addition complexes on treatment with halogens, and sulphides bearing α-methylene hydrogens usually react further to form α-halosulphides. This type of reaction can be classified as a Pummerer reaction of halosulphonium salts (equation 246).

Not only chlorine[424–427] and bromine[426,428], but also sulphuryl chloride[425,429–433], 3-iodopyridine–chlorine complex[434], NCS[435–442], and NBS[439,443] can serve as

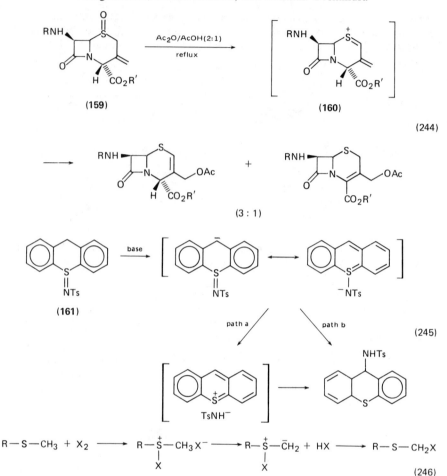

halogenating agents and afford α-halosulphides. Sometimes further halogenations, depending on the reaction conditions, are known to occur, giving di- or trihalogenosulphides[425,427,430–433]. Some investigations have presented strong evidence that the rate-determining step is proton removal. Tuleen and Marcum, for example, showed that in the chlorination of α-d_1-benzyl phenyl sulphide with NCS, the kinetic isotope effect was about 5.3–5.9[438] (equation 247).

$$\text{Ph}-\text{CHD}-\text{S}-\text{Ph} + \text{NCS} \longrightarrow \text{Ph}-\underset{|}{\overset{}{\text{CD}}}-\text{S}-\text{Ph} + \text{Ph}-\underset{|}{\overset{}{\text{CH}}}-\text{S}-\text{Ph} \qquad (247)$$
$$\qquad\qquad\qquad\qquad\qquad\qquad\qquad\quad \text{Cl} \qquad\qquad\quad \text{Cl}$$

In the reaction of α,α-d_2-thiolane with halogens in methanol to form the α-halothiolane. the competitive isotope effect was 5.1 with chlorine and 3.6 with bromine[426] (equation 248).

Involvement of the halosulphonium ylide intermediate was confirmed by a recent spectroscopic identification of the ylide **162** in the reaction of 1,3-dithiane with sulphuryl chloride[444] (equation 249).

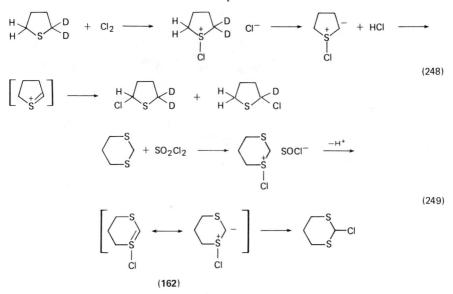

(248)

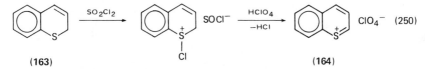

(249)

(162)

In a special case, a sulphur-stabilized carbonium ion was isolated in the reaction of the sulphide **163** with sulphuryl chloride, since the thiapyrilium salt produced **164** is aromatic[445] (equation 250).

(163)   (164)

(250)

Initial formation of the halosulphonium salt and the subsequent proton removal to form the halosulphonium ylide are similar to the usual Pummerer reaction of sulphoxides. The manner of the subsequent migration of halogen from sulphur to α-carbon, whether intramolecular or intermolecular, has not been fully investigated, however.

An interesting problem is the regioselectivity of the Pummerer α-halogenation of sulphides. Some data have suggested that migration of halogen takes place to the more alkylated α-carbon atom, contrary to the course of the usual Pummerer reaction of sulphoxides with acetic anhydride (see above). An example[441] is shown in equation 251. In this case, heterolytic cleavage of the S—Cl bond would be easy,

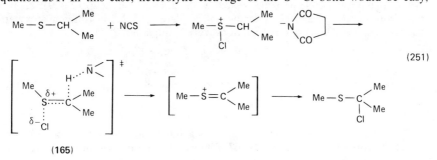

(251)

(165)

while the base available to remove the α-proton is weak. The situation resembles Saytzeff elimination of alkyl halides with bases[446]. Hence, the transition state **165** should be stabilized by α-alkyl substituents, since both π conjugation and a partial positive charge are developed at the α-carbon in the transition state.

In the reaction of sulphoxides with hydrochloric acid in the presence of molecular sieves which yields α-chlorosulphides, chlorination takes place preferentially at the α-carbon with the greater number of alkyl substituents[447]. This situation is analogous, since the initial step of the reaction between the sulphoxide and hydrochloric acid is presumably the formation of the corresponding chlorosulphonium salt (see Section IIIA2b), which subsquently undergoes elimination.

There are many other examples of α-halogenation of sulphoxides with various halogenating agents via the Pummerer reaction. Various carboxylic acid chlorides[284,378,429,436,448], sulphenyl[449,450], sulphinyl[451], and sulphonyl chlorides[276,452,453], thionyl chloride[429,436,454,455], boron trichloride[456], chloro-silanes[456–458], and phosphorus and phosphoryl chlorides[32,33,448,455,459] have been used successfully.

As in the abnormal Pummerer reaction of penam sulphoxide, ring enlargement of five-membered sulphur heterocycles appears to proceed by an abnormal Pummerer reaction via the chlorosulphonium salt (equations 252[460], 253[71], and 254[461]).

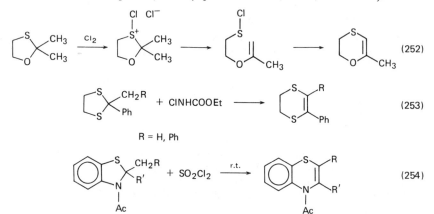

Dithioacetic acid is a fairly strong acid ($pK_1 = 2.55$) and has a strong nucleophilic thiol group. Therefore, it can add to nucleophilic alkenes and undergo Michael addition with electrophilic alkenes very readily[462]. It is a strong reducing agent and reduces sulphoxides and sulphimides to the sulphides even at low temperatures[462]. When phenyl vinyl sulphoxide reacts with dithioacetic acid (2 mol) at room temperature, an exothermic reaction takes place and 1,2-bis(dithioacetoxy)ethyl phenyl sulphide **166** is obtained quantitatively[462] (equation 255).

$$Ph-\underset{\underset{O}{\|}}{S}-CH=CH_2 + 2\ Me-\underset{\underset{S}{\|}}{C}-SH \longrightarrow Ph-\underset{\underset{S-C-Me}{\underset{\|}{S}}}{S}-CH-CH_2-S-\underset{\underset{S}{\|}}{C}-Me + H_2O \quad (255)$$

(**166**)

This reaction is considered to proceed via initial protonation at the sulphoxide oxygen to form a sulphonium salt **167**, followed by addition of dithioacetate to give the Pummerer intermediate **168**, which is eventually attacked by another molecule of dithioactic acid to yield the final product[462] (equation 256).

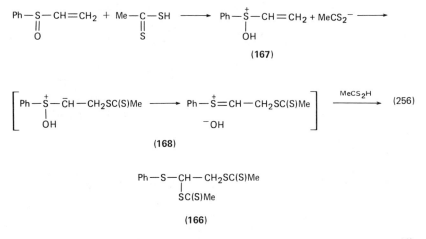

(167)

(256)

(168)

(166)

Similar Pummerer addition reactions are known, as shown in equations 257[463], 258[464], and 259[465].

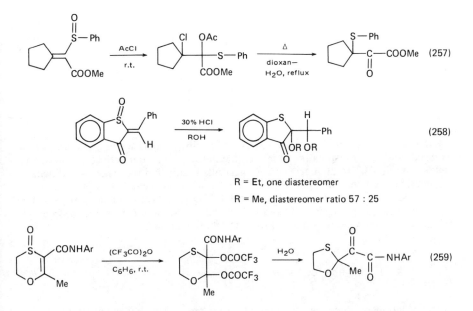

(257)

(258)

R = Et, one diastereomer

R = Me, diastereomer ratio 57 : 25

(259)

Addition of methoxide ion to phenyl vinyl methoxysulphonium salt results in the formation of the Pummerer product, into which 8.3% of deuterium was incorporated on reaction of the alkoxysulphonium salt in methanol-d_1. This implies that protonation of the ylide **169** to afford the β-substituted alkoxysulphonium salt **170** is also involved as a side-reaction[2,3] (equation 260).

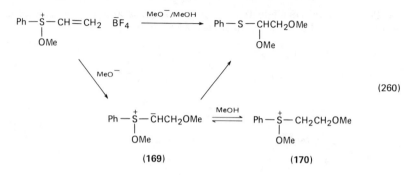

$$(260)$$

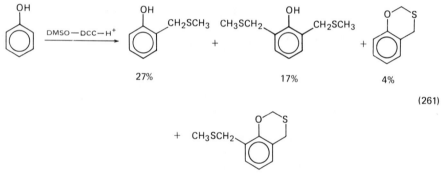

(169)   (170)

2. Sommelet–Hauser rearrangement

When phenol is treated with DMSO–DCC in the presence of acid, a Sommelet–Hauser type of rearrangement reaction occurs[466,467] to give four products, as shown in equation 261.

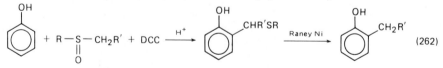

27%   17%   4%

$$(261)$$

+ CH₃SCH₂─[structure]

4%

Other alkyl-substituted phenols and *o*- and *p*-nitrophenols were also found to be oxidized to the corresponding 2-methylthiomethylphenol derivatives in moderate yields on similar treatment[466,468]. Other alkyl sulphoxides, such as dibenzyl, tetramethylene, benzyl methyl, methyl phenyl, and *t*-butyl methyl sulphoxide, were also found to be subject to the Sommelet–Hauser rearrangement, although the reaction proceeded sluggishly[468] (equation 262). α-Naphthol gave the bis(methylthiomethyl) compound as a major product[466,467] (equation 263).

[equation 262 structures]

$$(262)$$

The reaction products can be quantitatively reduced with Raney nickel to give 2-alkyl-substituted phenols[466], while 1,3-oxathiane derivatives were generally formed as minor products.

The mechanism of the reaction is suggested to involve attack of DMSO on protonated DCC to give an alkoxysulphonium salt **171**, which is then attacked by phenol at the sulphur atom to give the phenoxysulphonium salt **172**. Subsequent

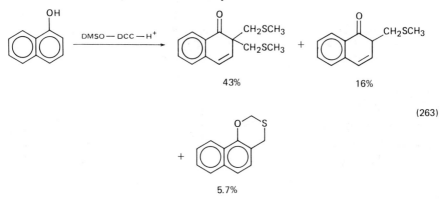

(263)

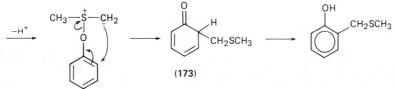

5.7%

α-proton abstraction from the salt **172** and attack of the ylide carbanion on the *ortho* position of phenol yields the dienone derivative **173**, which eventually affords the rearrangement product by re-aromatization[468,469] (equation 264).

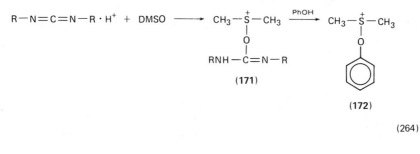

(171)

(172)

(264)

In the reaction of 2,6-disubstituted phenols, the intermediate dienone derivative **174** can be isolated generally in high yields, and then converted into the *p*-methylthiomethyl phenol derivative **175** quantitatively on treatment with trifluoroacetic acid or concentrated hydrochloric acid[466,470] (equation 265).

Transfer of the methylthiomethyl group from the intermediate **174** to the product **175** was found to be intermolecular in a cross-over experiment. When the dienone **174** was treated with five-fold excess of 2,6-dimethylphenol in the presence of a trace amount of trifluoroacetic acid, the reaction occurred spontaneously to give the cross-over product 4-methylthiomethyl-2,6-dimethylphenol in 33% yield, together with the **175** in 67% yield[470] (equation 265).

Formation of a 1,3-benzoxathiane derivative was explained tentatively as shown in equation 266, since 2,4,6-trichlorophenol gives **176** in 42% yield by the same treatment[470]. Possible involvement of an intermediate such as **177** was also suggested[417,467].

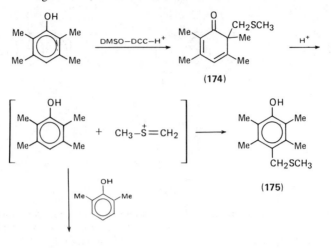

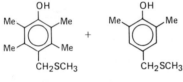

(265)

67%   33%

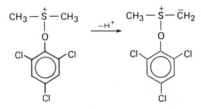

(266)

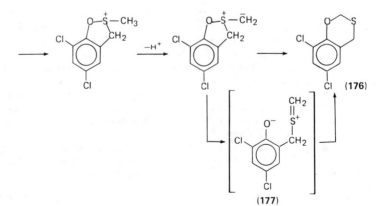

(177)

Methylthiomethylation of phenols with DMSO has also been performed in the presence of DCC–pyridinium trifluoroacetate[416,467,469], acetic anhydride[470-473], trifluoroacetic anhydride[416], phenyl chlorocarbonate[469], or pyrididine–SO$_3$ complex[474], and by extensive heating with DMSO[475,476].

ortho-Methylthiomethylation of phenols was also found to occur on treatment with either the azasulphonium salt or the chlorosulphonium salt (equation 267). Several *o*- and *p*-substituted phenols were thus converted into the corresponding 2-methylthiomethyl phenol derivatives in 49–73% yields[477].

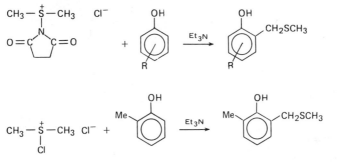

(267)

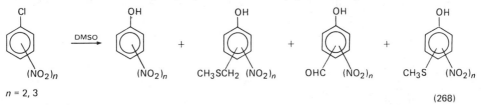

Reaction of dimethyl sulphoxide with dinitro- and trinitrochlorobenzene was reported to afford the corresponding phenol derivatives, together with methylthiomethyl- and formyl-substituted phenols and thioanisoles as side-products, via initial formation of aryloxysulphonium salts[478] (equation 268).

Cl
⬡
(NO$_2$)$_n$
→ DMSO

OH
⬡
(NO$_2$)$_n$
+

OH
⬡
CH$_3$SCH$_2$ (NO$_2$)$_n$
+

OH
⬡
OHC (NO$_2$)$_n$
+

OH
⬡
CH$_3$S (NO$_2$)$_n$

n = 2, 3

(268)

Phenols and naphthols are known to give *o*- and *p*-dimethylsulphoniophenols or -naphthols by treatment either with dimethyl sulphoxide in the presence of acid[479] (equation 269), with an azasulphonium salt[59] (equation 270), or with the chlorosulphonium salt[76,480] (equation 271).

It was shown by Gassman's group and independently by Johnson's group that the azasulphonium salt undergoes the Sommelet–Hauser rearrangement. Johnson *et al.* reported that when the azasulphonium salt, prepared by treatment of the adduct between dimethyl sulphide and *N*-chlorobenzotriazole with aniline was allowed to react with NaH in THR, the Moffatt–Pfitzner reaction product was obtained quantitatively[28] (equation 272).

Gassman *et al.* reported an alternative route for the Sommelet–Hauser rearrangement of the azasulphonium salt. When the salt, prepared by the reaction between *N*-chloro-*N*-*t*-butylaniline and alkyl sulphide, was treated with bases such

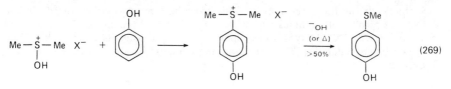

(269)

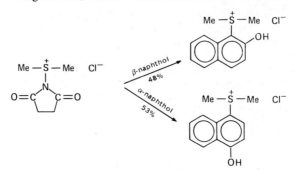

(270)

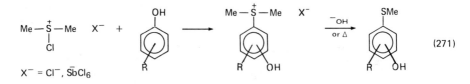

(271)

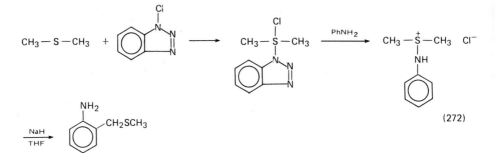

(272)

as methoxide in methanol or diethylamine, 2-methylthiomethyl-N-t-butylaniline was obtained quantitatively[73]. Subsequently, Gassman and coworkers carried out an extensive study of the Sommelet–Hauser rearrangement of azasulphonium salts.

The general procedure involves the five steps (equation 273): (1) mono-N-chlorination of the aniline with a suitable halogenating agent; (2) conversion of the N-chloroaniline into an azasulphonium salt by reaction with a dialkyl sulphide; (3) treatment of the azasulphonium salt with base to yield an azasulphonium ylide **178**; (4) Sommelet–Hauser rearrangement of the ylide to produce a substituted dienoneimine **179**; and (5) hydrogen transfer resulting in re-aromatization of the dienoneimine to give the o-alkylthioalkyl aniline. Raney nickel reduction then produces the $ortho$-alkylated aniline. Yields range from good to excellent[481,482].

Synthetically useful extensions of this procedure have been reported: indole derivatives result from anilines and β-ketosulphides[483–486] (equation 274), oxindoles from anilines and β-alkoxycarbonylsulphides[487,489], carbostyrils from anilines and γ-alkoxycarbonylsulphides[490], and 2-aminopyridines[491] are alkylated at the 3-position.

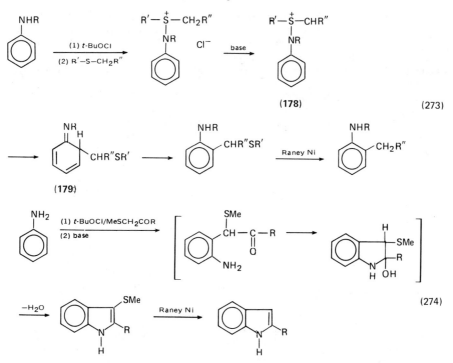

(178) (273)

(179)

(274)

Treatment of anilines with chlorine complexes of β-ketosulphides, β-carboalkoxysulphides and dialkyl sulphides similarly gives indoles, oxindoles, and alkylated anilines in good yields[492,493]. Phenol was converted into 2-methyl-3-methylthiobenzofuran in 12% yield by treatment with NCS and methylthioacetone[494]. Selective *ortho*-formylation of anilines[495] and phenols[496] can be similarly performed with 21–50% overall yields (equation 275).

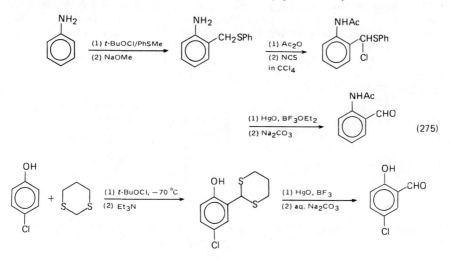

(275)

A major limitation to this synthetic procedure is the tendency for alkene formation when the alkyl group bonded to the sulphur atom of the azasulphonium salt bears β-hydrogens. Formation of alkene presumably proceeds via E_i elimination of the sulphimide formed by *in situ* deprotonation of the azasulphonium salt[482]. However, alkene formation can be prevented by the use of acetic anhydride or methyl chloroformate[497].

When *p*-toluenesulphonanilides were treated with the DMSO–DCC–H⁺ system, the compounds **180** and **181** were obtained in 24% and 27% yields, respectively, presumably via the ylide **182**. The anilide with two *o*-methyl groups gave the *N*-methylthiomethylsulphonamide in 21% yield at half-completion of the reaction[89] (equation 276).

(182)

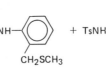

(276)

(180)   (181)

The Moffatt–Pfitzner reaction also takes place by treatment of sulphimides with base in aprotic solvents or without base in protic media[498]. The rate of the rearrangement was found to depend on the ease of proton abstraction from the S-alkyl group ($k_H/k_D = 2.5$–3.3 for X = Cl), and to decrease when electron-withdrawing substituents were present in the aniline. The rate of rearrangement of the azasulphonium ylide was also found to be greater than that of its re-protonation[499] (equation 277).

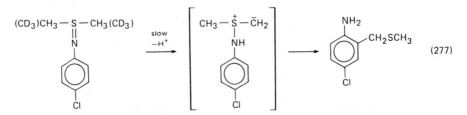

(277)

Sulphimides **183** and **184** gave the respective rearranged products **185** and **186** stereoselectively, via a suprafacial [2,3]-sigmatropic shift[500] (equation 278).

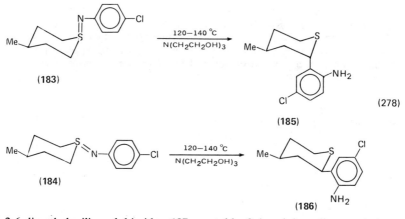

(183)

(185)

(278)

(184)

(186)

With 2,6-dimethylanilinosulphimides **187**, unstable 2,4-cyclohexadienone imines **188** were successfully isolated[501,502]. When an optically active sulphimide **187** was used, a partially optically active product **188** was obtained[503] (equation 279).

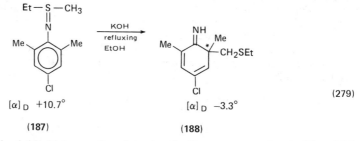

$[\alpha]_D$ +10.7°

(187)

$[\alpha]_D$ −3.3°

(188)

(279)

Diethyl *N*-aryl sulphimides were found to give the Pummerer product in 98% yield on treatment with triethylamine[501] (equation 280).

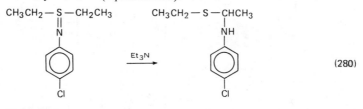

(280)

E. Kornblum Oxidation

When alkyl halides are heated with dimethyl sulphoxide, alkylation takes place either at sulphur or oxygen, depending on the alkylating agent. Only with methyl halides is extensive formation of the *S*-alkylated product observed[6,7]; with others, the reaction appears to proceed via an unstable alkoxysulphonium salt, which on further reaction affords an aldehyde and dimethyl sulphide[5,504,505] (equation 281).

$$RCH_2X + Me-\underset{\underset{O}{\|}}{S}-Me \xrightarrow[NaHCO_3]{100-160\ °C} \left[Me-\overset{+}{\underset{\underset{OCH_2R}{|}}{S}}-Me \quad X^- \right] \longrightarrow RCHO + MeSMe$$

(281)

This reaction can therefore be used as a covenient procedure for the preparation of aldehydes. Since Kornblum et al. showed that treatment of p-bromophenacyl bromide with DMSO gives phenyl glyoxal in a good yields[506], this reaction has been used extensively[507] (equation 282). The reaction can be applied not only to alkyl halides, but also to alkyl tosylates and other similar compounds with or without a base such as $NaHCO_3$[508–516].

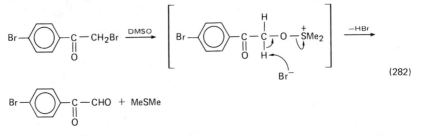

(282)

Simple alkyl chlorides and bromides are inert to DMSO even at relatively high temperatures, but are converted into the corresponding aldehydes in the presence of $AgBF_4$[517] or $AgClO_4$[518]. $AgNO_3$ cannot be used as a catalyst because of the nucleophilicity of nitrate ion[519]. Simple alkyl halides can be oxidized with DMSO after intial conversion of alkyl halides into the corresponding alkyl tosylates by treatment with silver tosylate[507].

The reaction has been shown to proceed through initial formation of the alkoxysulphonium salt. In the subsequent step, there are two conceivable routes (equation 283).

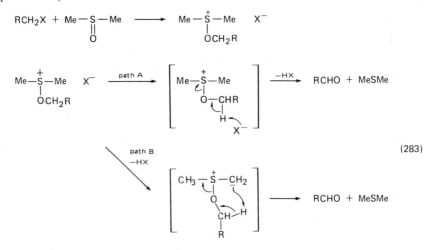

(283)

The use of DMSO-d_6 for this reaction revealed that the reaction of primary halides such as isobutyl derivatives apparently proceeds via path B, and results in the formation of CD_3SCD_2H, while that of the p-bromophenacyl compound proceeds via path A and dimethyl sulphide-d_6 is recovered intact[17,520] (equation 284).

Johnson and Phillips made similar observations on the reaction proceeding via path B[2,3] (equation 285). They also found that diphenyl methoxysulphonium tetrafluoroborate gave no formaldehyde on treatment with sodium methoxide in

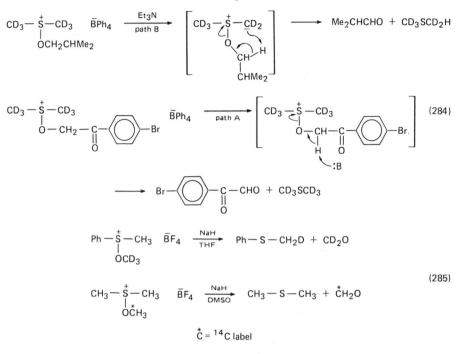

(284)

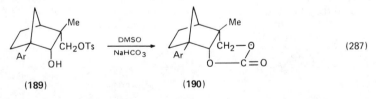

(285)

$$\overset{*}{C} = {}^{14}C \text{ label}$$

methanol, but was converted into diphenyl sulphoxide and presumably dimethyl ether. Thus, in order for oxidation to occur, the alkoxysulphonium salt seems to have to bear a hydrogen α to the sulphur atom[3] (equation 286).

Ph—$\overset{+}{S}$—Ph $\bar{B}F_4$ $\xrightarrow[\text{MeOH}]{\text{NaOMe}}$ Ph—$\underset{O}{\overset{\|}{S}}$—Ph + [MeOMe]   (286)
  |
  OMe

DMSO–NaHCO$_3$ mixture is known to oxidize both primary and secondary alkyl tosylates to the corresponding carbonyl compounds. The tosylate **189** is an interesting exception. It gives the cyclic carbonate **190**, indicating that bicarbonate anion is a better nucleophile than DMSO[521,522] (equation 287).

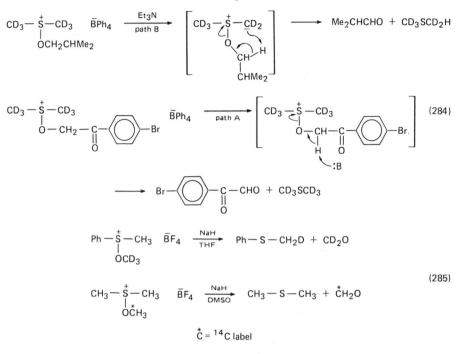

(287)

(189)   (190)

Oxidation of secondary halides and tosylates by this reaction is not generally useful for synthetic purposes, since elimination to alkene competes and in some cases becomes the major reaction[510–512,523,524].

α-Oxo secondary halides, however, were oxidized to the enolized α-diketones[512] in satisfactory yields. Carbonyl-substituted active methylene compounds were also oxidized to di- or tricarbonyl compounds on heating with DMSO in the presence of either hydrogen bromide[525] or iodine[526]. The reaction is considered to proceed through halogenation at the α-position of the carbonyl group and subsequent Kornblum oxidation. α-Picoline derivatives also gave the corresponding aldehydes in 30–36% yields[527] (equation 288).

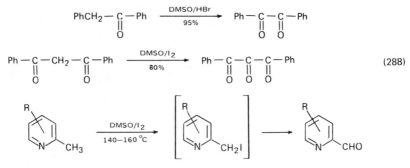

(288)

Reaction of 2-bromocyclohexanone with DMSO in the absence of base afforded 1-bromo-2-hydroxycyclohex-1-ene-3-one in 34% yield, presumably via further bromination of the diketone initially formed[528] (equation 289).

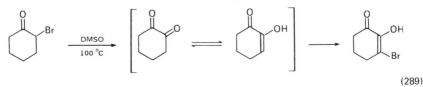

(289)

When epoxides were dissolved in anhydrous DMSO and treated with a catalytic amount of born trifuoride, the α-hydroxyketones were obtained in fairly good yields. The reaction was presumed to proceed as shown in equation 290[529]. DMSO has also

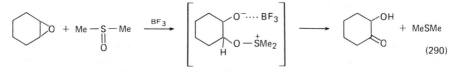

(290)

been employed for the oxidation of a variety of epoxides to α-hydroxyaldehydes and α-hydroxyketones in the presence of acid catalysts[530–534]. Swern and coworkers isolated the salts **191** from the reaction of epoxides with DMSO in the presence of a strong acid[20,21]. This was then converted into the α-hydroxyketone by treatment with base[21] (equation 291).

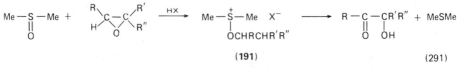

(191)   (291)

1,4-Cyclohexadiene dioxide was oxidized with DMSO but further dehydration took place to give catechol in 75% yield[535] (equation 292).

$$\text{O}\diamond\diamond\text{O} + \text{Me-S-Me} \xrightarrow[125\,°C]{BF_3} \text{catechol}$$

(292)

Benzyl alcohols were shown by Traynelis *et al.* to be selectively oxidized to the corresponding aldehydes by refluxing the alcohols in DMSO in a stream of air[536]; benzyl alcohol was converted into benzaldehyde (80%), and cinnamyl alcohol into cinnamaldehyde (60%) without the formation of any carboxylic acid. Primary and secondary saturated alcohols gave the corresponding carbonyl compounds in *ca.* 25% yields[537]. A possible explanation is that the reaction involves nucleophilic attack of the alcohol on sulphoxide sulphur (equation 293). Rate enhancement by

$$ArCH_2OH + Me-\underset{O}{\overset{\parallel}{S}}-Me \longrightarrow \left[ArCH_2O \overset{Me}{\underset{H}{\cdots}} \overset{Me}{\underset{Me}{S=O}} \right] \longrightarrow \left[ArCH-O-\overset{+}{S}Me_2 \atop H \right] \quad {}^-OH \quad (293)$$

$$\longrightarrow ArCHO + MeSMe + H_2O$$

electron-releasing *p*-substituents in benzyl alcohols may be rationalized by this mechanism. It is hard to explain, however, why the reaction should be facilitated by the passage of a stream of air.

Direct oxidation of related alcohols is known, and addition of a bivalent metal ion such as mercury II generally facilitates the oxidation of benzyl alcohol, cyclohexanol, and *sec*-butanol[538]. Oxidation of benzyl alcohol with DMSO is also facilitated by strong acids[539].

Tertiary alcohols are also dehydrated in good yields on refluxing in DMSO[359].

Aziridines are oxidized by treatment with DMSO, as shown in equations 294[540] and 295[541].

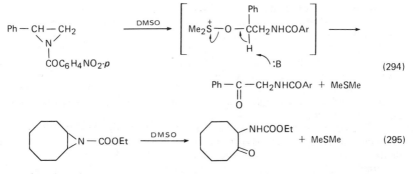

$$(294)$$

$$(295)$$

Base-catalysed oxidation of 2,5-dimethylquinolyl acetate by DMSO is an interesting reaction[542]. It is presumed to proceed as shown in equation 296.

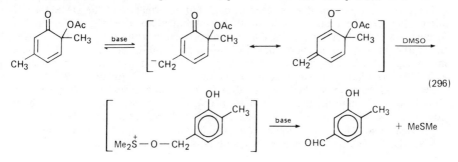

$$(296)$$

There are a few other examples of the Kornblum oxidation. Ketene and ketene imine were shown to give the corresponding α-hydroxy acid derivatives in DMSO–H_2O solution under acidic conditions[543]; α-chlorooximes also gave benzoic acids on heating with DMSO at 150°C[544] and benzylic N-tosyl sulphimides gave benzaldehyde nearly quantitatively on heating with DMSO at 180°C[545].

Generally, drastic conditions are necessary for direct oxidation of alcohols, with DMSO and the yields are generally low. On addition of a strong electrophile such as DCC to the reaction, DMSO is initially converted into a more reactive alkoxysulphonium salt **192**, which is easily attacked by alcohols to afford the alkoxysulphonium salt **193**. The resultant salt **193** undergoes ready elimination in base to give the corresponding carbonyl compound and dimethyl sulphide. Overall the reaction is a ready oxidation of alcohols to the corresponding carbonyl derivatives under mild conditions (equation 297).

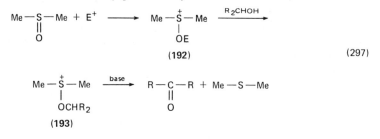

(297)

E^+ = electrophile

This modified procedure of DMSO oxidation has been widely used in the syntheses of carbohydrates, nucleosides, nucleotides, and alkaloids. Several oxidation systems of DMSO with different electrophiles have been developed and interesting results have been obtained. The following are well known electrophile–DMSO systems: DCC (Pfitzner–Moffatt)[546–548], acetic anhydride (Albright–Goldman)[549,550], P_2O_5 (Onodera)[551–553], SO_3-pyridine method (Parikh–Doering)[554], ketenimine–ynamine method (Harmon)[555], mercury(II) acetate[538], chlorine (Corey–Kim)[556], triflate[35], trifluoroacetic anhydride[557–559], methanesulphonic anhydride method[560], and oxalyl chloride[561].

A detailed mechanistic study has been carried out on the DMSO–DCC system. ^{18}O-labelled DMSO the oxygen atom of DMSO was transferred to DCC. A deuterium tracer experiment showed that butanol-1,1-d_2 gave butyl aldehyde-d_1 and dimethyl sulphide-d_1 while DMSO-d_6 gave dicyclohexylurea-d_1 and dimethyl sulphide-d_5 [562,563]. These observations suggest a mechanism which involves the alkoxysulphonium intermediate (equation 298).

The reaction is presumed to proceed through initial formation of the DMSO–DCC adduct **194**. On the attack of alcohol at the sulphur atom of **194**, the ylide **197** is formed and subsequent intramolecular proton abstraction within **197** eventually gives the carbonyl compound and dimethyl sulphide. For formation of the ylide intermediate **197** two mechanisms were proposed. One is the concerted mechanism as in **195**; the other is the stepwise route via a sulphurane intermediate **196**[563].

Sweat and Epstein studied the DMSO–DCC oxidation system using tritiated cholestanol and found the incorporation of tritium into dimethyl sulphide to be about 30–40%, suggesting that path A is favoured over path B[564] (equation 299).

Albright and Goldman showed that oxidation of ^{18}O-labelled alcohol with DMSO–DCC gave the aldehyde with retention of the isotope[550].

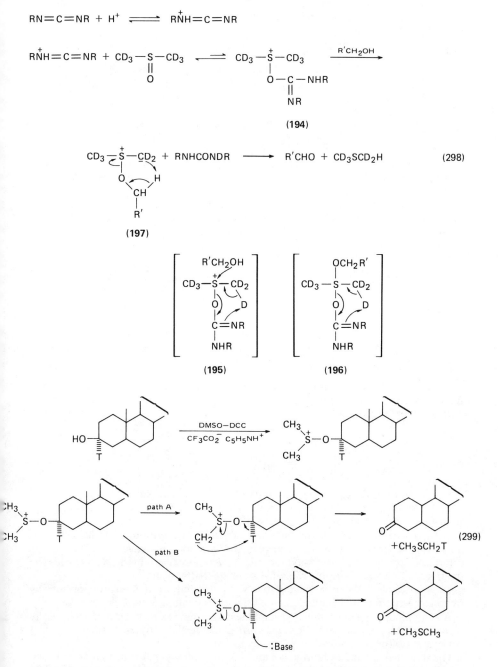

Barton *et al.* found that hydroxyl compounds were initially converted into their chloroformate esters, which were treated with DMSO and then triethylamine at room temperature, resulting eventually in formation of the carbonyl compounds in 65–70% yields[565] (equation 300).

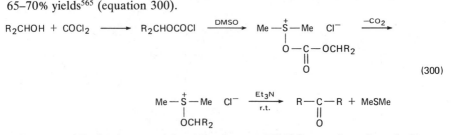

$$(300)$$

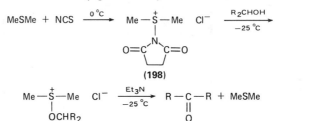

Barton and Forbes improved the chloroformate–DMSO procedure through the use of propylene oxide as an acid scavenger[566]. The application of Barton's original procedure in a prostaglandin synthesis has also been reported[567].

Corey and Kim found that the azasulphonium salt **198**, prepared *in situ* from dimethyl sulphide and NCS, when treated with an alcohol in the presence of triethylamine, caused oxidation to the carbonyl compound via initial conversion to the alkoxysulphonium salt[568,569] (equation 301).

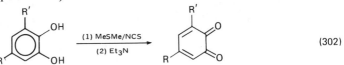

$$(301)$$

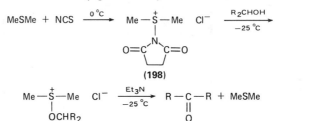

Catechols and hydroquinone were shown to be oxidized with the dimethyl sulphide–NCS system to the quinones in quantitative yields after addition of triethylamine[60] (equation 302).

$$(302)$$

A deuterium tracer experiment using the hexadeuterated dimethyl azasulphonium salt **199** showed that dimethyl sulphide–NCS oxidation proceeded through the alkoxysulphonium ylide **200**[570] (equation 303).

In the absence of base in the dimethyl sulphide–NCS system, alkyl halides were found to be produced regioselectively; the alcohol **201** gave only the allylic chloride[571] (equation 304). Alkyl bromides were obtained in 80–90% yields by treatment of alcohols with the dimethyl sulphide–NBS system[571].

Allyl alcohols **202** were quantitatively chlorinated at the allylic position by treatment with dimethyl sulphide and NCS[572,573] (equation 305).

An equimolar adduct of dimethyl sulphide and bromine can brominate an optically active secondary alcohol and afford the corresponding alkyl bromide with inversion of configuration (optical yield 91%)[574] (equation 306).

The dimethyl sulphide–chlorine complex and the methyl phenyl sulphide–chlorine complex oxidize 1,2-diols to α-hydroxyketones in good yields[575] (equation 307). The reaction is considered to proceed via initial formation of the alkoxysulphonium salt

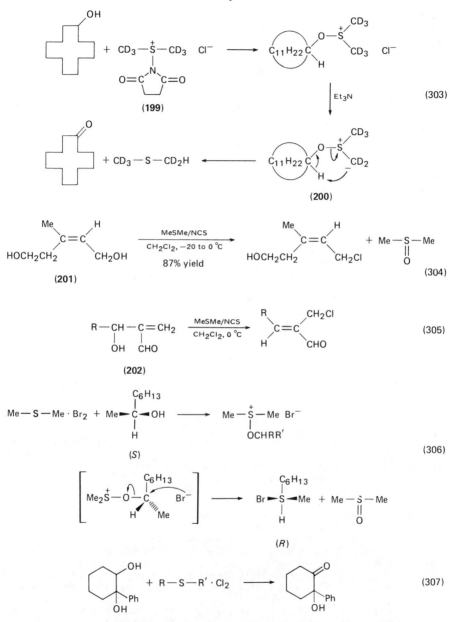

as in oxidation with the dimethyl sulphide–NCS system. The methyl phenyl sulphide–chlorine complex was also found to oxidize epoxides to α-chlorocarbonyl compounds by treatment with triethylamine, and on treatment with aqueous NaHCO₃ the α-chloro-alcohol was obtained quantitatively via hydrolysis of the intermediate β-chloroalkoxysulphonium salt[488].

660   Shigeru Oae, Tatsuo Numata, and Toshiaki Yoshimura

Polymer-bound thioanisole–chlorine complex **203** has been recommended for effective oxidation of sensitive alcohols to labile aldehydes in high yields[576] (equation 308).

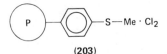

$$\text{(308)}$$

(203)

IV. REFERENCES

1. H. Meerwein, E. Battenberg, H. Gold, E. Pfeil and G. Willfang, *J. Prakt. Chem.*, **154**, 83 (1939); *Chem. Abstr.*, **34**, 2325 (1940).
2. C. R. Johnson and W. G. Phillips, *Tetrahedron Lett.*, 2101 (1965).
3. C. R. Johnson and W. G. Phillips, *J. Org. Chem.*, **32**, 1926 (1967).
4. S. Kabuss, *Angew. Chem.*, **78**, 714 (1966).
5. S. G. Smith and S. Winstein, *Tetrahedron*, **3**, 317 (1958).
6. R. Kuhn, *Angew. Chem.*, **69**, 570 (1957).
7. R. Kuhn and H. Trischmann, *Justus Liebigs Ann. Chem.*, **611**, 117 (1958).
8. N. J. Leonard and C. R. Johnson, *J. Amer. Chem. Soc.*, **84**, 3701 (1962).
9. A. N. Nesmeyanov, L. S. Isaeva and T. P. Tolstaya, *Dokl. Akad. Nauk SSSR*, **151**, 1339 (1963); *Chem. Abstr.*, **59**, 13859h (1963).
10. H. Meerwein, V. Hederich and K. Wunderlich, *Arch. Pharm.*, **291**, 541 (1958); *Chem. Abstr.*, **54**, 5427b (1960).
11. R. M. Acheson and J. K. Stubbs, *J. Chem. Soc. Perkin Trans. I*, 899 (1972).
12. M. Kobayashi, K. Kamiyama, H. Minato, Y. Oishi, Y. Takada and Y. Hattori, *J. Chem. Soc. Chem. Commun.*, 1577 (1971).
13. K. Kamiyama, H. Minato and M. Kobayashi, *Bull. Chem. Soc. Jap.*, **46**, 3895 (1973).
14. H. D. Durst, J. W. Zubrick and G. R. Kieczykowsi, *Tetrahedron Lett.*, 1777 (1974).
15. H. Minato, K. Yamaguchi, K. Okuma and M. Kobayashi, *Bull. Chem. Soc. Jap.*, **49**, 2590 (1976).
16. M. G. Ahmed, R. W. Alder, G. H. James, M. L. Sinnott and M. C. Whiting, *J. Chem. Soc. Chem. Commun.*, 1533 (1968).
17. K. Torssell, *Acta Chem. Scand.*, **21**, 1 (1967).
18. G. Natus and E. J. Goethals, *Bull. Soc. Chim. Belg.*, **74**, 450 (1965); *Chem. Abstr.*, **64**, 3339d (1966).
19. M. A. Khuddus and D. Swern, *Tetrahedron Lett.*, 411 (1971).
20. M. A. Khuddus and D. Swern, *J. Amer. Chem. Soc.*, **95**, 8393 (1973).
21. T. M. Santosusso and D. Swern, *J. Org. Chem.*, **40**, 2764 (1975).
22. Y. Hara and M. Matsuda, *J. Chem. Soc. Chem. Commun.*, 919 (1974).
23. H. Meerwein, K.-F. Zenner and R. Gipp, *Justus Liebigs Ann. Chem.*, **688**, 67 (1965).
24. W. Warthmann and A. Schmidt, *Chem. Ber.*, **108**, 520 (1975).
25. C. R. Johnson and M. P. Jones, *J. Org. Chem.*, **32**, 2014 (1967).
26. C. R. Johnson and J. J. Rigau, *J. Amer. Chem. Soc.*, **91**, 5398 (1969).
27. C. R. Johnson and D. McCant, Jr., unpublished result cited in reference 25.
28. C. R. Johnson, C. C. Bacon and W. D. Kingsbury, *Tetrahedron Lett.*, 501 (1972).
29. L. J. Adzima, C. C. Chiang, I. C. Paul and J. C. Martin, *J. Amer. Chem. Soc.*, **100**, 953 (1978).
30. J. C. Martin and R. J. Arhart, *J. Amer. Chem. Soc.*, **93**, 2339, 2341 (1971).
31. L. J. Kaplan and J. C. Martin, *J. Amer. Chem. Soc.*, **95**, 793 (1973).
32. R. Rätz and O. J. Sweeting, *Tetradedron Lett.*, 529 (1963).
33. R. Rätz and O. J. Sweeting, *J. Org. Chem.*, **28**, 1612 (1963).
34. A. K. Sharma and D. Swern, *Tetrahedron Lett.*, 1503 (1974).
35. J. B. Hendrickson and S. M. Schwartzman, *Tetrahedron Lett.*, 273 (1975).
36. H. Minato, K. Yamaguchi and M. Kobayashi, *Chem. Lett.*, 307 (1975).

37. H. Minato, K. Okuma and M. Kobayashi, *J. Org. Chem.*, **43**, 652 (1978).
38. H. Minato, K. Yamaguchi and M. Kobayashi, *Chem. Lett.*, 991 (1975).
39. R. Appel and W. Büchner, *Angew. Chem.*, **71**, 701 (1959).
40. R. Appel and W. Büchner, *Chem. Ber.*, **95**, 849 (1962).
41. Y. Tamura, K. Sumoto, J. Minamikawa and M. Ikeda, *Tetrahedron Lett.*, 4137 (1972).
42. Y. Tamura, H. Matsushima, M. Ikeda and K. Sumoto, *Synthesis*, 277 (1974).
43. Y. Tamura, H. Matsushima, J. Minamikawa, M. Ikeda and K. Sumoto, *Tetrahedron*, **31**, 3035 (1975).
44. Y. Tamura, J. Minamikawa and M. Ikeda, *Synthesis*, 1 (1977).
45. P. Stoss and G. Satzinger, *Tetrahedron Lett.*, 1973 (1974).
46. P. Stoss and G. Satzinger, *Chem. Ber.*, **111**, 1453 (1978).
47. Y. Tamura, J. Minamikawa, K. Sumoto, S. Fujii and M. Ikeda, *J. Org. Chem.*, **38**, 1239 (1973).
48. S. Oae and F. Yamamoto, *Tetrahedron Lett.*, 5143 (1973).
49. F. Yamamoto and S. Oae, *Bull. Chem. Soc. Jap.*, **48**, 77 (1975).
50. R. Appel, H. W. Fehlhaber, D. Hänssgen and R. Schöllhorn, *Chem. Ber.*, **99**, 3108 (1966).
51. E. P. Gosselink and R. G. Laughlin, *US Pat.*, 3,645,359; *Chem. Abstr.*, **76**, 139928w (1972).
52. Sandoz Ltd., *Belg. Pat.*, 667,056; *Chem. Abstr.*, **65**, 17092h (1966).
53. J. A. Cogliano and G. L. Braude, *J. Org. Chem.*, **29**, 1397 (1964).
54. R. G. Laughlin and W. Yellin, *J. Amer. Chem. Soc.*, **89**, 2435 (1967).
55. R. Appel and W. Büchner, *Chem. Ber.*, **95**, 2220 (1962).
56. M. Haake, *Tetrahedron Lett.*, 4449 (1970).
57. N. Furukawa, T. Omata and S. Oae, *J. Chem. Soc. Chem. Commun.*, 590 (1973).
58. T. Yoshimura, N. Furukawa, T. Akasaka and S. Oae, *Tetrahedron*, **33**, 1061 (1977).
59. E. Vilsmaier and W. Sprügel, *Tetrahedron Lett.*, 625 (1972).
60. J. P. Marino and A. Schwartz, *J. Chem. Soc. Chem. Commun.*, 812 (1974).
61. M. V. Likhosherstov, *Zh. Obshch. Khim.*, **17**, 1478 (1947); *Chem. Abstr.*, **43**, 172d (1949).
62. H. Kise, G. F. Whitfield and D. Swern, *Tetrahdron Lett.*, 1761 (1971).
63. H. Kise, G. F. Whitfield and D. Swern, *J. Org. Chem.*, **37**, 1121 (1972).
64. G. F. Whitfield, H. S. Beilan, D. Saika and D. Swern, *Tetrahdron Lett.*, 3543 (1970).
65. G. F. Whitfield, H. S. Beilan, D. Saika and D. Swern, *J. Org. Chem.*, **39**, 2148 (1974).
66. S. Oae, T. Masuda, K. Tsujihara and N. Furukawa, *Bull. Chem. Soc. Jap.*, **45**, 3586 (1972).
67. T. Fuchigami and K. Odo, *Chem. Lett.*, 247 (1974).
68. T. Fuchigami and K. Odo, *Bull. Chem. Soc. Jap.*, **50**, 1793 (1977).
69. A. J. Papa, *J. Org. Chem.*, **35**, 2837 (1970).
70. T. L. Gilchrist, C. J. Moody and C. W. Rees, *J. Chem. Soc. Perkin Trans. I*, 1964 (1975).
71. H. Yoshino, Y. Kawazoe and T. Taguchi, *Synthesis*, 713 (1974).
72. A. Heesing and G. Imsieke, *Chem. Ber.*, **107**, 1536 (1974).
73. P. G. Gassman, G. Gruetzmacher and R. H. Smith, *Tetrahedron Lett.*, 497 (1972).
74. Y. Ueno and M. Okawara, *Bull. Chem. Soc. Jap.*, **47**, 1033 (1974).
75. F. Knoll, M.-F. Müller-Kalben and R. Appel, *Chem. Ber.*, **104**, 3716 (1971).
76. R. Neidlein and B. Stackebrandt, *Justus Liebigs Ann. Chem.*, 914 (1977).
77. A. D. Dawson and D. Swern, *J. Org. Chem.*, **42**, 592 (1977).
78. K. Tomita, A. Terada and R. Tachikawa, *Heterocycles*, **4**, 729, 733 (1976).
79. E. Vilsmaier, E. Sprügel and K. Gagel, *Tetrahedron Lett.*, 2475 (1974).
80. P. K. Claus, W. Rieder, P. Hofbauer and E. Vilsmaier, *Tetrahedron*, **31**, 505 (1975).
81. D. Swern, I. Ikeda and G. F. Whitfield, *Tetrahedron Lett.*, 2635 (1972).
82. J. A. Franz and J. C. Martin, *J. Amer. Chem. Soc.*, **97**, 583 (1975).
83. J. C. Martin and J. A. Franz, *J. Amer. Chem. Soc.*, **97**, 6137 (1975).
84. T. Kitazume and J. M. Shreeve, *J. Amer. Chem. Soc.*, **100**, 985 (1978).
85. P. K. Claus, P. Hofbauer and W. Rieder, *Tetrahedron Lett.*, 3319 (1974).
86. A. K. Sharma, T. Ku, A. D. Dawson and D. Swern, *J. Org. Chem.*, **40**, 2758 (1975).

87. D. S. Tarbell and C. Weaver, *J. Amer. Chem. Soc.*, **63**, 2939 (1941).
88. U. Lerch and J. G. Moffatt, *J. Org. Chem.*, **36**, 3391 (1971).
89. U. Lerch and J. G. Moffatt, *J. Org. Chem.*, **36**, 3686 (1971).
90. U. Lerch and J. G. Moffatt, *J. Org. Chem.*, **36**, 3861 (1971).
91. H. Berger, R. Gall, M. Thiel, W. Voemel and W. Sauer, *Ger. Offen.*, 2,147,013; *Chem. Abstr.*, **78**, 159647y (1973).
92. T. E. Varkey, G. F. Whitfield and D. Swern, *J. Org. Chem.*, **39**, 3365 (1974).
93. R. Oda and S. Takashima, *Nippon Kagaku Zasshi*, **82**, 1423 (1961); *Chem. Abstr.*, **59**, 3802b (1963).
94. R. S. Glass and J. R. Duchek, *J. Amer. Chem. Soc.*, **98**, 965 (1976).
95. G. Kresze and M. Rössert, *Angew. Chem., Int. Ed. Engl.*, **17**, 64 (1978).
96. A. Le Berre, C. Renault and P. Giraudeau, *Bull. Soc. Chim. Fr.*, 3245 (1971).
97. I. Kapovits, F. Ruff and A. Kucsman, *Tetrahedron*, **28**, 4413 (1972).
98. N. Furukawa, T. Yoshimura, T. Omata and S. Oae, *Chem. Ind. (London)*, 702 (1974).
99. N. Furukawa, T. Omata, T. Yoshimura, T Aida and S. Oae, *Tetrahedron Lett.*, 1619 (1972).
100. T. Yoshimura, T. Omata, N. Furukawa and S. Oae, *J. Org. Chem.*, **41**, 1728 (1976).
101. R. Appel, W. Büchner and E. Guth, *Justus Liebigs Ann. Chem.*, **618**, 53 (1958).
102. C. R. Johnson, J. J. Rigau, M. Haake, D. McCants, Jr., J. E. Keiser and A. Gertsema, *Tetrahedron Lett.*, 3719 (1968).
103. D. Darwish and S. K. Datta, *Tetrahedron*, **30**, 1155 (1974).
104. P. Manya, A. Sekera and P. Rumpf, *Bull. Soc. Chim. Fr.*, 286 (1971).
105. B. K. Bandlish, S. R. Mani and H. J. Shine, *J. Org. Chem.*, **42**, 1538 (1977).
106. R. Appel and W. Büchner, *Chem. Ber.*, **95**, 855 (1962).
107. M. Becke-Goehring and H. P. Latscha, *Angew. Chem.*, **74**, 695 (1962).
108. P. Y. Blanc, *Experientia*, **21**, 308 (1965). *Chem. Abstr.*, **63**, 5520a (1965).
109. N. Furukawa, T. Yoshimura and S. Oae, *Tetrahedron Lett.*, 2113 (1973).
110. R. Appel and G. Büchler, *Justus Liebigs Ann. Chem.*, **684**, 112 (1965).
111. H. J. Shine and J. J. Silter, *J. Amer. Chem. Soc.*, **94**, 1026 (1972).
112. B. K. Bandlish, A. G. Padilla and H. J. Shine, *J. Org. Chem.*, **40**, 2590 (1975).
113. S. R. Mani and H. J. Shine, *J. Org. Chem.*, **40**, 2756 (1975).
114. J. L. Richards and D. S. Tarbell, *J. Org. Chem.*, **35**, 2079 (1970).
115. N. E. Heimer and L. Field, *J. Org. Chem.*, **35**, 3012 (1970).
116. M. Haake and H. Benack, *Synthesis*, 308 (1976).
117. W. Warthmann and A. Schmidt, *Z. Anorg. Allg. Chem.*, **418**, 57 (1975); *Chem. Abstr.*, **84**, 25306b (1976).
118. D. J. Pettitt and G. K. Helmkamp, *J. Org. Chem.*, **28**, 2932 (1963).
119. G. K. Helmkamp, D. C. Owsley, W. M. Barnes and H. N. Cassey, *J. Amer. Chem. Soc.*, **90**, 1635 (1968).
120. G. K. Helmkamp, H. N. Cassey, B. A. Olsen and D. J. Pettitt, *J. Org. Chem.*, **30**, 933 (1965).
121. N. E. Hester and G. K. Helmkamp, *J. Org. Chem.*, **38**, 461 (1973).
122. R. F. Hudson and F. Filippini, *J. Chem. Soc. Chem. Commun.*, 726 (1972).
123. G. Capozzi, O. DeLucchi, V. Lucchini and G. Modena, *Synthesis*, 677 (1976).
124. H. Minato, T. Miura and M. Kobayashi, *Chem. Lett.*, 1055 (1975).
125. G. Capozzi, V. Lucchini, G. Modena and F. Rivetti, *J. Chem. Soc. Perkin Trans. II*, 900 (1975).
126. R. Weiss and C. Schlierf, *Synthesis*, 323 (1976).
127. P. Dubs and R. Stüssi, *7th Int. Symp. Organosulfur Chem., Hamburg (1976)*.
128. W. K. Musker and P. B. Roush, *J. Amer. Chem. Soc.*, **98**, 6745 (1976).
129. W. K. Musker and T. L. Wolford, *J. Amer. Chem. Soc.*, **98**, 3055 (1976).
130. H. Minato, K. Yamaguchi, T. Miura and M. Kobayashi, *Chem. Lett.*, 593 (1976).
131. M. E. Peach, *Can. J. Chem.*, **47**, 1675 (1969).
132. Th. Zincke and W. Frohneberg, *Chem. Ber.*, **42**, 2721 (1909).
133. F. Boberg, G. Winter and G. R. Schultze, *Chem. Ber.*, **89**, 1160 (1956).
134. K. Fries and W. Vogt, *Justus Liebigs Ann. Chem.*, **381**, 337 (1911).
135. J. Böeseken, *Rec. Trav. Chim. Pays Bas*, **29**, 315 (1910); *Chem. Abstr.*, **5**, 678 (1911).

136. E. Bourgeois and A. Abraham, *Rec. Trav. Chim. Pays Bas*, **30**, 407 (1911); *Chem. Abstr.*, **6**, 623 (1912).
137. Th. Zincke and A. Dahm, *Chem. Ber.*, **45**, 3457 (1912).
138. Th. Zincke and O. Krüger, *Chem. Ber.*, **45**, 3468 (1912).
139. H. Böhme and E. Boll, *Z. Anorg. Allg. Chem.*, **290**, 17 (1957); *Chem. Abstr.*, **51**, 13739d (1957).
140. N. C. Baenziger, R. E. Buckles, R. J. Maner and T. D. Simpson, *J. Amer. Chem. Soc.*, **91**, 5749 (1969).
141. G. Allegra, G. E. Wilson, Jr., E. Benedetti, C. Pedone and R. Albert, *J. Amer. Chem. Soc.*, **92**, 4002 (1970).
142. Chr. Römming, *Acta Chem. Scand.*, **14**, 2145 (1960).
143. G. Y. Chao and J. D. McCullough, *Acta Crystallogr.*, **13**, 727 (1960); *Chem. Abstr.*, **55**, 55a (1961).
144. G. E. Wilson, Jr. and M. M. Y. Chang, *Tetrahedron Lett.*, 875 (1971).
145. G. E. Wilson, Jr., and M. M. Y. Chang, *J. Amer. Chem. Soc.*, **96**, 7533 (1974).
146. J. B. Lambert, D. A. Netzel, H.-N. Sun and K. K. Lilianstrom, *J. Amer. Chem. Soc.*, **98**, 3778 (1976).
147. M. J. Shapiro, *J. Org. Chem.*, **43**, 742 (1978).
148. S. Santini, G. Reichenbach and U. Mazzucato, *J. Chem. Soc. Perkin Trans. II*, 494 (1974).
149. N. Furukawa, F. Takahashi, T. Akasaka and S. Oae, *Chem. Lett.*, 143 (1977).
150. B. C. Menon and D. Darwish, *Tetrahedron Lett.*, 4119 (1973).
151. N. Furukawa, K. Harada and S. Oae, *Tetrahedron Lett.*, 1377 (1972).
152. M. Moriyama, N. Furukawa, T. Numata and S. Oae, *J. Chem. Soc. Perkin Trans. II*, 1783 (1977).
153. D. R. Rayner, A. J. Gordon and K. Mislow, *J. Amer. Chem. Soc.*, **90**, 4854 (1968).
154. D. Darwish and R. L. Tomilson, *J. Amer. Chem. Soc.*, **90**, 5938 (1968).
155. R. Scartazzini and K. Mislow, *Tetrahedron Lett.*, 2719 (1967).
156. D. Darwish, S. H. Hui and R. Tomilson, *J. Amer. Chem. Soc.*, **90**, 5631 (1968).
157. K. R. Brower and T. Wu, *J. Amer. Chem. Soc.*, **92**, 5303 (1970).
158. A. Garbesi, N. Corsi and A. Fava, *Helv. Chim. Acta*, **53**, 1499 (1970); cf. B. M. Trost, W. L. Schinski and I. B. Mantz, *J. Amer. Chem. Soc.*, **91**, 4320 (1969).
159. M. Moriyama, N. Furukawa, T. Numata and S. Oae, *Chem. Lett.*, 275 (1976).
160. R. E. Cook, M. D. Glick, J. J. Rigau and C. R. Johnson, *J. Amer. Chem. Soc.*, **93**, 924 (1971).
161. A. M. Griffin and G. M. Sheldrick, *Acta Crystallogr. B*, **31**, 893 (1975).
162. Y. Nishikawa, Y. Matsuura, M. Kakudo, T. Akasaka, N. Furukawa and S. Oae, *Chem. Lett.*, 447 (1978).
163. R. Tang and K. Mislow, *J. Amer. Chem. Soc.*, **91**, 5644 (1969).
164. J. Day and D. J. Cram, *J. Amer. Chem. Soc.*, **87**, 4398 (1965).
165. D. R. Rayner, D. M. von Schriltz, J. Day and D. J. Cram, *J. Amer. Chem. Soc.*, **90**, 2721 (1968).
166. D. J. Cram, J. Day, D. R. Rayner, D. M. von Schriltz, D. J. Duchamp and D. C. Garwood, *J. Amer. Chem. Soc.*, **92**, 7369 (1970).
167. S. Oae, M. Yokoyama, M. Kise and N. Furukawa, *Tetrahedron Lett.*, 4131 (1968).
168. B. W. Christensen, *J. Chem. Soc. Chem. Commun.*, 597 (1971).
169. B. W. Christensen and A. Kjaer, *J. Chem. Soc. Chem. Commun.*, 934 (1969).
170. M. Mikołajczyk and J. Drabowicz, *J. Chem. Soc. Chem. Commun.*, 775 (1974).
171. Unpublished data from these laboratories.
172. S. Oae, Y. Ohnishi, S. Kozuka and W. Tagaki, *Bull. Chem. Soc. Jap.*, **39**, 364 (1966).
173. U. Miotti, G. Modena and L. Sedea, *J. Chem. Soc. B*, 802 (1970).
174. S. Ahmed and J. L Wardell, *Tetrahedron Lett.*, 2363 (1972).
175. W. Tagaki, K. Kikukawa, K. Ando and S. Oae, *Chem. Ind. (London)*, 1624 (1964).
176. R. Harville and S. F. Reed, Jr., *J. Org. Chem.*, **33**, 3976 (1968).
177. K. M. More and J. Wemple, *Synthesis*, 791 (1977).
178. D. L. Tuleen and P. J. Smith, *J. Tenn. Acad. Sci.*, **46**, 17 (1971); *Chem. Abstr.*, **74**, 87536d (1971).

179. J. Benes, *Coll. Czech. Chem. Commun.*, **28**, 1171 (1963).
180. W. D. Kingsbury and C. R. Johnson, *J. Chem. Soc. Chem. Commun.*, 365 (1969).
181. Y. Sato, N. Kunieda and M. Kinoshita, *Chem. Lett.*, 1023 (1972).
182. Y. Sato, N. Kunieda and M. Kinoshita, *Chem. Lett.*, 563 (1976).
183. T. Higuchi and K.-H. Gensch, *J. Amer. Chem. Soc.*, **88**, 3874, 5486 (1966).
184. K. H. Gensch, I. H. Pitman and T. Higuchi, *J. Amer. Chem. Soc.*, **90**, 2096 (1968).
185. G. Barbieri, M. Cinquini, S. Colonna and F. Montanari, *J. Chem. Soc. C*, 659 (1968).
186. M. Cinquini, S. Colonna and F. Taddei, *Boll. Sci. Fac. Chim. Ind. Bologna*, **27**, 231 (1969); *Chem. Abstr.*, **72**, 89951t (1970).
187. K. C. Schreiber and V. P. Fernandez, *J. Org. Chem.*, **26**, 2910 (1961).
188. J. P. A. Castrillon and H. H. Szmant, *J. Org. Chem.*, **32**, 976 (1967).
189. N. Fujii, T. Sasaki, S. Funakoshi, H. Irie and H. Yajima, *Chem. Pharm. Bull.*, **26**, 650 (1978).
190. M. Hojo and R. Masuda, *Tetrahedron Lett.*, 613 (1976).
191. C. Walling and M. J. Mintz, *J. Org. Chem.*, **32**, 1286 (1967).
192. L. Skatteböl, B. Boulette and S. Solomon, *J. Org. Chem.*, **32**, 3111 (1967).
193. K. Kikukawa, W. Tagaki, N. Kunieda and S. Oae, *Bull. Chem. Soc. Jap.*, **42**, 831 (1969).
194. M. Kinoshita, Y. Sato and N. Kunieda, *Chem. Lett.*, 377 (1974).
195. M. Moriyama, S. Oae, T. Numata and N. Furukawa, *Chem. Ind. (London)*, 163 (1976).
196. M. Moriyama, T. Yoshimura, N. Furukawa, T. Numata and S. Oae, *Tetrahedron*, **32**, 3003 (1976).
197. T. Higuchi, I. H. Pitman and K.-H. Gensch, *J. Amer. Chem. Soc.*, **88**, 5676 (1966).
198. N. J. Leonard and W. L. Rippie, *J. Org. Chem.*, **28**, 1957 (1963).
199. C. R. Johnson, *J. Amer. Chem. Soc.*, **85**, 1020 (1963).
200. C. R. Johnson and D. McCants, Jr., *J. Amer. Chem. Soc.*, **87**, 5404 (1965).
201. J. J. Rigau, C. C. Bacon and C. R. Johnson, *J. Org. Chem.*, **35**, 3655 (1970).
202. K. K. Andersen, S. Colonna and C. J. M. Stirling, *J. Chem. Soc. Chem. Commun.*, 645 (1973).
203. K. K. Andersen, M. Cinquini, S. Colonna and F. L. Pilar, *J. Org. Chem.*, **40**, 3780 (1975).
204. M. Kishi and T. Komeno, *Int. J. Sulfur Chem.*, *A*, **2**, 1 (1972).
205. K. K. Andersen, R. L. Caret and I. Karup-Nielsen, *J. Amer. Chem. Soc.*, **96**, 8026 (1974).
206. K. K. Andrsen, *J. Chem. Soc. Chem. Commun.*, 1051 (1971).
207. K. K. Andersen, R. L. Caret and D. L. Ladd., *J. Org. Chem.*, **41**, 3096 (1976).
208. K. K. Andersen and N. E. Papanikolaou, *Tetrahedron Lett.*, 5445 (1966).
209. K. K. Andersen, M. Cinquini and N. E. Papanikolaou, *J. Org. Chem.*, **35**, 706 (1970).
210. S. Oae and Y. H. Khim, *Bull. Chem. Soc. Jap.*, **42**, 3528 (1969).
211. K. Tsumori, H. Minato and M. Kobayashi, *Bull. Chem. Soc. Jap.*, **46**, 3503 (1973).
212. C. R. Johnson and W. G. Phillips, *J. Org. Chem.*, **32**, 3233 (1967).
213. K. K. Andersen, W. H. Edmonds, J. B. Biasotti and R. A. Strecker, *J. Org. Chem.*, **31**, 2859 (1966).
214. P. Haake and R. D. Cook, *Tetrahedron Lett.*, 427 (1968).
215. S. Oae, K. Sakai and N. Kunieda, *Bull. Chem. Soc. Jap.*, **42**, 1964 (1969).
216. D. Landini, G. Modena, G. Scorrano and F. Taddei, *J. Amer. Chem. Soc.*, **91**, 6703 (1969).
217. U. Quintily and G. Scorrano, *J. Chem. Soc. Chem. Commun.*, 260 (1971).
218. R. Curci, F. Di Furia, A. Levi, V. Lucchini and G. Scorrano, *J. Chem. Soc. Perkin Trans. II*, 341 (1975).
219. R. P. Fedoezzhina, E. P. Buchikhin, E. A. Kanevskii and A. I. Zarubin, *Zh. Obshch. Khim.*, **44**, 877, 1351 (1974); *Chem. Abstr.*, **81**, 12897h, 62982c (1974).
220. G. A. Olah, A. T. Ku and J. A. Olah, *J. Org. Chem.*, **35**, 3904 (1970).
221. G. Gatti, A. Levi, V. Lucchini, G. Modena and G. Scorrano, *J. Chem. Soc. Chem. Commun.*, 251 (1973); see also Q Appleton, L. Bernander and G. Olofsson, *Tetrahedron*, **27**, 5921 (1971).
222. G. A. Olah, D. J. Donovan, H. C. Lin, H. Mayr, P. Andreozzi and G. Klopman, *J. Org. Chem.*, **43**, 2268 (1978).
223. G. Modena, *Int. J. Sulfur Chem.*, *C*, **7**, 95 (1972).
224. G. Scorrano, *Acc. Chem. Res.*, **6**, 132 (1973).
225. R. J. Gillespie and R. C. Passerini, *J. Chem. Soc.*, 3850 (1956).

226. S. Oae, T. Kitao and Y. Kitaoka, *Chem. Ind. (London)*, 291 (1961).
227. S. Oae, T. Kitao and Y. Kitaoka, *Bull. Chem. Soc. Jap.*, **38**, 543 (1965).
228. S. Oae, T. Kitao, Y. Kitaoka and S. Kawamura, *Bull. Chem. Soc. Jap.*, **38**, 546 (1965).
229. S. Oae and N. Kunieda, *Bull. Chem. Soc. Jap.*, **41**, 696 (1968).
230. N. Kunieda and S. Oae, *Bull. Chem. Soc. Jap.*, **46**, 1745 (1973).
231. S. Oae, M. Moriyama, T. Numata and N. Kunieda, *Bull. Chem. Soc. Jap.*, **47**, 179 (1974).
232. N. Kunieda and S. Oae, *Bull. Chem. Soc. Jap.*, **42**, 1324 (1969).
233. N. Kunieda and S. Oae, *Bull. Chem. Soc. Jap.*, **41**, 1025 (1968).
234. S. Oae, M. Yokoyama and M. Kise, *Bull. Chem. Soc. Jap.*, **41**, 1221 (1968).
235. N. Kunieda, K. Sakai and S. Oae, *Bull. Chem. Soc. Jap.*, **42**, 1090 (1969).
236. C. C. Addison and J. C. Sheldon, *J. Chem. Soc.*, 2705 (1956).
237. G. Modena, U. Quintily and G. Scorrano, *J. Amer. Chem. Soc.*, **94**, 202 (1972).
238. P. Bonvicini, A. Levi and G. Scorrano, *Gazz. Chim. Ital*, **104**, 1 (1974); *Chem. Abstr.*, **81**, 36968h (1974).
239. K. Mislow, T. Simmons, J. T. Melilio and A. L. Ternay, Jr., *J. Amer. Chem. Soc.*, **86**, 1452 (1964).
240. H. Yoshida, T. Numata and S. Oae, *Bull. Chem. Soc. Jap.*, **44**, 2875 (1971).
241. I. Ookuni and A. Fry, *J. Org. Chem.*, **36**, 4097 (1971).
242. W. Tagaki, K. Kikukawa, N. Kunieda and S. Oae, *Bull. Chem. Soc. Jap.*, **39**, 614 (1966).
243. D. Landini, F. Montanari, G. Modena and G. Scorrano, *J. Chem. Soc. Chem. Commun.*, 86 (1968); 3 (1969).
244. D. Landini, G. Modena, F. Montanari and G. Scorrano, *J. Amer. Chem. Soc.*, **92**, 7168 (1970).
245. N. Kunieda, T. Numata and S. Oae, *Phosphorus Sulfur*, **3**, 1 (1977).
246. H. Kwart and H. Omura, *J. Amer. Chem. Soc.*, **93**, 7250 (1971).
247. E. Casadevall and M. M. Bouisset, *Tetradedron Lett.*, 2023 (1975).
248. G. E. Hartzell and J. N. Paige, *J. Amer. Chem. Soc.*, **88**, 2616 (1966).
249. G. E. Manser, A. D. Mesure and J. G. Tillett, *Tetrahedron Lett.*, 3153 (1968).
250. K. Kondo, A. Negishi and I. Ojima, *J. Amer. Chem. Soc.*, **94**, 5786 (1972).
251. K. Kondo, A. Negishi and G. Tsuchihashi, *Tetrahedron Lett.*, 3173 (1969).
252. M. Kishi, S. Ishihara and T. Komeno, *Tetrahedron*, **30**, 2135 (1974).
253. Th. Zincke and J. Baeumer, *Justus Liebigs Ann. Chem.*, **416**, 86 (1918).
254. E. N. Karaulova and G. D. Gal'pern, *Zh. Obshch. Khim.*, **29**, 3033 (1959); *Chem. Abstr.*, **54**, 12096d (1960).
255. D. Landini, F. Montanari, H. Hogeveen and G. Maccagnani, *Tetrahedron Lett.*, 2691 (1964).
256. R. A. Strecker and K. K. Andersen, *J. Org. Chem.*, **33**, 2234 (1968).
257. G. Modena, G. Scorrano, D. Landini and F. Montanari, *Tetrahedron Lett.*, 3309 (1966).
258. J. H. Krueger, *Inorg. Chem.*, **5**, 132 (1966).
259. D. Landini, G. Modena, U. Quintily and G. Scorrano, *J. Chem. Soc. B*, 2041 (1971).
260. H. Hogeveen and F. Montanari, *Gazz. Chim. Ital.*, **94**, 176 (1964); *Chem. Abstr.*, **61**, 3691g (1964).
261. S. Allenmark, *Acta Chem. Scand.*, **20**, 910 (1966).
262. L. Sagramora, A. Garbesi and A. Fava, *Helv. Chim. Acta*, **55**, 675 (1972).
263. S. Tamagaki, M. Mizuno, H. Yoshida, H. Hirota and S. Oae, *Bull. Chem. Soc. Jap.*, **44**, 2456 (1971).
264. R. Curci, F. Di Furia, A. Levi and G. Scorrano, *J. Chem. Soc. Perkin Trans. II*, 408 (1975).
265. M. Gazdar and S. Smiles, *J. Chem. Soc.*, **97**, 2248 (1910).
266. K. Fries and W. Vogt, *Chem. Ber.*, **44**, 756 (1911).
267. H. Gilman and D. R. Swayampati, *J. Amer. Chem. Soc.*, **77**, 5944 (1955).
268. T. Aida, N. Furukawa and S. Oae, *Tetrahedron Lett.*, 3853 (1973).
269. M. Mikołajczyk and M. Para, *Bull. Acad. Pol. Sci. Ser. Sci. Chim.*, **16**, 295 (1968); *Chem. Abstr.*, **69**, 106818n (1968).
270. S. Oae, T. Yagihara and T. Okabe, *Tetrahedron*, **28**, 3203 (1972).
271. M. Mikołajczyk, *Chem. Ind. (London)*, 2059 (1966).
272. M. Mikołajczyk, *Angew. Chem.*, **78**, 393 (1966).

274. A. Nakanishi and S. Oae, *Chem. Ind. (London)*, 960 (1971).
274. S. Oae, A. Nakanishi and N. Tsujimoto, *Tetrahedron*, **28**, 2981 (1972).
275. M. Mikołajczyk and M. Para, *J. Chem. Soc. Commun.*, 1192 (1969).
276. A. Senning, *J. Chem. Soc. Chem. Commun.*, 64 (1967).
277. T. Numata, K. Ikura, Y. Shimano and S. Oae, *Org. Prep. Proced. Int.*, **8**, 119 (1976).
278. N. P. Volynskii, G. D. Gal'pern and V. V. Smolyaninov, *Neftekhimiya*, **1**, 473 (1961); *Chem. Abstr.*, **57**, 16510h (1962).
279. J. S. Grossert, W. R. Handstaff and R. F. Langler, *Can. J. Chem.*, **55**, 421 (1977).
280. J. Drabowicz and S. Oae, *Synthesis*, 404 (1977).
281. J. Drabowicz and S. Oae, *Chem. Lett.*, 767 (1977).
282. R. Tanikaga, K. Nakayama, K. Tanaka and A. Kaji, *Chem. Lett.*, 395 (1977).
283. T. Numata and S. Oae, *Chem. Ind. (London)*, 277 (1973).
284. R. Michelot and B. Tchoubar, *Bull. Soc. Chim. Fr.*, 3039 (1966).
285. T. M. Balthazor and J. C. Martin, *J. Amer. Chem. Soc.*, **97**, 5634 (1975).
286. P. Huszthy, I. Kapovits and A. Kucsman, *Tetrahedron Lett.*, 1853 (1978).
287. S. Oae and M. Kise, *Tetrahedron Lett.*, 1409 (1967).
288. S. Oae and M. Kise, *Bull. Chem. Soc. Jap.*, **43**, 1416 (1970).
289. M. Kise and S. Oae, *Bull. Chem. Soc. Jap.*, **43**, 1804 (1970).
290. E. Jonsson, *Acta Chem. Scand.*, **21**, 1277 (1967).
291. E. Jonsson, *Tetrahedron Lett.*, 3675 (1967).
292. J. Drabowicz and S. Oae, *Tetrahedron*, **34**, 63 (1978).,
293. M. Mikołajczyk, B. Bujnicki and J. Drabowicz, *Bull. Acad. Pol. Sci. Ser. Sci. Chim.*, **25**, 267 (1977); *Chem. Abstr.*, **87**, 167423y (1977).
294. M. Haake and H. Benack, *Synthesis*, 310 (1976).
295. K. H. Schlingensief and K. Hartke, *Tetrahedron Lett.*, 1269 (1977).
296. D. S. Tarbell and C. Weaver, *J. Amer. Chem. Soc.*, **63**, 2939 (1941).
297. G. Kresze and B. Wustrow, *Chem. Ber.*, **95**, 2652 (1962).
298. I. Kapovits, F. Ruff and A. Kucsman, *Tetrahedron*, **28**, 4405 (1972).
299. H. J. Shine and K. Kim, *Tetrahdron Lett.*, 99 (1974).
300. J. H. Krueger, *J. Amer. Chem. Soc.*, **91**, 4974 (1969).
301. J. H. Krueger and R. J. Kiyokane, *Inorg. Chem.*, **13**, 2522 (1974).
302. C. Dell'Erba, G. Guanti, G. Leandri and G. Poluzzi Corallo, *Int. J. Sulfur Chem.*, **3**, 261 (1973).
303. G. Guanti, G. Garbarino, C. Dell'Erba and G. Leandri, *Gazz. Chim. Ital.*, **105**, 849 (1975); *Chem. Abstr.*, **84**, 42794b (1976).
304. S. Oae, T. Aida and N. Furukawa, *Chem. Parm. Bull.*, **23**, 3011 (1975).
305. B. H. Nicolet and I. D. Willard, *Science*, **53**, 217 (1921).
306. F. G. Mann and W. J. Pope, *J. Chem. Soc.*, **121**, 1052 (1922).
307. A. Kucsman, F. Ruff, I. Kapovits and J. G. Fischer, *Tetrahedron*, **22**, 1843 (1966).
308. K. Tsujihara, N. Furukawa, K. Oae and S. Oae, *Bull. Chem. Soc. Jap.*, **42**, 2631 (1969).
309. F. Ruff and A. Kucsman, *J. Chem. Soc. Perkin Trans. II*, 509 (1975).
310. F. Ruff, K. Komoto, N. Furukawa and S. Oae, *Tetrahedron*, **32**, 2763 (1976).
311. S. Allenmark, *Acta Chem. Scand.*, **19**, 1, 2075 (1965).
312. S. Allenmark and H. Johnsson, *Acta Chem. Scand.*, **21**, 1672 (1967); **23**, 2902 (1969).
313. S. Allenmark and C.-E. Hagberg, *Acta Chem. Scand.*, **22**, 1461, 1694 (1968); **24**, 2225 (1970).
314. H. Johnsson and S. Allenmark, *Chem. Scr.*, **8**, 216, 223 (1975); *Chem. Abstr.*, **84**, 104763b, 120817c (1976).
315. C. E. Hagberg and S. Allenmark, *Chem. Scr.*, **5**, 13 (1974); *Chem. Abstr.*, **80**, 132521g (1974).
316. T. Numata, K. Sakai, M. Kise, N. Kunieda and S. Oae, *Chem. Ind. (London)*, 576 (1971).
317. T. Numata, K. Sakai, M. Kise, N. Kunieda and S. Oae, *Int. J. Sulfur Chem.*, A, **1**, 1 (1971).
318. D. Landini, F. Rolla and G. Torre, *Int. J. Sulfur Chem.*, A, **2**, 43 (1972).
319. D. Landini and F. Rolla, *J. Chem. Soc. Perkin Trans. II*, 1317 (1972).
320. D. Landini, A. M. Maia and F. Rolla, *J. Chem. Soc. Perkin Trans. II*, 1288 (1976).

321. W. Tagaki, M. Ochiai and S. Oae, *Tetrahedron Lett.*, 6131 (1968).
322. T. Numata and S. Oae, *Int. J. Sulfur Chem., A*, **1**, 6 (1971).
323. J. T. Doi and W. K. Musker, *J. Amer. Chem. Soc.*, **100**, 3533 (1978).
324. J. C. Martin and J. J. Uebel, *J. Amer. Chem. Soc.*, **86**, 2936 (1964).
325. H. Hogeveen, G. Maccagnani and F. Montanari, *J. Chem. Soc. C*, 1585 (1966).
326. M. Cinquini, S. Colonna and F. Montanari, *J. Chem. Soc. C*, 1213 (1967).
327. M. Cinquini, S. Colonna and F. Montanari, *J. Chem. Soc. C*, 572 (1970).
328. K. W. Buck, A. B. Foster, A. R. Perry and J. M. Webber, *J. Chem. Soc. Chem. Commun.*, 433 (1965).
329. D. N. Jones, M. J. Green, M. A. Saeed and R. D. Whitehouse, *J. Chem. Soc. C*, 1362 (1968).
330. F. Montanari, R. Danieli, H. Hogeveen and G. Maccagnani, *Tetrahedron Lett.*, 2685 (1964).
331. M. Cinquini, S. Colonna and F. Montanari, *Tetrahedron Lett.*, 3181 (1966).
332. M. Cinquini and S. Colonna, *Boll. Sci. Fac. Chim. Ind. Bologna*, **27**, 157 (1969); *Chem. Abstr.*, **72**, 120734a (1970).
333. S. Oae, *J. Amer. Chem. Soc.*, **78**, 4030 (1956).
334. G. Barbieri, M. Cinquini and S. Colonna, *Boll. Sci. Fac. Chim. Ind. Bologna*, **26**, 309 (1968); *Chem. Abstr.*, **70**, 105593u (1969).
335. J. Klein and H. Stollar, *Tetrahedron*, **30**, 2541 (1974).
336. O. Bohman and S. Allenmark, *Tetrahedron Lett.*, 405 (1973); *Chem. Scr.*, **4**, 202 (1973); *Chem. Abstr.*, **80**, 69970k (1974).
337. J. L. Kice and N. A. Favstritsky, *J. Amer. Chem. Soc.*, **91**, 1751 (1969).
338. S. H. Smallcombe and M. C. Caserio, *J. Amer. Chem. Soc.*, **93**, 5826 (1971).
339. H. Minato, T. Miura, F. Takagi and M. Kobayashi, *Chem. Lett.*, 211 (1975).
340. J. K. Kim and M. C. Caserio, *J. Amer. Chem. Soc.*, **96**, 1930 (1974).
341. H. Minato, T. Miura and M. Kobayashi, *Chem. Lett.*, 701 (1975).
342. P. Dubs and R. Stuessi, *Helv. Chim. Acta*, **59**, 1307 (1976).
343. G. K. Helmkamp, B. A. Olsen and D. J. Pettitt, *J. Org. Chem.*, **30**, 676 (1965).
344. G. K. Helmkamp, B. A. Olsen and J. R. Koskinen, *J. Org. Chem.*, **30**, 1623 (1965).
345. N. E. Hester, G. K. Helmkamp and G. L. Alford, *Int. J. Sulfur Chem., A*, **1**, 65 (1971).
346. G. Capozzi, O. De Lucchi, V. Lucchini and G. Modena, *Tetrahedron Lett.*, 2603 (1975).
347. G. Capozzi, O. De Lucchi, V. Lucchini and G. Modena, *J. Chem. Soc. Chem. Commun.*, 248 (1975).
348. G. Capozzi, V. Lucchini, G. Modena and P. Scrimin, *Tetrahedron Lett.*, 911 (1977).
349. R. Weiss, C. Schlierf and K. Schloter, *J. Amer. Chem. Soc.*, **98**, 4668 (1976).
350. N. Furukawa, S. Oae and T. Yoshimura, *Synthesis*, 30 (1976).
351. T. Yoshimura, T. Akasaka, N. Furukawa and S. Oae, *Heterocycles*, **7**, 287 (1977).
352. Y. Tamura, H. Matsushima, M. Ikeda and K. Sumoto, *Tetrahedron*, **32**, 431 (1976).
353. M. Edwards, T. L. Gilchrist and C. W. Rees, unpublished results cited in T. L. Gilchrist and C. J. Moody, *Chem. Rev.*, **77**, 409 (1977).
354. Y. Tamura, K. Sumoto, H. Matsushima, H. Taniguchi and M. Ikeda, *J. Org. Chem.*, **38**, 4324 (1973).
355. R. Annunziata, M. Cinquini and S. Colonna, *J. Chem. Soc. Perkin Trans. I*, 404 (1975).
356. R. Annunziata, M. Cinquini and S. Colonna, *J. Chem. Soc. Perkin Trans. I*, 1231 (1973).
357. E. M. Arnett and V. M. DePalma, *J. Amer. Chem. Soc.*, **99**, 5828 (1977).
358. J. C. Martin, J. A. Franz and R. J. Arhart, *J. Amer. Chem. Soc.*, **96**, 4604 (1974).
359. V. J. Traynelis, W. L. Hergenrother, H. T. Hanson and J. A. Valicenti, *J. Org. Chem.*, **29**, 123 (1964).
360. L. Horner and P. Kaiser, *Justus Liebigs Ann. Chem.*, **626**, 19 (1959).
361. R. Pummerer, *Chem. Ber.*, **43**, 1401 (1910).
362. L. Horner, *Justus Liebigs Ann. Chem.*, **631**, 198 (1960).
363. S. Oae, T. Kitao, S. Kawamura and Y. Kitaoka, *Tetrahedron*, **19**, 817 (1963).
364. C. R. Johnson, J. C. Sharp and W. G. Phillips, *Tetrahedron Lett.*, 5299 (1967).
365. C. R. Johnson and W. G. Phillips, *J. Amer. Chem. Soc.*, **91**, 682 (1969).
366. S. Oae, T. Kitao and S. Kawamura, *Tetrahedron*, **19**, 1783 (1963).
367. S. Iwanami, S. Arita and K. Takeshita, *Yuki Goseikagaku Kyokaishi*, **26**, 375 (1968).

368. T. Numata and S. Oae, *Chem. Ind. (London)*, 726 (1972).
369. S. Oae and T. Numata, *Tetrahedron*, **30**, 2641 (1974).
370. T. Numata and S. Oae, *Tetrahedron*, **32**, 2699 (1976).
371. S. Oae and M. Kise, *Tetrahedron Lett.*, 2261 (1968).
372. M. Kise and S. Oae, *Bull. Chem. Soc. Jap.*, **43**, 1426 (1970).
373. Y. Okamoto, R. Homsany and T. Yano, *Tetrahedron Lett.*, 2529 (1972).
374. Y. Okamoto, K. L. Chellappa and R. Homsany, *J. Org. Chem.*, **38**, 3172 (1973).
375. S. Oae and N. Furukawa, *Tetrahedron*, **33**, 2359 (1977).
376. E. Vilsmaier, *7th Int. Symp. Organosulfur Chem., Hamburg (1976)*.
377. G. E. Wilson, Jr., and C. J. Strong, *J. Org. Chem.*, **37**, 2376 (1972).
378. M. Cocivera, V. Malateska, K. W. Woo and A. Effio, *J. Org. Chem.*, **43**, 1140 (1978).
379. D. A. Davenport, D. B. Moss, J. E. Rhodes and J. A. Walsh, *J. Org. Chem.*, **34**, 3353 (1969).
380. T. Numata and S. Oae, unpublished results.
381. D. N. Jones, E. Helmy and R. D. Whitehouse, *J. Chem. Soc. Perkin Trans. I*, 1329 (1972).
382. T. Numata and S. Oae, *Tetrahedron Lett.*, 1337 (1977).
383. H. Kobayashi, N. Furukawa, T. Aida, K. Tsujihara and S. Oae, *Tetrahedron Lett.*, 3109 (1971).
384. N. Furukawa, T. Masuda, M. Yakushiji and S. Oae, *Bull. Chem. Soc. Jap.*, **47**, 2247 (1974).
385. N. Furukawa, S. Oae and T. Masuda, *Chem. Ind. (London)*, 396 (1975).
386. N. Furukawa, S. Oae and T. Yoshimura, *Phosphorus Sulfur*, **3**, 277 (1977).
387. T. Yagihara and S. Oae, *Tetrahedron*, **28**, 2759 (1972).
388. B. Stridsberg and S. Allenmark, *Acta Chem. Scand., B*, **28**, 591 (1974).
389. B. Stridsberg and S. Allenmark, *Acta Chem. Scand., B*, **30**, 219 (1976).
390. S. Oae and K. Ogino, *Heterocycles*, **6**, 583 (1977).
391. T. Numata, O. Itoh and S. Oae, *Chem. Lett.*, 909 (1977).
392. T. Numata, O. Itoh and S. Oae, *Tetrahedron Lett.*, 1869 (1979).
393. M. Mikołajczyk, A. Zatorski, S. Grzejszczak, B. Costisella and W. Midura, *J. Org. Chem.*, **43**, 2518 (1978).
394. T. Masuda, T. Numata, N. Furukawa and S. Oae, *Chem. Lett.*, 903 (1977).
395. T. Masuda, T. Numata, N. Furukawa and S. Oae, *J. Chem. Soc. Perkin Trans. II*, 1302 (1978).
396. U. Schöllkopf, E. Ruban, P. Tonne and K. Riedel, *Tetrahedron Lett.*, 5077 (1970).
397. S. Glue, I. T. Kay and M. R. Kipps, *J. Chem. Soc. Chem. Commun.*, 1158 (1970).
398. D. H. Bremner and M. M. Campbell, *J. Chem. Soc. Chem. Commun.*, 538 (1976).
399. D. H. Bremner and M. M. Campbell, *J. Chem. Soc. Perkin Trans. I*, 2298 (1977).
400. J. E. McCormick and R. S. McElhinney, *J. Chem. Soc. Chem. Commun.*, 171 (1969).
401. J. E. McCormick and R. S. McElhinney, *J. Chem. Soc. Perkin Trans. I*, 2533 (1976).
402. A. G. Brook and D. G. Anderson, *Can. J. Chem.*, **46**, 2115 (1968).
403. F. A. Carey and O. Hernandez, *J. Org. Chem.*, **38**, 2670 (1973).
404. F. A. Carey, O. D. Dailey, Jr., O. Hernandez and J. R. Tucker, *J. Org. Chem.*, **41**, 3975 (1976).
405. E. Vedejs and M. Mullins, *Tetrahedron Lett.*, 2017 (1975).
406. E. Vilsmaier, K. H. Dittrich and W. Sprügel, *Tetrahedron Lett.*, 3601 (1974).
407. E. Vilsmaier and R. Bayer, *Synthesis*, 46 (1976).
408. P. G. Gassman and R. J. Balchunis, *Tetrahedron Lett.*, 2235 (1977).
409. E. Vilsmaier, R. Bayer, I. Laengenfelder and U. Welz, *Chem. Ber.*, **111**, 1136 (1978).
410. E. Vilsmaier, R. Bayer, U. Welz and K.-H. Dittrich, *Chem. Ber.*, **111**, 1147 (1970).
411. K. W. Ratts and A. N. Yao, *J. Org. Chem.*, **33**, 70 (1968).
412. A. Terada and Y. Kishida, *Chem. Pharm. Bull*, **18**, 505 (1970).
413. T. Numata, O. Itoh and S. Oae, unpublished results.
414. T. Masuda, T. Numata, N. Furukawa and S. Oae, *Chem. Lett.*, 745 (1977).
415. D. K. Bates, *J. Org. Chem.*, **42**, 3452 (1977).
416. Y. Hiraki, M. Kamiya, R. Tanikaga, N. Ono and A. Kaji, *Bull. Chem. Soc. Jap.*, **50**, 447 (1977).

417. R. A. Olofson and J. P. Marino, *Tetrahedron*, **27**, 4195 (1971).
418. W. E. Parham and M. D. Bhavsar, *J. Org. Chem.*, **28**, 2686 (1963). cf. W. E. Parham and L. D. Edwards, *J. Org. Chem.*, **33**, 4150 (1968).
419. N. Kunieda, Y. Fujiwara and M. Kinoshita, *37th Annual Meeting of the Japanese Chemical Society*, Abs. 3B40 (1978).
420. T. Numata, O. Itoh and S. Oae, *Tetrahedron Lett.*, 161 (1979).
421. G. A. Koppel and L. J. McShane, *J. Amer. Chem. Soc.*, **100**, 288 (1978).
422. Y. Tamura, Y. Nishikawa, K. Sumoto, M. Ikeda, M. Murase and M. Kise, *J. Org. Chem.*, **42**, 3226 (1977).
423. C. A. Maryanoff, K. S. Hayes and K. Mislow, *J. Amer. Chem. Soc.*, **99**, 4412 (1977).
424. H. Böhme, H. Fisher and R. Frank, *Justus Liebigs Ann. Chem.*, **563**, 54 (1949).
425. H. Böhme and H.-J. Gran, *Justus Liebigs Ann. Chem.*, **577**, 68 (1952).
426. G. E. Wilson, Jr. and R. Albert, *J. Org. Chem.*, **38**, 2160 (1973).
427. F. Boberg, *Justus Liebigs Ann. Chem.*, **679**, 107 (1964).
428. F. Boberg, G. Winter and G. R. Schultze, *Chem. Ber.*, **89**, 1160 (1956).
429. F. G. Bordwell and B. M. Pitt, *J. Amer. Chem. Soc.*, **77**, 572 (1955).
430. W. E. Truce, G. H. Birum and E. T. McBee, *J. Amer. Chem. Soc.*, **74**, 3594 (1952).
431. H. Böhme and H.-J. Gran, *Justus Liebigs Ann. Chem.*, **581**, 133 (1953).
432. L. A. Paquette, *J. Amer. Chem. Soc.*, **86**, 4085 (1964).
433. L. A. Paquette, L. S. Wittenbrook and K. Schreiber, *J. Org. Chem.*, **33**, 1080 (1968).
434. E. Vilsmaier and W. Sprügel, *Justus Liebigs Ann. Chem.*, **749**, 62 (1971).
435. E. Vilsmaier and W. Sprügel, *Justus Liebigs Ann. Chem.*, **747**, 151 (1971).
436. T. Masuda, N. Furukawa and S. Oae, *Chem. Lett.*, 1103 (1977).
437. D. L. Tuleen and T. B. Stephens, *Chem. Ind. (London)*, 1555 (1966).
438. D. L. Tuleen and V. C. Marcum, *J. Org. Chem.*, **32**, 204 (1967).
439. D. L. Tuleen and D. N. Buchanan, *J. Org. Chem.*, **32**, 495 (1967).
440. D. L. Tuleen, *J. Org. Chem.*, **32**, 4006 (1967).
441. D. L. Tuleen and T. B. Stephens, *J. Org. Chem.*, **34**, 31 (1969).
442. W. H. Koster, J. E. Dolfini, B. Toeplitz and J. Z. Gougoutas, *J. Org. Chem.*, **43**, 79 (1978).
443. W. Tagaki, K. Kikukawa, K. Ando and S. Oae, *Chem. Ind. (London)*, 1624 (1964).
444. C. G. Kruse, N. L. J. M. Broekhof, A. Wijsman and A. van der Gen, *Tetrahedron Lett.*, 885 (1977).
445. A. Lüttringhaus and N. Engelhard, *Chem. Ber.*, **93**, 1525 (1960).
446. L. A. Paquette, R. E. Wingard, Jr, J. C. Philips, G. L. Thompson, L. K. Read and J. Clardy, *J. Amer. Chem. Soc.*, **93**, 4508 (1971).
447. R. H. Rynbrandt, *Tetrahedron Lett.*, 3553 (1971).
448. E. H. Amonoo-Neizer, S. K. Ray, R. A. Shaw and B. C. Smith, *J. Chem. Soc.*, 6250 (1965).
449. R. Oda and Y. Hayashi, *Tetrahedron Lett.*, 2181, 3141 (1967).
450. M. Oki and K. Kobayashi, *Bull. Chem. Soc. Jap.*, **43**, 1223 (1970).
451. G. A. Russell and E. T. Sabourin, *J. Org. Chem.*, **34**, 2336 (1969).
452. R. E. Boyle, *J. Org. Chem.*, **31**, 3880 (1966).
453. M. Hojo and Z. Yoshida, *J. Amer. Chem. Soc.*, **90**, 4496 (1968).
454. F. Loth and A. Michaelis, *Chem. Ber.*, **27**, 2540 (1894).
455. C. W. Bird, *J. Chem. Soc. C*, 1230 (1968).
456. M. F. Lappert and J. K. Smith, *J. Chem. Soc.*, 3224 (1961).
457. T. H. Chan, A. Meilnyk and D. N. Harpp, *Tetrahedron Lett.*, 201 (1969).
458. K. Naumann, G. Zon and K. Mislow, *J. Amer. Chem. Soc.*, **91**, 2788, 7012 (1969).
459. A. Michaelis and B. Godchaux, *Chem. Ber.*, **24**, 757 (1891).
460. G. E. Wilson, Jr., *J. Amer. Chem. Soc.*, **87**, 3785 (1965).
461. F. Chioccara, G. Prota, R. A. Nicolaus and E. Novellino, *Synthesis*, 876 (1977); see also F. Chioccara, V. Mangiacapra, E. Novellino and G. Prota, *J. Chem. Soc. Chem. Commun.*, 863 (1977).
462. S. Oae, T. Yagihara and T. Okabe, *Tetrahedron*, **28**, 3203 (1972).
463. H. Kosugi, H. Uda and S. Yamagiwa, *J. Chem. Soc. Chem. Commun.*, 71 (1976).
464. L. S. S. Reamonn and W. I. O'Sullivan, *J. Chem. Soc. Chem. Commun.*, 642 (1976).

465. R. R. King, R. Greenhalgh and W. D. Marshall, *J. Org. Chem.,* **43**, 1262 (1978).
466. M. G. Burdon and J. G. Moffatt, *J. Amer. Chem. Soc.,* **87**, 4656 (1965).
467. K. E. Pfitzner, J. P. Marino and R. A. Olofson, *J. Amer. Chem. Soc.,* **87**, 4658 (1965).
468. M. G. Burdon and J. G. Moffatt, *J. Amer. Chem. Soc.,* **88**, 5855 (1966).
469. J. P. Marino, K. E. Pfitzner and R. A. Olofson, *Tetrahedron,* **27**, 4181 (1971).
470. M. G. Burdon and J. G. Moffatt, *J. Amer. Chem. Soc.,* **89**, 4725 (1967).
471. Y. Hayashi and R. Oda, *J. Org. Chem.,* **32**, 457 (1967).
472. G. R. Pettit and T. H. Brown, *Can. J. Chem.,* **45**, 1306 (1967).
473. P. Claus, *Monatsh. Chem.,* **99**, 1034 (1968); *Chem. Abstr.,* **69**, 18768r (1968).
474. P. Claus, *Monatsh. Chem.,* **102**, 913 (1971); *Chem. Abstr.,* **75**, 88242n (1971).
475. P. Claus, N. Vavra and P. Schilling, *Monatsh. Chem.,* **102**, 1072 (1971); *Chem. Abstr.,* **75**, 129439c (1971).
476. J. Doucet and A. Robert, *C. R. Acad. Sci. Ser. C,* **272**, 1562 (1971); *Chem. Abstr.,* **75**, 34781u (1971).
477. P. G. Gassman and D. R. Amick, *Tetrahedron Lett.,* 889 (1974).
478. M. E. C. Biffin and D. B. Paul, *Aust. J. Chem.,* **27**, 277 (1974).
479. E. Goethals and P. de Radzitzky, *Bull. Soc. Chim. Belg.* **73**, 546 (1964); *Chem. Abstr.,* **61**, 10614a (1964).
480. P. Claus and W. Rieder, *Tetrahedron Lett.,* 3879 (1972).
481. P. G. Gassman and G. Gruetzmacher, *J. Amer. Chem. Soc.,* **95**, 588 (1973).
482. P. G. Gassman and G. Gruetzmacher, *J. Amer. Chem. Soc.,* **96**, 5487 (1974).
483. P. G. Gassman and T. J. van Bergen, *J. Amer. Chem. Soc.,* **95**, 590 (1973).
484. P. G. Gassman and T. J. van Bergen, *J. Amer. Chem. Soc.,* **95**, 591 (1973).
485. P. G. Gassman, D. P. Gilbert and T. J. van Bergen, *J. Chem. Soc. Chem. Commun.,* 201 (1974).
486. P. G. Gassman, T. J. van Bergen, D. P. Gilbert and B. W. Cue, Jr., *J. Amer. Chem. Soc.,* **96**, 5495 (1974).
487. P. G. Gassmana and T. J. van Bergen, *J. Amer. Chem. Soc.,* **95**, 2718 (1973).
488. H. Nakai and M. Kurono, *Chem. Lett.,* 995 (1977).
489. P. G. Gassman and T. J. van Bergen, *J. Amer. Chem. Soc.,* **96**, 5508 (1974).
490. P. G. Gassman and R. L. Parton, *J. Chem. Soc. Chem. Commun.,* 694 (1977).
491. P. G. Gassman and C. T. Huang, *J. Amer. Chem. Soc.,* **95**, 4453 (1973).
492. P. G. Gassman, T. J. van Bergen and G. Gruetzmacher, *J. Amer. Chem. Soc.,* **95**, 6508 (1973).
493. P. G. Gassman, G. Gruetzmacher and T. J. van Bergen, *J. Amer. Chem. Soc.,* **96**, 5512 (1974).
494. P. G. Gassman and D. R. Amick, *Synth. Commun.,* **5**, 325 (1975).
495. P. G. Gassman and H. R. Drewes, *J. Amer. Chem. Soc.,* **96**, 3002 (1974).
496. P. G. Gassman and D. R. Amick, *Tetrahedron Lett.,* 3463 (1974).
497. P. G. Gassman and R. L. Parton, *Tetrahedron Lett.,* 2055 (1977).
498. P. Claus and W. Vycudilik, *Tetrahedron Lett.,* 3607 (1968).
499. P. Claus and W. Rieder, *Monatsh. Chem.,* **103**, 1163 (1972); *Chem. Abstr.,* **78**, 15188b (1973).
500. P. K. Claus, W. Rieder and F. W. Vierhapper, *Tetrahedron Lett.,* 1335 (1976).
501. P. Claus, W. Vycudilik and W. Rieder, *Monatsh. Chem.,* **102**, 1571 (1971); *Chem. Abstr.,* **76**, 13539b (1972).
502. P. Claus and W. Vycudilik, *Monatsh. Chem.,* **101**, 396 (1970); *Chem. Abstr.,* **72**, 132225y (1970).
503. P. K. Claus, H. A. Schwarz, W. Rieder and W. Vycudilik, *Phosphorus Sulfur,* **1**, 11 (1976).
504. H. R. Nace and J. J. Monagle, *J. Org. Chem.,* **24**, 1792 (1959).
505. I. M. Hunsberger and J. M. Tien, *Chem. Ind. (London),* 88 (1959).
506. N. Kornblum, J. W. Powers, G. J. Anderson, W. J. Jones, H. O. Larson, O. Levand and W. M. Weaver, *J. Amer. Chem. Soc.,* **79**, 6562 (1957).
507. N. Kornblum, W. J. Jones and G. J. Anderson, *J. Amer. Chem. Soc.,* **81**, 4113 (1959).
508. A. P. Johnson and A. Pelter, *J. Chem. Soc.,* 520 (1964).
509. R. T. Major and H.-J. Hess, *J. Org. Chem.,* **23**, 1563 (1958).

510. D. N. Jones and M. A. Saeed, *J. Chem. Soc.*, 4657 (1963).
511. R. N. Iacona, A. T. Rowland and H. R. Nace, *J. Org. Chem.*, **29**, 3495 (1964).
512. H. R. Nace and R. N. Iacona, *J. Org. Chem.*, **29**, 3498 (1964).
513. M. M. Baizer, *J. Org. Chem.*, **25**, 670 (1960).
514. M. M. Robison, W. G. Pierson, R. A. Lucas, I. Hsu and R. L. Dziemian, *J. Org. Chem.*, **28**, 768 (1963).
515. Y. Morisawa and K. Tanabe, *Chem. Pharm. Bull.*, **17**, 1212 (1969).
516. C. H. Snyder, P. L. Gendler and H.-H. Chang, *Synthesis*, 655 (1971).
517. B. Ganem and R. K. Boeckman, Jr., *Tetrahedron Lett.*, 917 (1974).
518. W. W. Epstein and J. Ollinger, *J. Chem. Soc. Chem. Commun.*, 1338 (1970).
519. N. Kornblum and D. E. Hardies, *J. Amer. Chem. Soc.*, **88**, 1707 (1966).
520. K. Torssell, *Tetrahedron Lett.*, 4445 (1966).
521. N. Bosworth and P. D. Magnus, *J. Chem. Soc. Chem. Commun.*, 257 (1972).
522. N. Bosworth, P. Magnus and R. Moore, *J. Chem. Soc. Perkin Trans. I*, 2694 (1973).
523. F.-X. Jarreau, B. Tchoubar and R. Goutarel, *Bull. Soc. Chim. Fr.*, 887 (1962).
524. H. R. Nace, *J. Amer. Chem. Soc.*, **81**, 5428 (1959).
525. E. Schipper, M. Cinnamon, L. Rascher, Y. H. Chiang and W. Oroshnik, *Tetrahedron Lett.*, 6201 (1968).
526. N. Furukawa, T. Akasaka, T. Aida and S. Oae, *J. Chem. Soc. Perkin Trans. I*, 372 (1977).
527. A. Markovac, C. L. Stevens, A. B. Ash and B. E. Hackley, Jr., *J. Org. Chem.*, **35**, 841 (1970).
528. K. Sato, S. Suzuki and Y. Kojima, *J. Org. Chem.*, **32**, 339 (1967).
529. T. Cohen and T. Tsuji, *J. Org. Chem.*, **26**, 1681 (1961).
530. S. M. Osman and G. A. Qazi, *Fette, Seifen, Anstrichm.*, **77**, 106 (1975); *Chem. Abstr.*, **82**, 169963v (1975).
531. E. Brousse and D. Lefort, *C.R. Acad. Sci.*, **261**, 1990 (1965).
532. T. Tsuji, *Tetrahedron Lett.*, 2413 (1966).
533. G. Hanisch and G. Henseke, *Chem. Ber.*, **101**, 2074 (1968).
534. T. M. Santosusso and D. Swern, *Tetrahedron Lett.*, 4261 (1968).
535. B. McKague, *Can. J. Chem.*, **49**, 2447 (1971).
536. V. J. Traynelis, W. L. Hergenrother, J. R. Livingston and J. A. Valicenti, *J. Org. Chem.*, **27**, 2377 (1962).
537. V. J. Traynelis and W. L. Hergenrother, *J. Amer. Chem. Soc.*, **86**, 298 (1964).
538. J. M. Tien, H.-J. Tien and J.-S. Ting, *Tetrahedron Lett.*, 1483 (1969).
539. T. M. Santosusso and D. Swern, *Tetrahedron Lett.*, 4255 (1974).
540. H. W. Heine and T. Newton, *Tetrahedron Lett.*, 1859 (1967).
541. S. Fujita, H. Hiyama and H. Nozaki, *Tetrahedron*, **26**, 4347 (1970).
542. J. Leitich and F. Wessely, *Monatsh. Chem.*, **95**, 129 (1964); *Chem. Abstr.*, **61** 601g (1964).
543. W. H. W. Lunn, *J. Org. Chem.*, **30**, 2925 (1965).
544. M. E. C. Biffin and D. B. Paul, *Tetrahedron Lett.*, 3849 (1971).
545. T. Aida, N. Furukawa and S. Oae, *J. Chem. Soc. Perkin Trans. II*, 1432 (1976).
546. K. E. Pfitzner and J. G. Moffatt, *J. Amer. Chem. Soc.*, **85**, 3027 (1963).
547. K. E. Pfitzner and J. G. Moffatt, *J. Amer. Chem. Soc.*, **87**, 5661 (1965).
548. K. E. Pfitzner and J. G. Moffatt, *J. Amer. Chem. Soc.*, **87**, 5670 (1965).
549. J. D. Albright and L. Goldman, *J. Amer. Chem. Soc.*, **87**, 4214 (1965).
550. J. D. Albright and L. Goldman, *J. Amer. Chem. Soc.*, **89**, 2416 (1967).
551. K. Onodera, S. Hirano and N. Kashimura, *J. Amer. Chem. Soc.*, **87**, 4651 (1965).
552. K. Onodera, S. Hirano, N. Kashimura and T. Yajima, *Tetrahedron Lett.*, 4327 (1965).
553. K. Onodera, S. Hirano, N. Kashimura, F. Masuda, T. Yajima and N. Miyazaki, *J. Org. Chem.*, **31**, 1291 (1966).
554. J. R. Parikh and W. von E. Doering, *J. Amer. Chem. Soc.*, **89**, 5505 (1967).
555. R. E. Harmon, C. V. Zenarosa and S. K. Gupta, *J. Org. Chem.*, **35**, 1936 (1970).
556. E. J. Corey and C. U. Kim, *Tetrahedron Lett.*, 919 (1973).
557. K. Omura, A. K. Sharma and D. Swern, *J. Org. Chem.*, **41**, 957 (1976).
558. S. L. Huang, K. Omura and D. Swern, *J. Org. Chem.*, **41**, 3329 (1976).

559. S. L. Huang, K. Omura and D. Swern, *Synthesis*, 297 (1978).
560. J. D. Albright, *J. Org. Chem.*, **39**, 1977 (1974).
561. A. J. Mancuso, S.-L. Huang and D. Swern, *J. Org. Chem.*, **43**, 2480 (1978); see also K. Omura and D. Swern, *Tetrahedron*, **34**, 1651 (1978).
562. A. H. Fenselau and J. G. Moffatt, *J. Amer. Chem. Soc.*, **88**, 1762 (1966).
563. J. G. Moffatt, *J. Org. Chem.*, **36**, 1909 (1971).
564. F. W. Sweat and W. W. Epstein, *J. Org. Chem.*, **32**, 835 (1967).
565. D. H. R. Barton, B. J. Garner and R. H. Wightman, *J. Chem. Soc.*, 1855 (1964).
566. D. H. R. Barton and C. P. Forbes, *J. Chem. Soc. Perkin Trans. I*, 1614 (1975).
567. N. Finch, J. J. Fitt and I. H. S. Hsu, *J. Org. Chem.*, **40**, 206 (1975).
568. E. J. Corey and C. U. Kim, *J. Amer. Chem. Soc.*, **94**, 7586 (1972).
569. E. J. Corey and C. U. Kim, *J. Org. Chem.*, **38**, 1233 (1973).
570. J. P. McCormick, *Tetrahedron Lett.*, 1701 (1974).
571. E. J. Corey, C. U. Kim and M. Takeda, *Tetrahedron Lett.*, 4339 (1972).
572. J.-C. Depezay and Y. Le Merrer, *Tetrahedron Lett.*, 2751 (1974).
573. J.-C. Depezay and Y. Le Merrer, *Tetrahedron Lett.*, 2755 (1974).
574. N. Furukawa, T. Inoue, T. Aida and S. Oae, *J. Chem. Soc. Chem. Commun.*, 212 (1973).
575. E. J. Corey and C. U. Kim, *Tetrahedron Lett.*, 287 (1974).
576. G. A. Crosby, N. M. Weinshenker and H.-S. Uh, *J. Amer. Chem. Soc.*, **97**, 2232 (1975).

The Chemistry of the Sulphonium Group
Edited by C. J. M. Stirling and S. Patai
© 1981   John Wiley & Sons Ltd

CHAPTER **16**

Synthetic applications of sulphonium salts and sulphonium ylides

ERIC BLOCK

Department of Chemistry, University of Missouri-St. Louis, St. Louis, Missouri 63121, USA

I. INTRODUCTION

Sulphonium salts and the ylides derived from them have become increasingly useful and important in organic synthesis since the demonstration in 1962 by Corey and Chaykovsky of the easy preparation and highly selective methylene transfer capabilities of dimethylsulphonium and dimethyloxosulphonium methylides ($Me_2\overset{+}{S}\overset{-}{C}H_2$ and $Me_2\overset{+}{S}(O)\overset{-}{C}H_2$, respectively[1]. Other chapters in this book deal with mechanistic aspects of reactions of sulphonium and heterosulphonium salts and ylides and introduce a number of synthetic uses of these reagents. This chapter will concentrate on recent applications (to the end of 1978) of these sulphonium reagents in the synthesis of a wide range of organic molecules. Coverage will not be encylopaediac because of the availability of other recent reviews of this subject[2].

II. SYNTHETIC APPLICATIONS OF SULPHONIUM SALTS

A. Alkylation: Inter- and Intramolecular

The best known synthetic application of sulphonium salts involves their deprotonation to sulphonium ylides which then undergo various reactions (to be considered in Section III) such as carbonyl addition followed by intramolecular displacement of sulphide in the resulting sulphonium salt betaine (equation 1), a

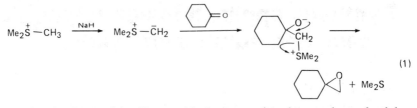

$$\tag{1}$$

reaction already discussed in Chapter 12. In contrasting the reactions of sulphur and of phosphorus ylides (the Wittig reaction) with ketones, the superior leaving-group ability of sulphides compared with phosphines is in part responsible for the different paths taken in these two reactions[2a]. Sulphonium salts are in fact excellent alkylating agents, undergoing ready S_N2 attack at carbon by a variety of nucleophiles in intermolecular processes as well as in intramolecular processes such as shown in equation 1.

In living systems S-adenosyl-L-methionine (**1**) serves as the alkyl donor in enzyme catalysed transmethylation reactions (Chapter 17); intramolecular displacement of thiomethyl adenosine is also possible (equation 2)[3]. It has been known for some

$$\text{Adenosine}-\overset{+}{\underset{\text{Me}}{S}}-CH_2CH_2\overset{-}{C}H(NH_3{}^+) \longrightarrow \underset{NH_3{}^+}{\text{(ring)}}=O \ + \ \text{Adenosine}-S-Me \tag{2}$$

(1)

time that N- and O-methylation occur on treatment of various amino and hydroxyl compounds with sulphur ylides (equations 3–6)[4]. These reactions are thought to involve deprotonation followed by alkylation of the heteroanion by the

$$Me_2\overset{+}{S}(O)\overset{-}{C}H_2 \ + \ PhCH=N-NHPh \longrightarrow PhCH=N-\overset{-}{N}Ph \ \overset{+}{+} \ Me-\overset{+}{S}(O)Me_2 \longrightarrow$$
$$PhCH=N-N(Me)Ph \ + \ Me_2SO \tag{3}$$
$$86\%$$

$$p\text{-}NO_2C_6H_4CO_2H \ + \ Me_2\overset{+}{S}(O)\overset{-}{C}H_2 \longrightarrow p\text{-}NO_2C_6H_4CO_2Me \tag{4}$$
$$69\%$$

$$\tag{5}$$

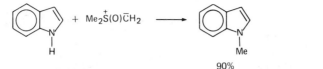

$$(6)$$

90%

simultaneously formed sulphonium or oxosulphonium salt (equation 3). Trimethylsulphonium hydroxide ('MSH') can also function as a useful methylating agent and model for S-adenosylmethionine (equation 7)[5]. In connection with the

$$Me_3S^+OH^- + RXH \xrightarrow{70-120\ °C} RXMe + H_2O + Me_2S \qquad (7)$$

examples given in Table 1 involving methylation by 'MSH', it can be noted that the reagent is cheap and easily produced, separation and purification of products are simple, and the reactions are rapid and nearly quantitative. The methylation apparently works best when RXH has a pK_a less than 12.

As previously indicated, intramolecular displacement of sulphide is a common reaction of sulphonium salts. Additional examples are given in equations 8[6], 9[7], 10[8], 11[9], 12[2], 13[10], and 14[14], illustrating both synthetic and mechanistic aspects of displacements involving oxygen or carbon nucleophiles.

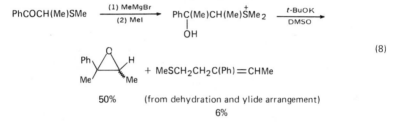

$$(8)$$

50%   (from dehydration and ylide arrangement)
6%

TABLE 1. Methylation with trimethyl sulphonium hydroxide, $Me_3S^+OH^-$[5]

Reactant	Conditions	Product	Yield (%)
$HOCH_2CH_2SH$	100°C, 20 min	$HOCH_2CH_2SMe$	100
Me_3CCO_2H	120°C, 20 min	Me_3CCO_2Me	100
	110°C, 30 min		84
	70°C, 20 min		100
	100°C, 5 min		90

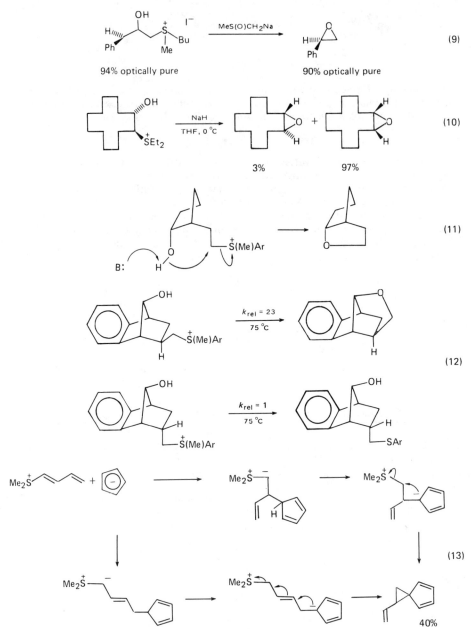

B. Synthesis of Alkyl Halides

Reaction of sulphonium salts with halide ion, the reverse of the standard method for the preparation of sulphonium salts from sulphides and alkyl halides, can in fact

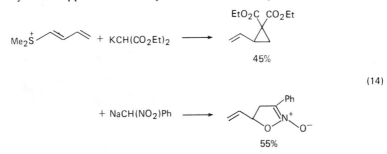

(14)

be developed into useful syntheses of both of these latter classes of compounds if it is recognized that interchange of alkyl groups can occur (equation 15). A procedure

$$R_2S + R'Br \; \overset{\Delta}{\rightleftharpoons} \; R_2\overset{+}{S}R' \; Br^- \; \overset{\Delta}{\rightleftharpoons} \; RBr + RSR'$$

(15)

$$RSR' + R'Br \; \overset{\Delta}{\rightleftharpoons} \; R'_2\overset{+}{S}R \; Br^- \; \overset{\Delta}{\rightleftharpoons} \; RBr + R'_2S$$

for homologation of alkyl halides involves *C*-alkylation of an α-thio-carbanion with an alkyl halide followed by 'one-pot' *S*-alkylation with methyl iodide and displacement by iodide to give the homologated alkyl halide (equation 16)[11]. This

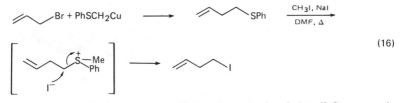

(16)

sequence is a key step in the stereospecific total synthesis of the *dl*-C_{18} cecropia juvenile hormone (equation 17)[12]. A novel variant of this reaction involves the use

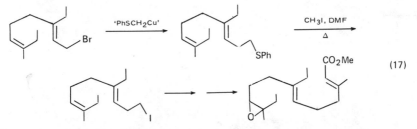

(17)

of polymeric phenylthiomethyllithium as a re-usable homologation reagent (equation 18)[13]. Other recent synthetic applications are summarized in equations 19[14], 20[15a], and 21[15b]. In the last reaction (equation 21) it is necessary to employ the more easily alkylated thiomethyl group rather than the thiophenyl group to avoid secondary decomposition processes resulting in loss of iodine[15b]. Finally, it is noted that allylic hydroxyl functions can be selectively replaced by chloride in the presence of non-allylic alcohol units if the allylic oxygen is activated as an oxysulphonium group (equation 22)[15c].

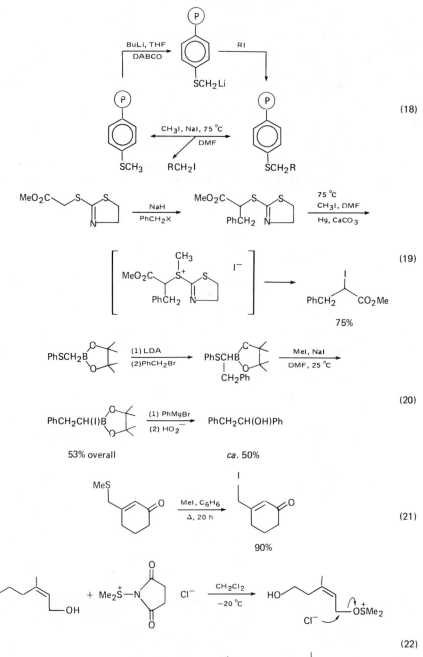

(18)

(19)

(20)

53% overall   *ca.* 50%

(21)

90%

(22)

C. Synthesis of Sulphides

The sulphonium salt ligand interchange process depicted in general terms in equation 16 forms the basis for the unusual sulphide syntheses shown in equation 23[16] (one step is a vinylogous reaction) and equation 24[17]. Even more remarkable

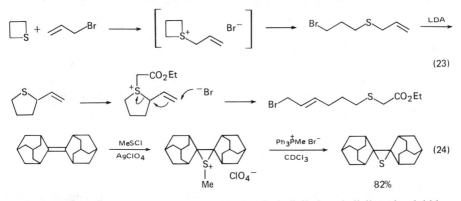

(23)

(24)

is the synthesis of symmetrical and unsymmetrical dialkyl and dialkenyl sulphides under neutral conditions by the simple expedient of heating a neat mixture of sulphide and halide[18]. When the sulphide is dimethyl sulphide, the by-product is the trimethylsulphonium salt halide, which is simply removed by filtration. Applications of this novel reaction are given in Table 2, and the mechanism for the formation of symmetrical sulphides, R_2S, from RBr and Me_2S is shown in Scheme 1.

A second sulphonium salt-mediated sulphide synthesis is the conversion of sulphoxides into sulphides via oxysulphonium salts[21]. An example of this reaction is the synthesis of 2,3,5-trithiahexane, a naturally occurring component of cabbage,

TABLE 2. Sulphide synthesis via the sulphonium salt ligand interchange route

Halide	Sulphide[a]	Product	Yield (%)	Reference
Br—◇(=O)O (bromo-γ-butyrolactone)	Me_2S	MeS—◇(=O)O	96	19
ICH_2CN	Me_2S	$MeSCH_2CN$	88	20
Br—C(=CO_2Me)CO_2Me	Me_2S	MeS—C(=CO_2Me)CO_2Me	100	18
CH_2Br_2	Me_2S	$(MeS)_2CH_2$	65	18
EtO_2CCH_2Br	Me_2S	EtO_2CCH_2SMe	89	18
EtO_2CCH_2Br	Et_2S	EtO_2CCH_2SEt	95	18
$PhCH_2Br$	Me_2S	$PhCH_2SMe$	100	18
EtO_2CCH_2Br	Me_2S[b]	$(EtO_2CCH_2)_2S$	96	18

[a]Unless otherwise indicated, a four-fold molar excess of sulphide was used.
[b]Molar ratio of halide to sulphide = 2:3.

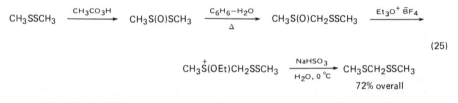

broccoli and cauliflower[22], through deoxygenation of 2,3,5-trithiahexane 5-oxide (equation 25)[23].

$$CH_3SSCH_3 \xrightarrow{CH_3CO_3H} CH_3S(O)SCH_3 \xrightarrow[\Delta]{C_6H_6-H_2O} CH_3S(O)CH_2SSCH_3 \xrightarrow{Et_3O^+ \bar{B}F_4}$$

(25)

$$CH_3\overset{+}{S}(OEt)CH_2SSCH_3 \xrightarrow[H_2O, 0\,°C]{NaHSO_3} CH_3SCH_2SSCH_3$$
72% overall

III. SYNTHETIC APPLICATIONS OF SULPHONIUM YLIDES

This section will illustrate the use of sulphur ylides in the synthesis of a wide range of organic structures such as cyclophanes, heterocycles, terpenes, monosaccharides, gibberellic acid, prostaglandins, steroids, phenols, anilines, olefins, and carbonyl compounds, among others. The generation of sulphur ylides through deprotonation of sulphonium salts and reaction of sulphur compounds with carbenes, benzynes, and other reagents has been reviewed in Chapter 12 and elsewhere[2] and will not be explicitly considered here. For purposes of organization, synthetic applications will be considered in the context of the type of intermediate or reaction involved, e.g. thiocarbonyl ylides, carbonyl addition, Michael addition ('cyclopropanation'), alkylation and other displacements, elimination (α and α,β'), fragmentation, and rearrangement ([1,2]-, [1,5]-, and [2,3]-sigmatropic)[21].

A. Synthetic Methods Involving Thiocarbonyl Ylides

Thiocarbonyl ylides (2) may be viewed as 1,3-dipoles in which two trivalent carbon atoms are bonded to sulphur[24]. Two important approaches to these

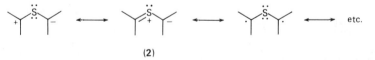

(2)

interesting, reactive species include fragmentation of Δ³-1,3,4-thiadiazolines (available in good yields from ketones or thioketones as shown in equation 26) and photocyclization of divinyl sulphides (equation 27)[2a]. One of the characteristic reactions of thiocarbonyl ylides is closure to the valence tautomeric thiiranes, an electrocyclic process which occurs in a conrotatory manner (equation 28 and 29)[25]. Thiocarbonyl ylides also undergo stereospecific cycloaddition with

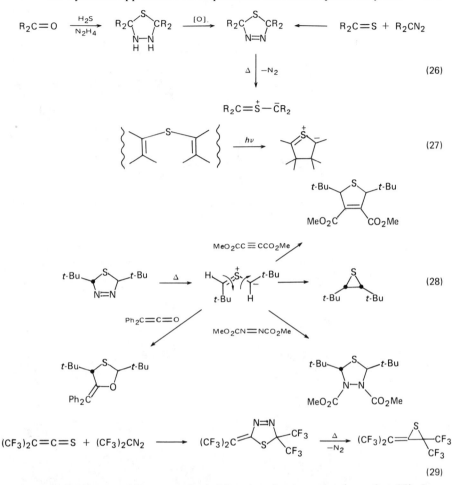

(26)

(27)

(28)

(29)

1,3-dipolarophiles providing a synthesis of various heterocycles (equation 28). Iron carbonyl complexes of thiocarbonyl ylides (or related delocalized species) can undergo Diels–Alder reactions, as seen in equation 30[26].

(30)

Since thiiranes can be readily desulphurized to olefins, Δ^3-1,3,4-thiadiazolines can serve as precursors to olefins, either in a two-step process with isolation of thiirane (equation 31[27]), or via a single 'two-fold extrusion' process wherein the Δ^3-1,3,4-thiadiazoline is heated with a phosphine and sulphur extruded directly (equation 32)[28]. A number of examples of the 'two-fold extrusion' approach to olefin synthesis are given in Table 3. A complication in the synthesis of highly hindered olefins by the 'two-fold extrusion' route is the alternative decomposition of Δ^3-1,3,4-thiadiazolines to thiones and diazo compounds (a retro 1,3-dipolar

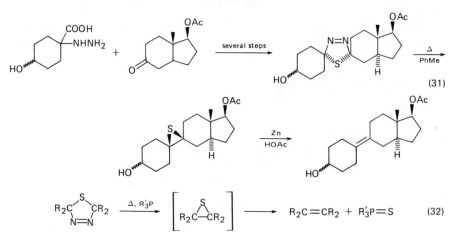

(31)

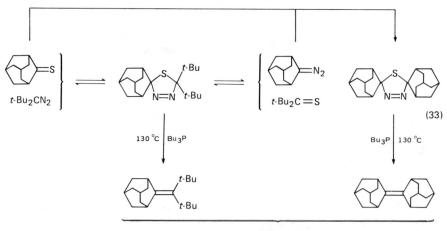

(32)

addition process), as shown in equation 33[30]. A consequence of the latter reaction is the formation of more than one olefin from unsymmetrically substituted Δ^3-1,3,4-thiadiazolines.

(33)

42% total

Cyclic thiocarbonyl ylides, formed through photocyclization of divinyl sulphides, can undergo [1,4]-suprafacial hydrogen shifts. When this hydrogen shift is followed by desulphurization the result is a useful C—C bond-forming reaction known as 'heteroatom directed photoarylation' and illustrated by the reactions in equations 34–36[31].

B. Carbonyl Addition: Oxirane Synthesis

Whereas phosphorus ylides react with aldehydes and ketones to give olefins via the well known Wittig reaction, sulphur ylides affored oxiranes under analogous conditions, e.g. equation 1. In view of the importance of oxiranes as synthetic intermediates and the ready availability of carbonyl compounds, oxirane formation

TABLE 3. Olefin synthesis via two-fold extrusion

Ketone (or thioketone)	Olefin product	Overall yield (%)	Reference
		42	28
		69	29a
		65	29b
		20	29c
		90[b]	29d
		50	29e
$t\text{-Bu}_2C{=}S^a$	$t\text{-Bu}_2C{=}CPh_2$	68[b]	28
		73	29f
		70	29g
		82	29h

[a]Co-reactant is diphenyldiazomethane.
[b]From cis-Δ^3-1,3,4-thiadiazoline.
[c]Co-reactant is 2-diazopropane.
[d]Co-reactant is analogous diazo compound.

using sulphur ylides is a very useful procedure that has been studied in great detail both from a mechanistic and a preparative standpoint (see Chapter 12 for a thorough discussion). The generality of the reaction is demonstrated by the lack of interference from enol ethers, acetals, amides, nitriles, divalent sulphur, and in some cases esters and hydroxyl and amino groups[2d]. Representative of synthetic

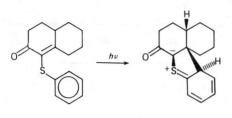
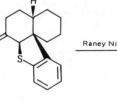

(34)

55%

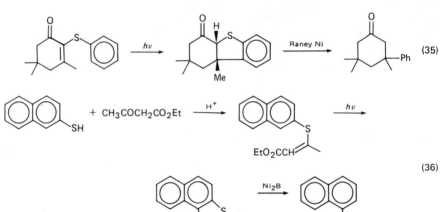

(35)

(36)

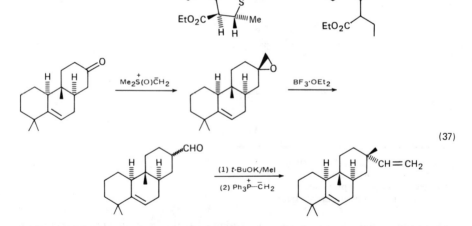

(37)

applications of this reaction are synthesis of *dl*-rimuene (equation 37)[32], *R*-styrene oxide (equation 38)[33] [^{14}C]squalene oxide (equation 39)[34], acorenone B (equation 40)[35], projected dodecahedrane precursor **3** (equation 41)[36], and steroid derivative **4** (equation 42)[37]. The last three reactions employ cyclopropyl sulphur ylides in the

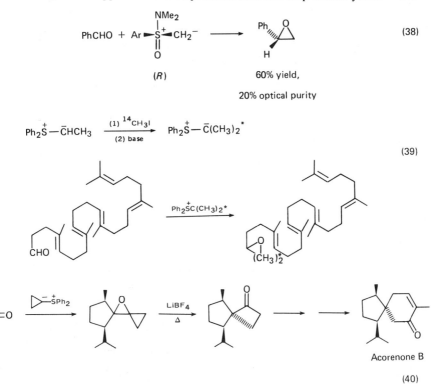

PhCHO + Ar—$\overset{\overset{NMe_2}{|}}{\underset{\underset{O}{\|}}{\overset{+}{S}}}$—$CH_2^-$ ⟶   (38)

(R)   60% yield,

20% optical purity

$Ph_2\overset{+}{S}$—$\bar{C}HCH_3$ $\xrightarrow[\text{(2) base}]{\text{(1) }^{14}CH_3I}$ $Ph_2\overset{+}{S}$—$\bar{C}(CH_3)_2^*$

(39)

(40)

Acorenone B

process termed 'spiroannelation'. An interesting oxirane synthesis has been recently described using a polymeric sulphonium ylide (equation 43)[38]. The advantages of this process, which employs a three-phase system with a phase transfer catalyst, are ease of isolation of product from polymeric reagent, avoidance of bad odours associated with volatile sulphides, and regeneration of polymeric reagent with no loss of activity[38]. One instance of intramolecular oxirane formation has been published recently (equation 44)[39]; a previous attempt to achieve this reaction led instead to rearranged products (equation 45)[40].

C. Michael Addition: Cyclopropane Synthesis

Sulphonium and oxosulphonium ylides readily add to such Michael acceptors as α,β-unsaturated ketones, esters, nitriles, isonitriles, sulphones, sulphoxides, sulphonamides, sulphonates, and nitro compounds to afford cyclopropanes[2d], e.g. equations 46[41], 47[42], 48[43], 49[44], 50[45], and 51[46]. All of these reactions are nucleophilic cyclopropanations, which proceed best with electron-deficient olefins (see Chapter 12 for mechanistic details). Whereas sulphonium ylides such as diphenyl sulphonium methylide are normally unreactive toward olefins such as cis- and trans-2-octene, it has been discovered that cyclopropanation does occur in the presence of copper salts and that these reactions are stereospecific with retention (equation 52[47] and 53[48]). Ylide 5 (equation 53) is a stable crystalline solid which can be stored at room temperature without special precautions.[48]

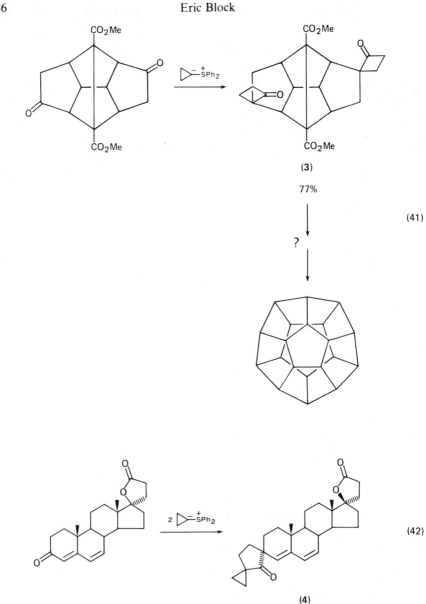

(3)

77%

(41)

?

(42)

(4)

D. Alkylation and Other Displacements

Sulphur ylides participate in a variety of displacement reactions, in keeping with their carbanionic character (e.g. equations 54[49] and 55[50]). Sufficiently stabilized sulphur ylides show reduced reactivity in alkylation reactions and in certain instances can even function as protecting groups, as illustrated in equation 56[51].

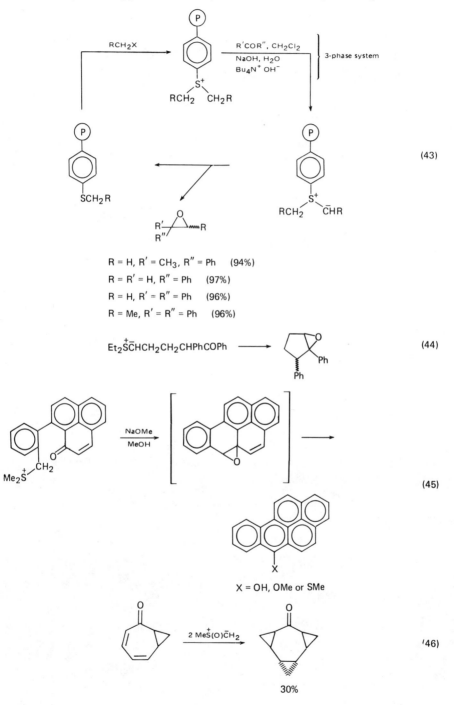

R = H, R′ = CH₃, R″ = Ph   (94%)

R = R′ = H, R″ = Ph   (97%)

R = H, R′ = R″ = Ph   (96%)

R = Me, R′ = R″ = Ph   (96%)

(43)

(44)

(45)

X = OH, OMe or SMe

30%

(46)

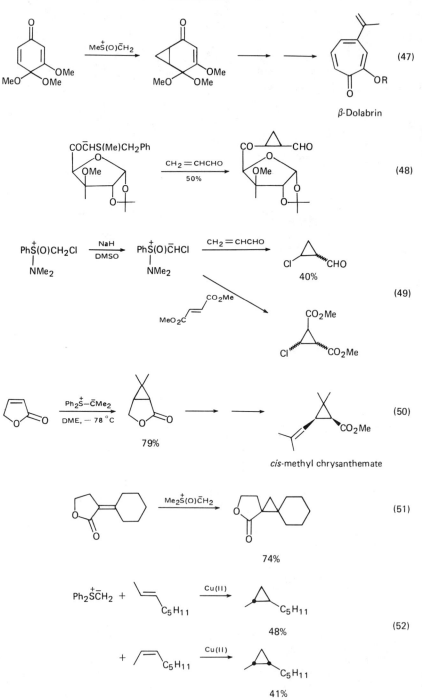

(47)

β-Dolabrin

(48)

(49)

(50)

cis-methyl chrysanthemate

(51)

74%

(52)

48%

41%

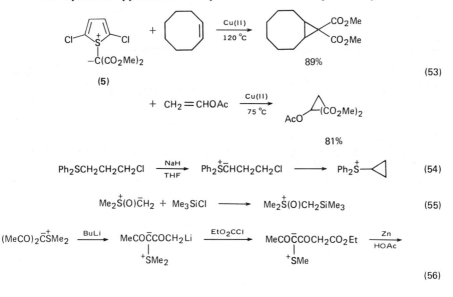

(53)

81%

$$Ph_2SCH_2CH_2CH_2Cl \xrightarrow[THF]{NaH} Ph_2\overset{+-}{S}CHCH_2CH_2Cl \longrightarrow Ph_2\overset{+}{S}-\triangleleft \qquad (54)$$

$$Me_2\overset{+}{S}(O)\overset{-}{C}H_2 + Me_3SiCl \longrightarrow Me_2\overset{+}{S}(O)CH_2SiMe_3 \qquad (55)$$

$$(MeCO)_2\overset{-+}{C}SMe_2 \xrightarrow{BuLi} \underset{\underset{SMe_2}{|}}{MeCO\overset{-}{C}COCH_2Li} \xrightarrow{EtO_2CCl} \underset{\underset{SMe}{|}}{MeCO\overset{-}{C}COCH_2CO_2Et} \xrightarrow[HOAc]{Zn}$$

(56)

$$MeCOCH_2COCH_2CO_2Et$$

E. α-Elimination

A limited number of examples of α-elimination reactions of sulphur ylides have been reported. These eliminations produce carbenes which undergo such characteristic processes as Wolff rearrangement, insertion into C—H bonds, addition to C=C double bonds and formation of dimers. Two examples of synthetic applications are given in equations 57[52] and 58[53].

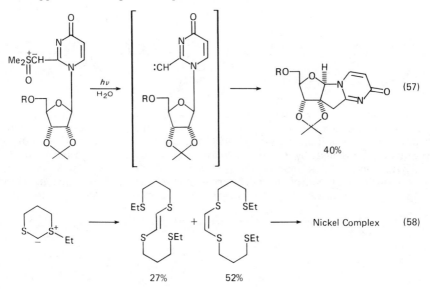

(57)

40%

(58)

27%   52%

F. α,β′-Elimination

The general form of an α,β′-elimination reaction of a sulphonium ylide is shown in equation 59. In the case where $X = CR_2$ (**6a**) the process of equation 59 may be

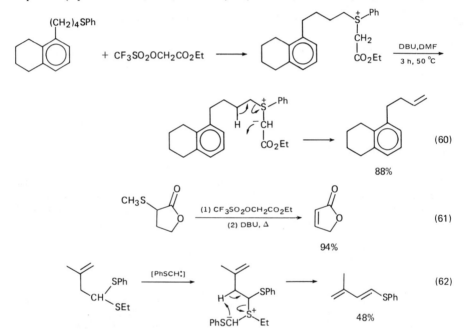

$$RSCH_3 + \quad C{=}X \qquad (59)$$

(6a) X = CR_2

(6b) X = O

an undesired side-reaction in the preparation and utilization of sulphonium ylides bearing β-hydrogen atoms. A few synthetic applications of this reaction have been reported (equations 60[54], 61[54], and 62[55]). α,β′-Elimination of oxysulphonium ylides

(60)

88%

(61)

94%

(62)

48%

(equation 59, **6b**) is a key step in a variety of procedures for selective oxidation of alcohols to aldehydes and ketones[2a]. These useful procedures stem from the discovery of Kornblum *et al.* in 1959 that alkyl tosylates could be oxidized to carbonyl compounds by dimethyl sulphoxide (DMSO) at elevated temperatures via oxysulphonium salts intermediates (equation 63[56]). These oxidations can be performed with the alcohols under very mild conditions, provided that an activating agent (generally a Lewis acid) is present. Table 4 provides a detailed comparison of

$$CH_3(CH_2)_7OTs + (CH_3)_2SO \xrightarrow[\text{NaHCO}_3,\ 150\ ^\circ C]{} CH_3(CH_2)_6CH_2O{-}\overset{+}{S}(CH_3)_2 \longrightarrow$$

(63)

$$CH_3(CH_2)_6CHO$$

78%

TABLE 4. Oxidation of $n\text{-}C_9H_{19}CH_2OH$ to $C_9H_{19}CHO$ (A) and of $n\text{-}C_6H_{13}CH(OH)CH_3$ to $n\text{-}C_6H_{13}COCH_3$ (B) by activated DMSO[57]

Activator[a]	Solvent	Temperature (°C)	Time (h)	Yield of A (%)	Yield of B (%)
$(CH_3CO)_2O$	DMSO	25	27	27	30
$Py \cdot SO_3$	DMSO	25	0.5	91	93
$Me_2S\text{-}Cl^b$	$PhCH_3$	-25	1.5	94	95
$(CH_3SO_2)_2O$	HMPA	-15	0.25	69	84
Cyanuric chloride	$HMPA\text{—}CH_2Cl_2$	-15	0.5	73	82
CH_3SO_2Cl	$HMPA\text{—}CH_2Cl_2$	20	0.75	62	77
$p\text{-}CH_3C_6H_4SO_2Cl$	$HMPA\text{—}CH_2Cl_2$	5	1.25	72	90
$(CF_3CO)_2O$	CH_2Cl_2	-50	0.5	56	78
$SOCl_2$	CH_2Cl_2	-60	0.25	76	88
$(COCl)_2$	CH_2Cl_2	-60	0.25	97	98
PCl_3	CH_2Cl_2	-30	0.25	45	59
$POCl_3$	CH_2Cl_2	-30	0.25	43	52
CH_3COBr	CH_2Cl_2	-60	0.25	58	70
CH_3COCl	CH_2Cl_2	-20	0.25	40	57
$PhCOCl$	CH_2Cl_2	-60	0.25	97	—
$(CH_3SO_2)_2O^c$	$HMPA^d$	-15	0.5	94	—
Cyanuric chloride[c]	$HMPA^d$	-15	0.5	95	—

[a]Et_3N also required.
[b]No DMSO used.
[c]$i\text{-}Pr_2NEt$ employed instead of Et_3N.
[d]HMPA = hexamethylphosphoramide.

yields for the oxidation of n-decanol and octan-2-ol by DMSO combined with various activating agents[57]. DMSO or dimethyl sulphide derived oxidants have been employed in the synthesis of gibberellic acid (equation 64[58]), betalains (water-soluble, nitrogeneous plant pigments; equation 65[59]), monosaccharides (equations 66[60] and 67[61]), prostaglandins (equations 68[62]), steroids (equation 69[63]), and a variety of simpler structures (equations 70[64] and 71[65]). In equation 68 an insoluble carbodiimide linked to a cross-linked polystyrene matrix (symbolized as Ⓟ) is used with excellent results as the DMSO activator. The oxidation shown in

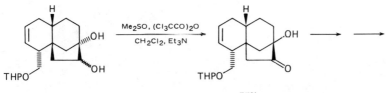

75%

(64)

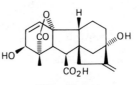

Gibberellic acid

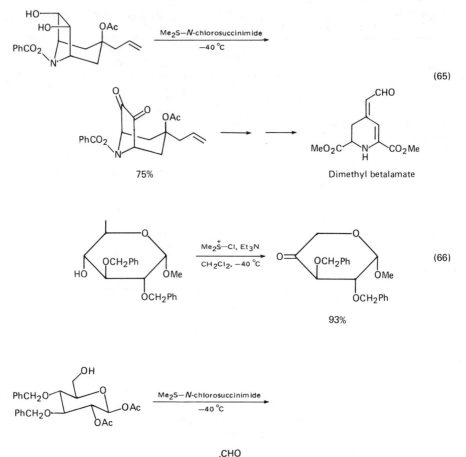

(65)

75%   Dimethyl betalamate

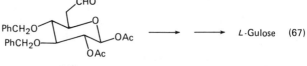

(66)

93%

(67)

66%   L-Gulose

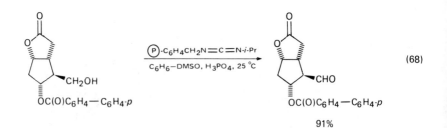

(68)

91%

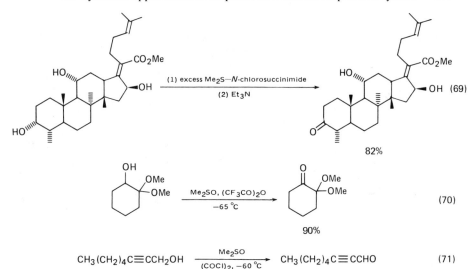

(69)

82%

(70)

90%

$$CH_3(CH_2)_4C\equiv CCH_2OH \xrightarrow[\text{(COCl)}_2, -60\,°C]{Me_2SO} CH_3(CH_2)_4C\equiv CCHO \qquad (71)$$

95%

equation 69 is notable in that only one of the three hydroxyl groups has been oxidized even with excess of reagent.

G. Fragmentations

For want of a better description, a limited number of ylide reactions may be categorized as 'fragmentations'[2a]. A single example, representing a novel means of stereospecifically desulphurizing thiiranes, is given in equation 72[66].

(72)

72%; > 99.5% cis

H. Rearrangements

Sulphur ylides are known to undergo [1,2]-sigmatropic (Stevens), [1,4]-sigmatropic, [1,5]-sigmatropic, and [2,3]-sigmatropic (Sommelet–Hauser) rearrangements[2a]. The first and last 'name reactions' have been studied extensively with respect to mechanism and are of demonstrated synthetic utility. Some recent applications of the Stevens ([1,2]-sigmatropic) rearrangement in the synthesis of cyclophanes are presented in equations 73[67], 74[68], and 75[69], while equation 76[70] provides an example of a synthetic application involving a [1,5]-sigmatropic rearrangement.

Although the first example of a Sommelet–Hauser ([2,3]-sigmatropic) rearrangement of a sulphur ylide was reported as early as 1938[71], only in recent years has the reaction proved to be of synthetic importance[2a]. Examples of

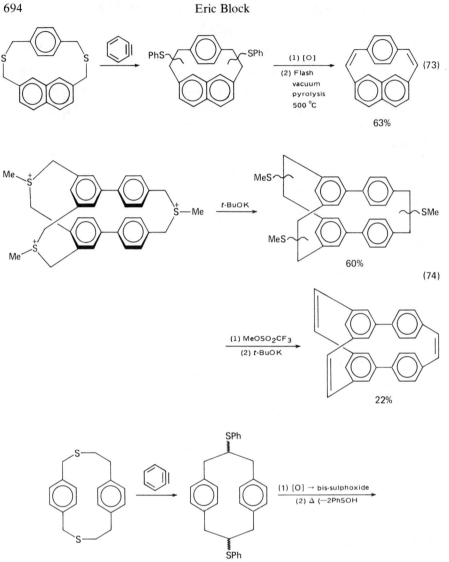

(73)

63%

(74)

60%

22%

29%

(75)

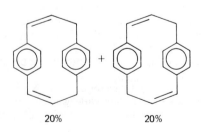

20%   20%

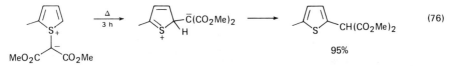

(76)

95%

syntheses involving this rearrangement are the syntheses of yomogi alcohol (equation 77[72]) and artemisia ketone (equation 78[73]), the procedure for allylic functionalization with overall retention of double bond position (equation 79[74]),

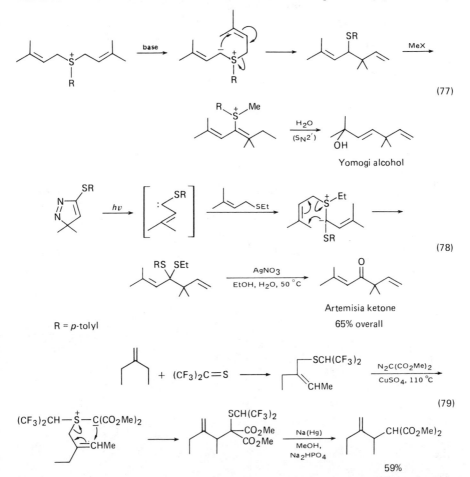

(77)

Yomogi alcohol

(78)

Artemisia ketone
65% overall

R = *p*-tolyl

(79)

59%

and the procedure for easily repeatable multi-carbon ring expansion ('ring-growing reaction') involving ylides derived from α-vinyl heterocycles (equation 80 and Table 5)[75–79]. The Sommelet–Hauser rearrangement also has been utilized for the *ortho* functionalization of anilines, phenols, and related aromatic compounds, as illustrated by equations 81[80], 82[81], and 83[81] and Scheme 2[82]. The rearrangement of ylides derived from *N*-arylazasulphonium salts provides a simple high-yield conversion of anilines into *ortho*-alkylated anilines, indoles, oxindoles, and

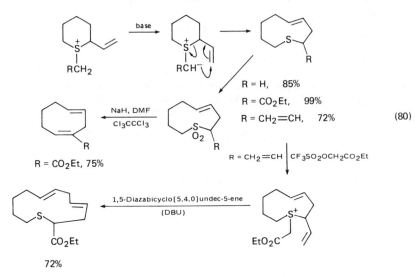

(80)

TABLE 5. Multi-carbon ring expansion via [2,3]-sigmatropic shifts of sulphonium ylides

Substrate	Base	Products and yields (%)		Reference
(structure, PF₆⁻, Me)	t-BuOK	77%	13%	77
(structure, PF₆⁻, Et)	t-BuOK	25%	37%	77
(structure, Br⁻)	KOH	45%	26%	79
(structure, CF₃SO₃⁻, CO₂Et)	t-BuOK	80%		78

TABLE 5. *continued*

Substrate	Base	Products and yields (%)	Reference
HO₂C / CO₂Et	—	S—CO₂Et, CO₂Et 53%	78
PF₆ CH₂Ph	NaOH	Ph 88% 7%	77
PF₆ Me	t-BuOK	85%	77
CF₃SO₃⁻	DBU	72% 3%	75
CF₃SO₃⁻ CO₂Et	DBU	9 CO₂Et 99%	75
PF₆ Me	t-BuOK	10 85%	77
CF₃SO₃⁻ CO₂Et	BDU	10 CO₂Et 91%	75

TABLE 5. *continued*

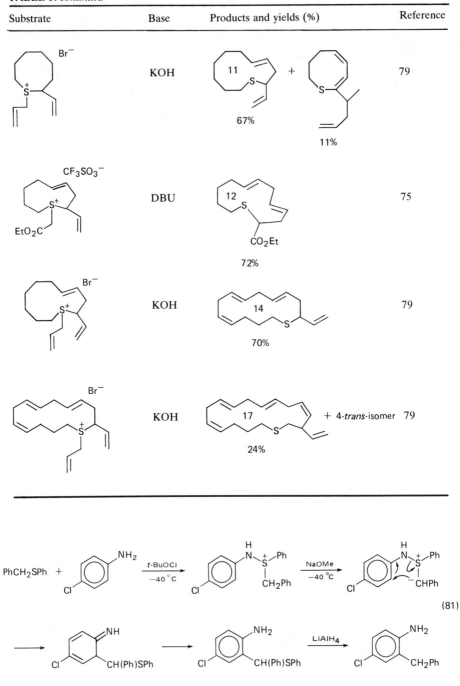

Substrate	Base	Products and yields (%)	Reference
	KOH	11 + 67% 11%	79
	DBU	12 72%	75
	KOH	14 70%	79
	KOH	17 + 4-*trans*-isomer 24%	79

(81)

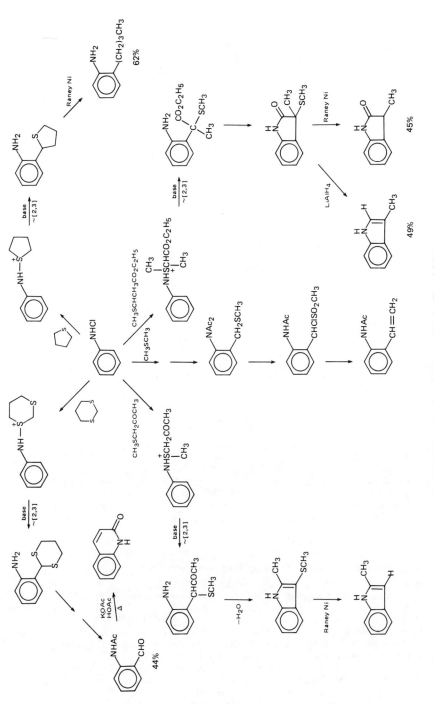

SCHEME 2. Azasulphonium route to substituted anilines and N-heterocycles

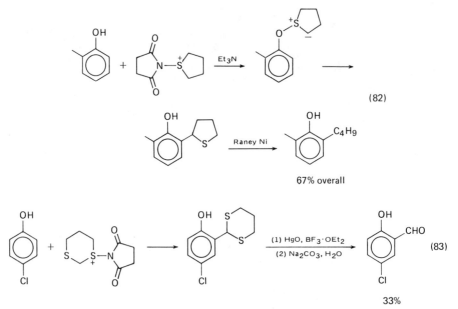

(82)

67% overall

(83)

33%

aminopyridines, as well as permitting *ortho*-formylation of anilines and aminopyridines. These reactions can be carried out over a wide temperature range (down to −78°C) with mild reagents and in the presence of a wide range of substituents. Little charge is built up on the aromatic ring so both electron-withdrawing and electron-donating substituents can be tolerated.

IV. REFERENCES

1. E. J. Corey and M. Chaykovsky, *J. Amer. Chem. Soc.*, **84**, 867, 3782 (1962).
2. (a) E. Block, *Reactions of Organosulfur Compounds*, Academic Press, New York, 1978; (b) E. Block, in *Organic Compounds of Sulphur, Selenium and Tellurium*, (Ed. D. R. Hogg), Specialist Periodical Reports, Vols. IV (1977) and V (1979), Chemical Society, London; (c) A. W. Johnson, in *Organic Compounds of Sulphur, Selenium and Tellurium* (Ed. D. H. Reid), Specialist Periodical Reports, Vols. I (1970), II (1973) and III (1975), Chemical Society, London; (d) B. M. Trost and L. S. Melvin, Jr., *Sulfur Ylides*, Academic Press, New York, 1975; (e) E. Block, *Aldrichim. Acta*, **11** (3), 51 (1978); (f) L. Field, *Synthesis*, 713 (1978); (g) C. J. M. Stirling, in *The Organic Chemistry of Sulfur*, (Ed. S. Oae), Plenum Press, New York, 1977; (h) C. R. Johnson, in *Comprehensive Organic Chemistry*, Vol. III (Ed. D. N. Jones), Pergamon Press, Oxford, 1979.
3. T. Irie and H. Tanida, *J. Org. Chem.*, **44**, 325 (1979).
4. (a) H. Metzger, H. König and K. Seelert, *Tetrahedron Lett.*, 867 (1964); (b) T. Kunieda and B. Witkop, *J. Org. Chem.*, **35**, 3981 (1970); (c) S. Tanimoto, T. Yamadera, T. Sugimoto and M. Okano, *Bull. Chem. Soc. Japan*, **52**, 627 (1979).
5. K. Yamauchi, T. Tanabe and M. Kinoshita, *J. Org. Chem.*, **44**, 683 (1979); see also K. Yamauchi, K. Nakamura and M. Kinoshita, *Tetrahedron Lett.*, 1787 (1979).
6. S. Kano, T. Yokomatsu and S. Shibuya, *Tetrahedron Lett.*, 4125 (1978).
7. C. R. Johnson, C. W. Schroeck and J. R. Shanklin, *J. Amer. Chem. Soc.*, **95**, 7424 (1973); see also W. H. Pirkle and P. L. Rinaldi, *J. Org. Chem.*, **44**, 1025 (1979).
8. J. M. Townsend and K. B. Sharpless, *Tetrahedron Lett.*, 3313 (1972).
9. J. K. Coward, R. Lok and O. Takagi, *J. Amer. Chem. Soc.*, **98**, 1057 (1976).
10. H. Braun and G. Huber, *Tetrahedron Lett.*, 2121 (1976).

11. E. J. Corey and M. Jautelat, *Tetrahedron Lett.*, 5787 (1968).
12. E. J. Corey, J. A. Katzenellenbogen, N. W. Gilman, S. A. Roman and B. W. Erickson, *J. Amer. Chem. Soc.*, **90**, 5618 (1968).
13. G. A. Crosby and M. Kato, *J. Amer. Chem. Soc.*, **99**, 278 (1977).
14. K. Hirai, Y. Iwano and Y. Kishida, *Tetrahedron Lett.*, 2677 (1977).
15. (a) D. S. Matteson and K. Arne, *J. Amer. Chem. Soc.*, **100**, 1325 (1978); (b) P. M. Wege, R. D Clark and C. H. Heathcock, *J. Org. Chem.*, **41**, 3144 (1976); (c) E. J. Corey and C. V. Kim, *J. Amer. Chem. Soc.*, **94**, 7586 (1972).
16. E. Vedejs, J. P. Hagen, B. L. Roach and K. L. Spear, *J. Org. Chem.*, **43**, 1185 (1978).
17. J. Bolster and R. M. Kellogg, *J. Chem. Soc. Chem. Commun.*, 630 (1978).
18. A. J. H. Labuschagne, J. S. Malherbe, C. J. Meyer and D. F. Schneider, *J. Chem. Soc. Perkin Trans. I*, 955 (1978).
19. B. M. Trost and H. C. Arndt, *J. Org. Chem.*, **38**, 3140 (1973).
20. B. M. Trost and L. S. Melvin, Jr., *Sulfur Ylides*, Academic Press, New York, 1975, p. 7.
21. C. R. Johnson, C. C. Bacon and J. J. Rigau, *J. Org. Chem.*, **37**, 919 (1972).
22. R. G. Buttery, D. G. Guadagni, L. C. Ling, R. M. Seifert and W. Lipton, *J. Agr. Food Chem.*, **24**, 829 (1976).
23. E. Block and J. O'Connor, *J. Amer. Chem. Soc.*, **96**, 3929 (1974).
24. R. M. Kellogg, *Tetrahedron*, **32**, 2165 (1976).
25. (a) J. Buter, S. Wassenaar and R. M. Kellogg, *J. Org. Chem.*, **37**, 4045 (1972); (b) W. J. Middleton, *J. Org. Chem.*, **34**, 3201 (1969).
26. T. Koyanagi, J. Hayami and A. Kaji, *Chem. Lett.*, 971 (1976).
27. D. J. Humphreys, C. E. Newall, G. H. Phillipps and G. A. Smith, *J. Chem. Soc. Perkin Trans. I*, 45 (1978).
28. D. H. R. Barton, F. S Guziec, Jr., and I. Shahak, *J. Chem. Soc. Perkin Trans, I*, 1794 (1974), and references cited therein.
29. (a) J. W. Everett and P. J Garratt, *J. Chem. Soc. Chem. Commun.*, 642 (1972); (b) A. P. Schaap and G. R. Faler, *J. Org. Chem.*, **38**, 3061 (1973); (c) H. Sauter, H. G. Horster and H. Prinzbach, *Angew. Chem. Int. Ed. Engl.*, **12**, 991 (1973); (d) R. M. Kellogg, M. Noteboom and J. M. Kaiser, *J. Org. Chem.*, **39**, 2573 (1975); (e) G. Seitz and H. Hoffmann, *Synthesis*, 20 (1977); (f) M. D. Bachi, O. Goldberg and A. Gross, *Tetrahedron Lett.*, 4167 (1978); (g) R. J. Bushby, M. D. Pollard and W. S. McDonald, *Tetrahedron Lett.*, 3851 (1978); (h) A. Krebs and W. Rüger, *Tetrahedron Lett.*, 1305 (1979).
30. F. Cordt, R. M. Frank and D. Lenoir, *Tetrahedron Lett.*, 505 (1979).
31. A. G. Schultz, W. Y. Fu, R. D. Lucci, B. G. Kurr, K. M. Lo and M. Boxer, *J. Amer. Chem. Soc.*, **100**, 2140 (1978).
32. R. E. Ireland and L. N. Mander, *Tetrahedron Lett.*, 3453 (1964).
33. C. R. Johnson and C. W. Schroeck, *J. Amer. Chem. Soc.*, **95**, 7418, 7424 (1973).
34. E. J. Corey, P. R. Ortiz de Montellano and H. Yamamoto, *J. Amer. Chem. Soc.*, **90**, 6254 (1968).
35. B. M. Trost, K. Hiroi and N. Holy, *J. Amer. Chem. Soc.*, **97**, 5873 (1975).
36. L. A. Paquette, M. J. Wyvratt, O. Schallner, D. F. Schneider, W. J. Begley and R. M. Blankenship, *J. Amer. Chem. Soc.*, **98**, 6744 (1976).
37. M. J. Green, H. J. Shue, A. T. McPhail and R. W. Miller, *Tetrahedron Lett.*, 2677 (1976).
38. M. J. Farrall, T. Durst and J. M. J. Frechet, *Tetrahedron Lett.*, 203 (1979).
39. P. Cazeau and B. Muckensturm, *Tetrahedron Lett.*, 1493 (1977).
40. M. S. Newman and L. F. Lee, *J. Org. Chem.*, **40**, 2650 (1975).
41. M. Oda, Y. Ito and Y. Kitahara, *Tetrahedron Lett.*, 977 (1978).
42. D. A. Evans, D. J. Hart and P. M. Koelsch, *J. Amer. Chem. Soc.*, **100**, 4593 (1978).
43. J. M. J. Tronchet and H. Eder, *Helv. Chim. Acta*, **58**, 1799 (1975).
44. H. G. Corkins, L. Veenstra and C. R. Johnson, *J. Org. Chem.*, **43**, 4233 (1978).
45. M. Servin, L. Hevesi and A. Krief, *Tetrahedron Lett.*, 3915 (1976).
46. T. Minami, M. Matsumoto, H. Suganuma and T. Agawa, *J. Org. Chem.*, **43**, 2149 (1978).
47. T. Cohen, G. Herman, T. M. Chapman and D. Kuhn, *J. Amer. Chem. Soc.*, **96**, 5627 (1974).
48. J. Cuffe, R. J. Gillespie and A. E. A. Porter, *J. Chem. Soc. Chem. Commun.*, 641 (1978).

49. B. M. Trost and L. S. Melvin, Jr., *Sulfur Ylides*, Academic Press, New York, 1975, p. 10.
50. H. Schmidbaur and W. Kapp, *Chem. Ber.*, **105**, 1203 (1972).
51. M. Yamamoto, *J. Chem. Soc. Chem. Commun.*, 289 (1975).
52. T. Kunieda and B. Witkop, *J. Amer. Chem. Soc.*, **93**, 3487 (1971).
53. S. R. Wilson and R. S. Myers, *Tetrahedron Lett.*, 3413 (1976).
54. E. Vedejs and D. A. Engler, *Tetrahedron Lett.*, 3487 (1976).
55. C. Huynh, V. Ratovelomanana and S. Julia, *Bull. Soc. Chim. Fr.*, 710 (1977).
56. N. Kornblum, W. J. Jones and G. J. Anderson, *J. Amer. Chem. Soc.*, **91**, 4113 (1959).
57. K. Omura and D. Swern, *Tetrahedron*, **34**, 1651 (1978).
58. E. J. Corey, R. L. Danheiser, S. Chandrasekran, P. Siret, G. E. Keck and J.-L. Gras, *J. Amer. Chem. Soc.*, **100**, 8031 (1978).
59. G. Buchi, H. Fliri and R. Shapiro, *J. Org. Chem.*, **43**, 4765 (1978).
60. P. Simon, J.-C. Ziegler and B. Gross, *Carbohydr. Res.*, **64**, 257 (1978).
61. D. K. Minster and S. M. Hecht, *J. Org. Chem.*, **43**, 3987 (1978).
62. N. M. Weinshenker and C.-M. Shen, *Tetrahedron Lett.*, 3285 (1972); for comparison with non-polymer-supported carbodiimide, see P. A. Grieco, Y. Yokoyama, G. P. Withers, F. J. Okuniewicz and C.-L. Wang, *J. Org. Chem.*, **43**, 4178 (1978).
63. S. S. Welankiwar and W. S. Murphy, *J. Chem. Soc. Perkins Trans. I*, 710 (1976).
64. C. Kowalski, X. Creary, A. J. Rollin and M. C. Burke, *J. Org. Chem.*, **43**, 2601 (1978).
65. A. J. Mancuso, S.-L. Huang and D. Swern, *J. Org. Chem.*, **43**, 2480 (1978).
66. Y. Hata, M. Watanabe, S. Inoue and S. Oae, *J. Amer. Chem. Soc.*, **97**, 2553 (1975).
67. J. R. Davy, M. N. Iskander and J. A. Reiss, *Tetrahedron Lett.*, 4085 (1978).
68. F. Vögtle and G. Steinhagen, *Chem. Ber.*, **111**, 205 (1978); see also F. Vögtle and N. Wester, *Justus Liebigs Ann. Chem.*, 545 (1978).
69. M. W. Haenel and A. Flatow, *Chem. Ber.*, **112**, 249 (1979).
70. R. J. Gillespie, A. E. A. Porter and W. E. Willmott, *J. Chem. Soc. Chem. Commun.*, 85 (1978).
71. L. A. Pinck and G. E. Hilbert, *J. Amer. Chem. Soc.*, **60**, 494 (1938).
72. B. M. Trost, P. Conway and W. G. Biddlecom, *J. Chem. Soc. Chem. Commun.*, 1639 (1971).
73. M. Franck-Neumann and J. J. Lohmann, *Tetrahedron Lett.*, 3729 (1978).
74. B. B. Snider and L. Füzesi, *Tetrahedron Lett.*, 877 (1978).
75. E. Vedejs, M. J. Arco, D. W. Powell, J. M. Renga and S. P. Singer, *J. Org. Chem.*, **43**, 4831 (1978).
76. E. Vedejs and S. P. Singer, *J. Org. Chem.*, **43**, 4884 (1978).
77. V. Ceré, C. Paolucci, S. Pollicino, E. Sandri and A. Fava, *J. Org. Chem.*, **43**, 4826 (1978).
78. E. Vedejs, J. P. Hagen, B. L. Roach and K. L. Spear, *J. Org. Chem.*, **43**, 1185 (1978).
79. R. Schmid and H. Schmid, *Helv. Chim. Acta*, **60**, 1361 (1977).
80. P. G. Gassman and H. R. Drewes, *J. Amer. Chem. Soc.*, **100**, 7600 (1978).
81. P. G. Gassman and D. R. Amick, *J. Amer. Chem. Soc.*, **100**, 7611 (1978).
82. P. G. Gassman and G. Gruetzmacher, *Org. Synth.*, **56**, 15 (1977); P. G. Gassman and T. J. van Bergen, *Org. Synth.*, **56**, 72 (1977); P. G. Gassman, G. Greutzmacher and T. J. van Bergen, *J. Amer. Chem. Soc.*, **96**, 5512 (1974), and preceding papers in series; see also reference 80.

The Chemistry of the Sulphonium Group
Edited by C. J. M. Stirling and S. Patai
© 1981 John Wiley & Sons Ltd

CHAPTER **17**

The biochemistry of sulphonium salts

G. A. MAW

Department of Biochemistry, Glasshouse Crops Research Institute, Littlehampton, Sussex, England

I. INTRODUCTION

Among the first instances of sulphonium compounds being viewed in a biological context was a report in 1872 that certain compounds of this class possess some of the pharmacological properties of the structurally analogous quaternary ammonium salts[1], a finding later confirmed by other workers[2,3]. Toennies[4] may be credited with being the first to suggest a biological role for sulphonium compounds, when in 1940 he proposed that a sulphonium derivative of methionine **1** might be formed as an intermediate in the conversion of the amino acid to cystine. Although this particular biochemical pathway has not been substantiated experimentally, Toennies' idea of the formation of sulphonium compounds from methionine has nevertheless proved correct with the discovery and isolation of several compounds of this type.

It was not until 1948 that the first sulphonium salt was established as occurring naturally. In that year Challenger and Simpson[5] described the isolation of a methylthetin from a marine alga. The term 'thetin' had been proposed by Brown and Letts[6] in 1878 to describe sulphonium compounds structurally analogous to the betaines. Although a methylsulphonium derivative of methionine had been synthesized by Toennies[4] and had been employed in a number of biochemical studies[7,8], the real emergence of a biochemical role for sulphonium salts coincided with the discovery of the algal thetin, when Du Vigneaud and coworkers[9,10] demonstrated the ability of methylthetins to act as nutritional sources of methyl groups for animals, and at about the same time Dubnoff and Borsook[11], working with liver preparations, showed that these compounds were involved in enzyme-catalysed methylation reactions.

The subsequent three decades have seen a most dramatic and diversified expansion of these early investigations, and the appearance of well over 1000 publications exploring and establishing the biochemistry of sulphonium compounds. Various aspects of the subject have also been discussed in depth in several reviews[12-19], and the reader is referred to these for further information. It is now clear that although there are few known naturally occurring compounds of this class, they occupy a key position in the intermediary metabolism of animals, plants, and microorganisms. In fact, one sulphonium derivative of the amino acid methionine, S-adenosylmethionine **4**, appears to be virtually as ubiquitous as adenosine triphosphate, and is implicated in almost every aspect of metabolism.

This chapter deals with the discovery of the naturally occurring sulphonium salts and with their distribution, biosynthesis, and involvement in various facets of metabolism. Particular attention will be paid to their participation in the biological transfer of methyl groups, which, like reactions such as phosphorylation, acylation, and oxidation–reduction, is a universally fundamental process. Recent work suggests that a second sulphonium derivative of methionine, S-methylmethionine, occasionally referred to as vitamin U, may have important medical applications, and the next few years may well see significant advances in our knowledge of the pharmacology of sulphonium salts, possibly coupled with their clinical use. Reference will also be made to sulphonium salts related structurally to the naturally occurring compounds, for example, the ethyl analogue of S-adenosylmethionine, the study of which has contributed substantially to our understanding of such widely differing topics as lipid and protein synthesis, carcinogenesis and mental disorders.

II. NATURAL OCCURRENCE

Despite the intensive examination of a wide range of living organisms, no more than four sulphonium compounds have so far been rigorously established as occurring naturally. These are the salts of 2-carboxyethyldimethylsulphonium (dimethyl-β-propiothetic) **2**, methioninemethylsulphonium (S-adenosyl-L-methionine) **3**, S-(5'-deoxyadenosyl-5')-L-methionine (S-adenosyl-L-methionine) **4**, and the decarboxylated derivative of S-adenosyl-L-methionine, namely S-adenosyl-(5')-3-methylthiopropylamine **5**. These compounds probably occur in combination with a variety of both inorganic and organic cations, and for simplicity have been represented below as their zwitterions.

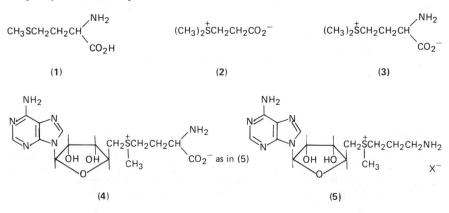

It would be premature to assume that other compounds of this type do not occur, but the evidence for their existence is indirect and, at best, far from convincing. The apparent paucity of sulphonium compounds in nature contrasts sharply with the widespread occurrence of a relatively large number of structurally analogous quaternary ammonium compounds, such as the betaines.

A. Dimethyl-β-propiothetin

This compound was the first sulphonium derivative to be isolated from natural sources in pure form. For some years Challenger of Leeds had been investigating the origin of the dimethyl sulphide reported by Haas[20] in 1935 to be readily evolved from the red marine algae *Polysiphonia fastigiata* and *P. nigrescens* when these seaweeds are exposed to air. In 1948 Challenger and Simpson[5] extracted the precursor of this dimethyl sulphide from *P. fastigiata* and characterized it unequivocally as dimethyl-β-propiothetin **2**. It was recognized that the evolution of dimethyl sulphide was the result of injury to the seaweed or of its exposure to abnormal conditions. Cantoni *et al.*[21] reported the concentration of the thetin in this alga to be as high as 0.04 M or, when estimated as its chloride, as 0.7% of the wet weight of the organism. In other algae the amount of thetin chloride present may vary from 0.07 to 2.9%[22].

Dimethyl-β-propiothetin was subsequently isolated from three other marine algae, *Enteromorpha intestinalis*, *Spongomorpha arcta*, and *Ulva lactuca*[23]. The thetin is readily cleaved by a non-enzymic reaction to dimethyl sulphide in the presence of cold concentrated sodium hydroxide solution, and this has been employed as a test for the presence of the compound in other organisms. In this way it has been detected in other marine and some freshwater algae[24], in several marine invertebrates, and has been established as being of widespread occurrence in various species of plankton[22]. The amount present in these last-mentioned organisms, when expressed as the thetin chloride, was calculated to amount to about 1% of the wet weight of the cells. Marine organisms of this type constitute the principal food of many fish and filter-feeding creatures, which may account for the finding of dimethyl-β-propiothetin in appreciable quantities in the tissues of the North Pacific salmon[25] and in various molluscs and crustacea[26]. Dimethyl sulphide is one of the main volatile flavour components of fresh Pacific oysters, and the thioether probably originates from the thetin contained in ingested algae[27]. This may also explain the presence of dimethyl sulphide in some freshly caught fish, such as the Labrador cod[28] and the Baltic herring[29].

A further evolution of dimethyl sulphide occurs when the above-mentioned marine and freshwater algae are heated with alkali; in addition, some algal species produce the thioether only as a result of this further treatment. It has not yet been established whether this indicates that a proportion of the thetin present is in a bound or less readily available form, or that dimethylsulphonium derivatives other than dimethyl-β-propiothetin are present.

B. *S*-Methyl-L-methionine

During a survey of other natural sources of dimethyl-β-propiothetin, Challenger and Hayward[30] found that asparagus tips evolved dimethyl sulphide, but this required prolonged boiling of the plant tissue with alkali, indicating the possible presence of a different sulphonium compound. The precursor of the thioether was isolated and found to be not the above thetin but a derivative of the amino acid L-methionine, namely *S*-methyl-L-methionine **3**. Evidence was also obtained for the occurrence of this second sulphonium compound in common bracken (*Pteridium aquilinum*) and in several species of *Equisitum* ('Horse's tail')[24]. Some months previously Shive and coworkers[31] had reported the isolation of *S*-methyl-L-methionine from cabbage, and the compound was detected in the foliage of a range of higher plants, including parsley, pepper, onion, lettuce, and turnip. Its presence has since been established in tomato foliage and fruit[32], potato[33], and green tea[34]. Structurally, *S*-methylmethionine is related to the algal thetin and can be regarded as α-aminodimethyl-γ-butyrothetin. It appears to be a compound characteristic of plants, so that its identification in cow's milk[35] is probably the result of its transference from the herbage of the diet.

Employing a procedure involving alcohol extraction and ion-exchange purification, followed by two-dimensional chromatography, Kovacheva[36]

TABLE 1. *S*-Methylmethionine content of various vegetables. (Data from Kovacheva[36]; reproduced with permission of Plenum Publishing Corporation, New York)

	S-Methylmethionine content	
Source	mg/100 g fresh tissue	mg/100 g dry weight of tissue
Cabbage leaves	7.5–12	123–197
Celery leaves	1.0–1.4	10.5–14.7
Celery roots	6.5–7.5	57.5–66.4
Radish	0.8–1.2	17.4–26.1
Turnip	3.0–4.0	54.5–72.7
Kohlrabi	7.5–8.5	81.0–91.8
Parsley	0.8–1.6	9.4–18.8
Green onion	0.6–1.0	8.0–13.3
Carrot	0.5–0.6	5.4–6.5
Potato	0.1–0.3	0.6–1.6
Tomato fruit, unripe	1.8–3.0	20.0–33.3
Tomato fruit, ripe	0.4–0.6	7.9–11.9

determined the content of *S*-methylmethionine in a range of vegetables (see Table 1). Cabbage leaves and kohlrabi were found to contain relatively high levels of the compound, corresponding to as much as 0.2% and 0.1% of the tissue dry weight, respectively. Another analytical method, employing electrophoretic rather than chromatographic separation, has been used to verify the presence of *S*-methylmethionine in various other plants, including radish, beetroot, plantain, and broad bean[37]. As will be discussed in Section V, *S*-methylmethionine has aroused considerable interest particularly in Japan and the USSR, on account of its purported wide spectrum of medicinal properties.

C. *S*-Adenosyl-L-methionine

The discovery of this compound was the culmination of research begun in the early 1930s on the physiological significance of the amino acid L-methionine. Growth studies made on rats had established the indispensibility of the *S*-methyl group of methionine in nutrition and metabolism, for it was found that methionine could not be replaced in the diet of growing rats by homocysteine **6** or homocystine **7** unless certain other compounds, including the quaternary

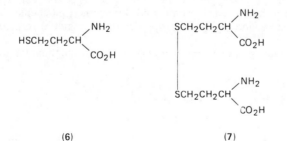

HSCH₂CH₂CH(NH₂)CO₂H

(6)

SCH₂CH₂CH(NH₂)CO₂H | SCH₂CH₂CH(NH₂)CO₂H

(7)

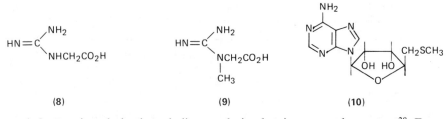

(8)   (9)   (10)

methylammonium derivatives choline or glycine betaine, were also present[38]. From this finding has developed the concept of 'transmethylation', or the biological transfer of methyl groups from one metabolite to another. The emergence of transmethylation as a fundamental biochemical process has been described in several reviews by Du Vigneaud[12,39], whose research school was responsible for setting it on a sound experimental basis. Detailed accounts of the topic can also be found in a number of other publications[13-19,40-42]. Methyl transfer reactions have subsequently become of paramount biological importance and the associated literature so ramified and extensive that they will be discussed in various sections only as far as the participation of sulphonium compounds is concerned.

In contrast to the now classical studies of Du Vigneaud on whole animals, experiments were being carried out by Borsook and Dubnoff at the cellular level. In 1940 they reported that the muscle constituent creatine 9 was formed from guanidinoacetic acid (glycocyamine) 8 in rat liver slices[43]. This methyl transfer reaction required aerobic conditions and was enhanced by the addition of methionine to the medium. Further, in the presence of guinea-pig liver suspensions, the formation of creatine was more than doubled by the addition of either adenosine triphosphate (ATP) or a source of this compound[44]. The assumption was made that the aerobic conditions were needed to ensure the continuous provision of ATP which was involved in some way in the methyl transfer reaction. Other workers also showed that nicotinamide could be methylated by methionine in liver slices, and this reaction was likewise essentially aerobic[45].

Following up these observations, Cantoni[46] showed that ATP also participated in the methylation of nicotinamide by methionine, but he made the fundamentally important observation that ATP in this reaction was not acting in its usual role as a phosphorylating agent, but that it brought about the enzymic conversion of methionine to a biochemically active form before the methyl group of the amino acid was transferred. This 'active methionine' could be produced from methionine and ATP in the presence of an enzyme system occurring in mammalian liver. The compound contained no phosphorus, and from a study of its properties Cantoni concluded that it was a sulphonium compound containing an S-adenosyl moiety, with the structure 4 given in the previous section[47]. The correctness of this formula was established when the compound was synthesized chemically and shown to have the chemical and biological properties of the material formed enzymically by rabbit liver[48,49].

S-Adenosylmethionine possesses a chiral centre at the sulphonium group:

R = $-CH_2CH_2CH(NH_2)CO_2H$

The α-carbon atom of the amino acid moiety is also chiral, so that if the asymmetry associated with the pentose ring is excluded, there are four diastereoisomers of biochemical significance. The initial chemical synthesis of S-adenosylmethionine from 5'-methylthioadenosine **10** and DL-2-amino-4-bromobutyric acid was assumed to result in a mixture of all four diastereoisomers, *viz.* (±)-S-adenosyl-DL-methionine. Subsequently, (±)-S-adenosyl-L-methionine and (±)-S-adenosyl-D-methionine have been prepared by chemical methylation of the appropriate S-adenosylhomocysteine isomers. These two isomers of S-adenosylmethionine are chemically identical with the 'active methionine' produced biosynthetically from rabbit liver or from baker's yeast, but they differ with respect to their optical rotations (see Table 2)[50].

Furthermore, the product obtained by chemical synthesis from S-adenosyl-L-homocysteine was found to possess only 50% of the activity of the biosynthetic product in certain enzymic reactions (see Section IV.C), indicating that the enzymes concerned differentiate between the configurations of groups attached to the sulphonium group. Mudd and coworkers[50] were able to prepare (+)-S-adenosyl-L-methionine from (±)-S-adenosyl-L-methionine by preferential enzymic utilization of the (−)-isomer and to show its distinct difference in optical activity from the biosynthetically produced (−)-isomer, which can be regarded as the naturally occurring diastereoisomer. This work appears to be the first biological resolution of a compound which is optically active at a sulphonium centre.

S-Adenosylmethionine appears to be a constituent of virtually all living tissue. It has been detected in a wide range of bacteria, fungi, and plants, as well as various tissues of higher animals. Much attention has been paid to its occurrence in yeasts, such as *Saccharomyces cerevisiae*, the cells of which normally contain about 0.3–0.8 μmol of S-adenosylmethionine per gram of fresh tissue. The sulphonium compound is thus present at concentrations of the same order of magnitude as nicotinamide–adenine dinucleotide (NAD) or ATP. However, when the culture medium is supplemented with L-methionine at concentrations up to 5 μmol/ml, the cellular content of S-adenosylmethionine is increased to more than 20 μmol/g fresh weight, and this can be further increased to as much as 50 μmol/g by the addition

TABLE 2. Optical rotations of diastereoisomers of S-adenosylmethionine. (Data from De La Haba *et al.*[50]; reproduced with permission of the American Chemical Society)

Compound	Source	Concentration (%)	Specific rotation (degrees) 589 nm	Specific rotation (degrees) 436 nm
(−)-S-Adenosyl-L-methionine	Methionine-activating enzyme of rabbit liver	1.78	+48.5	+100.5
	Methionine-activating enzyme of baker's yeast	0.92	+47.2	+101.2
(±)-S-Adenosyl-L-methionine	Chemical methylation of S-adenosyl-L-homocysteine	1.81	+52.2	+106
(+)-S-Adenosyl-L-methionine	Enzymic resolution with guanidinoacetate methyl-transferase	1.83	+57	+115
(±)-S-Adenosyl-D-methionine	Chemical methylation of S-adenosyl-D-homocysteine	1.29	+16	+33

of extra supplements[51–53]. The yeast *Candida utilis* contains up to 1.6 μmol of the sulphonium compound per gram of cells when grown on standard medium, and this can be increased to up to 20–40 μmol/g after supplementation with L-methionine. The 50–100-fold enrichment of yeast cells when grown in the presence of the parent amino acid, coupled with improved isolation procedures, has provided an excellent source of *S*-adenosylmethionine for metabolic studies and, indeed for the commercial production of the compound.

Methionine taken up by *S. cerevisiae* is converted into *S*-adenosylmethionine with extreme rapidity. In experiments with ³H-labelled L-methionine, half-maximal specific activity of the resulting *S*-adenosylmethionine was reached within 30 s of the addition of the amino acid to the culture medium[54]. It has been calculated that a logarithmically growing yeast cell contains about 2×10^6 molecules of the sulphonium compound. It becomes accumulated within the cell in the vacuole[51,53,55] where, presumably, it appears to be metabolically inert, acting as a reservoir to maintain the concentration of the compound in the surrounding cytoplasm for immediate metabolic functions.

In contrast to yeast cells, animal tissues contain appreciably lower amounts of *S*-adenosylmethionine. For example, values for the contents of rat brain and liver have been given as 30 and 83 nmol/g fresh tissue, respectively[56]. Using an improved and more specific isotope dilution method of analysis, Salvatore *et al.*[57] obtained contents ranging from 15 to 70 nmol/g fresh tissue in rabbit brain and liver, respectively, and some of their data are given in Table 3. Figures cited elsewhere for various animal tissues generally range from 40 to 140 nmol/g fresh weight[58].

These low values for a metabolite with such a varied and biochemically essential role imply that the compound has a very high metabolic turnover rate, and in this respect it is akin to the coenzymes NAD and flavin–adenine dinucleotide (FAD). Such low concentrations contrast with the much higher levels of dimethyl-β-propiothetin and *S*-methylmethionine found naturally. In algae, for example, the thetin is present at concentrations as much as 500–1000-times that of *S*-adenosylmethionine in animal tissues. This clearly indicates that the thetin and probably *S*-methylmethionine have different biochemical roles from the adenosyl

TABLE 3. *S*-Adenosylmethionine content of various mammalian tissues. (Data from Salvatore *et al.*[57]; reproduced with permission of Academic Press, New York)

Source	*S*-Adenosylmethionine content (nmol/g wet weight of tissue)
Rat liver	53
kidney	45
heart	50
spleen	48
Rabbit liver	70
kidney	20
lung	53
brain	15
skeletal muscle	30
Calf liver	55
Chicken liver	58
Eel liver	37
Human blood serum	30[a]

[a]Expressed as nmol/ml.

sulphonium compound, possibly concerned with osmotic regulation or the storage of essential metabolite precursors.

D. S-Adenosyl-(5′)-3-methylthiopropylamine

Recently, Ito and Nicol[59] found that the tapetum lucidum, or reflecting material of the eyes of a range of teleost fish, including the sea catfish (*Aris felis*), consists of acidic and basic components, and among the latter they isolated S-adenosyl-(5′)-3-methylthiopropylamine **5**, which is present together with its demethylated derivative, S-adenosyl-(5′)-3-thiopropylamine[60]. This sulphonium compound is the decarboxylation product of S-adenosylmethionine, and is present in relatively large amounts (0.9 µmol/g fresh weight of eye) compared with those of S-adenosylmethionine itself in other tissues (*ca.* 0.05 µmol/g fresh weight). This amount represents some 17% of the dry matter of the purified reflecting material, or 3.4% of the total dry matter of the eye.

Decarboxylated S-adenosylmethionine has already been recognized to be of special significance in other tissues as an intermediate in the conversion of the diamine putrescine (1,4-diaminobutane) to the bases spermidine **84** and spermine **85** (see Section IV.C.3), and is capable of transferring its methyl group enzymatically to homocysteine. Its precise role in the eye at such high concentrations has yet to be fully established, but one of its functions might be to make the tapetal pigment less soluble by forming a covalently bound complex.

E. Other Sulphonium Compounds

In 1895, Abel[61] reported that dog urine, when treated with hot alkali, gave rise to a volatile sulphur compound which he concluded was diethyl sulphide. A later reinvestigation by Challenger and coworkers[14,62], however, revealed that the evolved volatile material was more probably a mixture of methyl *n*-propyl and methyl *n*-butyl sulphides. These thioethers appear to be the cleavage products of two sulphonium salts present in the urine, with cations having the structures **11** and **12**. The nature of the third substituent group (or groups) still remains to be identified.

(11)   (12)

III. BIOSYNTHESIS

There is good evidence that at least one major pathway for the biosynthesis of each of the four known naturally occurring sulphonium compounds involves the amino acid methionine as an intermediate. It is therefore relevant to consider the formation of methionine, although for reasons of space it will be considered in outline only. More detailed accounts of this aspect of sulphur biochemistry can be found in numerous reviews[13,17,63,64].

A. L-Methionine

Microorganisms and plants are able to form methionine from the simplest materials – C_1 units, inorganic sulphate, and ammonia. The C_4 chain of the amino acid arises via the initial biosynthesis of homoserine **14** from aspartic acid **13**:

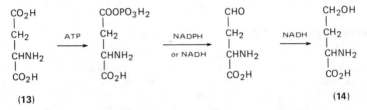

(13)   (14)

The sulphur of methionine originates from inorganic sulphate, which is initially reduced to hydrogen sulphide. This occurs by a multi-step pathway involving the preliminary activation of the sulphate ion by a two-stage interaction with ATP to form 3′-phosphoadenosine-5′-phosphosulphate **15**, generally abbreviated to

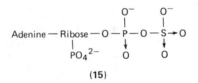

(15)

PAPS[65]. PAPS undergoes subsequent reduction by a series of reactions, first to a sulphite derivative and then to inorganic sulphide. The latter reacts with serine **16** in the presence of the enzyme serine sulphydrase to give cysteine **17**.

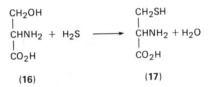

(16)   (17)

Interaction of the *O*-succinyl ester of homoserine **18** with cysteine occurs in the presence of the enzyme cystathionine synthase to form the intermediate thioether, cystathionine **19**. Hydrolytic fission of the C—S bond of the cysteinyl residue is

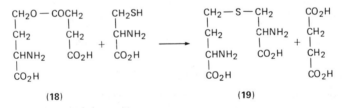

(18)   (19)

catalysed by the β-cleaving enzyme cystathionase, giving homocysteine **6** and pyruvate[66].

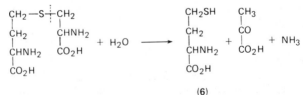

(6)

The final methylation of homocysteine is intrinsically an alkyl transfer reaction, but the methyl donor is the relatively complex structure $N_{(5)}$-methyltetrahydropteroyl monoglutamate **22** or trigulamate, generally abbreviated to $N_{(5)}$-methyltetra-hydrofolate, or further to 5-CH$_3$-H$_4$PteGlu. In outline, the methylation is as follows:

$$\text{Homocysteine} + 5\text{-CH}_3\text{-H}_4\text{PteGlu} \rightarrow \text{Methionine} + \text{H}_4\text{PteGlu} + \text{H}^+$$

the enzyme mediating the reaction being $N_{(5)}$-methyltetrahydrofolate:homocysteine methyltransferase. However, there appear to be two pathways for this final methyl transfer[67,68]. Both are related, one being dependent on the presence of a cobalamin (vitamin B$_{12}$) as the prosthetic group of the corresponding methyltransferase and requiring the additional presence of the cofactors FADH$_2$, NADH and the sulphonium compound S-adenosylmethionine **4**. The methyl donor may be either

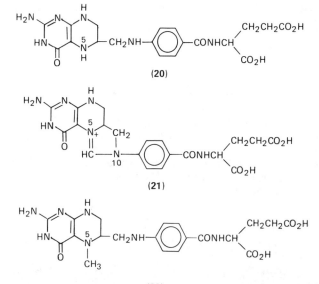

(20)

(21)

(22)

$N_{(5)}$-methyltetrahydrofolate mono- or triglutamate. The cobalamin is believed to act as an intermediate methyl carrier[69]:

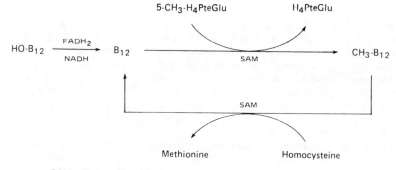

SAM = S-adenosylmethionine

This B_{12}-dependent methylation operates in most microorganisms and plants, and is of particular interest in that it is also present in mammalian tissues[70].

A second pathway occurs, often alongside the first, in many bacteria such as *Escherichia coli* and *Aerobacter aerogenes*, and has been detected in plants. It is simpler, mechanistically, does not require the participation of vitamin B_{12} and utilizes only $N_{(5)}$-methyltetrahydrofolate triglutamate as methyl donor.

The methyl group of $N_{(5)}$-methyltetrahydrofolate can initially be generated as a C_1 fragment from serine[71]. This is transferred to tetrahydrofolate **20** in the presence of serine transhydroxymethylase to give $N_{(5)},N_{(10)}$-methylenetetrahydrofolate **21**, which in turn is reduced to $N_{(5)}$-methyltetrahydrofolate **22**. The terminal methylation of homocysteine has also been shown to be brought

$$
\begin{array}{c}
CH_2OH \\
| \\
CHNH_2 \\
| \\
CO_2H
\end{array}
+ H_4PteGlu \longrightarrow
\begin{array}{c}
CH_2NH_2 \\
| \\
CO_2H
\end{array}
+ N_{(5)}, N_{(10)}\text{-Methylene-}H_4PteGlu
$$

$$
\downarrow
$$

$$
N_{(5)}\text{-}CH_3\text{-}H_4PteGlu
$$

about in bacteria, yeasts, and plants in the presence of *S*-adenosylmethionine **4** or *S*-methylmethionine **3** as methyl donors[15,72,73]. However, neither of these mechanisms can account for the *de novo* synthesis of methyl groups, and must be regarded purely as alternative pathways for the transfer of methyl groups generated via tetrahydrofolate.

An alternative pathway for methionine biosynthesis, not involving cystathionine as an intermediate, is possible, since *O*-succinylhomoserine has been shown to react enzymically with methanethiol[74]:

$$
\begin{array}{cc}
CH_2O - COCH_2 & \\
| \quad\quad | & \\
CH_2 \quad\quad CH_2 & \\
| \quad\quad | & + CH_3SH \longrightarrow \\
CHNH_2 \quad CO_2H & \\
| & \\
CO_2H &
\end{array}
\quad
\begin{array}{cc}
CH_2SCH_3 & CO_2H \\
| & | \\
CH_2 & CH_2 \\
| & + | \\
CHNH_2 & CH_2 \\
| & | \\
CO_2H & CO_2H
\end{array}
$$

Whether this transmethylation is of physiological significance, however, is not known.

In contrast to lower forms of life, higher animals and man are unable to achieve the total synthesis of methionine, and require the amino acid largely pre-formed. They can bring about partial synthesis from either the demethylated derivative, homocysteine, or its disulphide, homocystine, or the deaminated derivative, *viz.* the corresponding α-keto acid, $CH_3SCH_2CH_2COCO_2H$. Animals are unable to synthesize the homocysteinyl moiety because they lack the cystathionase capable of catalysing the β-cleavage of cystathionine. They possess instead a γ-cleaving cystathionase in the liver which splits the thioether to cysteine and α-ketobutyrate[66]. The latter enzyme is also present in most microorganisms and plants, which accounts for the ability of lower organisms to bring about the reversible interconversion of cysteine and methionine, whereas animals can only degrade methionine to cysteine.

Under normal circumstances, animals obtain their methionine in the diet, either directly from plant protein or from the flesh of herbivores who are themselves dependent upon plant protein as well as on the bacterial synthesis of methionine in the rumen. Dietary methionine can be replaced experimentally by homocysteine or homocystine, provided that a source of biologically available methyl groups is also present, for example choline in the form of phospholipids, or glycine betaine[12,19]. In practice, however, homocysteine and homocystine do not occur in significant amounts in the diet, although they may be involved in the intracellular utilization and turnover of pre-formed methyl groups originating from the diet.

B. Dimethyl-β-propiothetin

Methionine has been found to be an efficient precursor of dimethyl-β-propiothetin **2** in the marine alga *Ulva lactuca*, which contains relatively large amounts of the thetin[75]. Using L-methionine labelled with ^{35}S, with ^{14}C in the carbon chain and ^{14}C and ^{3}H in the methyl group. Greene[75] established that the methyl group and the sulphur of the amino acid are incorporated into both methyl groups and the sulphonium group of the thetin, also that the α-carbon of methionine is the precursor of the carboxyl-carbon of the thetin. The probable pathway of the biosynthesis is as follows:

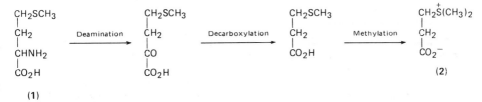

(1)

The source of the methyl group for the terminal methylation is thought to be $N_{(5)}$-methyltetrahydrofolate, as discussed previously.

C. *S*-Methyl-L-methionine

The biosynthesis of *S*-methyl-L-methionine **3** has been shown to occur in a variety of plant species, including oat, tobacco, pumpkin, and onion[76–79]. When jack bean seedlings were grown in solutions containing ^{14}C- or ^{35}S-labelled L-methionine, up to 80% of the non-protein radioactivity was found to be in the form of *S*-methylmethionine, the sulphonium compound being located almost entirely in the roots[80]. The enzymic methylation of methionine has also been demonstrated in cell-free extracts of jack bean roots and hypocotyls[80], and also in apple tissue[81]. The reaction requires the presence of *S*-adenosylmethionine[80,82] and proceeds according to the equation

$$\text{Methionine} + S\text{-Adenosylmethionine} \rightarrow S\text{-Methylmethionine} \\ + S\text{-Adenosylhomocysteine}$$

The relatively high concentration in plants of *S*-methylmethionine compared with that of *S*-adenosylmethionine suggests that the former compound may have some important metabolic function. An interesting clue to this has come from recent work on the biosynthesis of ethylene in flowers of the plant Morning Glory (*Ipomoea tricolor*). Ethylene is an important plant growth regulator concerned with the ageing of tissues and in many if not all higher plants its major precursor is L-methionine. Tracer studies by Hanson and Kende[83] showed that

S-methylmethionine is actively synthesized from methionine in flower tissue from Morning Glory, but that during flower senescence, which is associated with the production of ethylene, there is more than a 10-fold increase in tissue methionine content with little change in the concentration of *S*-methylmethionine. The amino acid does not arise simply from the degradation of *S*-methylmethionine but is the result of an active methylation of homocysteine by the sulphonium compound. Hanson and Kende speculated that during senescence ethylene acts to increase the availability of homocysteine, either by directly stimulating its biosynthesis or by altering cellular compartmentation to release stored homocysteine. The homocysteine undergoes methylation by *S*-methylmethionine in the presence of a methyltransferase, forming methionine which can then generate more endogenous ethylene. This could be one explanation of the dramatic increase in ethylene production which occurs during the ripening of flowers and fruits, and would place *S*-methylmethionine as a source of ethylene as well as being a methyl donor in plants. Its significance in the latter role has also been suggested in connection with the methylation of cell-wall pectins[76].

D. *S*-Adenosyl-L-methionine

1. Biosynthesis from L-methionine

Following the discovery that ATP participates as an essential cofactor in those methyl transfer reactions in which methionine serves as the methyl donor[44,46], Cantoni showed that the role of ATP was to convert the amino acid into a metabolically active form. The activation is catalysed by an enzyme system found in extracts of rat, pig, beef, and rabbit liver, and one of the other products of the reaction is orthophosphate[46]. With the recognition of 'active methionine' as a sulphonium compound, and the elucidation of its structure and its total synthesis by Baddiley and coworkers[48,49], a more detailed examination has been made of its biosynthesis from L-methionine and ATP.

A novel feature of the reaction is that during the biosynthesis ATP is completely dephosphorylated, in contrast to its partial breakdown to adenosine diphosphate (ADP) in the majority of its metabolic reactions. A nucleophilic transfer of the 5′-deoxyadenosyl moiety of ATP to the methionine-S atom takes place without the involvement of ADP or adenylic acid (AA) as intermediates. Experiments with ^{32}P-labelled ATP indicated that the terminal phosphate is released as inorganic phosphate, the remaining two phosphate groups being split from the ribose moiety as pyrophosphate[84,85]. Later work showed that the activating enzymes from both yeast and liver form predominantly the (−)-diastereoisomer of *S*-adenosyl-L-methionine[50] (see Section II.C). The reaction can be formulated as follows:

L-Methionine + Adenosine triphosphate → (−)-*S*-Adenosyl-L-methionine
+ Pyrophosphate + Inorganic phosphate

The enzyme responsible for the reaction has been named methionine-activating enzyme, or *S*-adenosylmethionine synthetase, but is now referred to as ATP: L-methionine *S*-adenosyltransferase.

An important contribution to the study of this enzyme was the finding of Mudd and Mann[86] that enzyme-bound tripolyphosphate is an obligatory intermediate in the reaction, and that preparations of the adenosyltransferase possess tripolyphosphatase activity, enabling the enzyme to cleave tripolyphosphate in a specific manner[87,88]. The biosynthesis of S-adenosylmethionine (SAM) from methionine (M) has been represented as a four-stage process, involving the transient formation of a ternary complex between the substrates and the adenosyltransferase (represented as E):

$$E + ATP \xrightleftharpoons[(a)]{} E\cdots ATP \xrightleftharpoons[(b)]{M} M\cdots E\cdots ATP \xrightleftharpoons[(c)]{} SAM\cdots E\cdots PPP_i \xrightleftharpoons[(d)]{}$$

$$SAM + E + PP + P_i$$

Step (d) is catalysed by the tripolyphosphatase function of the enzyme[58,85]. Step (c) is thought to be rate limiting in both directions, and both S-adenosylmethionine and tripolyphosphate are more tightly bound to the enzyme than are ATP, methionine, pyrophosphate, or inorganic phosphate, thus facilitating the release of S-adenosylmethionine. The equilibrium of the overall reaction strongly favours synthesis of the sulphonium compound; it is irreversible with S-adenosylmethionine plus pyrophosphate and inorganic phosphate as reactants, and proceeds only to a limited extent with the sulphonium compound and tripolyphosphate[86].

The enzyme thus apparently has two catalytic functions, an adenosyltransferase activity and a tripolyphosphatase activity. It requires Mg^{2+} in relatively high concentrations, together with a monovalent cation, such as K^+ or NH_4^+, it is stimulated by S-adenosylmethionine at low concentrations, when the sulphonium compound is probably acting as an allosteric effector, but at high concentrations S-adenosylmethionine exhibits product inhibition[87]. The enzyme is widely distributed in most organisms, which accounts for the ubiquitous appearance of the sulphonium compound. It is present in practically all animal tissues, the liver being by far the richest source of the enzyme, and containing up to twenty times the concentration found in other tissues.

The adenosyltransferase shows an absolute specificity towards ATP, no other nucleoside triphosphates, including the closely related inosine triphosphate and 2′-deoxyadenosine triphosphate being significantly active as substrates. In fact, many common nucleoside triphosphates, such as those of inosine, guanine, cytosine, uridine, and thymine act as competitive inhibitors with respect to ATP. Tripolyphosphate and also tetrapolyphosphate ions are fairly powerful inhibitors in competition with ATP, indicating that the phosphate groups of ATP represent one point of attachment of the substrate to the adenosyltransferase[58].

The enzyme is less specific towards the amino acid substrate, however. The α-hydroxy analogue of L-methionine (2-hydroxy-4-methylthio-n-butyric acid) and the S-ethyl analogue, ethionine, are converted into the corresponding S-adenosylsulphonium derivatives, although at a slower rate[15]. The last mentioned compound, S-adenosylethionine, has proved of considerable biochemical interest and is discussed in more detail in Section IV.D. The enzyme can also utilize the D-isomer of methionine as substrate. Biosynthesis of S-adenosyl-D-methionine can be readily carried out in the presence of yeasts, C. utilis being more effective in this respect than S. cerevisiae[88]. A further substrate for the enzyme is the selenium analogue of methionine[89], an illustration of the parallelism between many aspects of sulphur and selenium biochemistry.

A number of other analogues of methionine, as well as closely related compounds, are competitive inhibitors of the enzyme. These include several C_6 acids, for example, DL-2-amino-trans-4-hexanoic acid (trans-crotylglycine) 23 and L-2-amino-5-chloro-trans-4-hexanoic acid 24. These analogues have been used to elucidate the complex kinetics of the adenosyltransferase reaction[58]. Two other

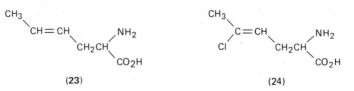

(23)   (24)

extremely effective inhibitors of the reaction are L-2-aminohex-4-ynoic acid **25** and 1-aminocyclopentane-1-carboxylic acid (cycloleucine) **26**.

(25)   (26)

In the animal there appear to be sex differences in the level of the adenosyltransferase in the liver, female animals having higher concentrations, and the administration of androgens causing a lowering of the enzyme level in castrated males. The amount of the enzyme present in tissues is believed to be considerably in excess of physiological requirements. Its activity is probably controlled largely by the availability of substrate methionine, although S-adenosylmethionine itself exerts control as an allosteric effector at low concentrations and through product inhibition at high concentrations[90]. In microorganisms, e.g. yeasts, addition of methionine to the growth medium represses methionine biosynthesis[91], but S-adenosylmethionine appears to play a more dominant role through feedback inhibition.

The ready formation of S-adenosylmethionine from methionine can explain some of the toxic effects produced by the administration of high doses of methionine to animals, since this causes an immediate rise in tissue methionine levels, particularly in the liver and, as a result of the ensuing S-adenosylmethionine biosynthesis, can induce an acute deficiency of ATP[92,93], with consequent effects on other metabolic pathways dependent upon the availability of ATP for phosphorylation reactions. This situation also occurs in yeast when the culture medium contains an excess of methionine. The growth inhibition which results from a deficiency of ATP can be largely circumvented by supplementation of the growth medium with adenine[94].

2. Other biosynthetic pathways

In addition to the pathway homocysteine $\xrightarrow{C_1}$ methionine $\xrightarrow{ATP}$ S-adenosylmethionine, at least two other mechanisms for the synthesis of S-adenosylmethionine are possible. There is evidence in the yeasts S. cerevisiae and C. utilis[95], as well as in mammalian liver[96] and many other animal tissues[140], for an enzyme-catalysed reaction of adenosine with L-homocysteine yielding S-adenosyl-L-homocysteine, which can then be methylated by $N_{(5)}$-methyltetrahydrofolate to S-adenosyl-methionine:

$$\text{Adenosine} + \text{L-Homocysteine} \rightleftharpoons \text{S-Adenosyl-L-homocysteine} + H_2O$$

The enzyme catalysing the initial reaction is S-adenosyl-L-homocysteine hydrolase, but the K_m of the reaction is such as to greatly favour the reaction in the reverse direction, namely, the interaction to give the thioether.

Isotopic labelling studies have shown that 5'-methylthioadenosine 10 can also act as a precursor of S-adenosylmethionine in *C. utilis*[97], and it has been proposed that this nucleoside reacts with a C_4 amino acid precursor, although the nature of this reactant is as yet unknown. An alternative pathway is possible in bacteria, which contain a nucleosidase capable of splitting off the adenine moiety of 5'-methylthioadenosine, leaving 5'-methylthioribose[98]:

$$\text{5'-Methylthioadenosine} + H_2O \rightarrow \text{Adenine} + \text{5'-Methylthioribose}$$

A cleavage of 5'-methylthioribose to methanethiol, by analogy with the established enzymic breakdown of S-ribosyl-L-homocysteine to L-homocysteine[99] (see Section IV.C.1), could be followed by interaction of the thiol with O-succinylhomoserine to give methionine, as described earlier (see Section III.A), the amino acid then being activated by ATP:

$$\text{5'-Methylthioribose} \rightarrow \text{Methanethiol} + C_5\text{-fragment}$$
$$\text{Methanethiol} + O\text{-succinylhomoserine} \rightarrow \text{Methionine} + \text{Succinic acid}$$
$$\text{Methionine} + \text{ATP} \rightarrow S\text{-Adenosylmethionine} + PP + P_i$$

The formation of S-adenosylmethionine from 5'-methylthioadenosine in all organisms, and that from homocysteine and S-adenosylhomocysteine in animals, are mechanisms for the regeneration of the sulphonium compound, rather than its *de novo* synthesis, since they are in effect recycling its metabolic breakdown products. Nevertheless, by conserving essential parts of the S-adenosylmethionine molecule for repeated use, they could be of considerable importance in the sulphur economy of organisms.

E. S-Adenosyl-(5')-3-methylthiopropylamine

The biosynthesis of this recently discovered sulphonium compound 5 does not appear to have been studied in any detail in the tissues from which it has been isolated. However, a natural assumption would be that the compound arises from L-methionine via S-adenosyl-L-methionine, which is then decarboxylated, although alternative pathways may be operative. The decarboxylation of S-adenosyl-methionine is also an important step in the biogenesis of polyamines, and will be described in more detail in Section IV.C.3.

IV. METABOLISM

The initial step in the metabolism of most sulphonium compounds is almost invariably an enzyme-catalysed C—S bond fission, leading to the release of a thioether. However, when an amino acid moiety is attached to the sulphonium pole, modifications of this part of the molecule, e.g. decarboxylation or deamination, may first occur. The presence of different sulphur-linked groups can result in the formation of different thioethers, depending on which C—S bond is broken, and the cleavage of a particular C—S bond usually requires the mediation of a specific enzyme. This accounts for the different pathways in the catabolism of the naturally occurring sulphonium salts and related compounds of this class.

A. Simple Thetins and Related Compounds

1. Cleavage to thioethers

The production of dimethyl sulphide by marine algae, which led to the discovery and isolation of dimethyl-β-propiothetin, is an example of such a C—S bond cleavage. The reaction was studied by Cantoni et al.[21] in cell-free extracts of *P. fastigiata* and shown to be analogous to the non-enzymic decomposition of the thetin by alkali, acrylic acid being the other product:

$$(CH_3)_2\overset{+}{S}CH_2CH_2CO_2^- \longrightarrow (CH_3)_2S + CH_2 = CHCO_2H$$

The cleavage enzyme shows a fairly high degree of substrate specificity in that the lower homologue, dimethylacetothetin, $(CH_3)_2\overset{+}{S}CH_2CO_2^-$ (originally referred to as dimethylthetin), is attacked only slowly and S-methylmethionine and S-adenosylmethionine not at all.

This mechanism may also be the first step in the bacterial fermentation of dimethyl-β-propiothetin by a mud-dwelling *Clostridium* which can utilize the thetin as its sole source of carbon for growth[100]. The overall stoichiometry of the reaction is

$$3(CH_3)_2\overset{+}{S}CH_2CH_2CO_2^- + 2H_2O \longrightarrow 2C_2H_5CO_2H$$
$$+ CH_3CO_2H + 3(CH_3)_2S + CO_2 + 3H^+$$

The organism contains the enzyme acrylylcoenzyme A aminase, which metabolises the acrylic acid formed initially.

2. Alkyl group transfer

A second type of C—S bond cleavage is feasible in the case of the thetins, leading to the formation of an alkylthiocarboxylic acid and the loss of one methyl group. Like the first reaction, this has proved to be an enzyme-catalysed pathway for the metabolism of thetins. Of greater significance is that it is an example of the widespread reaction of the naturally-occurring sulphonium salts, referred to in Section IV.C.1 as transmethylation, or methyl group transfer, involving the participation of other specific substrate molecules as acceptors of the sulphonium-methyl group. This topic will also be considered in detail in connection with the metabolism of S-adenosylmethionine, as this sulphonium compound undergoes an extremely large number of metabolically important methyl transfer reactions.

The existence of this type of C—S bond fission became apparent from the animal nutrition studies of Du Vigneaud and coworkers[9,10,39], who found that when rats were fed on a diet in which the amino acid methionine was replaced by its demethylated derivative, homocysteine (in the form of its disulphide, homocystine), the animals failed to grow, lost weight, and developed fatty livers and haemorrhagic damage to the kidneys. Alleviation of these symptoms occurred when one of a small group of compounds was included in the 'methyl-deficient' diet, including choline, glycine betaine, its sulphur analogue, dimethylacetothetin, and also dimethyl-β-propiothetin. Diethylacetothetin, lacking the essential methyl groups, failed to support the growth of the animals, whereas the mixed thetin, ethylmethylacetothetin, was partially effective. The natural assumption was that the active compounds were supplying metabolically essential methyl groups for the conversion of homocysteine to methionine.

The enzymic nature of the methyl transfer reaction was demonstrated by Dubnoff and Borsook[11] using suspensions of mammalian liver and kidney, dimethylacetothetin and dimethyl-β-propiothetin acting as methyl donors to L-homocysteine. The reaction involving the acetothetin is as follows:

$$(CH_3)_2\overset{+}{S}CH_2CO_2^- + HSCH_2CH_2CH\begin{array}{c} NH_2 \\ CO_2H \end{array} \longrightarrow$$

$$CH_3SCH_2CH_2CH\begin{array}{c} NH_2 \\ CO_2H \end{array} + CH_3SCH_2CO_2^- + H^+$$

Since a proton is liberated (in contrast to the analogous methylation of homocysteine by glycine betaine), it has been possible to follow the methyl transfer manometrically in a bicarbonate buffer system in terms of the CO_2 released[101].

The enzyme catalysing the reaction, thetin:L-homocysteine methyltransferase, is present in the liver and to some extent the kidneys of mammals, including man[102]. It has been identified in algae[103] but appears to be absent from various plants examined and from yeasts. The enzyme has a specific requirement for L-homocysteine as the methyl acceptor, although the D-isomer is slightly active in this respect, but it can use a number of methylsulphonium salts as methyl donors, as shown for the rat liver enzyme in Table 4[104]. The naturally occurring dimethyl-β-propiothetin and S-methylmethionine can act as substrates, but the most active methyl donor in this system is dimethylacetothetin. Thetin:homocysteine methyltransferase has been purified 400-fold from horse liver, where it represents some 1% of the total protein present[105].

Glycine betaine has also been found to act as a methyl donor for the enzyme, although its substrate activity is only 0.3% of that of dimethylacetothetin[105].

TABLE 4. Activity of various sulphonium halides as substrates of thetin-homocysteine methyltransferase. (Data from Maw[104]; reproduced with permission of the Biochemical Society, London)

Compound[a]	Relative activity[b]
Dimethylacetothetin chloride	100
Ethylmethylacetothetin chloride	85.0
Diethylacetothetin chloride	0
Dimethyl-α-propiothetin bromide	51.9
Dimethyl-β-propiothetin bromide	24.9
Ethylmethyl-β-propiothetin bromide	33.0
Dimethyl-γ-butyrothetin bromide	23.6
Methionine methylsulphonium chloride	17.5
Sulphocholine iodide	0
Trimethylsulphonium chloride	16.5
Triethylsulphonium chloride	0
Ethyldimethylsulphonium iodide	6.6
Butyldimethylsulphonium iodide	4.0

[a]Substrates: DL-homocysteine, 0.03 M; sulphonium halides, 0.1 M.
[b]Activities expressed as percentages of the initial reaction rate obtained with dimethylacetothetin chloride.

Moreover, a second methyltransferase, betaine:L-homocysteine methyltransferase, again located in the liver and kidney, likewise uses not only glycine betaine as methyl donor but also the above-mentioned thetins[106]. However, the possibility that these two enzymes are identical was eliminated by Cantoni and coworkers[107], who showed that the thetin and betaine methyltransferases possess different stabilities and physicochemical characteristics and are distributed differently within the cell Furthermore, in the regenerating liver tissue of rats which had been partially hepatectomized, the concentration of the thetin enzyme falls rapidly to 40% of its original activity, owing to its negligible synthesis and its further dilution by other newly synthesized proteins. In contrast, the concentration of the betaine methyltransferase remains at a level close to the initial, normal value, indicating its active synthesis along with other metabolically important enzymes.

Thetin:homocysteine methyltransferase thus represents something of a biochemical enigma. It is a distinct entity, present in the liver in relatively large amounts, and with a turnover number considerably higher than that of other methyltransferases, yet its apparently preferred substrate is neither a constituent of the diet nor a normal metabolite. If the enzyme does have a physiological function, it may well be unconnected with methionine synthesis or even methyl group transfer. The betaine methyltransferase, although present in tissues at a lower concentration, is now regarded as the principal mammalian enzyme concerned with the mobilization of methyl groups supplied by the diet in the form of choline, after its oxidation to glycine betaine.

3. Oxidation

When dimethylacetothetin or dimethyl-β-propiothetin is administered to rats, about 30% of the methyl-carbon appears as CO_2 in the respired air[108] and up to 60% of the sulphur is excreted as urinary inorganic sulphate[109]. The rate of oxidation of the thetin-methyl groups is similar to that of the methyl groups of betaine, and it is probable that they undergo transfer prior to their oxidation.

Oxidation of the sulphur of sulphonium salts is not a general catabolic pathway, for ethylmethylacetothetin forms sulphate to a lesser extent than does dimethylacetothetin, and diethylacetothetin is not oxidized at all in this way. Trimethyl-, diethylmethyl-, and triethylsulphonium salts are also not converted into inorganic sulphate to any significant extent. The oxidation of the sulphur of dimethylacetothetin and dimethyl-β-propiothetin is unlikely to involve the preliminary cleavage of these compounds to dimethyl sulphide, as this thioether is not further oxidized to sulphate in the rat[109]. However, the respective demethylation products of dimethylacetothetin and ethylmethylacetothetin, namely methylthioacetic and ethylthioacetic acids, form sulphate to approximately the same extent as the parent thetins, suggesting that methyl transfer is an initial and obligatory step in the metabolism of the sulphur of the thetins. This has been confirmed by in vitro experiments, when it was shown that dimethylacetothetin was not oxidized to sulphate by rat liver or kidney slices unless homocysteine was present as the specific methyl acceptor, whereas methylthioacetic acid, the demethylation product of the thetin, was oxidized in the absence of homocysteine[110].

4. Metabolism of sulphocholine

Sulphocholine (2-hydroxyethyldimethylsulphonium salt) 27 is of interest in being the sulphur analogue of choline, a component of phospholipids and the major source of dietary methyl groups for the animal. When fed to rats on a methyl-deficient diet sulphocholine, unlike choline itself, failed to promote growth of the animals and was toxic, unless present at low levels in the diet[111]. However, it could prevent some of the symptoms of methyl deficiency, namely the weight loss characteristic of animals maintained on methyl-deficient diets and also the development of fatty livers and haemorrhagic damage to the kidneys. This is because it may be replacing choline in at least some of its metabolic roles, and it could even make tissue choline available for methyl transfer reactions by a 'sparing' action. In support of this, evidence was obtained for the incorporation of sulphocholine into the liver phospholipids of animals fed the compound. This was subsequently confirmed[112] with the aid of sulphocholine labelled with ^{14}C and ^{35}S. Use of the labelled compound also indicated that sulphocholine forms a number of metabolites in the animal, including the demethylated product, 2-methylthioethanol, and the corresponding sulphoxide and sulphone. Two further metabolites excreted in the urine of animals fed the compound were N-(S-methylthioacetyl)glycine **28** and inorganic sulphate.

$$(CH_3)_2\overset{+}{S}CH_2CH_2OH \quad X^- \qquad\qquad CH_3SCH_2CONHCH_2CO_2H$$

$$\text{(27)} \qquad\qquad\qquad\qquad\qquad\qquad \text{(28)}$$

In vitro experiments have shown that sulphocholine can act as a methyl donor to homocysteine, but that unlike the methylation of homocysteine by dimethylacetothetin or dimethyl-β-propiothetin, the reaction requires aerobic conditions[113]. This strongly suggests that sulphocholine requires to be oxidized to dimethylacetothetin as an initial step in the methyl transfer. In this respect sulphocholine closely parallels choline, which is not itself a direct methyl donor, but requires to be first oxidized to glycine betaine, the natural substrate of betaine:homocysteine methyltransferase[114]. However, the oxidation of sulphocholine to dimethylacetothetin does not appear to take place fast enough to provide an adequate supply of labile methyl groups in the growing animal.

5. Metabolism of trimethylsulphonium salts

Only one type of C—S bond cleavage is possible in this type of compound, leading to the formation of dimethyl sulphide and loss of a methyl group. Trimethylsulphonium chloride acts as a substrate for the thetin:homocysteine methyltransferase of rat liver[104] (see Table 4), having 17% of the activity of dimethylacetothetin as a methyl donor. It has also proved to be a more useful substrate than the thetin itself in studies on the distribution of the enzyme[102]. Administration of the compound to rats resulted in a negligible increase in the excretion of inorganic sulphate[109], since its catabolite, dimethyl sulphide, is largely excreted unchanged or converted into the corresponding sulphone.

An interesting finding has been that a soil bacterium, a strain of *Pseudomonas*, is capable of growing on trimethylsulphonium salts as its sole source of carbon. The organism is able to utilise methyl-carbon for cellular syntheses, while the dimethyl sulphide formed is released. The bacterium contains an enzyme, trimethylsulphonium:tetrahydrofolate methyltransferase, which catalyses the cleavage in the presence of tetrahydrofolate as methyl acceptor[115]:

$$(CH_3)_3\overset{+}{S} + H_4PteGlu \rightarrow (CH_3)_2S + 5\text{-}CH_3\text{-}H_4PteGlu + H^+$$

The $N_{(5)}$-methyltetrahydrofolate formed is known to be an important precursor of

724   G. A. Maw

methionine (see Section III.A), the methyl group of which can be transferred to a
variety of metabolites via the intermediate formation of S-adenosylmethionine.
Surprisingly, the trimethylsulphonium methyltransferase is substrate specific and is
unable to utilize S-adenosylmethionine or dimethyl-β-propiothetin as methyl
donors.

B. S-Methyl-L-methionine

1. Cleavage to dimethyl sulphide

As in the case of dimethyl-β-propiothetin, there are two possible pathways of
C—S bond fission for S-methylmethionine **3**. There is good evidence for an
enzymic mechanism for the removal of the amino acid moiety with the formation of
dimethyl sulphide. This is a feature of various bacteria which are able to use the
sulphonium compound as their sole source of carbon for growth[116,117].
Methioninesulphonium lyase has been suggested as a trivial name for the enzyme
responsible, and the second product of the reaction is homoserine **30**, although, as
in the non-enzymic degradation of S-methylmethionine by alkali, α-amino-γ-
butyrolactone **29** may be formed as an intermediate:

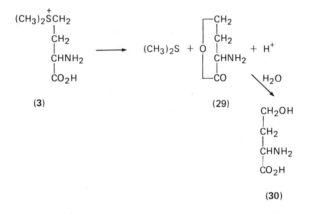

The enzyme also appears to be present in plants, e.g. cabbage[118] and onion[119], and
is distinct from the enzyme in algae which cleaves dimethyl-β-propiothetin.

Enzymic fission of S-methylmethionine to dimethyl sulphide can be brought
about during the brewing of beer by yeasts which act on the S-methylmethionine
originating from germinating barley. The dimethyl sulphide so formed contributes
to beer flavour, but its concentration can occasionally be raised to undesirable
levels through the further action of spoilage bacteria[120]. The ready formation of the
thioether from S-methylmethionine, particularly as a result of heating under mild
alkaline conditions, accounts in part for the presence of dimethyl sulphide as a
constituent of the flavour volatiles of many cooked foods[121], e.g. cabbage, sweet
corn[122], and canned tomato juice[123].

2. Methyl group transfer

Early nutritional studies on animals and microorganisms made with the aim of
identifying the biochemical significance of S-methylmethionine indicated that the

sulphonium compound can act in lieu of methionine itself[7,124-126]. This would be explicable on the basis of the loss of one methyl group from the sulphonium compound to form the parent amino acid. Detailed investigations by Shapiro[73,127] established that not only does this reaction occur enzymically, but that the methyl group of S-methylmethionine is transferred specifically to L-homocysteine, the net result being the production of *two* molecules of methionine:

$$(CH_3)_2\overset{+}{S}CH_2CH_2CH\overset{NH_2}{\underset{CO_2^-}{}} + HSCH_2CH_2CH\overset{NH_2}{\underset{CO_2H}{}} \longrightarrow 2CH_3SCH_2CH_2CH\overset{NH_2}{\underset{CO_2H}{}}$$

However, at least for yeasts and probably other microorganisms, there is no strong case for assuming the existence of a specific S-methylmethionine:homocysteine methyltransferase. Further work by Shapiro and coworkers[128,129] established that in these organisms S-methylmethionine is, in fact, a highly effective substrate for S-adenosylmethionine:homocysteine methyltransferase, the latter enzyme being so called because S-adenosylmethionine is normally its natural methyl donor substrate. The characteristics of this methyltransferase are described in Section IV.C.1.

S-Methylmethionine methyltransferase activity has been observed in mammalian tissues, such as pig liver[130], but here again it is unnecessary to invoke the operation of a specific enzyme, since S-methylmethionine has been shown to act as a substrate for both the thetin:homocysteine and betaine:homocysteine methyltransferase of liver[104,106], as well as S-adenosylmethionine:homocysteine methyltransferase[127]. This sulphonium compound is a precursor of methionine in many plant tissues, such as oat seedlings[76], pea coleoptiles[131], jack beans[132], and various seeds[133], where it could be acting as a substrate for the S-adenosylmethionine methyltransferase. However, more recent work by Allamong and Abrahamson[134] with winter wheat (*Triticum aestivum*) has revealed that S-methylmethionine methyltransferase activity is considerably higher in the dry seeds than is the accompanying S-adenosylmethionine methyltransferase activity, and the two activities change independently during germination, so that there may be a case for accepting the existence of two distinct enzymes in plants.

The demethylation of S-methylmethionine is the first step in the catabolism of this compound, which subsequently follows the well established pathways of methionine metabolism, including activation to S-adenosylmethionine, trans-sulphuration to cysteine and eventual oxidation to inorganic sulphate.

C. S-Adenosyl-L-methionine

S-Adenosylmethionine is susceptible to breakdown, both chemically and enzymically, at a number of sites in the molecule, these being represented in Figure 1.

Three different C—S bond cleavages are possible, namely at sites A, B, and C. Breakage of the ribose—S bond at A can take place to a limited extent in the presence of hot alkali, giving rise to methionine as the sulphur-containing product, although this is invariably preceded by hydrolysis at D to yield free adenine. So far, however, there is no well established biological counterpart of this alkaline fission, either in the form of an enzymic release of adenosine or, as would be more likely, an adenosyl group transfer. A reversal of the S-adenosylmethionine synthetase reaction might be regarded as illustrating this type of reaction, in that the enzyme

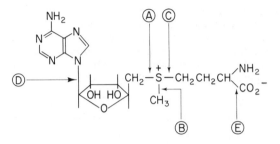

FIGURE 1. Sites of chemical and enzymic bond
fission in S-adenosylmethionine.

can catalyse the formation of adenosine triphosphate from S-adenosylmethionine
and polyphosphate[86], but whether such a reverse reaction occurs naturally is not yet
known. Some experiments with labelled S-adenosylmethionine incubated with
extracts of an *E. coli* mutant have suggested that the sulphonium compound can
transfer its adenosyl group to L-homocysteine by the following reaction[135]:

S-Adenosylhomocysteine + H⁺

However, this reaction does not seem to have been investigated further.

Not only can the remaining two C—S bond cleavages be brought about
chemically, but enzymes catalysing them also occur in cells. Loss of the S-methyl
group at B requires prolonged heating of the sulphonium compound with
concentrated hydriodic acid. The corresponding enzymic reaction takes place very
readily through the mediation of methyltransferases, provided a second substrate is
present to act as a methyl acceptor (see Section IV.C.1).

When S-adenosylmethionine is heated under mildly acidic conditions, splitting at
site C occurs with the formation of 5′-methylthioadenosine and α-amino-γ-butyrolactone, the latter being rapidly hydrolysed to homoserine. The corresponding
enzyme-catalysed reaction is a characteristic of many microorganisms (see Section
IV.C.4).

Two further reactions of the sulphonium compound can take place at sites other
than the sulphonium group. The adenine–ribose linkage is labile and, as mentioned
earlier, can be split by hot acid and also by hot alkali[136]. The reaction, analogous to
the alkaline decomposition, which forms adenine and S-ribosylmethionine, has not
been established with certainty as a biological process, although enzymic fission at
site D can occur in the presence of a specific nucleosidase, following a preliminary
cleavage of the S-adenosylmethionine molecule at either site A or B. Enzymic
decarboxylation of the amino acid moiety also occurs with the formation of
S-adenosyl-(5′)-3-thiopropylamine **5** (see Section IV.C.3), but this part of the
molecule does not act as a substrate for other amino acid-metabolizing enzymes,
such as the amino acid oxidases.

1. Methyl group transfer

The enzymic methyl group transfer reactions undergone by the methylthetins and by S-methylmethionine are confined to the S-methylation of L-homocysteine. S-Adenosylmethionine, on the other hand, is a remarkably versatile methyl donor, both in terms of the number of compounds and the types of group it is able to methylate. The great variety of O- and N-methylated compounds found naturally largely derive their methyl groups directly from S-adenosylmethionine. A great many C-methyl groups in natural products also originate from the same source. Methylation of inorganic compounds of arsenic, selenium, tellurium, mercury, etc., is achieved by many fungi, and the evidence so far available indicates that methionine is an efficient source of the methyl groups[136a]. It is therefore not unreasonable to suppose that S-adenosylmethionine may again be involved as the methyl donor.

An indication of the dominant role that S-adenosylmethionine plays in biological methylation is given by the fact that in 1977 the International Union of Biochemistry listed 47 methyltransferases utilizing the sulphonium compound as the specific methyl donor[137]. A more realistic estimate of the number of these enzymes must be several times greater than this figure, in view of the existence of hundreds of naturally occurring methylated compounds, coupled with the fact that many of the methyltransferases are also highly specific with respect to their methyl acceptor substrates. The number of individual methylation reactions mediated by S-adenosylmethionine is, in fact, so large that limitations of space will allow reference to only a selected number of the more important of these. Before considering these processes, it is, perhaps, interesting to reflect that soon after the discovery of S-adenosylmethionine, Cantoni[138] suggested that this compound might be the sole methyl donor in reactions other than methionine biosynthesis, and so far no reliable evidence has been put forward to contradict this idea.

The general mechanism of methyl group transfer from S-adenosylmethionine to an appropriate methyl acceptor (designated as I—X—H, where X is —S—, —N≡, —O—, etc.) is similar to that from dimethyl-β-propiothetin and S-methylmethionine in the methylation of homocysteine, and involves a nucleophilic attack by the methyl acceptor on the S-methyl group, which is probably displaced as a carbocation ion. A variant of the reaction is the methylation of methionine to S-methylmethionine with the formation of a second sulphonium group. On the other hand, C-methylation reactions generally involve an addition across a double bond. In S-, N-, and O-methylations the methyl group has been clearly established as being transferred intact to the methyl acceptor, but this is not necessarily the case with C-methylation reactions.

In the course of all methyl transfer reactions from S-adenosylmethionine, the sulphonium compound is converted into S-adenosyl-L-homocysteine[139]. This thioether is further metabolized by a number of pathways. Hydrolytic fission of the S-ribosyl linkage to yield L-homocysteine and adenosine can occur in the presence of S-adenosylhomocysteine hydrolase, provided that the reaction products are removed by further enzyme reactions. The hydrolase has been identified in rat liver[96] and in most other animal tissues[140], as well as in yeasts[95,141].

S-Adenosylhomocysteine, in contrast to S-adenosylmethionine, is a substrate for L-amino acid oxidase, being deaminated to S-adenosyl-γ-thio-α-ketobutyrate 31. This pathway has been demonstrated in bacteria[142,143] and in rat liver and kidney[144], and when S-adenosylhomocysteine is administered to rats, the above keto acid is the main metabolite excreted in the urine[145]. In the mould Aspergillus

$$\text{Adenosyl} - \text{SCH}_2\text{CH}_2\text{CH} \overset{\text{NH}_2}{\underset{\text{CO}_2\text{H}}{\diagup\diagdown}} \quad + \; O_2 \; + \; H_2O \quad \longrightarrow$$

$$\text{Adenosyl} - \text{SCH}_2\text{CH}_2\text{COCO}_2\text{H} \; + \; NH_3 \; + \; H_2O_2$$

(31)

oryzae, S-adenosylhomocysteine acts as a substrate for another deaminating enzyme, adenine deaminase, the product being S-inosylhomocysteine[146].

Duerre and Walker[98,147] found a third catabolic pathway to be operative in many bacteria, including *E. coli*, namely the cleavage of the glycosyl linkage in the presence of a specific nucleosidase to yield adenine and S-ribosyl-L-homocysteine:

S-Adenosyl-L-homocysteine + H_2O → Adenine + S-Ribosyl-L-homocysteine

S-Ribosylhomocysteine is further degraded by a bacterial lyase to give homocysteine and a C_5-fragment[148].

The catabolism of S-adenosylmethionine following its participation in methyl transfer reactions is summarised in Figure 2.

The homocysteine formed either directly from S-adenosylhomocysteine or via the formation of S-ribosylhomocysteine may undergo condensation with serine to yield cystathionine. This intermediate thioether can be cleaved by cystathionase, forming cysteine which may be further oxidized in the animal to inorganic sulphate. Alternatively, homocysteine may be methylated by betaine-, S-adenosylmethionine- or tetrahydrofolate methyltransferases to give methionine, which may be converted into S-adenosylmethionine by the methionine-activating enzyme.

This last pathway is of interest in that it provides a means of re-utilizing part of the S-adenosylmethionine molecule, and is one of several such pathways for recycling the sulphonium compound of potential importance in the metabolic

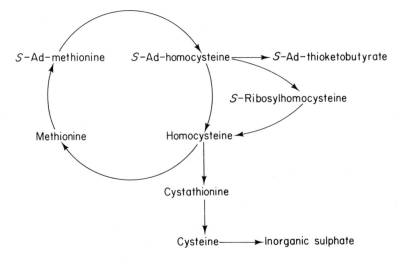

FIGURE 2.   Pathways of S-adenosylmethionine catabolism (Ad = adenosyl).

economy of cells. The ready turnover and regeneration of S-adenosylmethionine from its breakdown products accounts for its ability to take part in a variety of biological processes, despite its very low concentration in tissues, and in this respect it functions as a coenzyme in much the same way as NAD, FAD and ATP.

One highly significant finding is that S-adenosylhomocysteine has proved to be a potent inhibitor of practically all known methyl transfer reactions catalysed by S-adenosylmethionine-dependent methyltransferases[149–152]. Kinetic studies with a number of methyltransferases have shown that the inhibition produced by the thioether is competitive with respect to S-adenosylmethionine, and in many instances the K_i value for S-adenosylhomocysteine is lower than the K_m value for S-adenosylmethionine[150–152], indicating that the thioether has a greater affinity for the enzymes than has the methyl donor substrate.

Since the concentration of S-adenosylhomocysteine in tissues is generally comparable with that of S-adenosylmethionine[57], it is probable that the thioether plays a major role in the control of S-adenosylmethionine-mediated methylation reactions. In addition, the various enzymes concerned with the metabolism of S-adenosylhomocysteine, e.g. S-adenosylhomocysteine hydrolase, by affecting the local concentration of the thioether in tissues, would be expected to exert an extra regulatory role in methylation, as summarized in Figure 3.

a. S-*Methylation*. In studies on the biosynthesis of methionine, Shapiro and coworkers identified the enzyme S-adenosylmethionine:L-homocysteine methyltransferase, which catalyses the methylation of L-homocysteine. The enzyme has been purified over 500-fold from the yeast *S. cerevisiae*[128,129] and is also present in mammalian liver, in bacteria, including *E. coli* and *Aerobacter aerogenes*[127], and also in the yeast *Candida albicans*[153]. An examination of the enzyme from yeast showed that either L- or D-homocysteine can serve as the methyl acceptor. The enzyme has only moderate specificity with regard to the methyl donor substrate and can use either of the (−)- and (+)-diastereoisomers of S-adenosyl-L-methionine[154], although S-adenosyl-D-methionine is only weakly effective as a substrate. Other related compounds able to act as methyl donors are S-inosyl-L-methionine[149], S-methyl-L-methionine (which is comparable in activity to S-adenosyl-L-methionine)[128,129], S-adenosyl-2-methyl-DL-methionine[155], and S-adenosyl-(5′)-3-methylthiopropylamine[154]. S-Adenosylethionine has some activity as a substrate and is able to transfer its ethyl group to homocysteine[154]; similarly, the corresponding n-propyl analogue has weak but significant propyl-donating activity[156].

S-Adenosylmethionine also acts as the methyl donor for two other S-methyltransferases, one of these being found in the microsomes of a variety of

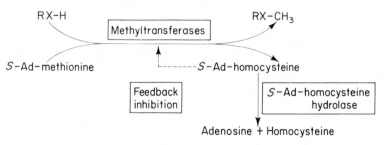
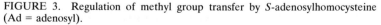

FIGURE 3. Regulation of methyl group transfer by S-adenosylhomocysteine (Ad = adenosyl).

animal tissues. This enzyme is capable of S-methylating a wide range of thiols many of which do not occur naturally, such as BAL (1-hydroxypropane-2,3-dithiol), 2-hydroxyethanethiol, thioloacetic acid, as well as methanethiol and hydrogen sulphide[157]. It is, however, unable to methylate homocysteine, cysteine, or glutathione, and may be identical with an S-adenosylmethionine-dependent S-methyltransferase which can methylate tetrahydrofurfuryl thiol and other related compounds, also present in the microsomes of liver and other organs[158].

Another S-methyltransferase requiring S-adenosylmethionine as methyl donor and apparently distinct from the above enzymes catalyses the methylation of a number of 2-, 6- or 8-thio-substituted purines, and to a lesser extent a number of 2-, 4- or 5-thio-substituted pyrimidines[159]. It is found in many animal tissues and is especially abundant in kidney, although of limited occurrence in microorganisms. The thiopurine nucleosides, rather than the free bases themselves, are the preferred substrates for this methyltransferase, which may have some detoxicating role. A further example of the participation of S-adenosylmethionine in S-methylation is the conversion of methionine to S-methylmethionine by a methyltransferase in many plant tissues (see Section III.C).

b. N-*Methylation*. S-Adenosylmethionine functions as the specific methyl donor in a large number of enzymic N-methylation reactions of amino groups and also of N in heterocyclic structures. Of major importance in the metabolism of vertebrates is the methylation of glycocyamine (guanidinoacetic acid) **8** to creatine **9**, the N-phosphoric acid derivative of the latter compound playing a key role in phosphate group transfer, muscular contraction, etc. This methylation was initially observed in rat liver slices[43,44] and was subsequently shown to take place in other tissues, e.g. muscle[160], as well as in extracts from wheat seedlings[161]. Cantoni and Vignos[162] achieved a partial purification from pig liver of the enzyme responsible, guanidinoacetate N-methyltransferase, and established its dependence on S-adenosylmethionine as the methyl donor. An assay of the enzyme based on the use of ^{14}C-labelled S-adenosylmethionine as substrate[163] showed it to be present in the liver and testes of a range of animal species but absent from various bacteria, yeasts, and plant seeds.

A second metabolically important N-methylation mediated by the sulphonium compound is the synthesis of choline from ethanolamine, which was first achieved *in vitro* in homogenates of mammalian muscle and liver[164]. These preparations could also convert dimethylethanolamine into choline. The involvement of S-adenosylmethionine as the methyl donor was established with the aid of the labelled sulphonium compound in rat liver homogenates[165], all three methyl groups of choline originating from S-adenosylmethionine[166]. Whole animal experiments had suggested that phosphatidylethanolamine rather than free ethanolamine is the true methyl acceptor[166], and the methylation pathway is now believed to be as follows:

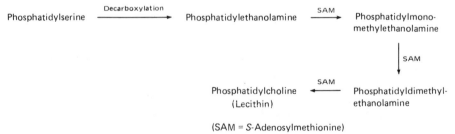

(SAM = S-Adenosylmethionine)

Phosphatidylserine is the precursor of phosphatidylethanolamine. The methylating enzyme, S-adenosylmethionine:phosphatidylethanolamine N-methyltransferase, is located in microsomes. It has been found in lung tissue[167], although it may have a physiologically insignificant role in this organ compared with its activity in liver. The stepwise methylation occurs in bacteria that accumulate lecithin[168], and there is evidence for its presence in yeast[169]. It is not clear whether a single methyltransferase brings about all three methylations; in fact, there are indications of the existence of two enzymes, one catalysing the last two steps[168].

A number of S-adenosylmethionine-dependent methyltransferases are now known to catalyse the N-methylation of several α-amino acids. Among these are a specific glycine N-methyltransferase, present in mammalian liver[170,170a], which forms sarcosine (N-methylglycine) **32**. In plants, the methylation of glycine goes to completion with the formation of glycine betaine **33**. The stepwise methylation of

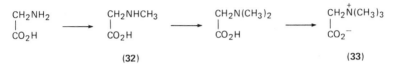

(32)   (33)

proline **34** to form stachydrine **36**, possibly via the intermediate formation of hygric acid **35**, which has been studied in alfalfa plant tissues[171], requires the

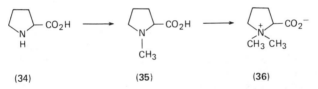

(34)   (35)   (36)

methionine-methyl group, thus suggesting the involvement of S-adenosylmethionine. There is also evidence that the sulphonium compound furnishes methyl groups for the conversion of lysine **37** to carnitine **40** in rats and in *Neurospora*[172,173]. Carnitine appears to function as a carrier of acyl groups from the cytoplasm to the mitochondria during the cellular oxidation of long-chain fatty acids, and probably arises from ε-N-trimethyllysine **38** after loss of carbon atoms 1 and 2 to give γ-butyrobetaine **39**, followed by hydroxylation. It now seems likely that it is not free lysine which is the initial methyl acceptor, but rather a specific lysine residue bound in protein[174].

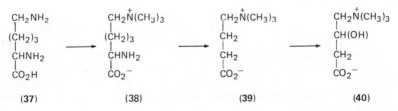

(37)   (38)   (39)   (40)

S-Adenosylmethionine is specifically involved in the N-methylation of numerous naturally occurring amines, many of which have marked pharmacological effects even though they may be present in tissues at very low concentrations. Axelrod and coworkers have contributed extensively to the characterization of the methyltransferases concerned, and have written a number of valuable reviews on biogenic amines and the role of methylation in their synthesis and metabolism[175–177].

Examples of such methylations include the formation of adrenaline **41** from noradrenaline in the adrenal medulla[178]. Adrenaline itself may be further methylated in the presence of S-adenosylmethionine to form N-methyladrenaline, a normal metabolite of the hormone in many animal species. Axelrod refers to the noradrenaline-methylating methyltransferase as phenylethanolamine N-methyltransferase[175,179], since the enzyme from several animal species can methylate a range of phenylethanolamines, although the same enzyme in other animals and man methylates only 2-hydroxyphenylethanolamines. This enzyme is also present in nervous tissues and the brain, suggesting that adrenaline may have an additional role as a neurotransmitter.

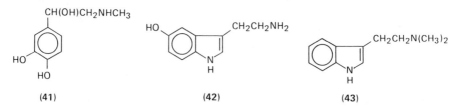

(41)   (42)   (43)

Animal tissues also exhibit a relatively non-specific S-adenosylmethionine-dependent amine N-methyltransferase activity catalysing the methylation of a range of indoleamines, such as tryptamine and serotonin (5-hydroxytryptamine, 5-HT) **42**, as well as many phenylethanolamines, including noradrenaline[180]. Many drugs, e.g. ephedrine, amphetamine, normorphine, and nornicotine, can likewise be methylated by this system. The N-methyltransferase(s) responsible occurs in lung, heart, and stomach tissue, but is most active in liver. Its apparent lack of activity in brain is due to the presence of potent naturally occurring inhibitors. When these are removed from brain extracts by dialysis, the enzyme present can readily methylate tryptamine first to N-methyltryptamine and then further to NN-dimethyltryptamine **43**[181], which produces hallucinogenic effects in man. It is of interest that this N-methyltransferase is also found in human blood platelets, so that its activity in patients with various mental disorders can be easily assayed by blood sampling. This has revealed that in schizophrenic subjects there is an abnormally high activity for converting tryptamine to the mono- and dimethyl derivatives, apparently due to a reduction in the tissue levels of the natural inhibitors of the enzyme[182].

Other workers have reported the presence in human, sheep, chick, and ox brain of an S-adenosylmethionine:indolethylamine methyltransferase, distinct from the above-mentioned non-specific enzyme, which acts on various indolethylamines but does not N-methylate phenylethylamines, such as noradrenaline. It is very active towards tryptamine and serotonin as methyl acceptors[183,184]. The latter compound is methylated to the NN-dimethyl derivative bufotenin **44** for which there is accumulating evidence that it plays a role in the onset of psychosis in schizophrenia.

The simple alkaloid hordenine **46** is widely distributed in plants, and was shown originally by Raoul[185] to be synthesized in barley seedlings from the amino acid tyrosine via the intermediate formation of tyramine **45**, the methyl groups being derived from methionine[186]. N-Methyltyramine is the first methylation product[187], and further methylation can occur in some plants, e.g. barley and certain cacti, to give the quaternary ammonium base candicine **47**[188]. Mann and Mudd[189] have identified a tyramine *N*-methyltransferase in the roots of germinating barley which requires *S*-adenosylmethionine in the formation of *N*-methyltyramine, and there is evidence for the participation of the enzyme in a further methylation to hordenine[190].

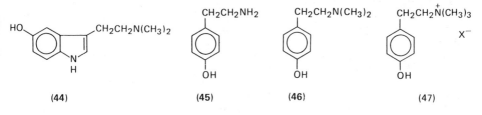

(44)   (45)   (46)   (47)

According to Mudd[191], plant tissues such as barley shoots contain an indoleamine methyltransferase distinct from that found in animal tissues, which can methylate 3-aminomethylindole **48** first to the 3-methylamino derivative and then to the naturally occurring dimethyl derivative, the alkaloid gramine **49**.

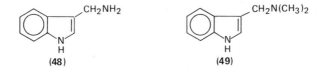

(48)   (49)

Another amino group *N*-methyltransferase has been partially purified from tobacco roots and methylates the base putrescine **83** to *N*-methylputrescine, the latter compound being a likely intermediate in the biosynthesis of nicotine[192].

The first heterocyclic *N*-methylation to be studied biochemically was the conversion of nicotinamide to its methochloride in rat liver slices[45,46]. This methylation is the main pathway for the metabolism of nicotinamide in animals and man[193]. The methyltransferase responsible is specific for the methyl acceptor and is unable to methylate nicotinic acid. The latter compound is, however, methylated to trigonelline **50** in the presence of a specific nicotinate *N*-methyltransferase, isolated from extracts of pea seedlings[194]. In muscle tissue, carnosine (β-alanyl-L-histidine) **52** can be methylated to the 1-methyl derivative, anserine **53**, by another specific *S*-adenosylmethionine-dependent methyltransferase[195].

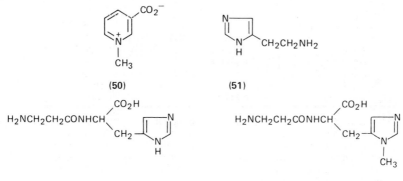

(50)   (51)

(52)   (53)

The nuclear methylation of histamine 51 to 1-methylhistamine is the principal pathway in the metabolism of this powerful vasopressor amine in the animal[196]. A specific imidazole N-methyltransferase catalysing the methylation was identified in mouse liver by Axelrod and coworkers[197] and subsequently partially purified from guinea pig brain[198] and liver[199]. It has also been found in the majority of animal tissues, including red blood cells, but the highest activity is present in the brain. Since 1-methylhistamine has negligible physiological activity compared with histamine itself, the enzymic methylation may operate to deactivate the parent amine.

With the aid of S-adenosylmethionine labelled with ^{14}C or ^{3}H in the methyl group it is possible to identify and determine minute quantities of histamine in tissue extracts[177]. The extracts are incubated with the labelled methyl donor in the presence of a purified preparation of histamine methyltransferase. The N-methylated derivative is then extracted and its radioactivity assayed by liquid scintillation counting, and in this way the detection of as little as 20 pg of histamine is possible. Other pharmacologically active amines, e.g. noradrenaline and serotonin, can be assayed in a similar manner using the appropriate N-methyltransferases, as indicated in Table 5.

There is abundant evidence for the existence of other N-methyltransferases in plants which are responsible for the formation of the nuclear N-methyl groups of many alkaloids and related compounds. In several instances tracer studies with whole plants or extracts of plant tissues have indicated methionine as the source of the methyl groups, so that the participation of S-adenosylmethionine, if not demonstrated directly, is strongly suspected[188]. Such alkaloids include codeine, thebaine, morphine, ricinine, and nicotine[200,201]. A recent confirmation of the role of S-adenosylmethionine as the methyl donor in these reactions is that poppy latex has been shown to contain N-methyltransferases which catalyse the methylation of the nor-precursors of these alkaloids in the presence of the sulphonium

TABLE 5. Use of S-adenosylmethionine and methyltransferases to measure biogenic amines and drugs in tissues. (Data from Axelrod[177]; reproduced with permission of Columbia University Press, New York)

Amine	Enzyme[a]	Sensitivity (pg)
Dopamine	COMT	100
Noradrenaline	COMT	20
Octopine	PNMT	25
Phenylethanolamine	PNMT	25
Phenylethylamine	PNMT (+ DBH)	200
Serotonin	HIOMT	20
N-Acetylserotonin	HIOMT	25
Histamine	HMT	20
Tryptamine	NMT	1000
Amphetamine	NMT	10,000

[a]COMT, catechol O-methyltransferase (from rat liver); PNMT, phenylethanolamine N-methyltransferase (from bovine adrenal); DBH, dopamine β-hydroxylase (from bovine adrenal); HIOMT, hydroxyindole O-methyltransferase (from bovine pineal); HMT, histamine N-methyltransferase (from guinea pig brain); NMT, non-specific N-methyltransferase (from rabbit lung).

compound[202]. This type of methylation is not confined to plants, however, and can be brought about by a methyltransferase in rabbit lung[203].

Suzuki[204] demonstrated that discs cut from tea leaves incorporated the methyl group of S-adenosylmethionine into caffeine. Subsequently, tea leaves were shown to contain two N-methyltransferases involved in the transfer of methyl groups from the sulphonium compound to 7-methylxanthine, giving first theobromine (3,7-dimethylxanthine) and then caffeine (1,3,7-trimethylxanthine)[205]. Methylation of 1-methylxanthine can also occur, giving theophylline (1,3-dimethylxanthine) and paraxanthine (1,7-dimethylxanthine).

S-Adenosylmethionine is also the methyl donor for the $N_{(3)}$-methylation of the naturally occurring bases adenine and $N_{(6)}$-methyladenine, as well as the purine analogues 7-methyladenine and 2,6-diaminopurine[206]. The specific methyltransferase catalysing these methylations is present in rabbit lung, kidney, and spleen. It is distinct from an aminopurine methyltransferase present in many bacteria, capable of methylating substituted purines to yield the corresponding 2-methylamino derivatives[207].

c. O-*Methylation*. An important pathway in the metabolism of catecholamines is the conversion of these compounds to their methoxy derivatives. This has been shown to require the action of the S-adenosylmethionine-dependent enzyme catechol O-methyltransferase[175,177,208]. Normally, methylation takes place on the 3-hydroxy group, but when other substituent groups are present in the ring a mixture of 3- and 4-methoxy derivatives may result. Catechol O-methyltransferase is widely distributed in plants, crustacea, and in animals, where it is a constituent of most tissues. Adrenaline **41** is methylated by the enzyme to metanephrine **54** the principal metabolite of the hormone in man[209]. Other substrates are noradrenaline and dopamine **55**. The amino acid dopa **56** is readily methylated

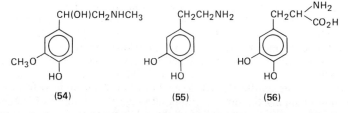

(54)   (55)   (56)

to 3-methoxydopa, and since the former compound is used medicinally in the treatment of Parkinson's disease, its administration in excessive amounts could lead to a depletion of S-adenosylmethionine levels in the body. As evidence of this, the injection of dopa into rats causes a 76% reduction in the level of the sulphonium compound in the brain and a 51% reduction in the adrenal medulla within a period of 45 min[210].

In the acute state of schizophrenia, S-adenosylmethionine levels in the brain are also markedly reduced as a result of utilization of the compound for methylation[211]. Whereas dopamine is normally metabolized to its 3-methoxy derivative, followed by further oxidation to 3-methoxy-4-hydroxyphenylacetic acid, in schizophrenic subjects both hydroxy groups of the drug are methylated to yield 3,4-dimethoxyphenylethylamine, which appears to be a characteristic constituent of the urine in this mental condition[212]. It is of interest that the 3,4-dimethoxy derivative is known to have properties akin to those of the hallucinogenic substance mescaline (3,4,5-trimethoxyphenylethylamine).

Catechol O-methyltransferase has a relatively low substrate specificity in that it can catalyse the methylation of a variety of other catechols, and also polyphenols and 5,6-dihydroxyindoles. The enzyme from human placenta accepts 2-hydroxyestrogens as substrates[213] and the methylation by S-adenosylmethionine of 2-hydroxyestradiol-17β has been demonstrated in the presence of the enzyme in human liver slices[214]. The rat liver enzyme has also been reported to catalyse the methylation of L-ascorbic acid to 2-methyl-L-ascorbic acid[215]. Other enzymes with catechol O-methyltransferase activity have been identified in various plants[216,217].

Another S-adenosylmethionine-dependent O-methyltransferase participates in the final step in the synthesis of melatonin **59** from tryptophan **57**, serotonin **42** and acetylserotonin **58** being intermediates in the pathway[218]. Melatonin is a

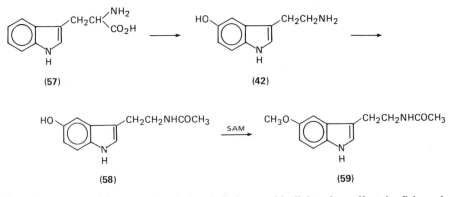

hormone secreted by the pineal gland. It has a skin-lightening effect in fish and amphibia and inhibits the activity of the gonads in mammals. The methylating enzyme is now referred to as acetylserotonin O-methyltransferase on account of its high substrate activity, although a few other 5-hydroxyindoles serve as methyl acceptors. The reported O-methylation of 5-hydroxy-NN-dimethyltryptamine in pineal gland tissue[219] may be due to the action of this enzyme.

A wide variety of O-methylation reactions have been observed in plants and microorganisms where, as with N-methylation reactions, the role of S-adenosylmethionine as the methyl donor has often been implied from the ability of methionine to supply the methyl group, although in a number of instances there is direct evidence for the participation of the sulphonium compound. Examples are the incorporation of the methionine-methyl group into the guiacyl, syringyl, veratryl, and trimethylgallyl residues of lignins in plants such as barley and tobacco[220,221], likewise the transfer of methyl groups from methionine in pectin synthesis in oat and maize tissues[76,222], as well as in the synthesis of alkaloids[200,201] and chalcone-type compounds[223]. Methionine was also shown to be a source of methyl groups for the formation of the 6-methoxy group of ubiquinone-9 in the organism *Euglena gracilis*[224], as was found some years earlier by Lawson and Glover to be the situation in rat liver and intestine[225]. More recently, the direct involvement of S-adenosylmethionine as the methyl donor in ubiquinone biosynthesis has been confirmed in rat liver mitochondria[226].

Goodwin and coworkers have provided unequivocal evidence for S-adenosylmethionine acting as the methyl donor in the biosynthesis of methoxy carotenoids in the bacterium *Rhodopseudomonas spheroides*[227], where methylation of tertiary hydroxy groups takes place. O-Methylation of anthocyanidins also requires the sulphonium compound[228].

High substrate specificity is a feature of the S-adenosylmethionine-dependent O-methyltransferase catalysing the methylation of inositol. Thus, two separate enzymes, myo-inositol 1-O-methyltransferase and myo-inositol 3-O-methyltransferase have been characterized in the pea plant[229].

A recently identified methyltransferase, caffeate 3-O-methyltransferase, has been partially purified by Poulton and Butt[230] from spinach beet leaves and from soybean cell cultures[231]. This enzyme catalyses the methylation of the *meta*-hydroxy group of caffeic acid by S-adenosylmethionine, forming ferulic acid **60**, a

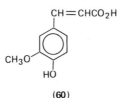

(60)

precursor of lignin. Although caffeic acid is the preferred substrate, other hydroxycinnamic acids, such as 5-hydroxyferulic acid and 3,4,5-trihydroxycinnamic acid, can also act as methyl acceptors. A second enzyme concerned in the metabolism of catechol-type residues is a 3'-4'-dihydroxyflavonoid 3-O-methyltransferase, which occurs together with the caffeate methyltransferase in soybean cultures[232]. It differs from the first enzyme in its stability and in the way its specific activity changes during the growth of the cell cultures. It is believed that the caffeate methyltransferase is part of the lignin biosynthesis pathway, whereas the other enzyme is concerned with flavonoid biosynthesis.

A further O-methyltransferase, present in the fungus *Lentinus lepidens*, catalyses the methylation by S-adenosylmethionine of esters of hydroxycinnamic acids, the free acids being inactive as substrates. An example of its action is the methylation of methyl coumarate (p-hydroxycinnamate) to methyl p-methoxycinnamate[233]. The fungal enzyme preparation was reported to convert cinnamic acid into its methyl ester, but it is not yet clear whether this was due to the presence of another methyltransferase. The methylation of free carboxylic acids to the corresponding methyl esters can occur enzymically, one biologically important example of this reaction being a step in the synthesis of chlorophyll a, namely, the methylation of magnesium-protoporphyrin IX by S-adenosylmethionine, whereby a carboxyl group in ring V is esterified. The reaction has been studied in the organism *Rh. spheroides*[234] and in wheat seedling tissue[235], and a partial purification of the enzyme responsible, magnesium-protoporphyrin methyltransferase, from wheat seedlings has been described by Ellsworth and Dullaghan[236]. An enzyme preparation from *Mycobacterium phlei* has been found to catalyse the conversion of several fatty acids to their methyl esters in the presence of S-adenosylmethionine, oleic acid being the most effective methyl acceptor[237].

Incorporation of the methyl group of the sulphonium compound into insect juvenile hormones has also been established, and the presence of an appropriate O-methyltransferase detected in extracts from the tobacco hornworm[238].

d. C-*Methylation*. Although some C-methyl groups in natural products arise by the reduction of structures such as $>C=CH_2$ and $\geq C-CH_2OH$, many are recognized to be formed as a result of methyl transfer reactions. As with O- and N-methylated compounds, methionine has been demonstrated as a source of the methyl groups, but in comparatively few instances has the direct participation of

S-adenosylmethionine been shown. Much still remains to be uncovered about this aspect of biological methylation, for example, whether there are only a few relatively non-specific *C*-methyltransferases, or whether a large number of substrate-specific enzymes exist.

Methyl transfer from methionine is implicated in the biosynthesis of ubiquinones, tocopherols[239], plastoquinones, C_{31}-triterpenes, tropolones[240], sterols, branched-chain fatty acids, and other secondary metabolites, such as antibiotics in fungi. All seven methyl groups of the corrin ring structure of cobalamins have been found to originate from methionine[241]. The *C*-methyl groups of certain minor purine and pyrimidine bases in DNA and RNA are apparently formed in the same way in microorganisms, plants, and animals. Earlier investigations of the nature of the process, using methionine labelled with ^{14}C and ^{2}H in the methyl group, revealed that whereas transfer of the intact methyl group sometimes occurs, loss of one or two hydrogen atoms during the transfer is not uncommon.

The mechanisms of *C*-methylation have been outlined in illuminating reviews by Lederer[242-244]. Intact methyl transfer is considered to involve the interaction of *S*-adenosylmethionine with a double bond activated by adjacent electron-releasing groups, for example

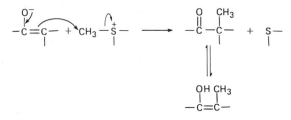

This mechanism was first proposed by Birch *et al.*[245], and shown to account for the biosynthesis of mycophenolic acid **61**[246]. A number of antibiotics contain

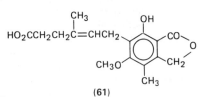

(61)

methyl-branched 6-deoxyhexoses, e.g. noviose **62** and mycarose **63**, in which the *C*-methyl groups are derived from methionine. Pape *et al.*[247] have provided

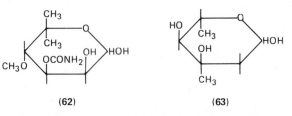

(62)   (63)

evidence for intact methyl group transfer in the formation of the mycarose residues in *Streptomyces*, probably by the above mechanism. Intact methyl group transfer is also believed to take place in the biosynthesis of various methylated amino acid residues found in antibiotics, e.g. N-methyl-C-methyl-L-isoleucine, 3-methyltryptophan, and 3-methylphenylalanine.

A second mechanism has been proposed to account for the partial transfer of the methyl group, involving an interaction at an isolated double bond. Initially, the formation of an intermediate carbocation ion may occur, the carbocation ion then

becoming stabilized in one of several ways:

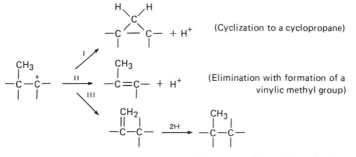

(Cyclization to a cyclopropane)

(Elimination with formation of a vinylic methyl group)

(Elimination followed by reduction)

Evidence for the operation of the type I reaction is the conversion of *cis*-vaccenic acid **64** into lactobacillic acid **65** by *Clostridium butyricum* in the presence of

$$CH_3(CH_2)_5CH=CH(CH_2)_9CO_2H \xrightarrow{SAM} CH_3(CH_2)_5CH\overset{\displaystyle H\ H}{\underset{\displaystyle C}{\diagup\diagdown}}CH(CH_2)_9CO_2H$$

(64)   (65)

S-adenosylmethionine as methyl donor[248,249]. The true substrate for this system appears to be a vaccenic acid residue in a phosphatidylethanolamine and not the free acid. The methylation of monounsaturated alkanoic acids to give other cyclopropane acids seems to be a feature of several bacteria[250].

The type III reaction explains the formation of tuberculostearic acid **68** from oleic acid **66** in the organism *Mycobacterium phlei*[251], and the expected intermediate, 10-methylenestearic acid **67**, has been isolated and characterized[252].

$$CH_3(CH_2)_7CH=CH(CH_2)_7CO_2H \longrightarrow CH_3(CH_2)_7\overset{\displaystyle CH_2}{\overset{\displaystyle \|}{C}}-CH_2(CH_2)_7CO_2H$$

(66)   (67)

$$CH_3(CH_2)_7\overset{\displaystyle CH_3}{\overset{\displaystyle |}{C}}-CH_2(CH_2)_7CO_2H$$

(68)

Since free tuberculostearic acid is not found, the true substrate is probably a phospholipid-bound oleic acid residue.

A better documented example of the type III *C*-methyl transfer is provided by the methylation of many plant and yeast sterols in the $C_{(24)}$ position. Studies of sterol biosynthesis begun about 25 years ago indicated that this methyl group is not derived from acetate as are the rest of the carbon atoms[253], and Parks[254] showed that it originates from the methyl group of *S*-adenosylmethionine. An *S*-adenosylmethionine-dependent $\Delta_{(24)}$-sterol methyltransferase was identified and characterized in the yeast *S. cerevisiae*, and its properties have been described in detail by Parks *et al*.[255].

The role of the enzyme may be illustrated by the methylation which it catalyses at the $C_{(24)}$–$C_{(25)}$ double bond of zymasterol **69** with the introduction of a methylene group to give fecosterol **70**. Reduction of this group then occurs, together with dehydrogenation in ring B to form ergosterol **71**. The sterol

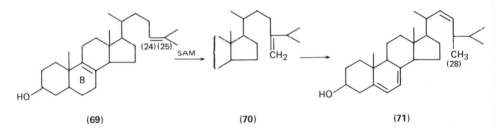

(69)   (70)   (71)

methyltransferase has been partially purified from yeast[256], where it is located in the mitochondria. Its reaction kinetics are complex and it is markedly inhibited by cations, especially monovalent cations, e.g. K^+ and Na^+, so that cation flux may exert a regulatory role in sterol biosynthesis in yeast.

More than eighty sterols have been characterized in such marine organisms as sponges and shellfish, many of these compounds being derived from the cholesterol skeleton, but having various modifications in the side chain at $C_{(24)}$. Again, *C*-methylation is implicated in the formation of the alkylated branched chains[257,258]. Frequently a cyclopropane ring occurs in the side chain, as in gorgosterol **72**, a

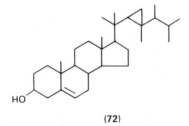

(72)

constituent of the gorgonian *Plexaura flexuosa*[259]. This ring probably results from a type I reaction, following the addition of the *S*-adenosylmethionine-methyl group to an olefinic double bond.

A double *C*-methylation was postulated by both Birch[260] and Nes and coworkers[261] for the biogenesis of ethylidene and ethyl side-chains found in phytosterols, and the mechanism has been confirmed for the formation of β-sitosterol[261], spinasterol, and fucosterol. This was established by feeding methionine labelled with ^{14}C in the methyl group to pea seedlings, the sterols in question being shown to have incorporated the label into both the $C_{(28)}$ and $C_{(29)}$ positions. The first methylation is believed to form a methylene derivative **73**, which acts as the acceptor for a second methyl group.

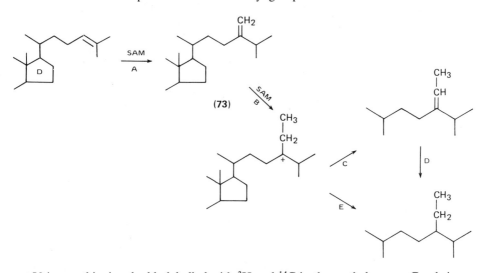

Using methionine doubly labelled with ^{3}H and ^{14}C in the methyl group, Goodwin and coworkers[262] concluded that the ethyl side-chain of sitosterol in maize and larch leaves contains only four H atoms derived from methionine-methyl groups, which would imply the biosynthetic pathway A → B → C → D. However, Lederer and coworkers[263], using ^{2}H-methyl-labelled methionine, found that in a slime mould five H atoms were incorporated, favouring the sequence A → B → E.

2. Methylation of macromolecules

a. *Polysaccharides.* O-Methoxy sugars occur widely in the form of polysaccharides or in conjugation with sterols, phenols, hydroxyalkanoic acids, etc., and include 2-O-methylrhamnose, 2-O-methylfucose, 2,4- and 3,4-di-O-methylrhamnose, and 2,3-di-O-methylglucose. Nucleic acids contain O-methylpentoses, and in yeast transfer-RNA some 1.5% of the D-ribose residues are O-methylated[264]. Hemicelluloses contain 4-O-methyl-D-glucuronic acid as a component, and O-methyl-D-galacturonic acid is present in pectins. Byerrum and coworkers[265] reported the incorporation of the methionine-methyl group into pectin in radish plants, and Kauss and Hassid[266] subsequently demonstrated with mung bean extracts that S-adenosylmethionine is the precursor of methyl ester groups in the polygalacturonate side-chains of pectin, the incorporation taking place at the macromolecular level. Kauss and Hassid also showed that particulate preparations of corn cob can transfer the methyl group of the sulphonium compound to yield 4-O-methylglucuronate residues in hemicullulose B[267].

Ballou and coworkers have investigated the methylation of polysaccharides in *Mycobacteria*, which are exceptionally rich in O-methyl sugars. They found that 6-O-methyl-D-glucose is a hydrolysis product of a lipid fraction of *Mycobacterium phlei*[268]. This methoxy sugar proved to be a constituent of a unique lipopolysaccharide, consisting of eighteen hexose units, the terminal position at the non-reducing end of the chain being occupied by a 3-*O*-methylglucose unit. Also part of the chain and starting at the fifth hexose unit from the non-reducing end is a sequence of ten 6-*O*-methyl-D-glucose units. It is believed that this lipopolysaccharide, having strongly hydrophilic and lipophilic properties, functions in some way at the interface between membranes and the cytoplasm in the organism[269].

The methylation of sugar residues is thought to play an important role in the biological function of polysaccharides by modifying properties such as solubility, ion-exchange and water-binding capacity, and the formation of cross-linkages[270]. The effect on a polymer such as pectin, when polar carboxyl groups are converted into neutral methyl ester groups, may be to alter cellular plasticity, for a firmer cell wall structure is generally associated with a low pectin methyl ester content, which in turn is related to a higher content of divalent ions, such as Ca^{2+}, bound to free carboxyl groups.

b. *Proteins.* *N*-Methylated amino acids have been known for some time to be minor components of proteins. These compounds are also present in small amounts in mammalian urine, and their excretion is regarded as a consequence of normal protein catabolism. Such amino acid derivatives include 3-*N*-methylhistidine, *N*-methyllysines, and *N*-methylarginines. The natural occurrence of these compounds has been reviewed in detail by Paik and Kim[271]. The possibility that the naturally occurring *N*-methylamino acids are first synthesized and then incorporated into proteins has now been discarded in favour of the concept of the introduction of methyl groups into preformed polypeptide structures. This has been clearly established on an experimental basis by several groups of workers[272-274], who achieved methylation of histones *in vitro* in the presence of *S*-adenosylmethionine.

Paik and Kim[271,275], among others, subsequently made a detailed study of the mechanism of protein methylation, and have characterized three classes of enzyme involved, which they termed for convenience protein methylases I, II, and III. Protein methylase I, also referred to as protein (arginine) methyltransferase, catalyses the methylation by *S*-adenosylmethionine of the guanidino moiety of arginine residues in histones and other proteins. It is present in many organs and has been partially purified from calf thymus. The main reaction product is mono-*N*-methylarginine, but the *NN*-dimethyl derivative is also formed.

An intriguing finding is that *N*-methylarginine occurs at a specific site, namely, residue 107, in a basic protein referred to as A1 protein, a major component of the myelin from animal and human brain[276]. This basic protein is thought to confer on myelin the ability to interact with lipid constituents in cell membranes. Protein methylase I can give rise to mono-, di- and tri-*N*-methylarginines, which are present in different combinations in the A1 protein from the brain of different animal species. This interesting aspect of protein methylation has been reviewed by Cantoni[174]. Its significance in relation to brain function is still a matter of speculation. The protein methylase I from rat liver cytoplasm can use the analogue *S*-adenosylethionine as a substrate as effectively as the methionine sulphonium compound in the alkylation of histone, and is inhibited by the reactive metabolites of various carcinogens in addition to ethionine, for example aminofluorene derivatives and acridine orange. There appears to be a correlation between the malfunction of histone produced by these compounds and the induction of neoplasia[277].

Protein methylase II, otherwise known as protein carboxymethyltransferase, was discovered during a study of protein methylase I[278]. This second methylase was found to catalyse the formation from protein substrates of an unstable product which readily breaks down with the release of methanol. A similar enzyme had been reported previously by Liss and Edelstein[279]. The enzyme catalyses the O-methylation of free carboxyl groups in proteins, forming methyl ester groups, which are easily hydrolysed back to the free acids under weakly acidic or alkaline conditions.

This O-methyltransferase is present in a wide variety of animal tissues, including brain, liver, heart, thymus, and pituitary, the testis being especially rich in the enzyme. Among the amino acid residues capable of being esterified are glutamic and aspartic acids. The enzyme can also methylate hormones from the anterior lobe of the pituitary, and by forming derivatives of the hormones from which they may be readily released, the methyltransferase may play a role in their storage and transport. It acts on internal, but not terminal, carboxyl groups and can methylate gelatin, pepsin, ovalbumin, pancreatic ribonuclease, and histones[280].

Kinetic studies on protein methylase II have shown that its K_m value for S-adenosylmethionine is of the order of 1×10^{-6} M, or one tenth to one fifteenth of both the concentration of the sulphonium compound in most tissues and also the corresponding K_m values for this substrate in the case of other methyltransferases, such as catechol-O-methyltransferase, acetylserotonin O-methyltransferase, and phenylethanolamine N-methyltransferase. This means that under normal circumstances the action of the protein methylase is not limited by the concentration of S-adenosylmethionine, but that when the concentration of the sulphonium compound does fall to critical levels, protein O-methylation may proceed in preference to the methylation of biogenic amines[275].

Unlike the first two protein methylases, which are soluble enzymes, protein methylase III, or protein (lysine) methyltransferase, is bound to cell nuclei, where it appears to be associated with chromatin. This enzyme mediates in the methylation by S-adenosylmethionine of the ε-amino groups of lysine residues in histones[281]. The mono-, di- and tri-N-methyl derivatives of the amino acid are formed, but there is some reason to believe that more than one lysine methyltransferase is involved. Like protein methylase II, it has a particularly low K_m value for the methyl donor substrate, indicating that its action might be favoured over that of other methyltransferases.

It is of interest that the specific activity of protein methylase III is higher in rapidly proliferating tissues, such as foetal or regenerating adult tissue, rapidly growing tumour tissue, etc., than in normal adult tissue. In fact, there appears to be a close correlation between the extent of methylation of protein–lysine residues and cell proliferation. The enzyme is present in most tissues, testis being a good source.

Much remains to be discovered about the biological roles of the protein methylases and to what extent their functions are interlinked. Of particular interest is the way the specific activity of protein methylase II in rat brain alters with the age of the animal, a 15-fold increase (on a cell-unit basis) taking place between birth and the attainment of a body weight of 120 g. This dramatic change is not exhibited by the two other protein methylases[282]. Protein methylase II, but not I or III, is present in circulating blood and does not show the marked changes in specific activity in proliferating or tumour tissue observed for protein methylase III. It has been speculated that protein methylase II may be involved in the processes of brain maturation, or it may operate as a regulator in brain function[278]. Furthermore, protein methylases I and III decrease greatly in activity in rat testes over the period of body weight development from 100 to 200 g, whereas protein methylase II activity first increases and then declines during this period. This suggests a possible link between protein methylation and sexual maturation.

c. *Nucleic acids.* The occurrence of 7-methylguanine **74** as a constituent of human urine was reported as long ago as 1898[283]. Subsequently, various other methylated purines and pyrimidines were identified as minor excretory products. However, about 60 years elapsed after the original discovery before it was recognized that small amounts of these methylated bases are normal components of nucleic acids from mammalian species and other organisms[284,285], the metabolic turnover of these nucleic acids accounting for the appearance of the bases in urine.

Extensive analytical studies have revealed that at least fifteen of these bases occur naturally in this manner, and that their distribution pattern varies with the species of the organism and the type of tissue from which the nucleic acids are obtained. Thus rat liver transfer RNA (tRNA) contains 7-methylguanine **74**, 1-methylguanine **75**, $N_{(2)}$-methylguanine **76**, $N_{(2)},N_{(2)}$-dimethylguanine **77**, 1-methyladenine **78**, 5-methyluracil (thymine) **79**, and 5-methylcytosine **80**. The tRNAs of other tissues, including tumour tissues, contain the same bases but in different proportions, while tRNA from the bacterium *E. coli* contains in addition 2-methyladenine **81** and $N_{(6)}$-methyladenine **82**[286]. An extensive literature on the composition of nucleic acids has developed, which has been summarized in valuable reviews by Salvatore and Cimino[287] and by Nau[288].

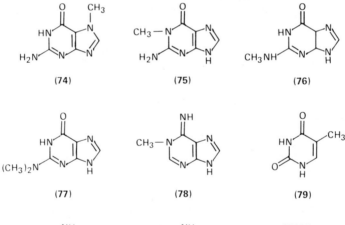

(74)   (75)   (76)

(77)   (78)   (79)

(80)   (81)   (82)

Figure 4 illustrates the distribution of methylated bases in a typical tRNA, consisting of about ninety nucleotide units. The average number of these bases present may be as low as two in some bacterial tRNAs, but up to ten in mammalian tRNAs. They are located mainly at nine sites in the molecule, comprising residues, 9, 10, 19, 29, 55, 64, 65, 70, and 74, although some are occasionally found at position 14, 35, 37, 40, 42, or 43.

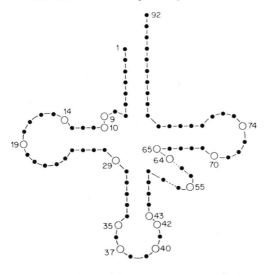

FIGURE 4. Methylation sites in tRNA.
O, Commonly methylated nucleoside residues;
●, residues occasionally methylated.

Borek and coworkers[289,290] have played a leading part in elucidating the biochemical mechanisms whereby these methylated bases arise. The situation parallels that found for polysaccharides and proteins, in that the methyl groups are introduced into selected purine and pyrimidine residues of nucleic acids at the macromolecular level and not into the free bases themselves[291]. Methionine was shown to be the source of the methyl groups of both the component bases of nucleic acids and likewise the free bases and nucleosides found in urine[292,293].

The study of tRNA methylation has made considerable advances as a result of the isolation of a mutant of *E. coli*, K12, exhibiting certain genetic lesions, which render it possible to separate the initial synthesis by the bacterium of 'nascent' or unmethylated tRNA (as well as ribosomal RNA and to some extent DNA) from the subsequent methylation step[294]. The mutant is unable to form methionine from the usual inorganic sulphur sources and requires the amino acid to be supplied in the culture medium for normal growth. In the absence of added methionine, tRNA is produced but is deficient in its normal complement of methylated bases. Under the appropriate conditions and in the presence of S-adenosylmethionine as the methyl donor, preparations of this submethylated tRNA from the *E. coli* mutant can be made to undergo further methylation *in vitro*, so that it can serve as a methyl acceptor for tRNA methyltransferases.

However, its role as a substrate is not limited to the enzymes from *E. coli* itself, since it is also an efficient methyl acceptor for other tRNA methyltransferases of plant, animal and viral origin. In fact, the methylation of this submethylated tRNA by heterologous tRNA methyltransferases (i.e. from species other than *E. coli*) may result in the formation of supermethylated tRNA, with more component bases being methylated than can be achieved with enzyme preparations from the bacterial mutant. In these circumstances nucleoside residues not normally methylated in *E. coli* tRNA become sites for the acceptance of methyl groups.

With the aid of the submethylated tRNA as a substrate, it has been demonstrated that a multiplicity of tRNA methyltransferases exist in mammalian tissues as well as in plant tissues and microorganisms. These enzymes show varying degrees of specificity towards individual purine or pyrimidine residues (base specificity) and different groups in the nucleoside molecule (group specificity), also towards the same nucleoside located at different points in the tRNA chain (site specificity) and towards tRNAs from different types of organism (species specificity).

Thus, in any one species, a uracil tRNA methyltransferase C-methylates a uracil moiety to yield 5-methyluracil, whereas at least three types of species- and site-specific guanine tRNA methyltransferase N-methylate guanine residues in the 1- or 5-position, or introduce one or two methyl groups in the 2-amino group. Similarly, adenine-specific tRNA methyltransferases C-methylate the purine moiety in the 2-position, N-methylate it in the 1-position, or form the mono- or dimethyl derivatives at the 6-amino group. In addition, other tRNA methyltransferases methylate the 2'-O- position of the ribose moiety in the nucleoside residues[295]. Detailed accounts of the distribution and properties of the tRNA methyltransferases may be found in a recent monograph on S-adenosylmethionine biochemistry[19].

The mechanism proposed for the C-methylation of a uridine residue by the enzyme transfer ribonucleate uracil-5 methyltransferase involves an interaction between S-adenosylmethionine and the reactive carbon–carbon double bond in the base[296]:

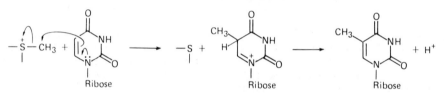

The methyl group is transferred intact, as shown by Law and coworkers[297] using methionine labelled with ^{2}H in the methyl group and added to the culture medium of *E. coli*.

The methylation of an adenosine to a 2-methyladenosine residue is a unique example of the enzymic C-methylation of a carbon–nitrogen double bond:

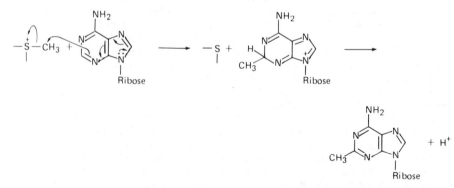

The tRNA methyltransferases are stimulated to various extents by monovalent or divalent cations, e.g. NH_4^+, Ca^{2+}, and Mg^{2+}, and also by physiological concentrations of the bases putrescine (83), spermidine (84) and spermine (85), which frequently occur in cells in close association with nucleic acids[298,299]. These enzymes are subject to product inhibition by S-adenosylhomocysteine and, moreover, they appear to be considerably more sensitive to the presence of this compound than are other methyltransferases, such as glycine and nicotinamide methyltransferases. Glycine methyltransferase generally occurs at much higher concentrations in animal tissues, so that not only may it compete more effectively with tRNA methyltransferases for the substrate S-adenosylmethionine, but it can also act as an efficient generator of inhibitory S-adenosylhomocysteine[300]. In this way it can exert a powerful regulatory effect on tRNA methylation. It is significant that in foetal tissue and that of newborn animals, where there is high tRNA methyltransferase activity, glycine methyltransferase activity is low or even absent[301]. Ageing in rats has been found to be associated with a rise in glycine methyltransferase activity, while at the same time there is a decrease in activity of tRNA methyltransferases, which would lead to the production of submethylated tRNAs[302]. Nicotinamide behaves as another natural inhibitor of tRNA methylation[303], and this may be because its methylation by nicotinamide N-methyltransferase competitively removes S-adenosylmethionine and, like glycine methyltransferase, produces inhibitory concentrations of S-adenosylhomocysteine.

Ribosomal RNA also contains a number of methylated bases, although it is methylated to only a quarter of the extent of tRNA. On the other hand, methylated bases have been reported as absent from messenger RNA[296]. Bacterial DNA contains 6-methyladenine as the predominant methylated base, whereas in plant and animal DNAs there is a preponderance of 5-methylcytosine[304]. The latter compound is the principal methylated base identifiable in DNA of various organs when rats are given methionine labelled in the methyl group[305], or when normal or leukaemic human white blood cells are incubated with the amino acid[306]. Methyltransferases are again implicated in the formation of the methylated bases, transferring methyl groups from S-adenosylmethionine to specific adenine or cytosine residues in DNA, depending on the species[307]. The methylation of DNA often occurs very shortly after the synthesis of the nascent DNA, sometimes within 2 min of replication, although there are situations where there is a delay of several hours in the onset of methylation[308]. S-Adenosylhomocysteine is a potent competitive inhibitor, reducing the methylation of DNA in isolated rat liver nuclei by as much as 50% when present at concentrations comparable with that of the methyl donor substrate[309].

The biological function of nucleic acid methylation is still far from clear, although several plausible suggestions have been put forward, based on the changes in physical and chemical properties conferred on the entire molecule or on its nucleoside residues, by the introduction of methyl groups into the component bases[291,296]. Through steric and electronic effects the presence of methyl groups will affect the configuration of the nucleic acid molecule as a whole by modifying the degree of ionization of the various polar groups, hydrogen bonding, etc. In DNA these effects would be expected to govern its base-pairing properties and could thereby control further DNA replication, transcription, and cell differentiation. Consequently, any divergence from the normal pattern of methylation, whether this results in the production of submethylated or supermethylated DNA, could lead to atypical base pairing and the formation of new nucleic acid chains lacking the conformations characteristic of the species.

Borek and coworkers[290,291] have proposed that an appropriate degree of methylation creates the maximum extent of organization within the nucleic acid structure and protects the replicating mechanism of the organism from the action of invading viruses. The location of methylated bases at specific sites within the tRNA molecule, as shown in Figure 4, further suggests that in any one tRNA the methylated bases comprise a code for the attachment of a particular amino acid, recognizable only by its appropriate activating enzyme, so serving to exclude other enzymes activating different amino acids. It is tempting to suggest that the submethylation of tRNA, resulting from a decrease in tRNA methyltransferase activity in senility, referred to previously, may account for the changes in protein synthesis and cellular organization associated with the ageing process.

There is accumulating evidence that aberrant methylation of nucleic acids is linked with the onset of malignancy, although it remains to be established whether the biochemical lesion is a cause or an effect of the cellular transformation[310,311]. Numerous reports indicate a pronounced elevation of tRNA methylating activity in a wide range of tumour tissues, amounting to a twofold to tenfold increase over that in normal adult tissues. The reason for this increase seems to be the release of the enzymic inhibition exerted by tissue constituents and competing methyltransferases, and to the appearance of tRNA methyltransferases with altered specificities[312]. Whatever the cause, the nett effect is an increased methylating ability with the formation of supermethylated tRNA, manifested by an increase in the urinary excretion of methylated bases. Thus, an increased excretion of 7-methylguanine by leukaemic patients and of other methylated bases by animals with various experimentally induced tumours has been observed[310]. Waalkes et al.[313] have also reported increased levels of methylated nucleosides in the urine of patients with various forms of cancer, and they have considered the feasibility of using the excretion of these compounds as a biochemical indicator in the diagnosis of malignancy[314].

The idea of a connection between nucleic acid methylation and cancer is strengthened by the fact that the well known liver carcinogen dimethylnitrosamine is metabolized to a powerful alkylating agent. Its administration to animals results in an increased methylation of DNA and RNA in liver and kidney tissue, with 7-methylguanine, normally a rare constituent, becoming one of the prevalent methylated bases[315]. Furthermore, dimethylnitrosamine induces increased tRNA methyltransferase activity in the kidney of rats fed the compound, as well as when added to tissue culture preparations[316,317]. The S-ethyl analogue of methionine, ethionine, which is also a liver carcinogen, has been reported to affect tRNA methyltransferase activities in the same way[318]. The mode of action of this compound is more complex, however, as discussed in Section IV.D.2.

The above account gives only a brief glimpse of the steadily unfolding story of nucleic acid methylation in which S-adenosylmethionine plays an all-important role. A further aspect worthy of mention concerns the mechanisms by which bacteria can combat foreign DNAs in the form of invading viruses. To prevent the duplication of these DNAs within the cell, the bacterial host can degrade them by a process termed restriction, or can further overcome their effects by modification. Restriction consists of the splitting of the foreign DNAs at certain points in the molecule by highly specific endonucleases which do not attack the organism's own DNA. These enzymes have a requirement for S-adenosylmethionine, possibly as an allosteric factor. Modification results from a methylation of DNA by specific S-adenosylmethionine-dependent methyltransferases, methyl groups being introduced into the 6-amino group of adenine in nucleotides at or near the points where the restriction endonucleases cleave the polynucleotide chains. The DNA of the host cells is methylated and thereby protected from attack by the restriction endonucleases, while foreign DNAs are cleaved before they can be methylated.

These cellular defence mechanisms, representing yet another intriguing facet of the biochemistry of S-adenosylmethionine, have recently been the subject of several reviews[319–321].

3. Decarboxylation and polyamine biosynthesis

Polyamines have long been known to be constituents of bacteria, as well as being present in plants and most animal tissues[322]. Commonly occurring compounds of this type include putrescine (1,4-diaminobutane) **83**, spermidine **84**, and spermine **85**. Strains of E. coli have been reported to contain 0.06 mmol of putrescine and 0.017 mmol of spermidine per gram dry weight of cells, which correspond to concentrations in the intracellular fluid of 19 and 6 mM, respectively[323]. In animals, spermine is the more commonly found polyamine, seminal fluid and the prostate gland being particularly rich in this compound[324]. The concentration of spermine in human seminal fluid is given as 12–14 mM and in human prostate as 0.023 mmol per gram of fresh tissue, brain and ovary containing only small amounts.

Polyamines are associated with nucleic acids in animal tissues; spermidine, for example, is bound to cell structural elements in amounts capable of neutralizing a large fraction of the phosphate groups of RNA[325]. Various roles have been attributed to polyamines, generally related to their interaction with polynucleotides. There is also abundant evidence for an elevated excretion of these compounds in some forms of cancer, and of abnormal levels of spermidine in the serum of cancer patients[326,327].

Important contributions to an understanding of the biosynthesis of these amines have been made by Tabor et al.[328], who showed that the carbon chain of putrescine is incorporated into spermidine in E. coli cells, precursors of putrescine being the amino acid ornithine, and agmatine, the decarboxylation product of arginine. Greene[329] had shown that in spermidine synthesis the propylamine moiety was derived from methionine, but the more immediate precursor of this group was established as the decarboxylated derivative of S-adenosylmethionine, namely S-adenosyl-(5′)-3-methylthiopropylamine **5**[330]. The transfer of the propylamino group is shown below, 5′-methylthioadenosine **10** being a further product of the reaction.

$$NH_2(CH_2)_4NH_2 \;+\; Adenosine\!-\!\overset{+}{\underset{\underset{CH_3}{|}}{S}}\!-\!CH_2CH_2CH_2NH_2 \longrightarrow$$

(83)   (5)

$$NH_2(CH_2)_4NH(CH_2)_3NH_2 \;+\; CH_3\!-\!S\!-\!Adenosine \;+\; H^+$$

(84)   (10)

Spermidine can itself act as the acceptor in a second propylamino group transfer from decarboxylated S-adenosylmethionine, yielding spermine and 5′-methylthioadenosine:

$$NH_2(CH_2)_4NH(CH_2)_3NH_2 \;+\; Adenosine\!-\!\overset{+}{\underset{\underset{CH_3}{|}}{S}}\!-\!CH_2CH_2CH_2NH_2 \longrightarrow$$

$$NH_2(CH_2)_3NH(CH_2)_4NH(CH_2)_3NH_2 \;+\; CH_3\!-\!S\!-\!Adenosine \;+\; H^+$$

(85)

These reactions have also been established as accounting for the presence of spermidine and spermine in the rat prostate gland[331,332].

The enzyme S-adenosylmethionine decarboxylase, which generates the initial propylamine donor, has been studied in some detail. It is widely distributed, with a K_m value for the substrate of 4×10^{-5} M, so that in most organs its activity is not limited by the concentration of available S-adenosylmethionine. The enzyme has also been identified in $E.\ coli$, yeasts, and plants[333]. It shows a high degree of specificity towards its substrate[333], being unable to decarboxylate the $(+)$-diastereoisomer of S-adenosylmethionine, its α-hydroxy analogue, or related compounds such as S-methylmethionine and S-inosylmethionine. Putrescine is an activator of the decarboxylase[331,334], while thiole and carbonyl reagents are inhibitors, the latter interacting with pyruvic acid bound to the enzyme as a cofactor[335].

Another potent inhibitor is an antileukaemic agent, methylglyoxal bis(guanylhydrazone) 86, which exerts antiproliferative effects on various experimentally induced tumours. This compound produces a 50% inhibition of the decarboxylase from human prostate at concentrations as low as 10^{-7} M, although the enzyme from $E.\ coli$ is relatively unaffected[336].

$$\underset{HN}{\overset{H_2N}{>}}CNHN = CHC = NNHC\underset{NH}{\overset{NH_2}{<}}$$
$$\underset{CH_3}{|}$$

(86)

Recent work on the distribution of S-adenosylmethionine decarboxylase activity in various mammalian tissues indicates that, through its participation in spermidine and spermine biosynthesis, it plays an important role in the development and function of the brain and is in some way linked to the action of hormones, in particular those concerned with lactation.

Decarboxylated S-adenosylmethionine has some activity as a methyl donor for the homocysteine S-methyltransferase of yeast[154]. It serves as a weak inhibitor of other methyltransferases, and this may mean that S-adenosylmethionine decarboxylase has a regulatory effect on methylation reactions, not only by removing the methyl donor substrate but also by actively depressing them.

Much less is known about spermidine synthase which catalyses the propylamino group transfer from decarboxylated S-adenosylmethionine[328,337]. In rat brain, at least, there is evidence that the second propylamino group transfer to form spermine is brought about by a separate enzyme[338].

4. Cleavage to 5'-methylthioadenosine

The third type of bond fission in S-adenosylmethionine (at site C, see Figure 1) leads to loss of the α-amino-n-butyryl moiety, yielding 5'-methylthioadenosine 10. This nucleoside has been recognized for a long time to be a minor constituent of yeast extracts[339,340], although its biochemical significance remained obscure until it was found to behave as a precursor of methionine in yeast[341], and its enzymic formation from S-adenosylmethionine was demonstrated in $Aerobacter\ aerogenes$[342] and in yeast[343].

The enzyme responsible for the formation of 5'-methylthioadenosine, S-adenosylmethionine lyase, catalyses the displacement of the $C_{(4)}$ group as α-amino-γ-butyrolactone 29, which is a hydrolysed non-enzymically to homoserine 30, the overall reaction being analogous to the decomposition of

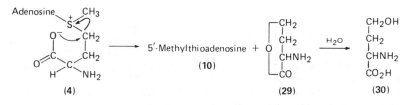

S-methylmethionine by a corresponding lyase (see Section IV.B.1). Mudd[343] proposed that the reaction proceeds by an intramolecular attack of the carboxyl ion on the electrophilic carbon atom adjacent to the sulphonium group. An alternative mechanism is possible, involving an elimination reaction rather than a substitution, and giving rise to 5′-methylthioadenosine and 2-amino-3-butenoic acid as initial products, by analogy with the known enzymic decomposition of dimethyl-β-propiothetin to dimethyl sulphide and acrylic acid (Section IV.A.1). However, Mudd[343] has reported that this latter reaction does not occur.

More recently, Nishimura et al.[344] have reported the presence of an enzyme in E. coli which catalyses the transfer of the α-amino-n-butyryl group of S-adenosylmethionine to uridine residues in tRNA, giving 3-(α-amino-n-butyryl)uridine. As in the above intramolecular reaction, 5′-methylthioadenosine would be an expected reaction product, although it does not appear to have been identified in this case.

In bacteria, 5′-methylthioadenosine is further metabolized by rapid enzymic hydrolysis to give adenine and 5′-methylthioribose[342], serving as the substrate for a nucleosidase which also acts on S-adenosylhomocysteine to form adenine and S-ribosylhomocysteine (Section IV.C.1)[98,147]. The hydrolysis of 5′-methylthioadenosine does not apparently take place readily in yeast, and for this reason might be expected to accumulate in the cells in appreciable quantities. That this does not happen is explained by the rapid utilization of the compound for the resynthesis of S-adenosylmethionine, as observed in the yeast Candida utilis[97,345]. As discussed in Section III.D.2, a regenerative pathway for the sulphonium compound in bacteria is possible, involving methanethiol as an intermediate. This is shown as the cycle in Figure 5, which provides a means of recovery of the utilizable fragments of the S-adenosylmethionine molecule additional to that already

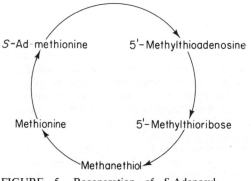

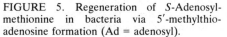

FIGURE 5. Regeneration of S-Adenosyl-methionine in bacteria via 5′-methylthio-adenosine formation (Ad = adenosyl).

described for animals and microorganisms in Figure 2, and which can also recycle the 5'-methylthioadenosine formed as a by-product of polyamine synthesis.

An interesting and related aspect of S-adenosylmethionine metabolism is the role of the sulphonium compound as a precursor of ethylene in plants[346], which it shares with S-methylmethionine (see Section III.C). Ethylene generated by plant organs is derived from carbon atoms 3 and 4 of methionine[347,348], the formation of the olefin being one of the main pathways of methionine metabolism in apple tissue[348]. To account for the continued production of ethylene from the relatively low concentrations of methionine in apple, Baur and Yang[349] suggested that a recycling of the sulphur of the amino acid must take place.

Incubation of methionine with pieces of apple tissue gave 5'-methylthioribose as the main product, together with smaller quantities of 5'-methylthioadenosine[350]. This takes place only in climacteric tissue, which also actively converts the amino acid into ethylene, but it does not occur in immature or preclimacteric tissue which is unable to produce ethylene. The tissue rapidly converts 5'-methylthioadenosine into 5'-methylthioribose, explaining the small amounts of the former compound found. Radioactive labelling experiments have subsequently shown that the thiomethyl group of methionine is transferred intact to the two metabolites. The most likely mechanism to explain these findings is that methionine is first activated to S-adenosylmethionine, the α-amino-n-butyryl moiety of which is the source of ethylene. The sulphonium compound is split to 5'-methylthioadenosine, which is then hydrolysed to the corresponding ribose derivative. Recycling of the sulphur fragments may then take place by the scheme shown in Figure 5.

In his most recent work, Yang[305a] has shown that the α-amino-n-butyryl moiety of S-adenosylmethionine gives rise to 1-aminocyclopropane-1-carboxylic acid, most probably as a result of a γ-elimination coupled with the elimination of an α-hydrogen atom, leading to the formation of a cyclopropane ring. This amino acid, already known for some time to be a constituent of ripe fruits, readily breaks down oxidatively to yield ethylene in a wide range of plant tissues, and would appear to be the immediate precursor of ethylene originating from methionine in plants[350b]. The conversion of S-adenosylmethionine to 1-aminocyclopropane-1-carboxylic acid is the rate-limiting step in ethylene biosynthesis by this pathway, and a further finding of great significance is that the plant hormone auxin, or indol-3-ylacetic acid, which stimulates ethylene production, does so by inducing the synthesis of the enzyme catalysing 1-aminocyclopropane-1-carboxylic acid formation[350c].

5. Other metabolic roles

An account of the biochemistry of S-adenosylmethionine would not be complete without reference to some of the many roles the compound plays in other aspects of metabolism but which have not as yet been fully explored. For example, there is accumulating evidence that S-adenosylmethionine participates in the synthesis of the vitamin biotin by virtue of its ability to act as an amino group donor. The conversion of 7-keto-8-aminopelargonic acid into 7,8-diaminopelargonic acid, a precursor of desthiobiotin, occurs by an aminotransferase reaction in which S-adenosylmethionine is a substrate[351–353].

The sulphonium compound is also involved as an activator in several complex enzymic reactions in which it is thought to act as an allosteric effector. These include the so-called phosphoroclastic reaction, or the conversion of pyruvate into acetyl phosphate and formate, catalysed by pyruvate–formate lyase[354,355], the action of lysine 2,3-diaminomutase, which catalyses the conversion of L-lysine into

L-β-lysine[356,357], and also the terminal step in the *de novo* synthesis of methionine, namely the methylation of homocysteine by $N_{(5)}$-methyltetrahydrofolate (**22**). These reactions are catalysed by enzymes containing bound iron or cobalt ions, and S-adenosylmethionine may be concerned in the stabilization of the metal ions within the prosthetic groups. In the pyruvate–formate lyase reaction, it appears that the sulphonium compound converts an inactive form of the lyase to the catalytically active form. It has been suggested that this takes place through the transient adenosylation of a further iron-dependent enzyme functioning as the converter enzyme[358].

The role of S-adenosylmethionine as an activator of cystathionine synthase is of potential significance in the control of sulphur utilization in the animal[359]. S-Adenosylmethionine in this situation may be acting as a precursor of S-adenosylhomocysteine, which Finkelstein et al.[360] have shown to activate partially purified preparations of cystathionine synthase from rat liver. Since the latter compound is an inhibitor of betaine:homocysteine methyltransferase in liver (a key enzyme in the mobilization of dietary methyl groups), it is capable of altering the partition of tissue homocysteine between methionine synthesis on the one hand and cysteine synthesis, leading to the formation of taurine and inorgnaic sulphate, on the other.

Another role of S-adenosylmethionine which needs further study is its connection with threonine metabolism, the sulphonium compound stimulating the breakdown of O-phosphothreonine to the free amino acid by an enzyme present in the leaves of sugar beet[361]. In addition to inhibiting methyltransferases indirectly by acting as a source of S-adenosylhomocysteine, S-adenosylmethionine also controls the action of a number of other enzyme systems. It interacts with pyridoxal 5-phosphate and, *in vitro* at least, can cause inhibition of pyridoxal 5-phosphate-requiring enzymes, such as tyrosine aminotransferase[362]. This type of action may contribute to the growth inhibition of yeast caused by the presence of high concentrations of methionine, when there is an over-production of S-adenosylmethionine.

D. S-Adenosylethionine

S-Adenosylmethionine and its parent amino acid, ethionine, are structural analogues of S-adenosylethionine and methionine, respectively. Some years ago apparently convincing evidence was presented for the presence of ethionine in at least four bacterial species[363], as well as for the synthesis of the labelled amino acid from radioactive inorganic sulphate in these organisms[364], and this gave grounds for believing that S-adenosylethionine might likewise be a normal cell constituent. However, the natural occurrence of ethionine has not so far been confirmed by other workers; indeed, Lederer and coworkers[365] have refuted its existence in bacteria. Ethionine and consequently S-adenosylethionine must still be regarded as artifacts, although this in no way diminishes their significance as research tools in metabolic studies.

Ethionine was first synthesized in 1938 by Dyer[366], who found the compound to be highly toxic to rats, inducing a cessation of growth, the development of fatty livers, and kidney damage, generally leading to the death of the animal. These effects were partially reversed by the subsequent addition of methionine or choline to the diet, and were prevented completely when both methionine and ethionine were fed together. This established that ethionine was acting as a metabolic antagonist of methionine, and at the time the inference was drawn that the metabolically essential methyl group supplied by dietary methionine cannot be replaced by the ethyl group. Further studies have shown that prolonged administration of ethionine to animals induces liver carcinoma[367]. Other effects produced by administration of the compound include changes in the lipid composition of the blood, degenerative changes in certain organs, such as the pancreas, also hypoglycaemia, and an inhibition of protein synthesis[368,369]. The sequence of events following an injection of ethionine in animals has been well described by Farber *et al.*[370], and commences with the development of an acute deficiency of ATP, then a marked inhibition of RNA synthesis, followed by an inhibition of protein synthesis. This in turn is followed by an accumulation of lipids in the liver, the release of triglyceride alkanoic acids being impaired.

The mode of action of ethionine is obviously more complex than was originally thought, and it seems likely that it exerts multiple biochemical effects in the whole organism through a number of different mechanisms. Certain of these effects may be attributed to the action of ethionine *per se*. Thus the compound is known to take part in protein synthesis and to be incorporated in place of methionine in polypeptide chain formation[371-374]. In *Bacillus subtilis* ethionine was shown to be capable of replacing 40% of the methionine residues in α-amylase[375]. This did not significantly alter either the physiochemical or the catalytic properties of the enzyme; nevertheless, it has yet to be established whether other ethionine-containing proteins behave in the same way as their normal counterparts.

Ethionine administration to animals causes marked decreases in the activities of various tissue enzymes, including choline and xanthine oxidases, threonine dehydrase, and glucose-6-phosphatase[368]. This may be partly a reflection of the ability of ethionine to inhibit protein synthesis (see Section IV.D.1), although there may be other contributory causes. Ethionine also inhibits tRNA methylation in rats and bacteria, leading to the production of submethylated tRNA[376,377].

Cleland[378] has observed a number of effects produced by ethionine on plant tissues. The ethyl compound strongly inhibited the transfer of methyl groups to the coleoptile cell-wall material of oats (*Avena sativa*), and counteracted the ability of auxin to enhance the rate of methylation. It also caused an immediate suppression of tissue elongation in this plant. Other effects of ethionine derive from its conversion into S-adenosylethionine, which is discussed in the next section.

1. Biosynthesis

Following the recognition of methionine S-adenosyltransferase as the enzyme responsible for the formation of S-adenosylmethionine from methionine, purified preparations of the enzyme were also found to act on ethionine as the substrate[84,379,380], although the reaction with ATP occurs at about half the rate of activation of methionine[381]. Proof that the two amino acids are substrates of the same enzyme is indicated by the ability of ethionine to act as a strong competitive inhibitor of S-adenosylmethionine synthesis by the rat liver enzyme[382].

Parks[383] was the first to isolate S-adenosylethionine, using as the source cultures of *S. cerevisiae* grown in the presence of L-ethionine. Yeasts, containing an active methionine S-adenosyltransferase, constitute an excellent organism for the preparation of this sulphonium compound[52,88]. *S. cerevisiae* gives high yields of S-adenosyl-L-ethionine (up to 100 μmol/g dry weight of cells) when supplied with L-ethionine at a concentration of 5 μmol/ml of culture medium. Under these conditions, *C. utilis* produces 30–40 μmol/g dry weight of cells. For the preparation of S-adenosyl-D-ethionine, the latter yeast is the preferred source, accumulating up to 35 μmol of the compound per gram dry weight of cells, or more than seven times that accumulated by *S. cerevisiae*.

When ethionine is injected into animals there is a rapid accumulation of S-adenosylethionine in the liver[384]. For example, Stekol[381] showed that three hours after an injection of 1.7 μmol of ethionine per gram body weight into young rats, the livers contained 3.1 μmol of the sulphonium compound per gram of fresh tissue, compared with the formation of 0.4 μmol of S-adenosylmethionine per gram of tissue from an equimolar amount of methionine. Six hours after the oral administration of ethionine to partially hepatectomised rats, the concentration of S-adenosylethionine in the liver may rise to as much as 30–40 times that of S-adenosylmethionine after the feeding of methionine[385]. These values indicate the rapid withdrawal of the naturally occurring S-adenosylmethionine for normal metabolic needs and the relatively slow metabolism of the ethyl analogue.

Exogenously applied ethionine, through its ready conversion into S-adenosylethionine, causes a marked fall in tissue ATP levels, particularly in the liver[386,387]. This 'ATP-trapping' action of ethionine, in addition to reducing the supply of ATP for S-adenosylmethionine synthesis, has adverse effects on other ATP-requiring processes, one of these being protein synthesis. This has been observed in animals, bacteria, and protozoa, and results in turn from a depression of the synthesis of RNA[370,388]. The administration of methionine does not seriously deplete tissue ATP levels, probably because of the rapid turnover of the S-adenosylmethionine produced.

Schlenk[389] has shown that small amounts of S-adenosylethionine accumulate in yeast when the cells are exposed to ethanethiol. S-Ethylcysteine and 5'-ethylthioadenosine are also precursors of the sulphonium compound in C. utilis, possibly as a result of being converted into ethanethiol[97,390].

2. Metabolism

The types of metabolic reaction undergone by S-adenosylethionine resemble those previously described for S-adenosylmethionine. There is no evidence, however, to indicate the existence of enzymes specific for the ethyl analogue. Mudd[343] has shown that it serves as a substrate for S-adenosylmethionine lyase, being split to 5'-ethylthioadenosine and α-amino-γ-butyrolactone, but at 30% of the rate of cleavage of S-adenosylmethionine. At high concentrations S-adenosylethionine acts as a competitive inhibitor of 5'-methylthioadenosine formation from S-adenosylmethionine. The ethyl compound is a poor substrate for S-adenosylmethionine decarboxylase[391], although as with S-adenosylmethionine, decarboxylation is markedly enhanced by putrescine.

There is accumulating evidence that S-adenosylethionine can act as a substrate for a number of methyltransferases, giving rise to ethylated metabolites. However, their formation is often appreciably slower than that of the naturally occurring methyl compounds[392]. Homocysteine S-methyltransferase is one of the least specific of the methyltransferases, and the enzyme from S. cerevisiae can utilize the ethyl analogue almost as effectively as S-adenosyl-L-methionine itself[129,383]. Ethionine is a product of the reaction:

S-Adenosylethionine + Homocysteine → S-Adenosylhomocysteine + Ethionine

S-Adenosylethionine also proved to be as effective as S-adenosylmethionine in the S-alkylation of foreign thiols, such as O-methylmercaptoethanol, by a methyltransferase from rat liver microsomes[157].

Experiments in which labelled ethionine was fed to animals established that the ethyl group can be transferred to form the ethyl analogues of choline and creatine[393]. The normal methylation reaction giving rise to choline is significantly depressed at the same time, owing to competition between methionine and ethionine for the same enzyme systems[394]. The ethyl analogue of choline is, incidentally, an inhibitor of growth in rats[395]. Presumptive evidence for S-adenosylethionine as an ethyl donor came from the finding of Tuppy and Dus[396] that, in pig liver homogenates, ethionine alkylates glycocyamine to N-ethylglycocyamine, provided that ATP is present. Later, Winnick and Winnick[195] were able to show directly that S-adenosylethionine N-ethylates histidine and carnosine in the presence of chick muscle extracts. The utilization of S-adenosylmethionine and S-adenosylethionine by common enzymic pathways was demonstrated by Stekol et al.[392], who showed that each sulphonium compound depresses the utilization of the other in the alkylation of the precursors of choline and creatine in rat liver preparations. This is further confirmed by the ability of S-adenosylethionine to inhibit the methylation by S-adenosylmethionine of lysine and arginine residues in histones, catalysed by protein methyltransferases from rat liver[397].

Numerous studies have also established that the ethyl group of ethionine administered to animals appears in RNA and DNA[398,399]. Again, S-adenosylethionine has been shown to be the direct donor[400] and to inhibit competitively the methylation of DNA by S-adenosylmethionine[385]. The injection of [^{14}C-*ethyl*]-labelled ethionine into rats leads to the appearance of the labelled ethyl group in DNA and RNA as 7-ethylguanine residues[381,401]. Pegg[402] produced evidence that the ethylation of tRNA which takes place in the livers of rats treated with ethionine is, at least in part, mediated by tRNA methyltransferases, although the distribution of the resulting ethylated bases is different from that found for the normal methylated bases. Thus, $N_{(2)}$-ethylguanine, $N_{(2)},N_{(2)}$-diethylguanine, $N_{(2)}$-ethyl-$N_{(2)}$-methylguanine, 7-ethylguanine, and two N-ethylpyrimidines were identified as constituent bases. However, whereas the normal methylation of tRNA gives rise to 1-methyladenine and 1-methylguanine as major components, the two corresponding ethyl derivatives are not found in tRNA from ethionine-treated animals. This selective ethylation of tRNA base residues confirms earlier findings by Rosen[403].

Experiments carried out *in vitro* with submethylated tRNA from E. coli K12 confirmed that rat liver methyltransferases catalyse the transfer of the ethyl group from S-adenosylethionine to yield residues of $N_{(2)}$-ethylguanine and traces of $N_{(2)},N_{(2)}$-diethylguanine, 7-ethylguanine, and ethylated pyrimidines, the rate of ethylation being only one twentieth of the rate of methylation by S-adenosylmethionine. Apparently S-adenosylethionine is not a substrate for the methyltransferases concerned in the formation of 1-methyladenine and 1-methylguanine residues.

A proportion of the ethylation associated with ethionine administration or the *in vitro* treatment of tRNA with S-adenosylethionine may, however, take place by a mechanism *not* involving the action of methyltransferases, and there is supporting evidence for this from other workers[401,404,405]. This alternative reaction is thought to account for the formation of 7-ethylguanine residues, and it is of particular interest that the $N_{(7)}$ position of guanine is also the preferred site for the action of other chemical alkylating agents, for example, dimethylnitrosamine (see Section IV.C.2), a number of which are carcinogenic[406].

S-Adenosylethionine acts *in vitro* as an effective competitive inhibitor of the methylation of submethylated tRNA from E. coli K12 by animal and bacterial methyltransferase[407]. Moreover, this is manifested as a preferential inhibition of the methylation of adenine and cytosine residues by S-adenosylmethionine, the uracil and guanine methyltransferases being more resistant to the inhibitor[408]. This contrasts with the effect of ethionine, also an inhibitor of tRNA methylation as referred to earlier, which appears to oppose selectively the methylation of uracil residues[409]. S-Adenosylethionine is also an inhibitor of the DNA modification methyltransferases and restriction endonucleases[321].

It is not yet possible to connect the carcinogenicity of ethionine with any one of the mechanisms by which it acts as an antagonist of methionine, for example its replacement of methionine in proteins, its ATP-trapping action, ethylation of bases in nucleic acids, or induction of submethylated nucleic acids. However, its biochemical behaviour does lend support to the hypothesis of Borek and Kerr[311] that aberrant methylation of nucleic acids is an important event in the development of tumours.

V. MEDICAL AND AGRICULTURAL ASPECTS

To complete this account of the biochemistry of sulphonium salts, mention must be made of their significance in other aspects of biology, for example pharmacology, therapeutics and agriculture. At present only a small number of these compounds appear to have applications in these areas, but they are a far from neglected group of substances, judging by the steady reference to them in journals and the patent literature. It will therefore not be surprising if more sulphonium compounds of practical value in a biological context emerge in the near future.

As mentioned in the Introduction, sulphonium compounds first aroused interest biologically through their pharmacological properties when they were found to mimic quaternary ammonium salts in exerting a curare-like action on skeletal muscle, causing paralysis[1-3]. The ability to block neuromuscular transmission by one of several mechanisms is a property of many quaternary ammonium salts including members of the polymethylenebis(trialkylammonium) series, such as hexamethonium and decamethonium, which are widely used as drugs. Their sulphonium analogues, hexamethylene-1,6-bis(dimethylsulphonium) and decamethylene-1,10-bis(dimethylsulphonium) halides, respectively, also act in a similar manner, although they are generally less potent[410,411]. With this type of compound the number of methylene groups present in the chain determines the nature of the pharmacological activity as well as its potency[412].

Other effects produced by quaternary ammonium salts and likewise by their sulphonium analogues are a reduction in blood pressure (due to their ability to block neural transmission in the sympathetic nervous system), antihistamine activity, and the inhibition of gastric and salivary secretion. Several monosulphonium salts, e.g., trimethylsulphonium, alkyldimethylsulphonium, and alkyldiethylsulphonium halides, show anticholinesterase, activity which increases appreciably in ascending the series CH_3 to C_5H_{11}. Replacement of methyl by benzyl gives an even greater increase. Among the monosulphonium compounds, 2-phenylethyldimethylsulphonium iodide shows a strong sympathomimetic effect. Two other compounds of this type are thiospasmin [2-(phenylcyclohexyl-acetoxy)ethyldimethylsulphonium iodide] **87** and hydroxythiospasmin **88**. Thio-

$$
\begin{array}{c}
C_6H_{11} \\
\diagdown \\
\hspace{1.5em}CHCOOCH_2CH_2\overset{+}{S}(CH_3)_2 \quad I^- \\
\diagup \\
C_6H_5
\end{array}
$$

(87)

$$
\begin{array}{c}
C_6H_{11} \\
\diagdown \\
\hspace{1.5em}C(OH)COOCH_2CH_2\overset{+}{S}(CH_3)_2 \quad I^- \\
\diagup \\
C_6H_5
\end{array}
$$

(88)

spasmin has strong spasmolytic properties, that is, it acts as a muscle relaxant, and appears to be more effective than currently available nitrogen-containing drugs used for this purpose[413]. The hydroxy derivative is less toxic than thiospasmin itself[414], and both compounds reduce gastric and salivary secretion in man.

Sulphocholine (2-hydroxyethyldimethylsulphonium salt) **27** is a substrate for choline acetyltransferase, and its *O*-acetyl derivative, the sulphonium analogue of acetylcholine, is hydrolysed by cholineesterase. Sulphocholine resembles choline in many of its pharmacological properties[415]. The ester, succinylbis-(sulphocholine) dibromide **89**, actively blocks neuromuscular transmission, but is longer acting than succinylcholine and is less toxic than its nitrogen analogue[416,417].

$$(CH_3)_2\overset{+}{S}CH_2CH_2OCOCH_2CH_2COOCH_2CH_2\overset{+}{S}(CH_3)_2 \quad 2Br^-$$

(89)

Several reports suggest that sulphonium compounds containing 2-chloroethyl groups have a cytostatic action in experimental animals when given by injection. A pronounced inhibitory effect on the development of Ehrlich carcinoma in mice, as well as of other tumours in rats and mice, has been claimed for tri(2-chloroethyl)sulphonium chloride and for alkyldi(2-chloroethyl)sulphonium halides[418,419].

Turning to the naturally occurring sulphonium compounds, *S*-adenosylmethionine exerts indirect pharmacological effects of a profound nature as a result of its participation in the biosynthesis and further methylation of biogenic amines, as discussed in Section IV.C.1. It has also been found to reduce the cholesterol content of the blood. This may be related to its ability to mobilize phospholipids through the rapid methylation of phosphatidylethanolamine to lecithin. *S*-Adenosylmethionine does have direct effects on tissues, although these have not yet been exploited to any extent. For example, when injected together with glucose into dogs, it shows adrenaline-like characteristics in raising the blood sugar level more than does glucose alone, and the level subsequently declines less rapidly. Tests on laboratory animals show *S*-adenosylmethionine to have anti-inflammatory and analgesic properties[420]. The aggregation of blood platelets which normally follows the addition of ADP to platelet-rich plasma was found to be completely inhibited by *S*-adenosylmethionine *in vitro*. The effect in the whole animal, although short-lived, was without side effects and could be of possible use for the prophylaxis and treatment of vascular thromboses[421].

Dimethyl-β-propiothetin exhibits no outstanding pharmacological properties, but its methyl ester has a high degree of activity in stimulating the parasympathetic nervous system. Thetin esters of the type $R'R''\overset{+}{S}CH_2CH_2COOR$ I^-, where R, R', and R'' are alkyl groups, are reported to show higher muscle relaxant and anti-acetylcholine activity than structurally analogous quaternary ammonium compounds.

In recent years there has been a spate of reports on the therapeutic properties of *S*-methylmethionine. It was found in 1955 that the development of experimentally induced gastric ulcers in guinea pigs could be prevented by the oral administration of extracts from cabbage leaves, and the factor responsible was named 'vitamin U'[422]. Further work[423] suggested that the anti-ulcer factor was identical with *S*-methylmethionine, already established as a constituent of cabbage[31]. *S*-Methylmethionine given by mouth to various animal species has since been shown to exert a preventive action against the development of gastric ulcers induced by pyloric ligation, the administration of arsenic compounds, food deprivation, the injection of histamine, or stress[424-428]. This property of the sulphonium salt is not shared by other closely related compounds; methionine in particular has no alleviating effect on such ulcers. The effective dosage rate for S-methylmethionine is 0.5–1 g of the chloride per kilogram of body weight per day over a 14-day period. At this level the compound reduces peptic activity without affecting the pH or the free acidity of the gastric juice[428]. It also enhances the rate of healing of ulcers produced by the ingestion of acetic acid, and its anti-inflammatory properties afford protection against lesions of the gastric mucosa caused by large doses of aspirin[429]. The anti-ulcer activity of S-methylmethionine has been the subject of several reviews[430,431], and a number of patents have been filed for its use in the treatment of gastric ulcers and other gastric disorders.

S-Methylmethionine is also claimed to stimulate skin regeneration and the healing of wounds in experimental animals[432,433]. When given by injection to guinea pigs it appeared to increase the formation of antibodies to protein allergen from horse serum, and given orally it was found to lower the permeability of skin capillaries and to potentiate the anti-inflammatory effects of small doses of aspirin[429]. Daily injections given to rats at a level of 28 mg of the chloride per kilogram of body weight are reported to raise the number of circulating blood platelets by 30% with no significant effect on the leucocyte count. Furthermore, intramuscular injections of the compound seem to be effective in the alleviation of acute neuralgia, arthritis, and rheumatism in patients[434]. Its ability to increase the survival of X-irradiated mice has been linked with a protective effect against degeneration of the intestinal mucosa resulting from radiation damage[435].

Another feature of the seemingly limitless therapeutic activity of S-methylmethionine, and one which may have some bearing on its anti-ulcer properties, is its effect on blood cholesterol levels. It has been found that the serum cholesterol content of rats injected with the compound are lower than those of control animals[436]. The blood bilirubin level is also lowered in humans[437]. When administered intravenously, S-methylmethionine has a preventive effect against dietary hypercholesteraemia in rabbits[438]. This seems to be a fairly specific action, for dimethyl-γ-butyrothetin, which lacks the α-amino group, is less effective and methionine has negligible activity. Doses of 20 mg of S-methylmethionine chloride per day given orally to rabbits fed for 30 days on a diet containing 0.4 g of cholesterol per day prevented the rise in blood cholesterol and the development of atherosclerosis. S-Methylmethionine is therefore a sulphonium salt which could have widespread clinical applications in the future.

So far, sulphonium compounds have made relatively little impact as agricultural or horticultural chemicals. A number of patents describe sulphonium salts with insecticidal activity, for example various arylmethylphenacylsulphonium tetrafluoroborates for the control of corn root worms (*Diabrotica* species), and benzyldimethylsulphonium and *p*-nitrobenzyldimethylsulphonium tetraphenylborates for use as nematocides. Metcalf and coworkers[439] found that the introduction of a sulphonium group into the insecticide *OO*-diethyl *S*-2-ethylthioethylphosphorothiolate by its conversion into the corresponding methosulphate increased the anticholinesterase activity 100-fold, and yielded a compound with high systemic activity against insects infesting the cotton plant.

In the field of fungicides, a number of ethylenebis-sulphonium bromides, e.g. compound **90**, are claimed to control soil-borne pathogens, such as *Pythium ultimum*, which attacks pea seedlings. 2-Hydroxyethylsulphonium methylsulphates of structure $RCH_3SCH_2CH_2OH$ $CH_3SO_4^-$, where $R = C_7-C_{10}$ *n*-alkyl, are also

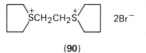

(**90**)

reported to be fungicides with low phytotoxicity and to control pea rust almost as effectively as the commonly used dithiocarbamate maneb.

Since a number of quaternary ammonium and phosphonium halides are widely used as plant growth regulators, it might be expected that sulphonium salts would show similar properties. Numerous patents, in fact, claim plant growth regulatory activity for alkyldimethylsulphonium salts, which are said to induce stimulation of leaf growth, hasten fruit ripening and abscission, and increase the flow of latex from rubber trees. Similarly, isopropyldimethylsulphonium bromide has been reported to increase the crop yield of wheat and barley when given as a spray treatment, while 2-bromoethyldimethylsulphonium bromide enhances the respiration of potato plants and aids in frost protection[440].

Several dialkylethylsulphonium iodides, including 2-chloroethyldimethylsulphonium and 2-hydroxyethyldimethylsulphonium iodides, when applied to the soil after sowing, have been claimed to reduce the height of wheat plants to a greater extent than the well known growth retardant chlormequat (2-chloroethyltrimethylammonium chloride). In a survey of the plant growth regulator activity of a wide range of onium salts, Wain and coworkers[441] tested the effects on wheat, pea, and dwarf bean seedlings of fourteen sulphonium bromides added to the soil or to the nutrient culture solution. Benzyldimethylsulphonium bromide and the corresponding mono- and dichlorobenzyl compounds reduced plant height, particularly in the case of wheat, but were phytotoxic and also suppressed root growth. Various benzyldi-n-butylsulphonium bromides at concentrations as low as 10^{-6} M also caused dwarfing of wheat plants and produced other abnormalities akin to those associated with the action of the growth retardant maleic hydrazide. Allyldimethylsulphonium bromide brought about a reduction in stem internode length with no serious side effects, although this compound was less effective on a molar basis than the corresponding quaternary ammonium analogue or chlormequat itself.

Among other miscellaneous effects of sulphonium compounds may be mentioned the possible use of triphenylsulphonium chloride as a soil conditioner. When applied to the soil surface it is reported to improve the tilth and to lead to an improved germination of lettuce, peas, oats, and other crops. When used as a foliar spray, it was found to aid control of iron chlorosis on the foliage of roses, pears, and lemons. The sulphonium compound appears to mobilize iron in the plant under conditions when it is in a form unavailable for chlorophyll synthesis[442].

VI REFERENCES

 1. A. Crum Brown and T. R. Fraser, *Trans. R. Soc. Edinburgh*, **7**, 663 (1872).
 2. H. R. Ing and W. M. Wright, *Proc. R. Soc. Ser. B*, **109**, 337 (1931).
 3. R. Hunt and R. R. Renshaw, *J. Pharmacol.*, **44**, 63 (1932).
 4. G. Toennies, *J. Biol. Chem.*, **132**, 455 (1940).
 5. F. Challenger and M. I. Simpson, *J. Chem. Soc.*, 1591 (1948).
 6. C. Brown and E. A. Letts, *Trans. R. Soc. Edinburgh*, **28**, 571 (1878).
 7. A. Bennett, *J. Biol. Chem.*, **141**, 573 (1941).
 8. P. Handler and M. L. C. Bernheim, *J. Biol. Chem.*, **150**, 335 (1943).
 9. V. Du Vigneaud, A. W. Moyer and J. P. Chandler, *J. Biol. Chem.*, **174**, 477 (1948).
10. G. A. Maw and V. Du Vigneaud, *J. Biol. Chem.*, **174**, 381 (1948); **176**, 1037 (1948).

11. J. W. Dubnoff and H. Borsook, *J. Biol. Chem.*, **176**, 789 (1948).
12. V. Du Vigneaud, *A Trail of Research in Sulphur Chemistry and Metabolism and Related Fields*, Cornell University Press, Ithaca, N.Y., 1952.
13. L. Young and G. A. Maw, *The Metabolism of Sulphur Compounds*, Methuen, London, 1958.
14. F. Challenger, *Aspects of the Organic Chemistry of Sulphur*, Butterworths, London, 1959.
15. S. K. Shapiro and F. Schlenk, *Adv. Enzymol.*, **22**, 237 (1960).
16. S. H. Mudd and G. L. Cantoni, in *Comprehensive Biochemistry*, Vol. 15 (Eds. M. Florkin and E. H. Stotz), Elsevier, Amsterdam, 1964, p. 1.
17. S. K. Shapiro and F. Schlenk (Editors), *Transmethylation and Methionine Biosynthesis*, University of Chicago Press, Chicago, 1965.
18. J. B. Lombardini and P. Talalay, *Adv. Enzyme Regulation*, **9**, 349 (1971).
19. F. Salvatore, E. Borek, V. Zappia, H. G. Williams-Ashman and F. Schlenk (Editors), *The Biochemistry of Adenosylmethionine*, Columbia University Press, New York, 1977.
20. P. Haas, *Biochem. J.*, **29**, 1258 (1935).
21. G. L. Cantoni, D. G. Anderson and E. Rosenthal, *J. Biol. Chem.*, **222**, 171 (1956).
22. R. G. Ackman, C. S. Tocher and J. McLachlan, *J. Fish. Res. Bd. Can.*, **23**, 357 (1966).
23. R. Bywood and F. Challenger, *Biochem. J.*, **53**, xxvi (1953).
24. F. Challenger, R. Bywood, P. Thomas and B. J. Hayward, *Arch. Biochem. Biophys.*, **69**, 514 (1957).
25. T. Motohiro, *Mem. Fac. Fish. Hokkaido Univ.*, **10**, 1 (1962).
26. R. G. Ackman and H. J. Hingley, *J. Fish. Res. Bd. Can.*, **25**, 267 (1968).
27. A. P. Ronald and W. A. B. Thomson, *J. Fish. Res. Bd. Can.*, **21**, 1481 (1964).
28. J. C. Sipos and R. G. Ackman, *J. Fish. Res. Bd. Can.*, **21**, 423 (1964).
29. B. Granroth and T. Hattula, *Finn. Chem. Lett.*, (6), 148 (1976).
30. F. Challenger and B. J. Hayward, *Chem. Ind. (London)*, 729 (1954).
31. R. A. McRorie, G. L. Sutherland, M. S. Lewis, A. D. Barton, M. R. Glazener and W. Shive, *J. Amer. Chem. Soc.*, **76**, 115 (1954).
32. F. F. Wong and J. F. Carson, *J. Agric. Food Chem.*, **14**, 247 (1966).
33. G. Werner, R. Hossli and H. Neukom, *Lebensm. Wiss. Technol.*, **2**, 145 (1969).
34. T. Kiribushi and T. Yamanishi, *Agric. Biol. Chem.*, **27**, 56 (1963).
35. T. W. Keenan and R. C. Lindsay, *J. Dairy Sci.*, **51**, 112 (1968).
36. E. G. Kovacheva, *Prikl. Biokhim. Mikrobiol.*, **10**, 129 (1974).
37. G. N. Khuchua and A. B. Stasyak, *Prikl. Biokhim. Mikrobiol.*, **11**, 914 (1975).
38. V. Du Vigneaud, J. P. Chandler, A. W. Moyer and D. M. Keppel, *J. Biol. Chem.*, **131**, 57 (1939).
39. V. Du Vigneaud and J. R. Rachele, in *Transmethylation and Methionine Biosynthesis* (Eds. S. K. Shapiro and F. Schlenk), University of Chicago Press, Chicago, 1965, p. 1.
40. F. Challenger, *Chem. Rev.*, **36**, 315 (1945).
41. F. Challenger, *Quart. Rev. Chem. Soc.*, **9**, 255 (1955).
42. S. J. Bach, *Biol. Rev. Cambridge Phil. Soc.*, **20**, 158 (1945).
43. H. Borsook and J. W. Dubnoff, *J. Biol. Chem.*, **132**, 559 (1940).
44. H. Borsook and J. W. Dubnoff, *J. Biol. Chem.*, **171**, 363 (1947).
45. W. A. Perlzweig, M. L. C. Bernheim and F. Bernheim, *J. Biol. Chem.*, **150**, 401 (1943).
46. G. L. Cantoni, *J. Biol. Chem.*, **189**, 203, 745 (1951).
47. G. L. Cantoni, *J. Biol. Chem.*, **204**, 403 (1953).
48. J. Baddiley, G. L. Cantoni and G. A. Jamieson, *J. Chem. Soc.*, 2662 (1953).
49. J. Baddiley and G. A. Jamieson, *Chem. Ind. (London)*, 375 (1954); *J. Chem. Soc.*, 4280 (1954); *J. Chem. Soc.*, 1085 (1955).
50. G. De La Haba, G. A. Jamieson, S. H. Mudd and H. H. Richards, *J. Amer. Chem. Soc.*, **81**, 3975 (1959).
51. F. Schlenk and R. E. DePalma, *J. Biol. Chem.*, **229**, 1037, 1051 (1957).
52. F. Schlenk, C. R. Zydek, D. J. Ehninger and J. L. Dainko, *Enzymologia*, **29**, 283 (1965).
53. F. Schlenk, in *Transmethylation and Methionine Biosynthesis*, (Eds. S. K. Shapiro and F. Schlenk), University of Chicago Press, Chicago, 1965, p. 48.

54. J. R. Warner, S. A. Morgan and R. W. Shulman, *J. Bacteriol.*, **125**, 887 (1976).
55. G. Svihla and F. Schlenk, *J. Bacteriol.*, **79**, 841 (1960).
56. G. E. Gaull and M. K. Gaitonde, *Biochem. J.*, **102**, 294, 7959 (1967).
57. F. Salvatore, R. Utili, V. Zappia and S. K. Shapiro, *Anal. Biochem.*, **41**, 16 (1971).
58. T.-C. Chou, A. W. Coulter, J. B. Lombardini, J. R. Sufrin and P. Talalay, in *The Biochemistry of Adenosylmethionine* (Eds. F. Salvatore, E. Borek, V. Zappia, H. G. Williams-Ashman and F. Schlenk), Columbia University Press, New York, 1977, p. 18.
59. S. Ito and J. A. C. Nicol, *Proc. R. Soc. Ser. B*, **190**, 33 (1975).
60. S. Ito and J. A. C. Nicol, *Biochem. J.*, **153**, 567 (1976).
61. J. J. Abel, *Hoppe-Seylers Z. Physiol. Chem.*, **20**, 253 (1895).
62. F. Challenger, D. Leaver and M. I. Whitaker, *Biochem. J.*, **56**, ii (1954).
63. S. P. Colowick and N. O. Kaplan (Editors), *Methods in Enzymology*, Vol. 17, Part B, Academic Press, New York, 1971.
64. D. E. Metzler, *Biochemistry: The Chemical Reactions of Living Cells*, Academic Press, New York, 1977.
65. J. A. Schiff and R. C. Hodson, *Ann. Rev. Plant. Physiol.*, **24**, 381 (1973).
66. C. Delavier-Klutchko and M. Flavin, *J. Biol. Chem.*, **240**, 2537 (1965).
67. H. P. Broquist and J. S. Trupin, *Ann. Rev. Biochem.*, **35**, 258 (1966).
68. F. Wagner, *Ann. Rev. Biochem.*, **35**, 425 (1966).
69. H. Weissbach and H. Dickerman, *Physiol. Rev.*, **45**, 80 (1965).
70. J. D. Finkelstein, W. E. Kyle and B. J. Harris, *Arch. Biochem. Biophys.*, **146**, 84 (1971).
71. J. L. Botsford and L. W. Parks, *J. Bacteriol.*, **97**, 1176 (1969).
72. S. K. Shapiro, A. Almenas and J. F. Thomson, *J. Biol. Chem.*, **240**, 2512 (1965).
73. S. K. Shapiro, *J. Bacteriol.*, **72**, 730 (1956).
74. M. Flavin and C. Slaughter, *Biochim. Biophys. Acta*, **132**, 400 (1967).
75. R. C. Greene, *J. Biol. Chem.*, **237**, 2251 (1962).
76. C. S. Sato, R. U. Byerrum, P. Albersheim and J. Bonner, *J. Biol. Chem.*, **233**, 128 (1958).
77. B. Ladešić and D. Keglević, *Arch. Biochem. Biophys.*, **111**, 653 (1965).
78. W. E. Splittstoesser and M. Mazelis, *Phytochemistry*, **6**, 39 (1967).
79. B. Granroth, *Ann. Acad. Sci. Fenn., Ser. A2*, No. 154 (1970).
80. R. C. Greene and N. B. Davis, *Biochim. Biophys. Acta*, **43**, 360 (1960).
81. A. H. Baur and S. F. Yang, *Phytochemistry*, **11**, 2503 (1972).
82. D. Karr, J. Tweto and P. Albersheim, *Arch. Biochem. Biophys.*, **121**, 732 (1967).
83. A. D. Hanson and H. Kende, *Plant Physiol.*, **57**, 528 (1976).
84. G. L. Cantoni and J. Durell, *J. Biol. Chem.*, **225**, 1033 (1957).
85. S. H. Mudd, in *Transmethylation and Methionine Biosynthesis* (Eds. S. K. Shapiro and F. Schlenk), University of Chicago Press, Chicago, 1965, p. 33.
86. S. H. Mudd and J. D. Mann, *J. Biol. Chem.*, **238**, 2164 (1963).
87. J. B. Lombardini, T.-C. Chou and P. Talalay, *Biochem. J.*, **135**, 43 (1973).
88. F. Schlenk, J. L. Dainko and S. M. Stanford, *Arch. Biochem. Biophys.*, **83**, 28 (1959).
89. S. H. Mudd and G. L. Cantoni, *Nature, Lond.*, **180**, 1052 (1957).
90. G. L. Cantoni, in *The Biochemistry of Adenosylmethionine* (Eds. F. Salvatore, E. Borek, V. Zappia, H. G. Williams-Ashman and F. Schlenk), Columbia University Press, New York, 1977, p. 557.
91. M. Masselot and H. De Robichon-Szulmajster, *Mol. Gen. Genet.*, **129**, 349 (1973).
92. R. Cox, J. T. Martin and H. Shinozuka, *Lab. Invest.*, **29**, 54 (1973).
93. A. Lescault and C. Laberge, *Union Med. Can.*, **103**, 456 (1974).
94. T. Takahashi, Y. Fujii and H. Takahashi, *Agric. Biol. Chem.*, **31**, 73 (1967).
95. J. A. Duerre and F. Schlenk, *Arch. Biochem. Biophys.*, **96**, 575 (1962).
96. G. De La Haba and G. L. Cantoni, *J. Biol. Chem.*, **234**, 603 (1959).
97. F. Schlenk and D. J. Ehninger, *Arch. Biochem. Biophys.*, **106**, 95 (1964).
98. J. A. Duerre and R. D. Walker, in *The Biochemistry of Adenosylmethionine* (Eds. F. Salvatore, E. Borek, V. Zappia, H. G. Williams-Ashman and F. Schlenk), Columbia University Press, New York, 1977, p. 43.
99. C. H. Miller and J. A. Duerre, *J. Biol. Chem.*, **243**, 92 (1968).
100. C. Wagner and E. R. Stadtman, *Arch. Biochem. Biophys.*, **98**, 331 (1962).

101. G. A. Maw, *Biochem. J.*, **63**, 116 (1956).
102. G. A. Maw, *Biochem. J.*, **72**, 602 (1959).
103. Y. Ishida and H. Kadota, *Nippon Suisan Gakkaishi*, **34**, 699 (1968).
104. G. A. Maw, *Biochem. J.*, **70**, 168 (1958).
105. J. Durell, D. G. Anderson and G. L. Cantoni, *Biochim. Biophys. Acta*, **26**, 270 (1957).
106. L.-E. Ericson, *Acta Chem. Scand.*, **14**, 2102, 2113, 2127 (1960).
107. W. A. Klee, H. H. Richards and G. L. Cantoni, *Biochim. Biophys. Acta*, **54**, 157 (1961).
108. M. F. Ferger and V. Du Vigneaud, *J. Biol. Chem.*, **185**, 53 (1950).
109. G. A. Maw, *Biochem. J.*, **55**, 42 (1953).
110. G. A. Maw, *Biochem. J.*, **58**, 665 (1954).
111. G. A. Maw and V. Du Vigneaud, *J. Biol. Chen.*, **176**, 1029 (1948).
112. K. S. Bjerve and J. Bremer, *Biochim. Biophys. Acta*, **176**, 570 (1969).
113. J. W. Dubnoff, *Arch. Biochem. Biophys.*, **22**, 478 (1949).
114. J. A. Muntz and J. Hurwitz, *J. Biol. Chem.*, **182**, 489 (1950).
115. C. Wagner, S. M. Lusty, H.-F. Kung and N. L. Rogers, *J. Biol. Chem.*, **241**, 1923 (1966); **242**, 1287 (1967).
116. K. Tanaka and K. Nakamura, *J. Biochem. (Tokyo)*, **56**, 172 (1964).
117. M. Mazelis, B. Levin and N. Mallinson, *Biochim. Biophys. Acta*, **105**, 106 (1965).
118. B. G. Lewis, C. M. Johnson and T. C. Broyer, *Biochim. Biophys. Acta*, **237**, 603 (1971).
119. T. Hattula and B. Granroth, *J. Sci. Food Agric.*, **25**, 1517 (1974).
120. H. J. Niefind and G. Spaeth, *Eur. Brew. Conv. Proc. Congr.*, **15**, 97 (1975).
121. R. Self, J. C. Casey and T. Swain, *Chem. Ind. (London)*, 863 (1963).
122. D. D. Bills and T. W. Keenan, *J. Agric. Food Chem.*, **16**, 643 (1968).
123. D. G. Guadagni, J. C. Miers and D. Venstrom, *Food Technol.*, **22**, 1003 (1968).
124. R. A. McRorie, M. R. Glazener, C. G. Skinner and W. Shive, *J. Biol. Chem.*, **211**, 489 (1954).
125. F. Schlenk and R. E. DePalma, *Arch. Biochem. Biophys.*, **57**, 266 (1955).
126. J. B. Ragland and J. L. Liverman, *Arch. Biochem. Biophys.*, **65**, 574 (1956).
127. S. K. Shapiro and D. A. Yphantis, *Biochim. Biophys. Acta*, **36**, 241 (1959).
128. S. K. Shapiro, D. A. Yphantis and A. Almenas, *J. Biol. Chem.*, **239**, 1551 (1964).
129. S. K. Shapiro, A. Almenas and J. F. Thomson, *J. Biol. Chem.*, **240**, 2512 (1965).
130. A. Stevens and W. Sakami, *J. Biol. Chem.*, **234**, 2063 (1959).
131. W. Dodd and E. A. Cossins, *Arch. Biochem. Biophys.*, **133**, 216 (1969).
132. L. Abrahamson and S. K. Shapiro, *Arch. Biochem. Biophys.*, **109**, 376 (1965).
133. J. E. Turner and S. K. Shapiro, *Biochim. Biophys. Acta*, **51**, 581 (1961).
134. B. D. Allamong and L. Abrahamson, *Bot. Gaz.*, **138**, 46 (1977).
135. M. Pfeffer and S. K. Shapiro, *Biochem. Biophys. Res. Commun.*, **9**, 405 (1962).
136. L. W. Parks and F. Schlenk, *J. Biol. Chem.*, **230**, 295 (1958).
   a. P. B. Dransfield and F. Challenger, *J. Chem. Soc.*, **1153** (1955).
137. *Chem. Abstr. Index Guide*, 424G (1977).
138. G. L. Cantoni, *Phosphorus Metab. Symp.*, **2**, 129 (1952).
139. G. L. Cantoni and E. Scarano, *J. Amer. Chem. Soc.*, **76**, 4744 (1954).
140. J. D. Finkelstein and B. Harris, *Arch. Biochem. Biophys.*, **159**, 160 (1973).
141. J. A. Duerre, *Arch. Biochem. Biophys.*, **124**, 422 (1968).
142. F. Schlenk and C. R. Zydek, *Biochem. Biophys. Res. Commun.*, **31**, 427 (1968).
143. S. S. Chen, J. Hudspeth-Walgate and J. A. Duerre, *Arch. Biochem. Biophys.*, **146**, 54 (1971).
144. C. H. Miller and J. A. Duerre, *J. Biol. Chem.*, **244**, 4273 (1969).
145. J. A. Duerre, C. H. Miller and G. G. Reams, *J. Biol. Chem.*, **244**, 107 (1969).
146. F. Schlenk, C. R. Zydek-Cwick and N. K. Hutson, *Arch. Biochem. Biophys.*, **142**, 144 (1971).
147. J. A. Duerre, *J. Biol. Chem.*, **237**, 3737 (1962).
148. C. H. Miller and J. A. Duerre, *J. Biol. Chem.*, **243**, 92 (1968).
149. V. Zappia, C. R. Zydek-Cwick and F. Schlenk, *J. Biol. Chem.*, **244**, 4499 (1969).
150. T. Deguchi and J. Barchas, *J. Biol. Chem.*, **246**, 3175 (1971).

151. J. K. Coward, M. d'Urso-Scott and W. D. Sweet, *Biochem. Pharmacol.*, **21**, 1200 (1972).
152. R. L. Lin, N. Narasimhachari and H. E. Himwich, *Biochem. Biophys. Res. Commun.*, **54**, 751 (1973).
153. D. N. Mardon and E. Balish, *Can. J. Microbiol.*, **17**, 795 (1971).
154. V. Zappia, C. R. Zydek-Cwick and F. Schlenk, *Biochim. Biophys. Acta*, **178**, 185 (1969).
155. K. D. Nakamura and F. Schlenk, *Arch. Biochem. Biophys.*, **177**, 170 (1976).
156. F. Schlenk and J. L. Dainko, *Biochim. Biophys. Acta*, **385**, 312 (1975).
157. J. Bremer and D. M. Greenberg, *Biochim. Biophys. Acta*, **46**, 217 (1961).
158. T. Fujita and Z. Suzuoki, *J. Biochem. (Tokyo)*, **74**, 717 (1973).
159. C. N. Remy, *J. Biol. Chem.*, **238**, 1078 (1963).
160. G. Steensholt, *Acta Physiol. Scand.*, **10**, 320 (1945).
161. H. K. Barrenscheen and T. Von Valyi-Nagy, *Z. Physiol. Chem.*, **277**, 97 (1942).
162. G. L. Cantoni and P. J. Vignos, *J. Biol. Chem.*, **209**, 647 (1954).
163. F. Salvatore and F. Schlenk, *Biochim. Biophys. Acta*, **59**, 700 (1962).
164. G. Steensholt, *Acta Physiol. Scand.*, **14**, 340 (1947).
165. J. A. Stekol, E. I. Anderson and S. Weiss, *J. Biol. Chem.*, **233**, 425 (1958).
166. J. Bremer and D. M. Greenberg, *Biochim. Biophys. Acta*, **35**, 287 (1959); **37**, 173 (1960); **46**, 205 (1961).
167. T. E. Morgan, *Biochim. Biophys. Acta*, **178**, 21 (1969).
168. H. Goldfine, *Ann. Rev. Biochem.*, **37**, 303 (1968).
169. R. Letters, *Biochim. Biophys. Acta*, **116**, 489 (1966).
170. J. Blumenstein and G. R. Williams, *Biochem. Biophys. Res. Commun.*, **3**, 259 (1960).
    a. J. E. Heady and S. J. Kerr, *J. Biol. Chem.*, **248**, 69 (1973).
171. G. Wiehler and L. Marion, *J. Biol. Chem.*, **231**, 799 (1958).
172. V. Tanphaichitr and H. P. Broquist, *J. Biol. Chem.*, **248**, 2176 (1973).
173. D. W. Horne and H. P. Broquist, *J. Biol. Chem.*, **248**, 2170 (1973).
174. G. L. Cantoni, *Ann. Rev. Biochem.*, **44**, 435 (1975).
175. J. Axelrod, in *Transmethylation and Methionine Biosynthesis* (Eds. S. K. Shapiro and F. Schlenk), University of Chicago Press, Chicago, 1965, p. 71.
176. P. B. Molinoff and J. Axelrod, *Ann. Rev. Biochem.*, **40**, 465 (1971).
177. J. Axelrod, in *The Biochemistry of Adenosylmethionine* (Eds. F. Salvatore, E. Borek, V. Zappia, H. G. Williams-Ashman and F. Schlenk), Columbia University Press, New York, 1977, p. 539.
178. N. Kirschner and M. Goodall, *Biochim. Biophys. Acta*, **24**, 658 (1957).
179. J. Axelrod, *J. Biol. Chem.*, **237**, 1657 (1962).
180. J. Axelrod, *J. Pharmacol. Exp. Ther.*, **138**, 28 (1962).
181. J. M. Saavedra and J. Axelrod, *Science*, **172**, 1365 (1972).
182. R. J. Wyatt, J. M. Saavedra and J. Axelrod, *Amer. J. Psychiat.*, **130**, 754 (1973).
183. M. Morgan and A. J. Mandell, *Science*, **165**, 492 (1969); *Nature New Biol.*, **230**, 85 (1971).
184. P. A. M. Eagles and M. Iqbal, *Brain Res.*, **80**, 177 (1974).
185. Y. Raoul, *C.R. Acad. Sci.*, **205**, 450 (1937).
186. E. Leete and L. Marion, *Can. J. Chem.*, **31**, 126 (1953).
187. J. Massicot and L. Marion, *Can. J. Chem.*, **35**, 1 (1957).
188. H. S. McKee, *Nitrogen Metabolism in Plants*, Clarendon Press, Oxford, 1962, Ch. 12, p. 358.
189. J. D. Mann and S. H. Mudd, *J. Biol. Chem.*, **238**, 381 (1963).
190. J. D. Mann, C. E. Steinhart and S. H. Mudd, *J. Biol. Chem.*, **238**, 676 (1963).
191. S. H. Mudd, *Nature, Lond.*, **189**, 489 (1961).
192. S. Mizusaki, Y. Tanabe, M. Noguchi and E. Tamaki, *Plant Cell Physiol.*, **12**, 633 (1971); **14**, 103 (1973).
193. P. Ellinger and M. M. Abdel Kader, *Biochem. J.*, **44**, 77 (1949).
194. J. G. Joshi and P. Handler, *J. Biol. Chem.*, **235**, 2981 (1960).
195. T. Winnick and R. E. Winnick, *Nature, Lond.*, **183**, 1466 (1959).

196. R. W. Schayer, S. A. Karjala, K. J. Davis and R. L. Smiley, *J. Biol. Chem.*, **221**, 307 (1956).
197. D. D. Brown, J. Axelrod and R. Tomchick, *Nature, Lond.*, **183**, 680 (1959).
198. D. D. Brown, R. Tomchick and J. Axelrod, *J. Biol. Chem.*, **234**, 2948 (1959).
199. A. Gustafsson and G. P. Forshell, *Acta Chem. Scand.*, **17**, 541 (1963).
200. A. R. Battersby and B. J. T. Harper, *Chem. Ind. (London)*, **365**, (1958).
201. L. J. Dewey, R. U. Byerrum and C. D. Ball, *J. Amer. Chem. Soc.*, **76**, 3997 (1954).
202. M. D. Antoun and M. F. Roberts, *Planta Med.*, **28**, 6 (1975).
203. J. Axelrod, *Life Sci.*, **1**, 29 (1962).
204. T. Suzuki, *FEBS Lett.*, **24**, 18 (1972).
205. T. Suzuki and E. Takahashi, *Biochem. J.*, **146**, 87 (1975).
206. J. Axelrod and J. Daly, *Biochim. Biophys. Acta*, **61**, 855 (1962).
207. C. N. Remy, *J. Biol. Chem.*, **234**, 1485 (1959).
208. J. Axelrod and R. Tomchick, *J. Biol. Chem.*, **233**, 702 (1958).
209. E. H. LaBrosse, J. Axelrod and S. S. Kety, *Science*, **128**, 593 (1958).
210. R. J. Wurtman, C. M. Rose, S. Matthysse, J. Stephenson and R. Baldessarini, *Science*, **169**, 395 (1970).
211. V. M. Andreoli and F. Maffei, *Lancet*, **ii** (7941) 922 (1975).
212. H. Thomas, *Med. Welt*, (8), 383 (1969).
213. R. Gugler and H. Breuer, *Symp. Deut. Ges. Endokrinol.*, **16**, 206 (1970).
214. H. Breuer and R. Knuppen, *Naturwissenschaften*, **47**, 280 (1960).
215. E. Blaschke and G. Hertting, *Biochem. Pharmacol.*, **20**, 1363 (1971).
216. J. D. Mann, H. M. Fales and S. H. Mudd, *J. Biol. Chem.*, **238**, 3820 (1963).
217. B. J. Finkle and R. F. Nelson, *Biochim. Biophys. Acta*, **78**, 747 (1963).
218. J. Axelrod and H. Weissbach, *J. Biol. Chem.*, **236**, 211 (1961).
219. R. B. Guchhait, *J. Neurochem.*, **26**, 187 (1976).
220. R. U. Byerrum, J. H. Flokstra, L. J. Dewey and C. D. Ball, *J. Biol. Chem.*, **210**, 633 (1954).
221. M. Shimada, H. Fushiki and T. Higuchi, *Mokuzai Gakkaishi*, **18**, 43 (1972).
222. R. Cleland, *Nature, Lond.*, **185**, 44 (1960).
223. P. M. Dewick, *Phytochemistry*, **14**, 983 (1975).
224. D. R. Threlfall, *Biochim. Biophys. Acta*, **280**, 472 (1972).
225. D. E. M. Lawson and J. Glover, *Biochem. Biophys. Res. Commun.*, **4**, 223 (1961).
226. B. L. Trumpower, R. M. Houser and R. E. Olson, *J. Biol. Chem.*, **249**, 3041 (1974).
227. R. K. Singh, G. Britton and T. W. Goodwin, *Biochem. J.*, **136**, 413 (1973).
228. D. Hess, *Z. Naturforsch.*, **19b**, 447 (1964).
229. F. Koller and O. Hoffmann-Ostenhof, *Hoppe-Seylers Z. Physiol. Chem.*, **357**, 1465 (1976).
230. J. E. Poulton and V. S. Butt, *Biochim. Biophys. Acta*, **403**, 301 (1975).
231. J. E. Poulton, K. Hahlbrock and H. Grisebach, *Arch. Biochem. Biophys.*, **176**, 449 (1976).
232. J. E. Poulton, H. Grisebach, J. Ebel, B. Schaller-Hekeler and K. Hahlbrock, *Arch. Biochem. Biophys.*, **173**, 301 (1976).
233. C.-K, Wat and G. H. N. Towers, *Phytochemistry*, **14**, 663 (1975).
234. G. H. Tait and K. D. Gibson, *Biochim. Biophys. Acta*, **52**, 614 (1961).
235. R. K. Ellsworth and L. A. St. Pierre, *Photosynthetica*, **9**, 340 (1975); **10**, 291 (1976).
236. R. K. Ellsworth and J. P. Dullaghan, *Biochim. Biophys. Acta*, **268**, 327 (1972).
237. Y. Akamatsu and J. H. Law, *J. Biol. Chem.*, **245**, 709 (1970).
238. D. Reibstein and J. H. Law, *Biochem. Biophys. Res. Commun.*, **55**, 266 (1973).
239. D. R. Threlfall, in *Vitamins and Hormones*, Vol. 29 (Eds. R. S. Harris and P. L. Munson), Academic Press, New York, 1971, p. 153.
240. G. Gold and R. Bentley, *Bioorg. Chem.*, **3**, 377 (1974).
241. A. R. Battersby, E. McDonald, R. Hollenstein, M. Ihara, F. Satoh and D. C. Williams, *J. Chem. Soc. Perkin Trans. I*, 166 (1977).
242. E. Lederer, *Isr. J. Med. Sci.*, **1**, 1129 (1965).
243. E. Lederer, *Quart. Rev. Chem. Soc.*, **23**, 453 (1969).

244. E. Lederer, in *The Biochemistry of Adenosylmethionine* (Eds. F. Salvatore, E. Borek, V. Zappia, H. G. Williams-Ashman and F. Schlenk), Columbia University Press, New York, 1977, p. 89.
245. A. J. Birch, D. Elliott and A. R. Penfold, *Aust. J. Chem.*, **7**, 169 (1954).
246. A. J. Birch, R. J. English, R. A. Massey-Wesdropp, M. Slaytor and H. Smith, *Proc. Chem. Soc.*, **204**, (1957).
247. H. Pape, R. Schmid, H. Grisebach and H. Achenbach, *Eur. J. Biochem.*, **10**, 479 (1969).
248. H. Zalkin, J. H. Law and H. Goldfine, *J. Biol. Chem.*, **238**, 1242 (1963).
249. A. E. Chung and J. H. Law, *Biochemistry*, **3**, 967 (1964).
250. L. A. Halper and S. J. Norton, *Biochem. Biophys. Res. Commun.*, **62**, 683 (1975).
251. Y. Akamatsu and J. H. Law, *Biochem. Biophys. Res. Commun.*, **33**, 172 (1968).
252. G. Jaureguiberry, M. Lenfant, R. Toubiana, R. Azerad and E. Lederer, *J. Chem. Soc. Chem. Commun.*, 855 (1966).
253. G. J. Alexander, A. M. Gold and E. Schwenk, *J. Amer. Chem. Soc.*, **79**, 2967 (1957).
254. L. W. Parks, *J. Amer. Chem. Soc.*, **80**, 2023 (1958).
255. L. W. Parks, R. B. Bailey and E. D. Thompson, in *The Biochemistry of Adenosylmethionine* (Eds. F. Salvatore, E. Borek, V. Zappia, H. G. Williams-Ashman and F. Schlenk), Columbia University Press, New York, 1977, p. 172.
256. R. B. Bailey, E. D. Thompson and L. W. Parks, *Biochim. Biophys. Acta*, **334**, 127 (1974).
257. P. De Luca, M. De Rosa, L. Minale and G. Sodano, *J. Chem. Soc. Perkin Trans. I*, 2132 (1972).
258. P. De Luca, M. De Rosa, L. Minale, R. Puliti, G. Sodano, F. Giordano and L. Mazzarella, *J. Chem. Soc. Chem. Commun.*, 825 (1973).
259. N. C. Ling, R. L. Hale and C. Djerassi, *J. Amer. Chem. Soc.*, **92**, 5281 (1970). 1963, p. 163.
261. M. Castle, G. A. Blondin and W. R. Nes, *J. Amer. Chem. Soc.*, **85**, 3306 (1963).
262. L. J. Goad, A. S. A. Hammam, A. Dennis and T. W. Goodwin, *Nature, Lond.*, **210**, 1322 (1966).
263. M. Lenfant, R. Ellouz, B. C. Das, E. Zissmann and E. Lederer, *Eur. J. Biochem.*, **7**, 159 (1969).
264. J. L. Starr and B. H. Sells, *Ann. Rev. Physiol.*, **49**, 623 (1969).
265. C. S. Sato, R. U. Byerrum and C. D. Ball, *J. Biol. Chem.*, **224**, 717 (1957).
266. H. Kauss and W. Z. Hassid, *J. Biol. Chem.*, **242**, 3449 (1967).
267. H. Kauss and W. Z. Hassid, *J. Biol. Chem.*, **242**, 1680 (1967).
268. Y. C. Lee and C. E. Ballou, *J. Biol. Chem.*, **239**, 3602 (1964).
269. W. L. Smith and C. E. Ballou, *J. Biol. Chem.*, **248**, 7118 (1973).
270. H. Deul and E. Stulz, *Adv. Enzymol.*, **20**, 341 (1958).
271. W. H. Paik and S. Kim, *Science*, **174**, 114 (1971).
272. D. G. Comb, N. Sarkar and C. J. Pinzino, *J. Biol. Chem.*, **241**, 1857 (1966).
273. A. M. Kaye and D. Sheratzky, *Biochim. Biophys. Acta*, **190**, 527 (1969).
274. W. K. Paik and S. Kim, *J. Biol. Chem.*, **245**, 88 (1970).
275. S. Kim, in *The Biochemistry of Adenosylmethionine* (Eds. F. Salvatore, E. Borek, V. Zappia, H. G. Williams-Ashman and F. Schlenk), Columbia University Press, New York, 1977, p. 415.
276. P. R. Carnegie, *Nature, Lond.*, **229**, 25 (1971); *Biochem. J.*, **123**, 57 (1971).
277. C. S. Baxter and P. Byvoet, *Cancer Res.*, **34**, 1418 (1974).
278. S. Kim, *Arch. Biochem. Biophys.*, **157**, 476 (1973); **161**, 652 (1973).
279. M. Liss and L. M. Edelstein, *Biochem. Biophys. Res. Commun.*, **26**, 497 (1967).
280. S. Kim and W. K. Paik, *J. Biol. Chem.*, **245**, 1806 (1970).
281. W. K. Paik, in *The Biochemistry of Adenosylmethionine* (Eds. F. Salvatore, E. Borek, V. Zappia, H. G. Williams-Ashman and F. Schlenk), Columbia University Press, New York, 1977, p. 401.
282. W. K. Paik and S. Kim, *Biochim. Biophys. Acta*, **313**, 181 (1973).
283. M. Kruger and G. Salomon, *Hoppe-Seylers Z. Physiol. Chem.*, **24**, 350 (1898).

284. J. W. Littlefield and D. B. Dunn, *Biochem. J.*, **70**, 642 (1958).
285. D. B. Dunn, *Biochim. Biophys. Acta*, **34**, 286 (1959); **38**, 176 (1960); **46**, 198 (1961).
286. M. Klagsbrun, *J. Biol. Chem.*, **248**, 2612 (1973).
287. F. Salvatore and F. Cimino, in *The Biochemistry of Adenosylmethionine* (Eds. F. Salvatore, E. Borek, V. Zappia, H. G. Williams-Ashman and F. Schlenk), Columbia University Press, New York, 1977, p. 187.
288. F. Nau, in *The Biochemistry of Adenosylmethionine* (Eds. F. Salvatore, E. Borek, V. Zappia, H. G. Williams-Ashman and F. Schlenk), Columbia University Press, New York, 1977, p. 258.
289. E. Fleissner and E. Borek, *Proc. Natl. Acad. Sci. USA*, **48**, 1199 (1962).
290. E. Borek and P. R. Srinivasan, *Ann. Rev. Biochem.*, **35**, 275 (1966).
291. P. R. Srinivasan and E. Borek, *Science*, **145**, 548 (1964).
292. L. R. Mandel and E. Borek, *Biochemistry*, **2**, 555, 560 (1963).
293. L. R. Mandel, P. R. Srinivasan and E. Borek, *Nature, Lond.*, **209**, 586 (1966).
294. E. Borek, A. Ryan and J. Rockenbach, *J. Bacteriol.*, **69**, 460 (1955).
295. R. H. Hall, *Biochemistry*, **3**, 876 (1964).
296. E. Borek and P. R. Srinivasan, in *Transmethylation and Methionine Biosynthesis* (Eds. S. K. Shapiro and F. Schlenk), University of Chicago Press, Chicago, 1965, p. 115.
297. B. E. Tropp, J. H. Law and J. M. Hayes, *Biochemistry*, **3**, 1837 (1964).
298. P. S. Leboy, *Biochemistry*, **9**, 1577 (1970).
299. A. E. Pegg, *Biochim. Biophys. Acta*, **232**, 630 (1971).
300. S. J. Kerr, *J. Biol. Chem.*, **247**, 4248 (1972).
301. S. J. Kerr and J. T. Heady, *Adv. Enzyme Regulation*, **12**, 1 (1974).
302. L. L. Mays, E. Borek and C. E. Finch, *Nature, Lond.*, **243**, 411 (1973).
303. R. M. Halpern, S. Chaney, B. C. Halpern and R. A. Smith, *Biochem. Biophys. Res. Commun.*, **42**, 602 (1971).
304. G. R. Wyatt, *Biochem. J.*, **48**, 584 (1951).
305. G. Seemayer, *Biochim. Biophys. Acta*, **224**, 10 (1970).
306. J. Malec, M. Wojnarowska and L. Kornacka, *Acta Biochim. Pol.*, **21**, 291 (1974).
307. R. H. Burdon and J. T. Douglas, *Nucleic Acids Res.*, **1**, 97 (1974).
308. J. W. Kappler, *J. Cell. Physiol.*, **75**, 21 (1970).
309. R. Cox, C. Prescott and C. C. Irving, *Biochim. Biophys. Acta*, **474**, 493 (1977).
310. V. M. Craddock, *Nature, Lond.*, **228**, 1264 (1970).
311. E. Borek and S. J. Kerr, *Adv. Cancer Res.*, **15**, 163 (1972).
312. A. Mittelman, R. H. Hall, D. S. Yohn and J. T. Grace, *Cancer Res.*, **27**, 1409 (1967).
313. T. P. Waalkes, C. W. Gehrke, R. W. Zumwalt, S. Y. Chang, D. B. Lakings, D. C. Tormey, D. L. Ahmann and C. G. Moertel, *Cancer*, **36**, 392 (1975).
314. T. P. Waalkes, C. W. Gehrke, W. A. Bleyer, R. W. Zumwalt, C. L. M. Olweny, K. C. Kuo, D. B. Lakings and S. A. Jacobs, *Cancer Chemother. Rep.*, **59**, 721 (1975).
315. P. N. Magee and E. Farber, *Biochem. J.*, **83**, 114 (1962).
316. B. W. Stewart and A. E. Pegg, *Biochim. Biophys. Acta*, **281**, 416 (1972).
317. R. Wilkinson and D. J. Pillinger, *Int. J. Cancer*, **8**, 401 (1971).
318. R. L. Hancock and P. I. Forrester, *Cancer Res.*, **33**, 1747 (1973).
319. M. Meselson, R. Yuan and J. Heywood, *Ann. Rev. Biochem.*, **41**, 447 (1972).
320. J. L. Marx, *Science*, **180**, 482 (1973).
321. S. Linn, B. Eskin, J. A. Lautenberger, D. Lackey and M. Kimball, in *The Biochemistry of Adenosylmethionine* (Eds. F. Salvatore, E. Borek, V. Zappia, H. G. Williams-Ashman and F. Schlenk), Columbia University Press, New York, 1977, p. 521.
322. H. Tabor, C. W. Tabor and S. M. Rosenthal, *Ann. Rev. Biochem.*, **30**, 579 (1961).
323. H. Tabor and C. W. Tabor, *Adv. Enzymol.*, **36**, 203 (1972).
324. H. G. Williams-Ashman, *Invest. Urol.*, **2**, 605 (1965).
325. S. S. Cohen, *Introduction to the Polyamines*, Prentice-Hall, Englewood Cliffs, N.J., 1971.
326. D. H. Russell, *Polyamines in Normal and Neoplastic Growth*, Raven Press, New York, 1973.
327. D. H. Russell and S. D. Russell, *Clin. Chem.*, **21**, 860 (1975).
328. H. Tabor, S. M. Rosenthal and C. W. Tabor, *J. Biol. Chem.*, **233**, 907 (1958).
329. R. C. Greene, *J. Amer. Chem. Soc.*, **79**, 3929 (1957).

330. H. Tabor, S. M. Rosenthal and C. W. Tabor, *J. Amer. Chem. Soc.*, **79**, 2978 (1957).
331. A. E. Pegg and H. G. Williams-Ashman, *J. Biol. Chem.*, **244**, 682 (1969).
332. H. G. Williams-Ashman, J. Janne, G. L. Coppoc, M. E. Geroch and A. Schenone, *Adv. Enzyme Regulation*, **10**, 225 (1972).
333. H. G. Williams-Ashman, A. Corti and G. L. Coppoc, in *The Biochemistry of Adenosylmethionine* (Eds. F. Salvatore, E. Borek, V. Zappia, H. G. Williams-Ashman and F. Schlenk). Columbia University Press, New York, 1977, p. 493.
334. V. Zappia, M. Cartenì-Farina and G. Della Pietra, *Biochem. J.*, **129**, 703 (1972).
335. R. B. Wickner, C. W. Tabor and H. Tabor, *J. Biol. Chem.*, **245**, 2132 (1970).
336. V. Zappia, M. Cartenì-Farina and P. Galletti, in *The Biochemistry of Adenosylmethionine*, (Eds. F. Salvatore, E. Borek, V. Zappia, H. G. Williams-Ashman and F. Schlenk), Columbia University Press, New York, 1977, p. 473.
337. W. H. Bowman, C. W. Tabor and H. Tabor, *J. Biol. Chem.*, **248**, 2480 (1973).
338. A. Raina and P. Hannonen, *Acta Chem. Scand.*, **24**, 3061 (1970); *FEBS Lett.*, **16**, 1 (1971).
339. J. A. Mandel and K. Dunham, *J. Biol. Chem.*, **11**, 85 (1912).
340. U. Suzuki, *J. Tokyo Chem. Soc.*, **34**, 1134 (1914).
341. R. L. Smith, E. E. Anderson, R. N. Overland and F. Schlenk, *Arch. Biochem. Biophys.*, **42**, 72 (1953).
342. S. K. Shapiro and A. N. Mather, *J. Biol. Chem.*, **233**, 631 (1958).
343. S. H. Mudd, *J. Biol. Chem.*, **234**, 87 (1959).
344. S. Nishimura, Y. Taya, Y. Kuchino and Z. Ohashi, *Biochem. Biophys. Res. Commun.*, **57**, 702 (1974).
345. F. Schlenk, C. R. Zydek-Cwick and J. L. Dainko, *Biochim. Biophys. Acta*, **320**, 357 (1973).
346. S. P. Burg, *Proc. Natl. Acad. Sci. USA*, **70**, 591 (1973).
347. Lieberman, A. Kunishi, L. W. Mapson and D. A. Wardale, *Plant Physiol.*, **41**, 376 (1966).
348. S. P. Burg and C. O. Clagett, *Biochem. Biophys. Res. Commun.*, **27**, 125 (1967).
349. A. H. Baur and S. F. Yang, *Phytochemistry*, **11**, 3207 (1972).
350. D. O. Adams and S. F. Yang, *Plant Physiol.*, **60**, 892 (1977).
     a. D. O. Adams and S. F. Yang, *Proc. Natl. Acad. Sci. USA*, **76**, 170 (1979).
     b. A. C. Cameron, C. A. L. Fenton, Y.-B. Yu, D. O. Adams and S. F. Yang, *HortSci.*, **14**, 178 (1979).
     c. Y.-B. Yu, D. O. Adams and S. F. Yang, *Plant Physiol.*, **63**, 589 (1979).
351. M. A. Eisenberg and G. L. Stoner, *J. Bacteriol.*, **108**, 1135 (1971).
352. G. L. Stoner and M. A. Eisenberg, *J. Biol. Chem.*, **250**, 4029 (1975).
353. Y. Izumi, K. Sato, Y. Tani and K. Ogata, *Agric. Biol. Chem.*, **39**, 175 (1975).
354. T. Chase and J. C. Rabinowitz, *J. Bacteriol.*, **96**, 1065 (1968).
355. H. Nakayama, G. G. Midwinter and L. O. Krampitz, *Arch. Biochem. Biophys.*, **143**, 526 (1971).
356. T. P. Chirpich, V. Zappia, R. N. Costilow and H. A. Barker, *J. Biol. Chem.*, **245**, 1778 (1970).
357. V. Zappia and F. Ayala, *Biochim. Biophys. Acta*, **268**, 573 (1972).
358. J. Knappe and T. Schmitt, *Biochem. Biophys. Res. Commun.*, **71**, 1110 (1976).
359. J. D. Finkelstein, W. E. Kyle, J. J. Martin and A. M. Pick, *Biochem. Biophys. Res. Commun.*, **66**, 81 (1975).
360. J. D. Finkelstein, W. E. Kyle and B. J. Harris, *Arch. Biochem. Biophys.*, **165**, 774 (1974).
361. J. T. Madison and J. F. Thompson, *Biochem. Biophys. Res. Commun.*, **71**, 684 (1976).
362. R. W. Trewyn, K. D. Nakamura, M. L. O'Connor and L. W. Parks, *Biochim. Biophys. Acta*, **327**, 336 (1973).
363. J. F. Fisher and M. F. Mallette, *J. Gen. Physiol.*, **45**, 1 (1961).
364. J. D. Loerch and M. F. Mallette, *Arch. Biochem. Biophys.*, **103**, 272 (1963).
365. V. R. Villanueva, M. Barbier, C. Gros and E. Lederer, *Biochim. Biophys. Acta*, **130**, 329 (1966).
366. H. M. Dyer, *J. Biol. Chem.*, **124**, 519 (1938).

367. E. Farber, *Adv. Cancer Res.*, **7**, 383 (1963).
368. J. A. Stekol, *Adv. Enzymol.*, **25**, 369 (1963).
369. G. A. Maw, *Ann. Rep. Chem. Soc.*, **63**, 639 (1966).
370. E. Farber, K. H. Shull, S. Villa-Trevino, B. Lombardi and M. Thomas, *Nature, Lond.*, **203**, 34 (1964).
371. M. V. Simpson, E. Farber and H. Tarver, *J. Biol. Chem.*, **182**, 81 (1950).
372. M. Levine and H. Tarver, *J. Biol. Chem.*, **192**, 835 (1951).
373. D. Gross and H. Tarver, *J. Biol. Chem.*, **217**, 169 (1955).
374. G. A. Maw, *Arch. Biochem. Biophys.*, **115**, 291 (1966).
375. A. Yoshida and M. Yamasaki, *Biochim. Biophys. Acta*, **34**, 158 (1959).
376. S. Villa-Trevino, K. H. Shull and E. Farber, *J. Biol. Chem.*, **241**, 4670 (1966).
377. E. Wainfan and F. A. Maschio, *Ann. N.Y. Acad. Sci.*, **255**, 567 (1975).
378. R. Cleland, *Plant Physiol.*, **35**, 585 (1960).
379. S. H. Mudd and G. L. Cantoni, *J. Biol. Chem.*, **231**, 481 (1958).
380. J. A. Stekol, S. Weiss, E. I. Anderson and M. Toporek, *Abstr. Mtg. Amer. Chem. Soc. New York*, 57C (1957).
381. J. A. Stekol, in *Transmethylation and Methionine Biosynthesis* (Eds. S. K. Shapiro and F. Schlenk), University of Chicago Press, Chicago, 1965, p. 231.
382. R. Cox and R. C. Smith, *Arch. Biochem. Biophys.*, **129**, 615 (1969).
383. L. W. Parks, *J. Biol. Chem.*, **232**, 169 (1958).
384. K. H. Shull, J. McConomy, M. Vogt, A. Castillo and E. Farber, *J. Biol. Chem.*, **241**, 5060 (1966).
385. R. Cox and C. C. Irving, *Cancer Res.*, **37**, 222 (1977).
386. J. A. Stekol, E. I. Anderson, P.-T. Hsu and S. Weiss, *Abstr. Mtg. Amer. Chem. Soc. Cincinnati, Ohio*, 4C (1955).
387. K. H. Shull, *J. Biol. Chem.*, **237**, PC1734 (1962).
388. S. Villa-Trevino, E. Farber, Th. Staehelin, F. O. Wettstein and W. Noll. *J. Biol. Chem.*, **239**, 3826 (1964).
389. F. Schlenk, *Arch. Biochem. Biophys.*, **69**, 67 (1957).
390. F. Schlenk and C. R. Zydek, *Arch. Biochem. Biophys.*, **123**, 438 (1968).
391. A. E. Pegg, *Biochim. Biophys. Acta*, **177**, 361 (1969).
392. J. A. Stekol, S. Weiss and C. Somerville, *Arch. Biochem. Biophys.*, **100**, 86 (1963).
393. J. A. Stekol and K. Weiss, *J. Biol. Chem.*, **185**, 577 (1950).
394. S. Simmonds, E. B. Keller, J. P. Chandler and V. Du Vigneaud, *J. Biol. Chem.*, **183**, 191 (1950).
395. J. A. Stekol and K. Weiss, *J. Biol. Chem.*, **185**, 585 (1950).
396. H. Tuppy and K. Dus, *Monatsh. Chem.*, **89**, 318 (1958).
397. C. S. Baxter and P. Byvoet, *Cancer Res.*, **34**, 1424 (1974).
398. E. Farber and P. N. Magee, *Biochem. J.*, **76**, 58P (1960).
399. J. A. Stekol, U. Mody and J. Perry, *J. Biol. Chem.*, **235**, PC59 (1960).
400. R. L. Hancock, *Cancer Res.*, **28**, 1223 (1968).
401. P. F. Swann, A. E. Pegg, A. Hawks, E. Farber and P. N. Magee, *Biochem. J.*, **123**, 175 (1971).
402. A. E. Pegg, *Biochem. J.*, **128**, 59 (1972).
403. L. Rosen, *Biochem. Biophys. Res. Commun.*, **33**, 546 (1968).
404. I. Merits, *Biochem. Biophys. Res. Commun.*, **10**, 254 (1963).
405. B. J. Ortwerth and G. D. Novelli, *Cancer Res.*, **29**, 380 (1969).
406. P. D. Lawley, *Progr. Nucleic Acids Res. Molec. Biol.*, **5**, 89 (1966).
407. B. G. Moore and R. C. Smith, *Can. J. Biochem.*, **47**, 561 (1969).
408. B. G. Moore, *Can. J. Biochem.*, **48**, 702 (1970).
409. J. S. Tscherne and E. Wainfan, *Nucleic Acids Res.*, **5**, 451 (1978).
410. D. Della Bella, *Naunyn-Schmiedebergs Arch. Exp. Pathol. Pharmakol.*, **226**, 335 (1955).
411. R. B. Barlow and J. R. Vane, *Brit. J. Pharmacol.*, **11**, 198 (1956).
412. D. Della Bella, W. Caliari and F. Rognoni, *Arch. Ital. Sci. Farmacol.*, **9**, 549 (1959).
413. M. Protiva and O. Exner, *Chem. Listy*, **47**, 213 (1953).
414. Z. Votava, Y. Metysh, I. Shramkova and M. Vanachek, *Farmakol. Toksikol. (Moscow)*, **20**, 35 (1957).

415. L. Frankenberg, G. Heimburger, C. Nilsson and B. Sörbo, *Eur. J. Pharmacol.*, **23**, 37 (1973).
416. D. Della Bella, R. Villani and G. F. Zuanazzi, *Naunyn-Schmiedebergs Arch. Exp. Pathol. Pharmakol.*, **229**, 536 (1956).
417. D. Della Bella, *Arch. Ital. Sci. Farmacol.*, **13**, 70 (1963).
418. A. E. Markitantova, *Tr. Inst. Eksp. Klin. Onkol. Akad. Med. Nauk SSSR*, **2**, 120 (1960).
419. A. Luettringhaus and H. Machatzke, *Arzneim.-Forsch.*, **13**, 366 (1963).
420. G. Stramentinoli, C. Pezzoli and E. Catto, *Minerva Med.*, **66**, 4434 (1975).
421. M. Cortellaro, *Minerva Med.*, **63**, 1854 (1972).
422. E. Adami, *Atti. Soc. Lombarda Sci. Med. Biol.*, **10**, 60 (1955).
423. Th. Bersin, A. Müller and E. Strehler, *Arzneim.-Forsch.*, **6**, 174 (1956).
424. G. G. Vinci, *Boll. Soc. Ital. Biol. Sper.*, **35**, 1672 (1959).
425. Y. Ishii and Y. Fujii, *Nippon Yakurigaku Zasshi*, **70**, 863 (1974).
426. I. V. Zaikonnikova and L. G. Urazaeva, in *Vitamin U (S-Metilmetionin), Prir. Svoistva. Primen* (Ed. V. N. Bukin), Nauka, Moscow, 1973, p. 25.
427. I. V. Zaikonnikova and L. G. Urazaeva, *Farmakol. Toksikol. (Moscow)*, **37**, 346 (1974).
428. A. O. Buchon, J. V. Pastor, A. B. Auban and J. E. Requera, *Rev. Esp. Enferm. Apar. Dig. Nutr.*, **47**, 495 (1976).
429. I. G. Urazaeva, *Farmakol. Toksikol. (Moscow)*, **39**, 316 (1976).
430. V. V. Efremov, in *Vitaminy* (Ed. M. I. Smirnov), Meditsina, Moscow, 1974, p. 487.
431. V. N. Bukin and G. N. Khuchua, *Usp. Biol. Khim.*, **10**, 184 (1969); in *Vitamin U (S-Metilmetionin), Prir. Svoistva, Primen*. (Ed. V. N. Bukin), Nauka, Moscow, 1973, p. 7.
432. U. Cucinotta, *Boll. Soc. Ital. Biol. Sper.*, **35**, 1142 (1959).
433. A. I. Bizyaev, in *Vitamin U (S-Metilmetionin), Prir. Svoistva, Primen*. (Ed. V. N. Bukin), Nauka, Moscow, 1973, p. 145.
434. I. Oodaira, T. Hasegawa and N. Mochizuki, *Shinyaku to Rinsho*, **5**, 193 (1956).
435. K. Kido, *Kansai Ika Daigaku Zasshi*, **25**, 104 (1973).
436. K. Nakamura and H. Ariyama, *Bull. Agric. Chem. Soc. Jap.*, **23**, 348 (1959).
437. B. Colombo, *Minerva Med.*, **50**, 2944 (1959).
438. K. Nakamura and H. Ariyama, *Tohoku J. Agric. Res.*, **11**, 273 (1960); **12**, 49, 383 (1961).
439. T. R. Fukuto, R. L. Metcalf, R. B. March and M. Maxon, *J. Amer. Chem. Soc.*, **77**, 3670 (1955).
440. R. P. Ivanova and K. S. Bokarev, *Fiziol. Rast.*, **17**, 58 (1970).
441. B. E. A. Knight, H. F. Taylor and R. L. Wain, *Ann. Appl. Biol.*, **63**, 211 (1969).
442. J. Antognini, R. Curtis and H. M. Day, in *Symposium on Metal Chelates in Plant Nutrition, Seattle*, Washington, (Ed. A. Wallace), University of California Press, Los Angeles, 1956, p. 61.

Author Index

This author index is designed to enable the reader to locate an author's name and work with the aid of the reference numbers appearing in the text. The page numbers are printed in normal type in ascending numerical order, followed by the reference numbers in parentheses. The numbers in *italics* refer to the pages on which the references are actually listed.

Subject Index